Spintronics

A sound understanding of magnetism, transport theory, spin relaxation mechanisms, and magnetization dynamics is necessary to engage in spintronics research. In this primer, special effort has been made to give straightforward explanations for these advanced concepts.

This book will be a valuable resource for graduate students in spintronics and related fields. Concepts of magnetism such as exchange interaction, spin-orbit coupling, spin canting, and magnetic anisotropy are introduced. Spin-dependent transport is described using both thermodynamics and Boltzmann's equation, including Berry curvature corrections. Spin relaxation phenomenology is accounted for with master equations for quantum spin systems coupled to a bath. Magnetic resonance principles are applied to describe spin waves in ferromagnets, cavity-mode coupling in antiferromagnets, and coherence phenomena relevant to spin qubits applications.

Key Features:

- A pedagogical approach to foundational concepts in spintronics with simple models that can be calculated to enhance understanding.
- Nineteen chapters, each beginning with a historical perspective and ending with an outlook on current research.
- 1200 references, ranging from landmark papers to frontline publications.

Jean-Philippe Ansermet is Professor Emeritus at École Polytechnique Fédérale de Lausanne (EPFL), where he pioneered experiments on giant magnetoresistance, current-induced magnetization switching, heat-driven spin torque, and nuclear magnetic resonance. He taught mechanics, thermodynamics, and spin dynamics for more than twenty years. A fellow of the American Physical Society and recipient of the 2022 Credit Suisse Teaching Award, he was an executive board member of the European Physical Society, president of the Swiss Physical Society, and teaching director at EPFL. He has authored or co-authored textbooks on mechanics and thermodynamics and published more than two hundred articles.

Spintronics
A Primer

Jean-Philippe Ansermet

CRC Press
Taylor & Francis Group
Boca Raton London New York

CRC Press is an imprint of the
Taylor & Francis Group, an **informa** business

Designed cover image: © Shutterstock_1895477707

First edition published 2025
by CRC Press
2385 NW Executive Center Drive, Suite 320, Boca Raton FL 33431

and by CRC Press
4 Park Square, Milton Park, Abingdon, Oxon, OX14 4RN

CRC Press is an imprint of Taylor & Francis Group, LLC

© 2025 Jean-Philippe Ansermet

ISBN: 978-1-032-43233-5 (hbk)
ISBN: 978-1-032-44022-4 (pbk)
ISBN: 978-1-003-37001-7 (ebk)

DOI: 10.1201/9781003370017

Typeset in CMR10
by KnowledgeWorks Global Ltd.

Publisher's note: This book has been prepared from camera-ready copy provided by the authors.

To my dedicated group members
who made this research field so much fun to explore

Contents

SECTION III   SPIN TRANSPORT

SECTION IV   SPIN RELAXATION

SECTION V   SPIN RESONANCE

Preface

When fundamental research leads to a widespread application within a few years, the scientific community takes notice. This was the case with the discovery of giant magnetoresistance (GMR). Investigation of newly formed metallic multilayers in which magnetic and non-magnetic layers alternated led to this finding. While cryogenic temperatures and magnetic fields in the tesla range were used initially, within ten years, the effect was observed at room temperature and GMR-based sensors were so sensitive as to detect the stray fields of magnetic bits in a hard disk.

The term "spintronics" began appearing in articles published at the turn of the twenty-first century. These early titles nicely express the enthusiasm for this emerging field. T. Studt boldly asked: *Will spintronics replace conventional electronics?*[1] S.A. Wolf echoed with: *Spintronics, electronics for the next millenium?*[2] The grounding of the field in nano-magnetism was affirmed by B. Heinrich: *Magnetic nanostructures. From physical principles to spintronics.*[3] Some authors readily linked this new field with research on spin qubits. Theoretical work by S. Das Sarma, J Fabian, X.D. Hu and I. Zutic yielded the link: *Spintronics: electron spin coherence, entanglement, and transport,*[4] while a stuby by G. Burhard, H.A. Engel, and D. Loss led them to propose: *Spintronics and quantum dots for quantum computing and quantum communication.*[5]

As researchers forged ahead in their efforts to develop magnetoresistive random-access memories, other scientists explored spin-dependent transport phenomena, inspired by the success story of GMR. Consequently, old problems were revisited, like conduction in manganites or tunnel junctions, leading to groundbreaking advancements. New phenomena were discovered, including spin torques of various origins, and spin-dependent transport revealed by the magnetic field dependence of heat and charge transport. Theoretical investigations coupled with experimental demonstrations unveiled instances of novel spin transport phenomena, including the spin Hall effect and the Hanle effect. Moreover, the spintronics community developed a keen interest in carbon nanostructures due to concerns regarding spin transport over extensive distances. Research on topological materials was another area of significance for the spintronics community, owing to the intrinsic spin-momentum locking feature in these materials.

The vast array of ideas and interconnected phenomena within the realm of spintronics can be overwhelming for newcomers to the field. The intricate blend of concepts pertaining to magnetism, mesoscopic physics and spin dynamics contributes to the challenge of approaching this specialized literature. This book aims to assist novices in the field of spintronics by providing introductions to foundational concepts, organized into four distinct

---

[1]Stud, T., *R&D Magazine*, 41(8), 14–16 (1999)

[2]Wolf, S.A., *J. Superconductivity*, 13(2), 195–199 (2000)

[3]Heinrich, B., *Can. J. Phys.*, 78(3), 161–199 (2000)

[4]Das Sarma, S., Fabian, J., Hu, X.D., Zutic, I, *Superlattices Microstruct.*, 27(5–6), 289–295 (2000)

[5]Burkard, G., Engel, H.A., Loss, D., *Fortschr. Phys.*, 48(9–11), 965–986 (2000)

parts:

- magnetism,
- transport,
- spin relaxation, and
- spin dynamics.

In an effort to facilitate comprehension, the book prioritizes the use of specific examples over general formulations whenever feasible. By adopting this approach, readers can develop a practical understanding that provides a solid foundation for delving into more advanced textbooks and specialized review articles.

Writing a book on an emerging field of research is a significant challenge as the pace of advancements often outstrips the time required to complete the writing process. To address this issue, each chapter of this book is structured as follows. Firstly, a concise **historical introduction** contextualizes the initial exploration of foundational concepts discussed within each chapter. These introductory sections do not strive to provide an exhaustive historical account but rather aim to captivate interest and encourage further exploration. Secondly, the main portion of each chapter delves into those fundamental notions that have withstood the test of time. These enduring concepts serve as the cornerstones of the spintronics field and provide a solid basis for understanding subsequent developments. Lastly, the concluding section of each chapter serves to ignite curiosity by highlighting recent advancements. This final section, purposefully titled "**Further Reading**," seeks to entice readers to engage in independent exploration and discover cutting-edge research and current trends within the field.

Regarding these last sections, it should be noted that the contents of this book were cultivated over a span of twenty years in conjunction with the development of a course on spin dynamics and spintronics. To enhance understanding and preparation, students in the associated course gave oral presentations on topics pertaining to the topics covered on the same day. This approach aimed to reinforce students' comprehension by integrating the lecture materials with the students' own investigations, often resulting in a sense of satisfaction upon witnessing the juxtaposition of foundational and contemporary perspectives. Thus, the final sections within the chapters of this book replace the traditional problem sets of introductory physics textbooks. The students' oral presentations at specific times are particularly suitable for advanced studies.

Acknowledgments

The diversity of research topics I could explore and the sophisticated experimental setups I could develop at École Polytechnique Fédérale de Lausanne (EPFL) were made possible thanks to its robust infrastructure and the wide range of competences accessible through collegial relationships and platforms. This environment enabled me to venture into the synthesis of magnetic nanowires, detection of charge- and heat-driven spin torques, and exploration of new magnetic resonance methodologies.

The University of Illinois at Urbana-Champaign (USA), holds a significant place in my academic journey. It is not only where I pursued my PhD and conducted my post-doctoral research under the guidance of the late Prof. C. P. Slichter, but it also served as a sanctuary where I could immerse myself in research away from administrative responsibilities. During my regular visits and a sabbatical stay, I had the distinct pleasure of continuing my collaboration with Prof. Slichter and had the opportunity to interact with distinguished experts in strongly correlated materials and spin-dependent transport.

At Beihang University (China), the Fert Beijing Institute provided an invaluable international platform for scientific exchanges and stimulating discussions that further broadened the scope of my research activities and helped me to gain insight into microelectronics applications of spintronics. To this day, despite geographical distance, Prof. Haiming Yu and I have maintained a close collaboration. Whenever possible, we facilitated the exchange of students pursuing an internship at EPFL, benefiting from a binational framework agreement. I am particularly indebted to Prof. Dapeng Yu, who sent Prof. Haiming Yu to conduct the experimental part of his PhD under my guidance. Moreover, Prof. Dapeng Yu graciously extended an invitation for me to establish connections in Shenzhen with SUSTech and the International Quantum Academy. Through this fruitful collaboration, along with Prof. Haiming Yu and his research group, I have had great opportunities to foster my interests in quantum materials and magnonics.

When I was building up my lecture notes on spin dynamics and spintronics, I made a point to include passages that I had worked out as a result of group meetings, discussions in the laboratory, or data analysis and modelling prepared for publications and dissertations. I am grateful for my coworkers' unwavering enthusiasm, focused approach, cooperative attitude, and eagerness to cultivate a comprehensive understanding of the phenomena that they observed.

A special acknowledgment is due to Dr. François Reuse, to whom I owe a debt of profound gratitude for his personal encouragement and support spanning three decades. Thanks to his theoretical expertise, I was able to recruit Dr. Sylvain Bréchet and invited him to collaborate with Dr. Reuse on investigating the thermodynamics of irreversible processes in systems where electromagnetic fields are state variables of the system. This collaboration led to the prediction of a heat-driven spin torque in magnetic insulators, which was subsequently experimentally verified within our research group. Dr. Reuse and Dr. Klaus Maschke, both seasoned theoreticians, made significant contributions as guest scientists in my research group. They developed master equations for spin systems coupled to a bath, exploring aspects such as spin relaxation, relaxation of N-electron systems, the quantum Boltzmann transport equation, and its classical limit. As soon as Dr. Bréchet joined my group, he and I collaborated on a textbook on thermodynamics. Our discussions encompassed a broad spectrum of topics, ranging from fundamental concepts to cutting-edge applications.

Dr. Jean-Marie Fürbringer introduced me to the art of generating figures using the LaTeX-based TikZ figure compiler. Recognizing my impatience, he kindly created sample figures for me, which encouraged me to persist in this endeavour. His persistence afforded me the opportunity to appreciate the intellectually stimulating thought processes involved in creating figures entirely by oneself. I am deeply grateful to him for providing me with such a valuable pedagogical tool.

I had the privilege of teaming up with my daughter, Lydia, who took on the role of editing assistant. Her innumerable suggestions have helped me to sharpen the text, straighten the punctuation, and clear the manuscript of typos. I would also like to extend my gratitude to the team of CRC Press, who assisted me in transforming my lecture notes into this book. I greatly appreciated the gentle emails that Rebecca Hogges-Davies sent me yearly until I had time to fully engage in writing this manuscript. I am thankful to Danny Kielty for his constructive inquiries regarding the editing process, and to Emma Brown for her assistance with copyright compliance.

Author

Jean-Philippe Ansermet graduated from Ecole Polytechnique Fédérale de Lausanne, with a degree in physics in 1980. During his doctoral and post-doctoral time at the University of Illinois at Urbana-Champaign, he developed nuclear spin resonance spectroscopy techniques that are relevant for heterogeneous catalysis research. He then worked for five years at the central research facilities of a Swiss chemical company on the dielectric properties of composites and the conductivity of polymer films containing charge transfer salts.

In 1992, he became professor of experimental physics at EPFL. His research focused primarily on spintronics and novel magnetic resonance methods, including sub-THz instrumentation. The latter activity led to a spinoff, Swissto12, now pioneering the development of lightweight, low-cost, geostationary satellites.

J.-Ph. Ansermet's laboratory studied giant magnetoresistance with current driven perpendicular to the interfaces of Co/Cu multilayers that were grown by electrodeposition in nanostructured templates. Using this nanostructure synthesis method, his group participated in the early experimental demonstration of charge-driven spin-transfer torques. His group also gave the first demonstration of heat-driven spin torque in conductive magnetic nanostructures. By developing the thermodynamics of continuous media where electric and magnetic fields are state variables, Ansermet's laboratory predicted a heat-driven spin torque in insulating ferromagnets and proceeded to demonstrate this effect experimentally.

Furthermore, J.-Ph. Ansermet's laboratory studied dynamic nuclear polarization (DNP) as a way of enhancing nuclear magnetic resonance signals, choosing strategies that could ultimately enhance nuclear spin polarization of molecules at electrode surfaces. A collaboration with the Swiss Plasma Center (SPC) led to the construction of a gyrotron which was used to excite the electron spin resonance of free radicals. The sub-THz test equipment developed in this framework was shown to be applicable for the excitation and detection of spin resonance in antiferromagnets, materials of special interest for ultra high frequency spintronics.

J.-Ph. Ansermet wrote a textbook on classical mechanics and co-authored a textbook on thermodynamics with Dr. S. Bréchet. J.-Ph. Ansermet was in charge of coordinating physics teaching for twelve year and received the 2022 EPFL Crédit Suisse award for this teaching commitment. He was member of the European Physical Society executive committee from 1993 to 1998, and president of the Swiss Physical Society from 2002 to 2006. He has been Fellow of the American Physical Society since 2011 and EPFL emeritus professor since 2022.

Section I

INTRODUCTION

1 Magnetoresistance

Albert Fert (born 1938)

Albert Fert studied conduction in ferromagnetic alloys. When the production of metallic multilayers became possible, his research led him to the discovery of giant magnetoresistance. While studying spin transport through multilayers, he closely examined the spin diffusion length, its value for different metals, and means of calculating it. In 2007, A. Fert and P. Gruenberg received the Nobel Prize in physics.

Spintronics: A primer
J.-Ph. Ansermet (https://orcid.org/0000-0002-1307-4864)

Chapter 1. Magnetoresistance

A variety of results are presented, pertaining to the effect of magnetization on charge current (magnetoresistance), and conversely, on the effect of charge current on magnetization (spin torque). This introduction will show that, in order to navigate in the spintronics literature, it is necessary to be acquainted with a few basic notions in the following four themes: magnetism, transport, relaxation, and resonance.

DOI: 10.1201/9781003370017-1

1.1 HISTORICAL INTRODUCTION

1.1.1 SPINTRONICS

The emerging field of spintronics thrives on the many possibilities offered by the prospect of using the spin of electrons in fast, nano-sized electronic devices. The first successes had to do with the effect of magnetization on the flow of conduction electron, leading to *giant magnetoresistance* (GMR) (Fig. 1.1).[1,2] Within 10 years, all reading heads of hard disks were equipped with GMR devices. The fabrication of high-quality tunnel junctions was also achieved,[3] and their promising use in magnetic memories was immediately recognized.[4]

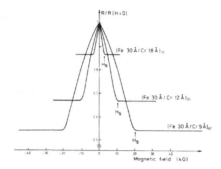

Figure 1.1   First report of giant magnetoresistance (GMR).[1]

The converse effect, i.e., the effect of spin-polarized current on magnetization was also discovered.[5, 6] This effect was first called "current-induced magnetization switching" (CIMS),[1] but it is usually referred to as *spin transfer torque*. This effect opened up a new possibility for writing selectively magnetoresistive memory bits.[7,8] This spin-polarized current forced the magnetization of spin valves into self-oscillation. Thus, ultra-small giga-hertz oscillators called *spin torque oscillators* (STO), could be designed based on this principle.[9]

GMR in metallic multilayers sparked the creation of a new field of research called *spin-tronics*. It is an amazing story to ponder when discussing the interplay of science and society. IBM certainly felt this way and asked on its website, "What is 10 years to you?," with the answer: "To us, it is enough time for a revolution!"

In the early 1980s, it became possible to make *metallic superlattices*. We can assume that physicists asked themselves what would become of the oscillatory response of non-magnetic metals in the vicinity of magnetic impurities (see RKKY coupling in Chapter 5), if instead of isolated magnetic atoms, there were magnetic layers separated by normal metals. This magnetic coupling was studied and found to alternate between ferromagnetic and anti-ferromagnetic interactions. At some point, groups in Paris (France) and Aachen (Germany) had the good idea of measuring the resistance of these thin films. The resistance was found to vary by about 30%. In this introduction, we will see that Fert and others studied spin transport in materials with a random distribution of scattering centers. These magnetic multilayers were fantastic opportunities to test their models with samples of controlled microstructure.

Permalloy, an alloy of iron and nickel, was known to have its resistance change by about 4% when a sufficiently large magnetic field was applied. This material was used in the reading heads of hard disks at the time GMR was discovered. The new measurements (Fig. 1.1) were carried out in a research environment at 4 K and fields of several teslas. It

---

[1]A table of abbreviation is given at the end of the book.

took research and development worldwide only 10 years to transform this discovery into a useful material. Thus, 10 years after this discovery in rather extreme conditions (cryogenic temperatures and very high magnetic fields), the reading heads of all hard disks contained a GMR sensor. What a beautiful example to illustrate the impact of physics on society!

1.1.2 ENGINEERING GMR MATERIALS

Controlling the magnetic configuration

As mentioned above, superlattices or multilayers of a magnetic metal alternating with a non-magnetic metal initially drove GMR sensor developments. One way of getting the antiparallel alignment was to insert a spacer layer just thin enough to produce a magnetic coupling mediated by that spacer, i.e., the RKKY coupling, that favors an antiparallel configuration of the two layers magnetization vectors (Fig. 1.2a). In a structure called a *spin valve* (Fig. 1.2b), one layer has its magnetization pinned, while the other layer is essentially free to reverse. The pinning mechanism that is commonly used relies on an interfacial effect, called *exchange biasing*, between a magnetic layer and an antiferromagnetic layer. This effect has been known since the 1950s.[10] The use of exchange biasing in spin valves has since sparked further research.[11, 12]

Magnetic composite materials

Granular materials with ferromagnetic ultra-fine grains in a non-magnetic matrix (Fig. 1.2c) were also shown to present GMR.[13, 14] A remarkable result was obtained with a combination of cobalt clusters and permalloy layers spaced with silver (Fig. 1.2d). Thus, a magnetoresistive sensitivity of 6.5% per oersted was achieved.[15]

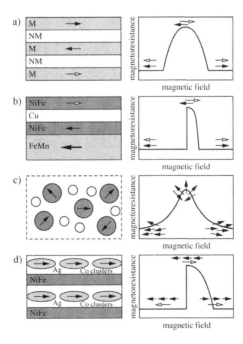

Figure 1.2 Nanostructures for GMR. From top to bottom: a) multilayers make use of an antiferromagnetic coupling among the layers; b) spin valve bi-layer structures have one layer pinned magnetically; c) granular materials; and d) hybrids. (Adapted from Doudin and Ansermet [16].)

Perpendicular spin transport

A simple argument, to be seen later on, is sufficient to predict GMR. However, some of the technical issues required extensive research. Here are a few properties of spin transport that had to be understood and controlled:

-   How far can a spin travel before it flips?
-   What is spin-dependent: the scattering potential itself, the scattering probability, or the density of the states?
-   Does the interface produce spin flips?
-   Is the transmission of spin at an interface spin-dependent?
-   What is the relative importance of bulk and interface scattering in producing GMR?

When the current runs essentially parallel to the interfaces of a multilayer, we speak of ***current-in-plane GMR*** (CIP-GMR). Since thin films were produced at first, CIP-GMR was the geometry of the earlier measurements. When the current runs perpendicular to the interfaces, ***current-perpendicular-to-the-plane GMR*** (CPP-GMR) is measured. CPP-GMR was particularly useful in shedding light on the issues listed above.[17]

CPP-GMR also has a practical advantage over CIP-GMR. In CPP-GMR, layers can be thicker than in CIP-GMR. This is relevant for practical applications, which must consider the ease of manufacturing and the reliability of the materials. The reason for this difference in thickness requirement is as follows. On the one hand, when the current runs parallel to the layers, the electrons experience two layers only if their separation is of the order of the electron mean free path. On the other hand, when the current is perpendicular to the layers, the spins undergo a diffusion process that extends over distances much longer than the electron mean free path since spin flip events are rare compared to electron collisions. Indeed, electron spin resonance experiments of the 1960s and 1970s determined that electrons can travel a long way without flipping their spin.

Nanostructuring for CPP-GMR

CPP-GMR poses some experimental challenges. A small square cut out of a thin film of a magnetic multilayer, sized in the sub-millimeter range, has a resistance in the nano-ohm range. In the early days of GMR, few groups around the world could accomplish such measurements.[18, 19] Some groups relied on advanced lithographic techniques to reduce the surface area and hence bring the resistance to more accessible values.[20] This nanostructuring demands quite an extensive process.[21] Furthermore, once pillars are very thin, the current density is no longer uniform throughout the column. Furthermore, the field sensitivity is reduced because of stray fields and possibly, unwanted magnetic anisotropies (notion introduced in Chapter 4 arising at the surface of the pillars). Both of these effects would cause the resistance transition to take place over a large field window.

Instead of using advanced lithography in order to obtain a thin column, some groups have produced multilayers in the form of wires by electrodepositing multilayers in pores of nanoporous membranes.[22–24] (Fig. 1.3a) Others have used grooved substrates and a straight deposition (Fig 1.3b) to measure GMR with current-at-an-angle (CAP) [25] or deposited the layers sideways, so as to have a structure in which the current is mostly perpendicular to the layers.[26] (Fig. 1.3c)

1.1.3   TUNNEL MAGNETORESISTANCE

Around the time that Slonczewski developed a model for spin-dependent tunneling between two ferromagnetic electrodes,[27] a major effort was launched to make spin valves with tunnel barriers.

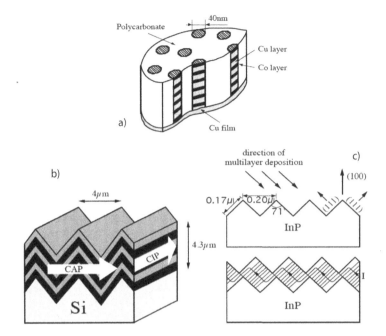

Figure 1.3 Special ways of measuring spin transport perpendicular to the planes: a) multilayered nanowires in nanoporous membranes; b) straight deposition on grooved surfaces; and c) oblique deposition on grooved substrates.

In the early days of tunnel barrier studies, Jullière argued that tunneling between two magnetic layers ought to depend on the relative orientation of the magnetization of the two layers.[28] In the 1980s, Tedrow, Meservey, and Moodera were famous for their studies on tunneling, in particular between magnets and superconductors.[29] In the 1990s, Moodera managed to make excellent aluminum oxide barriers and found sharp transitions and large relative changes in tunnel resistance.[3,30] This effect has been referred to as *tunnel magnetoresistance* (TMR).

The first TMR junctions were made of two ferromagnetic layers separated by an insulator. TMR devices became highly developed because of their potential application as *magnetoresistive random access memories* (MRAM). A TMR device looks like the stack shown in Fig. 1.4. Early research on TMR was done almost exclusively with Al_2O_3 barriers. Inspired by a calculation of Butler's group,[31] Yuasa et al. demonstrated very large TMR using MgO.[32] Thus, MgO has become a top candidate in the development of MRAMs.[33]

Granular materials composed of ferromagnetic metal particles in an insulating matrix also present a form of TMR.[34] There are also materials called *half-metallic ferromagnets* (HMFM), in which only one spin sub-band crosses the Fermi level.[35] CrO_2 is an example. TMR was found when such clusters of CrO_2 are embedded in a matrix of Cr_2O_3 matrix (Fig. 1.5).[36]

1.1.4 COLOSSAL MAGNETORESISTANCE AND HALF-METALLIC FERROMAGNETS

The phenomenon referred to nowadays as *colossal magnetoresistance* (CMR) is linked to GMR for historical reasons.[37] The perovskites that display CMR were already studied in the 1950s.[38] In the 1990s, the race to find any kind of "giant" magnetic field-dependent resistance was under way. For example, in 1994, a group showed that it was possible to

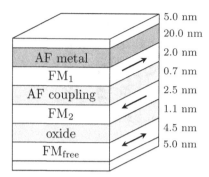

Figure 1.4 In a typical stack forming a magnetic bit, an artificial antiferromagnetically-coupled bilayer is pinned by a thick antiferromagnetic layer. A TMR device is formed by adding an oxide layer and a ferromagnetic layer, which is "free"; i.e., it has a very low switching field. The layer marked "AF coupling" is typically made of ruthenium and the thickness is chosen to produce an antiferromagnetic coupling.

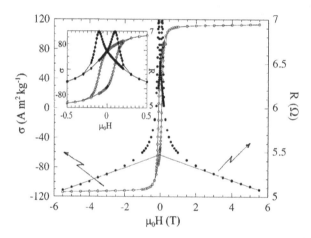

Figure 1.5 TMR in a granular material composed of CrO_2 nanoparticles embedded in a Cr_2O_3 matrix.

"tune" the structure of ***manganite perovskites*** so that the metal-insulator transition, which depends on magnetic field, would occur around room temperature.[39] In terms of the magnetoresistance ratio that is commonly used for anisotropic magnetoresistance (AMR) or GMR, namely, the relative change in resistance with respect to its value at zero field, this transition would amount to a change in resistance of 10,000% or more. Hence, the authors coined this phenomenon CMR.

A connection can be made between spin transport and CMR, in the sense that CMR has to do with the ***double exchange*** (DE) mechanism. This is a magnetic coupling among localized moments mediated by electron hoping between these sites. It was first introduced by C. Zener.[40] P. W. Anderson and H. Hasegawa explained the concept by considering an electron hopping between the two sites of a diatomic molecule in which both atoms carry a magnetic moment (see §2.4). The hopping probability is dependent on the relative orientation of the local moments, owing to the interaction between the electron and the local moment.[41]

However, as early as 1999, a review article warned that the complete explanation of CMR involves additional mechanisms.[42] A strong electron-phonon coupling and orbital ordering are also involved.[43–46] Near the phase transition, spin disorder scattering causes a resistance increase (see §1.2.4), and applying a magnetic field decreases the spin disorder. The DE mechanism alone is insufficient to account for the insulator character of the high-temperature phase. Jahn-Teller distortions (see §3.6) come into play.

Materials that are simpler in structure than manganites present an apparently similar complexity in their magnetic and transport properties. Mauger, for example, treated the case of **europium chalcogenides**. In order to account for ferromagnetism in this material, he proposed a model of ferromagnetism mediated by indirect exchange via conduction electrons, taking into account that the spin bands are split by the coupling to the magnetic lattice.[47–49]

Some materials are **HMFM**.[50] By definition, they are substances in which only one spin sub-band crosses the Fermi level. This means that electrons of only one spin orientation take part in conduction. Depending on the composition of the material, it is the majority band or the minority band that crosses the Fermi level. Some of the so-called Heusler alloys are half-metals. Manganite and the perovskite lanthanum strontium manganese oxide (LSMO) also have this property. These materials have been considered in spintronics application. A complete survey of the practical understanding of these materials and the state of the art in their synthesis can be found in the critical review by Coey, Viret, and von Molnar.[35]

Given their electronic spin structure, HMFM are expected to give rise to very strong TMR. Indeed, when they are used as the magnetic layers in tunnel junctions, a very large TMR is observed.[51] However, it is an open question whether the band structure of these materials determines their transport properties.[52] Spin-polarized spectroscopy is a more direct means of demonstrating the half-metallic character of these materials.[53,54]

1.1.5   EFFECT OF CURRENT ON MAGNETIZATION

Effects seen thus far, such as GMR or TMR, refer to the effect of a magnetic configuration on the ability of the material to transport electric current. In this section, we highlight major results concerning the converse effect: the effect of spin-polarized currents on magnetization.

In 1996, Berger [56] and Slonczewski [57] independently predicted that a spin-polarized current, injected into a magnetic layer, would flip the magnetization of this layer, if the current exceeded some threshold, estimated by both at about 10^7 A/cm^2. Initial experimental verifications appeared soon after.[6,58,59] This effect was called **CIMS**. The first observations were of a quasi-static sort. Either the magnetization following an impulse of current was detected,[8,55] or hysteresis loops were observed as a function of current (Fig. 1.6), as was first demonstrated by the groups of Buhrman and Ralph at Cornell University.[60]

CIMS was immediately perceived as a very important new field of research for two reasons. First, ensuring a physical description of the effective underlying mechanism was a challenge to the basic understanding of transport in magnetic materials. Second, this effect pointed to a possible efficient control of MRAMs.[7,8,61]

In the early years of the discovery of a spin torque acting on magnetization, Tsoi et al. discovered, using point contacts, that a strong current density could excite the magnetization dynamics of multilayers (Fig. 1.7).[58] Two-level fluctuations were observed in spin valves that could be interpreted with a charge-induced spin current model.[62–64] Over time, it became clear that a constant current could excite and maintain the precession of the magnetization. The first report showed broad self-resonances in spin valve nano-pillars (Fig. 1.8).[9] Devices where these self-resonances are excited are referred to as STO.

Later publications demonstrated much sharper resonances.[65] This improvement came about by perfecting the nanostructuring of the spin valves.[66] Instead of a zero-frequency

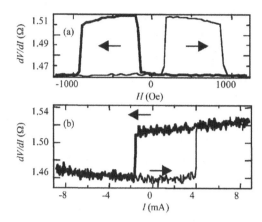

Figure 1.6 Magnetoresistance of a Co-Cu-Co nanopillar, plotted as the differential resistance as a function of field (top) or current (bottom).[55]

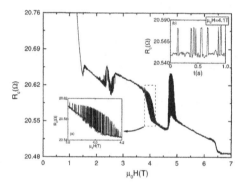

Figure 1.7 Magnetic excitation at a point contact to a magnetic multilayer.[58]

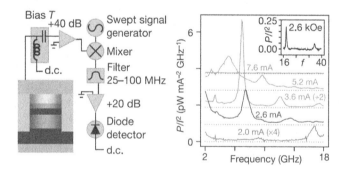

Figure 1.8 Principle of detection (left) and power spectrum (right) of the voltage generated by the excitation of the free layer of a spin valve by DC current.[64]

excitation, it also became possible to resonantly excite the free layer of a spin valve by CIMS.[67] The same was also achieved using MgO tunnel barriers. This experiment was truly remarkable, as the high current density was driven through a tunnel barrier and, furthermore, the magnetoresistance was of the order of 300%.[68] Whenever a current at frequency f is driven through a spin-valve, the magnetization oscillates at the same frequency. As a consequence, the voltage across the device is composed of two frequency components, one at frequency $2f$ and one at zero frequency. The existence of this zero-frequency component is often referred to as the ***spin-diode effect***, although the device is by no means a spin filter.

1.2 MAGNETORESISTANCE PHENOMENOLOGY

In this section, we introduce simple phenomenological models that account for various spin-dependent transport phenomena, namely: spin mixing, GMR, TMR, spin-disorder scattering, and the interaction between charge current and magnetization dynamics. In Chapter 8, we will show how to use the thermodynamics of irreversible processes to describe spin-dependent transport. In Chapter 9, we will analyze spin-dependent transport using the Boltzmann transport equation. In Chapter 15, we will see that thermodynamics can also describe the interaction between a charge-driven spin current and magnetization.

1.2.1 THE TWO-CURRENT MODEL

Mott introduced the idea of ***spin-up*** and ***spin-down currents*** in ferromagnets.[69, 70] He reasoned that, in a ferromagnet, we can assume spins that are polarized parallel or antiparallel to the magnetization will have different resistivities. This is often referred to as the ***Mott picture***. This model is based on an approximate model of the electronic structure of a transition-metal ferromagnet. The so-called "sp-bands"[2] are assumed to carry the current. The conduction electrons scatter into "d-bands" that are split by exchange (see Chap. 2). One of the d-bands is full, while the other has unoccupied states at the Fermi level. (Fig. 1.9). Hence, in working out the scattering amplitudes, the density of the final states may, roughly speaking, favor or forbid scattering (Fig. 1.10). Therefore, we write the

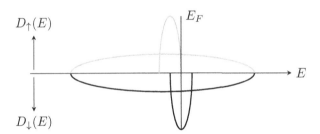

Figure 1.9 Schematics of the density of states for transition metals. The d-bands have a narrow (in energy) density of states, whereas the density of states of sp-bands is much broader. AT $T = 0$, states below E_F are occupied, and all states above E_F are empty.

resistivities of spins parallel or antiparallel to the magnetization as,

$$\rho_\uparrow = \rho(1 - \beta) \qquad \rho_\downarrow = \rho(1 + \beta) \tag{1.1}$$

where β is called the ***spin asymmetry*** of the resistivity.

The two-current model was used by Fert and Campbell to elucidate the departure from Matthiessen's law [73, 74] in ferromagnets doped with impurities, as shown below.[75–78]

According to ***Matthiessen's law***, the total scattering rate is given by the sum of the scattering rates corresponding to various scattering processes.[72, 79] Thus, we write an

---

[2]Terminology of the tight-binding method; see, e.g., [71, 72]. We will allude to this method in §13.2.3.

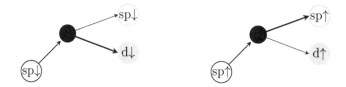

Figure 1.10 Left: a minority sp-electron scatters preferentially into the d-band and, hence contributes very little to the conduction. Right: the majority sp-electron cannot scatter into a filled d-band, so it scatters preferentially into the sp-band.

effective relaxation rate as,

$$\frac{1}{\tau} = \frac{1}{\tau_1} + \frac{1}{\tau_2} \tag{1.2}$$

As a consequence, according to the simple Drude model of resistivity, the resistivities due to these various processes are added as,

$$\rho = \rho_1 + \rho_2 \tag{1.3}$$

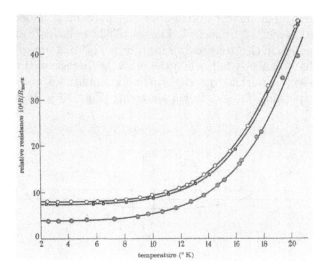

Figure 1.11 Resistance as a function of temperature for two samples of potassium doped with impurities.[80]

For example, the resistance of potassium doped with impurities was found to comprise two contributions, made evident by measuring the temperature-dependence resistivity as a function of doping (Fig. 1.11).[79] The low temperature limit is due to varying concentrations of impurities, while the temperature-dependent component is identical in all samples to that of the pure metal. Thus, Matthiessen's law may be used to estimate the resistivity of a sample when the temperature dependence of the pure metal is known.

When a ferromagnet is doped with impurities, deviations from Matthiessen's law may be observed. For example, Campbell, Fert, and Pomeroy reported the deviation shown in Fig. 1.12, which they wrote as:[75]

$$\Delta\rho(T) = \rho(T) - \rho_0 - \rho_P(T) \tag{1.4}$$

where ρ_0 is the low temperature limit of the resistivity and $\rho_P(T)$ is the temperature dependence of the pure metal. In order to account for this deviation, the two-current model

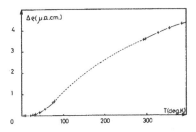

Figure 1.12   Deviation $\Delta\rho(T)$ from Matthiessen's law as a function of temperature.[75]

was applied by Fert and Campbell.[76] They assumed that, at low temperatures, the current would be carried in parallel by electrons with spin-up and spin-down electrons. The introduction of impurities would generate for each spin direction a residual resistivity $\rho_\uparrow(0)$ and $\rho_\downarrow(0)$. The overall resistivity would then be:

$$\rho(0) = \frac{\rho_\uparrow(0)\rho_\downarrow(0)}{\rho_\uparrow(0) + \rho_\downarrow(0)} \tag{1.5}$$

Now, having assumed this, we must ask ourselves how correct this might be. It turns out that, as the temperature rises, processes can mix momentum from one spin channel to the other. There are two processes known to produce this effect. One is the elastic scattering of electrons by spin waves. The other is a mutual spin-flip process between two interacting electrons. The description of these mechanisms is addressed later in this chapter.

We denote the average velocity in each channel by $\bar{v}_\uparrow$ and $\bar{v}_\downarrow$. In the Drude picture, we say that the electron is accelerated by the electric field E in between two collisions. We assume that this collision time depends on whether the electron has minority or majority spin. Thus, we distinguish $\tau_\uparrow$ and $\tau_\downarrow$. Furthermore, we include in our description processes that flip the spin without any momentum relaxation effect, called ***spin-mixing*** processes.

We write a detailed balance for each momentum channel. Thus, the momentum rate of change in each spin channel is given by spins leaving that channel or arriving from the other, at a rate $\tau_{\uparrow\downarrow}^{-1}$:

$$\begin{aligned}
\left\langle \frac{dp_\uparrow}{dt} \right\rangle &= \frac{m}{\tau_\uparrow}\bar{v}_\uparrow = -eE - \frac{m}{\tau_{\uparrow\downarrow}}\left(\bar{v}_\uparrow - \bar{v}_\downarrow\right) \\
\left\langle \frac{dp_\downarrow}{dt} \right\rangle &= \frac{m}{\tau_\downarrow}\bar{v}_\downarrow = -eE - \frac{m}{\tau_{\uparrow\downarrow}}\left(\bar{v}_\downarrow - \bar{v}_\uparrow\right)
\end{aligned} \tag{1.6}$$

For each spin channel, we define a current and a resistivity,

$$\begin{aligned}
i_\uparrow = ne\bar{v}_\uparrow \qquad & i_\downarrow = ne\bar{v}_\downarrow \\
\rho_\uparrow = \frac{m}{ne^2\tau_\uparrow} \qquad & \rho_\downarrow = \frac{m}{ne^2\tau_\downarrow}
\end{aligned} \tag{1.7}$$

It is convenient to define a ***spin-mixing resistivity***, given by,

$$\rho_{\uparrow\downarrow} = \frac{m}{ne^2\tau_{\uparrow\downarrow}} \tag{1.8}$$

Finally, the effective resistivity is calculated as,

$$\rho(T) = \frac{E}{i_\uparrow + i_\downarrow} \tag{1.9}$$

A few algebraic manipulations yield,[76]

$$\rho = \frac{\rho_\uparrow \rho_\downarrow + \rho_{\uparrow\downarrow}(\rho_\uparrow + \rho_\downarrow)}{\rho_\uparrow + \rho_\downarrow + 4\rho_{\uparrow\downarrow}} \tag{1.10}$$

We can write each spin channel resistivity in the spirit of Matthiessen's law as,

$$\begin{aligned} \rho_\uparrow(T) &= \rho_\uparrow(0) + \rho_{i\uparrow}(T) \\ \rho_\downarrow(T) &= \rho_\downarrow(0) + \rho_{i\downarrow}(T) \end{aligned} \tag{1.11}$$

where $\rho_{i\uparrow,\downarrow}(T)$ $(i = 1, \dots, 2)$ are meant as the intrinsic resistivity of each spin channel of the pure metal.

It is possible to obtain the spin-dependent scattering parameters based on the observed deviations from Matthiessen's law. Indeed, at low enough temperatures, we have,

$$\rho_\uparrow(0), \rho_\downarrow(0) >> \rho_{i\uparrow}(T), \rho_{i\downarrow}(T) \tag{1.12}$$

Therefore, to first order, we have,[76]

$$\rho = \rho(0) + \rho_i(T) + \rho_i(T)\frac{(\alpha - \mu)^2}{(1 + \alpha)^2} + \rho_{\uparrow\downarrow}\left(\frac{\alpha - 1}{\alpha + 1}\right)^2, \tag{1.13}$$

where,

$$\rho_i(T) = \frac{\rho_{i\uparrow}(T)\rho_{i\downarrow}(T)}{\rho_{i\uparrow}(T) + \rho_{i\downarrow}(T)} \tag{1.14}$$

Hence, the deviation from Matthiessen's law is given by,

$$\Delta\rho = \rho_i(T)\frac{(\alpha - \mu)^2}{(1 + \alpha)^2} + \rho_{\uparrow\downarrow}\left(\frac{\alpha - 1}{\alpha + 1}\right)^2 \tag{1.15}$$

In these formulas, the **spin asymmetries** are characterized by the coefficients,

$$\alpha = \frac{\rho_\downarrow(0)}{\rho_\uparrow(0)} \qquad \mu = \frac{\rho_{i\downarrow}(T)}{\rho_{i\uparrow}(T)} \tag{1.16}$$

Typical data are shown in Fig. 1.13. Fert and Campbell recognized, in the temperature-dependent part of their resistivity data for doped samples, a correction proportional to $\rho_i(T)$ plus a residual resistivity, which they attributed to $\rho_{\uparrow\downarrow}(T)$. The soundness of the model was tested by analyzing ternary alloys. Their results were summarized in a review paper.[78]

1.2.2 SPIN-VALVE GMR

Let us now consider a spin valve composed of two ferromagnetic layers separated by a non-magnetic metallic spacer. The magnetization is uniform in each layer and the magnetization of both layers can have either one of two magnetic configurations: parallel (P) or antiparallel (AP) alignment of the magnetization. Furthermore, spin-mixing and spin-flip processes are supposed to be negligible. In particular, electrons do not flip their spins going from one layer to the other (Fig. 1.14).

In the parallel (P) configuration, one spin orientation experiences $\rho_\uparrow$ in both layers, and the other spin orientation experiences $\rho_\downarrow$ in both layers. Using the notation of (1.1), the effective resistivity is given by:

$$\rho_P = \frac{2\rho_\uparrow \rho_\uparrow}{\rho_\uparrow + \rho_\uparrow} = \rho(1 - \beta^2) \tag{1.17}$$

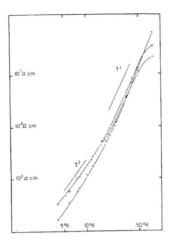

Figure 1.13 Low temperature resistivity variations. Square: ideal resistivity $\rho_i(T)$ of pure nickel. Circles: deviation $\Delta\rho(T)$ for nickel doped with 1400 ppm of Cr impurities. Crosses: same for 700 ppm Cr doping.[76]

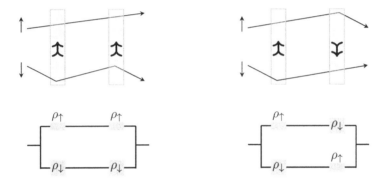

Figure 1.14 Top: magnetic configurations of two ferromagnetic layers separated by a non-magnetic spacer. Bottom: resistance network that expresses the spin-dependent resistivities (1.1).[81]

In the antiparallel (AP) configuration, each spin channel experiences $\rho_\uparrow$ and $\rho_\downarrow$ in series. The effective resistance is given by,

$$\rho_{AP} = \frac{\rho_\uparrow + \rho_\downarrow}{2} = \rho \tag{1.18}$$

The effective resistance is greater in the antiparallel (AP) configuration than in the parallel (P) configuration, no matter the sign of β. This model is a direct consequence of the two-current model, where the currents of both spin orientations are assumed to be separate.

The two-current model in the limit where no spin mixing takes place has been used by most authors to analyze GMR data. This is a reasonable model because scattering processes giving rise to spin mixing are rare.[1] However, Gijs et al. included the temperature-dependent term $\rho_{\uparrow\downarrow}(T)$ (1.8) in order to analyze successfully their measurements of the temperature dependence of CPP-GMR.[82]

1.2.3 JULLIÈRE'S TUNNEL MAGNETORESISTANCE

The experimentalist who wants to characterize a tunnel junction may refer to the work of Simmons.[83, 84] He showed that the current J is a linear function of the voltage V across a tunnel junction when $V \approx 0$, with,

$$J = J_L \sqrt{\bar{\varphi}} V \exp\left(-A\sqrt{\bar{\varphi}}\right), \qquad (1.19)$$

where

$$J_L = \frac{\sqrt{2m}}{\Delta s} \left(\frac{e}{\hbar}\right)^2 \quad \text{and} \quad A = \frac{4\pi \Delta s}{\hbar} \sqrt{2m}, \qquad (1.20)$$

m is the mass of the electron and $\hbar$ is Planck's constant, while Δs is the thickness of the barrier and $\bar{\varphi}$ is its height. In his derivation, Simmons assumed two identical electrodes with a density of states that is independent of momentum and energy.

Instead, Jullière considered two magnetic electrodes characterized by their electron spin polarization, P_1 and P_2. He defined the TMR ratio (TMR) as the relative change of conductance when the junction had parallel (P) or antiparallel (AP) magnetization. He considered the conductance of the tunnel junction in the parallel (P) and the antiparallel (AP) configurations to be proportional to the probability a of having an electron with majority spin orientation, and $1 - a$ when it is a minority spin. These probabilities are, respectively, a' and $1 - a'$ in the other electrode. Jullière assumed that no spin flip took place in the tunnel barrier. Thus, the conductance in the parallel (P) configuration is given by $G_P = aa' + (1-a)(1-a')$ whereas it is $G_{AP} = a(1-a') + (1-a)a'$ in the antiparallel (AP) configuration. The polarizations are $P = 2a - 1$ and $P' = 2a' - 1$. Straightforward algebra leads to,[28]

$$TMR = \frac{G_P - G_{AP}}{G_P + G_{AP}} = P_1 P_2 \qquad (1.21)$$

The spin polarizations P_1 and P_2 can be measured by tunneling from the material under examination into a superconductor.[85] The fundamental transport equations for tunneling into a BCS-type superconductor [3] were first established by Bardeen.[86, 87]

Slonczewski carried out a quantum mechanical calculation of spin tunnelling between two ferromagnets. He used a free-electron model for the conduction electrons, a rectangular barrier potential, and an internal exchange energy in the magnetic layers of the form:

$$-\boldsymbol{h} \cdot \boldsymbol{\sigma} \qquad (1.22)$$

The direction of the molecular field $\boldsymbol{h}$ in one layer differs by an angle θ from its value in the other layer, whereas the magnitude of $\boldsymbol{h}$ is the same in both layers. The junction conductance is found to vanish in the antiparallel (AP) configuration, provided only one spin band is present at the Fermi level. This would be the case when using HMFM. In the imperfect case where both spin bands cross the Fermi level, the junction conductance has the form:

$$G = G_{fbf} \left(1 + P_{fb}{}^2 \cos\theta\right), \qquad (1.23)$$

where

$$P_{fb} = \frac{(k_\uparrow - k_\downarrow)}{(k_\uparrow + k_\downarrow)} \frac{\left(\kappa^2 - k_\uparrow k_\downarrow\right)}{\left(\kappa^2 + k_\uparrow k_\downarrow\right)} \qquad (1.24)$$

Here, $k_\uparrow$ and $k_\downarrow$ are the spin up and spin down momentum values in the metals, and $i\kappa$ is the imaginary momentum in the barrier. G_{fbf} is proportional to $\exp\left(-2\kappa d\right)$ where d is the

---

[3]That is, a material where the superconductivity is the one described by the theory of Bardeen, Cooper, and Shrieffer.

barrier thickness. The first fraction in the expression of P_{fb} is a fractional spin polarization. The second factor of (1.24) is conceptually new compared to Jullière's model. It expresses a dependence of the effective relative polarization on the height of the barrier.

Slonczewski's model provides the basis for deriving two other effects besides TMR. First, an effective *junction exchange coupling* is found, of the form:

$$E(coupling) = -J_{\text{eff}} \cos \theta \qquad (1.25)$$

where the coupling constant J_{eff} varies in sign and magnitude depending on the barrier height. Second, Slonczewski derived a dynamical term that represents a damping for one sign of the applied voltage and excitation of oscillation of the magnetization for the other sign of the applied voltage.

Chui considered the difference in spin up and spin down chemical potentials $\Delta\mu$ at the ferromagnet/insulator interface. He found that $\Delta\mu$ depends on whether the magnetization vectors of the adjacent layers are parallel (P) or antiparallel (AP). He pointed out that these chemical potential differences result in a large bias dependence of the magnetoresistance.[88]

Vedyayev et al. pointed out the possibility of strongly enhanced TMR that could be obtained by adding to the tunnel structure a thin paramagnetic layer. Selective multiple reflections of either spin up or spin down electrons in this layer can result in very large MR effects at specific values of the thickness of the paramagnetic layer. This is the spin equivalent of an optical filter on a camera lens.[89]

Butler et al. cautioned against using the simple potential barrier models as discussed above. Their calculation of the electronic structure of spin-dependent structures such as Fe/Ge/Fe or Fe/GaAs/Fe indicated that the local density of states at the Fermi level at the interfacial metallic layer differs significantly from its value in the bulk of the metal. Furthermore, the band structure of the majority and minority spins in the semiconductor barrier differs significantly, although the barrier remains non-magnetic.[90] These considerations confirm the complications that Alvarado et al. detected earlier when using magnetic tips to study spin dependent tunneling.[91] They had argued that the hybridization of the wave function of the tip with the d-states of the substrates could lead to great variation in the spin-dependent current.

There can be impurities in a tunnel barrier, allowing for a two-step tunneling process.[92]. Doudin et al. found the following expression for the tunnel conductance $G(E)$ for a one-dimensional tunneling process via an impurity:[93]

$$G(E) = \frac{4e^2}{h} \frac{\Gamma_1 \Gamma_2}{(E - E_i)^2 + (\Gamma_1 + \Gamma_2)^2} \qquad (1.26)$$

where E_i is the energy of the impurity, and $\Gamma_1/\hbar$ and $\Gamma_2/\hbar$ are leak rates of electrons to the left and right electrodes. Off resonance, when $|E - E_i| \gg \Gamma_1 + \Gamma_2$, this formula is equivalent to Jullière's (1.21), when making the same assumptions about the spin polarization in the electrodes. On resonance, i.e., $E = E_i$, the TMR was shown to be of opposite sign.[93]

Zhang and Butler calculated the coherent tunneling in thin crystalline layers of MgO.[31, 94] They found that the majority-spin Bloch states with Δ_1 symmetry in the Fe electrodes decay as evanescent states with the same symmetry in MgO (Fig. 1.15). The Bloch states with Δ_5 and Δ_2 symmetries have both majority and minority spins. They decay very strongly in MgO.

Using scanning tunneling microscopy (STM), it is possible to detect spin-dependent tunneling using a spin-polarized tip.[95, 96]

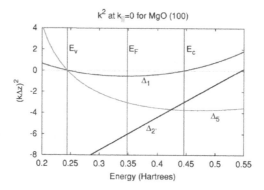

Figure 1.15 Decay rates (negative k^2) within MgO of Bloch states of Fe (100).[31]

1.2.4 SPIN-DISORDER SCATTERING

In the 1960s, the transport properties of rare earth metals and magnetic semiconductors were studied. In both cases, a peak in resistivity was observed at T_c. The magnetoresistance was quite large and was interpreted by many as arising from spin-disorder scattering. The case of metals was analyzed by P. G. de Gennes and J. Friedel.[97] As shown in the review article of C. Haas on magnetic semiconductors,[98] the general idea is that the spin of the conduction electrons scatters with localized magnetic spins owing to the exchange coupling between the two.

To clarify this concept, we follow de Gennes and Friedel. We assume that the conduction electrons can be treated as free electrons with an effective mass m. We assume the presence of a lattice of magnetic ions of spins S, concentration c, separated by a distance d, with Z nearest neighbors. The coupling between two localized moments is given by,

$$2J\boldsymbol{S}_R \cdot \boldsymbol{S}_{R'} \tag{1.27}$$

The coupling of the conduction electrons to the localized moments is assumed to be of the form :

$$\mathcal{H} = \sum_{R,p} G\,\delta\left(\boldsymbol{R} - \boldsymbol{r}_p\right)\left(\boldsymbol{S}_R \cdot \boldsymbol{s}_p\right) \tag{1.28}$$

In both cases, the transfer of energy from the conduction electron spins to the magnetization is negligible. In semiconductors, only small k values can be considered, whereas large k values must be included in metals, as did Friedel and de Gennes. However, it turns out that the theoretical treatment of Haas for small k does not change the overall picture much (see Fig. 1.16) and we follow the work of the former, since our objective here is simply to grasp the general physical picture.

De Gennes and Friedel apply the Born formula for scattering and get:

$$\frac{d\sigma}{d\Omega} = \left(\frac{m}{2\pi\hbar^2}\right)^2 \sum_{\alpha,\beta} p_\alpha \left| \langle\alpha| \sum_R \exp(i\boldsymbol{\kappa} \cdot \boldsymbol{R})\left(\boldsymbol{s}_p \cdot \boldsymbol{S}_R - \boldsymbol{s}_p \cdot \langle\boldsymbol{S}_R\rangle\right) |\beta\rangle \right|^2 \tag{1.29}$$

Here $|\alpha\rangle$ and $|\beta\rangle$ are the spin states of the overall system (electron plus localized moments), p_α the probability of finding the system in the state $|\alpha\rangle$ at thermal equilibrium, and $\hbar\boldsymbol{\kappa}$ the momentum change of the electron due to the collision. By summing over the final states,

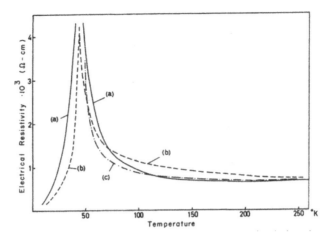

Figure 1.16 Resistivity at zero field of $Eu_{0.95}Gd_{0.05}Se$ (line a) compared with the theoretical prediction according to de Gennes and Friedel (curve b) and C.Haas (curve c).[98]

de Gennes and Friedel find,

$$
\frac{d\sigma}{d\Omega} = \left(\frac{m}{2\pi\hbar^2}\right)^2 \sum_{\alpha,\beta} p_\alpha \Big| \langle\alpha| \sum_{R,R'} \exp\left((i\boldsymbol{\kappa}\cdot(\boldsymbol{R}-\boldsymbol{R}'))\right) \\
G^2 \left(\boldsymbol{s}_p\cdot\boldsymbol{S}_R - \boldsymbol{s}_p\cdot\langle\boldsymbol{S}_R\rangle\right)\left(\boldsymbol{s}_p\cdot\boldsymbol{S}_{R'} - \boldsymbol{s}_p\cdot\langle\boldsymbol{S}_{R'}\rangle\right)|\alpha\rangle\Big|^2
$$

(1.30)

Thus, there is a thermal average of the expectation value of a correlation function that involves the electron spin operator and the spin of local moments at $\boldsymbol{R}$ and $\boldsymbol{R}'$. The average of the operators on the electron spin subspace yields a coefficient $\frac{1}{2}s_p(s_p+1)$ that multiplies a correlation function for the local moments:

$$
\langle\boldsymbol{S}_R\cdot\boldsymbol{S}_{R'}\rangle_T - \langle\boldsymbol{S}_R\rangle_T\langle\boldsymbol{S}_{R'}\rangle_T
$$

(1.31)

Given the fluctuation-dissipation theorem (see §12.5), there is a link between this correlation function and the susceptibility of the magnetic system. Thus, where a peak in susceptibility is found, one can likewise expect a peak in resistivity, since resistivity is directly linked to the scattering cross-section.

   If the moments were uncorrelated, then the scattering at one site would be independent of the moments at other sites. In particular, the scattering would be isotropic and the cross-section would also be isotropic. Then the relaxation time τ would be given by,

$$
\frac{1}{\tau} \propto 1 - \frac{\langle\boldsymbol{S}^2\rangle}{S(S+1)}
$$

(1.32)

This expression illustrates the connection between resistivity and the magnetic order parameter.

1.2.5   CURRENT-MAGNETIZATION INTERACTION: MAGNON DRAG

C. Haas [98] points out an important limitation to this approach: the electron occupation probability of the system at equilibrium in a given array of magnetic moments is used in the calculations. The effect of the current on the equilibrium magnetization and how this in turn changes the electron populations is not taken into account. This effect was expected by

Haas and others that he cites in his review and is referred to as ***magnon drag*** (see Chapter 9). Notably, this effect was introduced long before the predictions of Sloncewski and Berger regarding spin torque.[56,57] However, Berger in the 1970s worked out a phenomenological description of magnon drag and predicted its impact on thermoelectric power.[99]

Magnon drag was described in the framework of Boltzmann's transport equation.[100] A two-band model was applied to the calculation of the thermopower.[101] The effect of magnon drag on electron mass and mobility was also established.[102] Evidence for magnon drag was obtained in a few works on thermopower.[103] A strong interaction between a DC current and spin waves was also observed in bulk magnetic semiconductors.[104]

De Gennes and Friedel [97] estimated in a scattering process at a magnetic center, the energy transfer from the spin of incoming electrons to the magnetization. Their calculation showed that one should consider this type of inelastic scattering at low temperatures. In order to find a numerical estimate for this effect, they drew an analogy to magnetic neutron scattering.[105,106]

Clearly, these subtle effects cannot account for the magnetoresistance that was already observed in the 1960s, e.g., in doped EuSe (Fig. 1.17).[107] This type of effect requires a different mechanism and, nowadays, this kind of magnetoresistance is called "colossal".

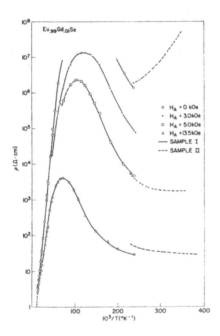

Figure 1.17   Resistivity of $Eu_{0.99}Gd_{0.01}Se$ as a function of temperature for several values of the applied field.[107]

A very large magnetoresistance can arise in the case where a strong correlation of conduction electron spins with the lattice magnetic moments leads to the formation of ***magnetic polarons***.[108–110].

1.3   FURTHER READINGS

Spin injection

A ***half-metallic ferromagnet*** has electrons of one spin orientation in a metallic state, while the other spin orientation has semiconductor or insulator characteristics. Evidence for the half-metallic character of magnetite (Fe_3O_4) was obtained by spin-resolved photo-electron spectroscopy.[111] Interfacial charge and polarization modulation were observed in thin film Fe_3O_4.[112] Furthermore, a **Schottky barrier** effect was detected in Fe_3O_4/ZnO and Fe_3O_4/Nb: SrTiO3 heterostructures. [113]

Spin injection from a ferromagnetic metal into a semiconductor is not possible, because of the great conductance mismatch between the materials in contact.[114,115] The injection must be carried out by tunneling. Let us mention two examples from the abundant literature. First, an alloy of cobalt, iron, and boron (CoFeB) could serve as a source of spin-polarized spins that are injected through an MgO tunnel barrier.[116] Second, magnetite (Fe_3O_4) could be used to inject spins into a GaAs quantum well.[117]

Injecting spin-polarized currents into a light-emitting diode has an effect on the polarization of the emitted light.[118] Hot electrons were injected into spin-valve photo-diodes.[119]

Proximity effect

We have seen that materials like Pt and Pd are close to satisfying the Stoner criterion (6.16) for ferromagnetism. Therefore, when these metals are deposited on a ferromagnet, we can expect these metals to become ferromagnetic because of a **proximity effect**. Transport measurement can reveal this induced ferromagnetism, typically by detecting the magnetoresistance induced by the underlying ferromagnet.[120–123]

Tunneling

Tunneling spectroscopy consists of measuring with high precision the $I(V)$ characteristics of a tunnel junction in order to study, for example, vibrational modes.[124] It's been used to study ferromagnetic dilute semiconductors.[125] Tunneling can also be used to study superconductors, e.g., the influence of strong spin orbit scattering.[126] Double-barrier junctions can reveal the quantized states of a quantum well.[127] More importantly for us, tunneling can be used to estimate the spin polarization of conduction electrons in metallic ferromagnets.[128–130]

Early attempts at injecting spins from a magnetic tip into GaAs showed that the effect could be marred because the electronic orbitals of the substrate and the tip may overlap.[91] The use of an antiferromagnetic tip was investigated.[131] The use of a CrO_2 tip gave clear spin-dependent tunneling.[95] **Spin-resolved STM** imaging became possible.[132–134]

Tunnel junctions can also be made with **organic** materials.[135,136] By photoemission spectroscopy, it was found that many organic materials maintain the high spin polarization of the substrate onto which they are deposited.[137] The mode of spin transport in the organic layer was debated.[138] The voltage-dependence of TMR was analyzed.[139]

Magnetic memories

In this chapter, we have referred to tunneling as a transport phenomenon. Of course, this aspect is fundamental, since TMR is a prime candidate for the manufacturing of magnetic memory bits in MRAM. When the size of the magnetic bit is in the nanometer range, coercivity becomes an issue. One way to overcome this issue is to make use of shape anisotropy.[140] It is possible to control the magnetic orientation of one layer of a spin valve or magnetic tunnel junction (MTJ) by driving a current through it. Spin-transfer torque is the method of choice, but it is also possible to make use of the spin-orbit torque effect.[141]

Making memories capable of logical operations would allow microprocessors to go beyond the von Neumann architecture. One possible option is to perform logic operations on current-driven magnetic domain walls.[142]

Spin-independent magnetoresistance: magneto-impedance

Since the early 1990s, there has been an intense interest in very large variations of the impedance of a magnetic wire. These effects arise in soft magnetic wires and ribbons at frequencies when the skin depth becomes small. Then, a small change in field is sufficient to alter the domain configuration and thus the differential susceptibility. Changes of impedance of the order of several hundred percent with as little as 1 mT can be obtained.[143] In [144], the impedance was measured up to the frequency range where ferromagnetic resonance occurs (Chapter 15).[144] Magneto-impedance may possibly have applications in magnetoresistive memories.[145–147]

Section II

MAGNETISM

2 Exchange

Walter Heitler (1904–1981)

In 1927, while working in Zurich with F. London, W. Heitler used the newly established quantum mechanics to investigate the electronic structure of the hydrogen molecule. In 1941, W. Heitler became a professor at the Dublin Institute for Advanced Studies, where E. Schroedinger served as director. By 1949, W. Heitler had transitioned to the University of Zurich, where he continued his career in theoretical physics.

Spintronics: A primer

J.-Ph. Ansermet (https://orcid.org/0000-0002-1307-4864)

Chapter 2. Exchange

Some fundamental notions of magnetism are defined: exchange, superexchange, spin-orbit coupling, the Zeeman interactions, and the Dzyaloshinskii-Moriya interaction. Exchange is introduced by way of a few molecular examples, the notion of spin-orbit coupling is derived from a non-relativistic argument, and the Dzyaloshinskii-Moriya interaction is inferred as a second-order effect of exchange and spin-orbit interactions.

DOI: 10.1201/9781003370017-2

2.1 HISTORICAL INTRODUCTION

The work of Langmuir in 1919, followed soon after by that of Lewis, Stoner, and Bohr, suggested that, in alkali metals, the electrons are clustered in shells with occupancies of 2, 8, or 18 electrons. In 1925, Pauli tried to explain spectroscopic data by assuming that four quantum numbers are needed to describe an electronic state, the principal quantum number n, the angular number l, another angular momentum $j = l \pm 1/2$, and finally m_j, the now-called magnetic number. In his 1925 paper,[148] Pauli wrote that the two-valuedness of this additional number *cannot be described from the classical point of view*. The following year, Uhlenbeck and Goudsmit discussed the Zeeman effect in optical spectra. They argued that the Zeeman effect could be accounted for provided that the gyromagnetic ratio (the ratio between the magnetic moment and the angular momentum) were twice as large for the spin as for the orbital angular momentum.[149,150] A series of papers sought to ascribe Pauli's new quantum number to the spinning of the electron. This turned out to be physically untenable, but the term ***spin*** remained.[151]

The Stern-Gerlach experiment was carried out in 1922. Stern's motivation was to measure a magnetic birefringence expected because of the orbital motion of the electron.[152] Its significance in demonstrating that there was a spin magnetic moment came only after Fraser realized in 1927 that the ground state orbital moment of silver, hydrogen, and sodium was zero.[153] This meant that Stern and Gerlach had measured the unpaired spin of the silver atoms used in their experiment.[152]

Based on the discovery of several elementary particles in the 1930s, Pauli was able to formulate the exclusion principle. It states that if we permute two particles of an N-particle wave function $\Psi(1, 2, ..., i, ..., j, ...N)$ representing the state of N identical particles of spin s, then the only possible wave functions are those that are transformed by this permutation into $(-1)^{(2s)} \Psi(1, 2, ..., i, ..., j, ...N)$. So, for example, electrons with $s = 1/2$ have states that are represented only by antisymmetric wave functions.

As we shall see, one of the consequences of the antisymmetrization of the wave function is the introduction of a so-called exchange energy term, as Heitler and London showed in a seminal paper seeking to describe the chemical bond using quantum mechanics.[154] Heitler tells the story of his coming up with the concept of exchange as follows.[155] *"I slept till very late in the morning, found I couldn't do work at all, had a quick lunch, went to sleep again in the afternoon and slept until five o'clock. When I woke up, I had clearly the picture before me of the two wave functions of two hydrogen molecules joined together with a plus and minus and with the exchange in it. So I was very excited, and I got up and thought it out. As soon as I was clear that the exchange did play a role, I called London up, and he came to me as quickly as possible. Meanwhile I had already started developing a sort of perturbation theory. We worked together then until rather late at night, and then by that time most of the paper was clear."*

The concept of exchange is ubiquitous in the field of magnetism. It is indeed very convenient to think in terms of exchange when working out the electronic structure of atoms, when deriving Hund's rules, when introducing the concept of Mott insulators or the Hubbard model, or when analyzing band ferromagnetism.

Naturally, the concept of exchange also appears in several areas of spintronics. Most notably, it is invoked in the s-d model, or when a spin-polarized current flows through a ferromagnet. While it is customary to speak of an exchange interaction as if it were an interaction between spins, in reality the strength of this coupling is typical of an electrostatic interaction, not a magnetic one. As we shall see in this chapter, the exchange interaction arises from an approach in which the state of the system is considered without imposing antisymmetrization of the wave function. Thus, the exchange is a correction that stems from the Pauli exclusion principle. In order to clarify the notion of exchange, several examples of increasing complexity are considered in this chapter (Fig. 2.1).

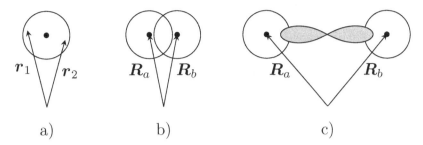

a)   b)   c)

Figure 2.1   a) Two electrons at positions r_1 and r_2 in two orbitals centered on one atom; b) two overlapping atomic orbitals centered on two atoms at positions R_a and R_b; and c) two atomic orbitals overlapping with the orbital of an intermediate atom.

2.2   INTRA-ATOMIC EXCHANGE

Consider two electrons coupled to a fixed positive charge Ze (Fig. 2.1, a).[156] We assume that we know the eigenstates of the Hamiltonian $\mathcal{H}_0$ describing the movement of one single electron around this positive charge, and we leave out the spin-orbit interaction (defined in §2.6). Thus, the eigenstates can be written as a product of a spatial wave function and a spinor function:

$$\varphi_i(r)|\sigma> \tag{2.1}$$

with i numbering the eigenstates. In the following, we assume that these eigenstates are orthogonal to one another. The spin states $|\sigma>$ can be up (u) or down (d) with respect to the spin quantization axis. We further assume that there are two low-lying degenerate states, φ_a and φ_b, that are much lower in energy than the other eigenstates. The Hamiltonian for two electrons around the nuclear charge Ze is given by:

$$\mathcal{H} = \mathcal{H}_0(r_1) + \mathcal{H}_0(r_2) + V(r_1, r_2) \tag{2.2}$$

where

$$V(r_1, r_2) = \frac{e^2}{4\pi\varepsilon_0|r_1 - r_2|} \tag{2.3}$$

$V(r_1, r_2)$ is the electrostatic potential of the electrons experiencing mutual Coulomb repulsions. We assume that the term $V(r_1, r_2)$ is a perturbation, i.e., it does not induce any significant admixture of the two ground states with the other states, that are assumed to have much higher energies. Because of the coupling term (2.3), we need to consider two-electron states. The **Pauli principle** requires that these states be antisymmetric. There are four such states that we can construct out of Slater determinants:

$$\Psi_1 = \frac{1}{\sqrt{2}} \begin{vmatrix} \varphi_a(1)u(1) & \varphi_a(2)u(2) \\ \varphi_b(1)u(1) & \varphi_b(2)u(2) \end{vmatrix} = \frac{1}{\sqrt{2}} \Big(\varphi_a(1)\varphi_b(2) - \varphi_a(2)\varphi_b(1) \Big) u(1)u(2)$$

$$\Psi_2 = \frac{1}{\sqrt{2}} \begin{vmatrix} \varphi_a(1)d(1) & \varphi_a(2)d(2) \\ \varphi_b(1)u(1) & \varphi_b(2)u(2) \end{vmatrix}$$

$$\Psi_3 = \frac{1}{\sqrt{2}} \begin{vmatrix} \varphi_a(1)u(1) & \varphi_a(2)u(2) \\ \varphi_b(1)d(1) & \varphi_b(2)d(2) \end{vmatrix} \tag{2.4}$$

$$\Psi_4 = \frac{1}{\sqrt{2}} \begin{vmatrix} \varphi_a(1)d(1) & \varphi_a(2)d(2) \\ \varphi_b(1)d(1) & \varphi_b(2)d(2) \end{vmatrix} = \frac{1}{\sqrt{2}} \Big(\varphi_a(1)\varphi_b(2) - \varphi_a(2)\varphi_b(1) \Big) d(1)d(2)$$

These definitions are harder to read than to generate. The determinants are constructed as follows. The wave functions are labeled with a and b, each column refers to one electron, and a spin function is attributed to each orbital function. Thus, we have for each electron the combinations $(a-u, b-u)$, $(a-d, b-u)$, $(a-u, b-d)$, and $(a-d, b-d)$. These four states are degenerate so long as $V(\boldsymbol{r}_1, \boldsymbol{r}_2)$ vanishes.

Let us now apply the method of perturbation theory to this manifold of two-electron states. That is, we need to diagonalize the matrix $\mathcal{H}$ of matrix elements $H_{ij} = \langle \Psi_i | \mathcal{H} | \Psi_j \rangle$. For example:

$$
\begin{aligned}
H_{11} &= \langle \Psi_1 | \mathcal{H} | \Psi_1 \rangle \\
&= \langle a | \mathcal{H}_0 | a \rangle + \langle b | \mathcal{H}_0 | b \rangle + \langle a1b2 | V(1,2) | a1b2 \rangle - \langle a1b2 | V(1,2) | a2b1 \rangle
\end{aligned} \tag{2.5}
$$

Here, we use the short-hand notation $|\varphi_a(1)\varphi_b(2)\rangle \equiv |a1b2\rangle$ and likewise for the other terms. The electrostatic interaction $V(1,2)$ between the electrons gives rise to two terms,

$$
\begin{aligned}
K_{ab} &= \langle a1b2 | V(1,2) | a1b2 \rangle \\
&= \int \int dr_1 dr_2 \varphi_a(1)^* \varphi_b(2)^* V(1,2) \varphi_a(1) \varphi_b(2) \\
J_{ab} &= \langle a1b2 | V(1,2) | a2b1 \rangle \\
&= \int \int dr_1 dr_2 \varphi_a(1)^* \varphi_b(2)^* V(1,2) \varphi_a(2) \varphi_b(1)
\end{aligned} \tag{2.6}
$$

The **exchange integral** J_{ab} can be seen as a repulsion energy when both electrons are in the state $\varphi_a(\boldsymbol{r})\varphi_b(\boldsymbol{r})^*$, therefore it must be positive.

We now proceed to examine the other matrix elements. We have the equality $H_{44} = H_{11}$ since the spin part has the same structure in Ψ_1 and Ψ_4. We use a notation that reflects the antisymmetry of Ψ_2 and Ψ_3:

$$
\begin{aligned}
H_{22} &= H_{33} = \langle a | \mathcal{H}_0 | a \rangle + \langle b | \mathcal{H}_0 | b \rangle + \langle a1b2 | V(1,2) | a1b2 \rangle \\
&= \langle a | \mathcal{H}_0 | a \rangle + \langle b | \mathcal{H}_0 | b \rangle + K_{ab}
\end{aligned} \tag{2.7}
$$

There is no J_{ab} term in these matrix elements because of the orthogonality of the spin states.[1] Finally, we have only two off-diagonal terms, $H_{23} = H_{32} = -J_{ab}$, as can readily be seen by writing out the developed expressions for these terms. We recall that these terms come about like this because of the orthogonality of the spin functions Ψ_2 and Ψ_3 and because the interaction $V(1,2)$ has no spin dependence.[2]

In summary, the matrix of $\mathcal{H}$ in the basis set (2.4) is given by,

$$
\mathcal{H} = \begin{pmatrix} E_0 + K_{ab} - J_{ab} & 0 & 0 & 0 \\ 0 & E_0 + K_{ab} & -J_{ab} & 0 \\ 0 & -J_{ab} & E_0 + K_{ab} & 0 \\ 0 & 0 & 0 & E_0 + K_{ab} - J_{ab} \end{pmatrix} \tag{2.8}
$$

where $E_0 = E_a + E_b$. We see that only the center block containing the second and third rows need to be diagonalized. That is, we need to diagonalize a matrix of the form,

$$
M = \begin{pmatrix} A & B \\ B & A \end{pmatrix} \tag{2.9}
$$

---

[1] When we analyze core polarization (§15.6.2), it will be useful to remember that the diagonal terms have no exchange contribution when the electron wave functions are built with spin functions of opposite directions.

[2] This will no longer be the case when we examine the Dzyaloshinskii-Moriya coupling at the end of this chapter.

Obviously, the eigenvectors are proportional to $(1, 1)$ and $(1, -1)$ with eigenvalues $A + B$ and $A - B$. Thus, the corresponding eigenstates are:

$$\Psi_2' = \frac{1}{\sqrt{2}}(\Psi_2 + \Psi_3) = (\varphi_a(1)\varphi_b(2) - \varphi_a(2)\varphi_b(1))\frac{1}{2}(d(1)u(2) + u(1)d(2))$$
$$\Psi_3' = \frac{1}{\sqrt{2}}(\Psi_2 - \Psi_3) = (\varphi_a(1)\varphi_b(2) + \varphi_a(2)\varphi_b(1))\frac{1}{2}(d(1)u(2) - u(1)d(2))$$

(2.10)

The states Ψ_1, Ψ_2', and Ψ_4 are called **triplet states**. They are eigenstates of $\left(\hat{S}_1 + \hat{S}_2\right)^2$ with the eigenvalue $S(S+1)$ where S=1. In this notation, the corresponding angular momentum operators are $S_1 = \hbar\hat{S}_1$ and likewise for S_2. The states Ψ_1, Ψ_2', and Ψ_4 are also eigenstates of $S_{1z} + S_{2z}$ with eigenvalues 1, 0, −1. The state Ψ_3' is called a **singlet**, for which $S = 0$. Thus, we find that:

- the triplet states have the energy $E_t = E_a + E_b + K_{ab} - J_{ab}$
- the singlet state has the energy $E_s = E_a + E_b + K_{ab} + J_{ab}$

We indicated that the intra-atomic exchange energy J_{ab} is positive, therefore the triplet states have the lowest energy. This is the origin of the **first Hund rule**. The first Hund rule states that the lowest energy state is the one with a maximum spin. We saw in working out the matrix $\mathcal{H}$ in (2.8) that the diagonal elements of the states Ψ_1 and Ψ_4 are those that contain the term $(-J_{ab})$ and they are states for which all spins are in the same direction. To the contrary, Ψ_2' and Ψ_3' have the single electron spin states mixed and do not contain $(-J_{ab})$. We can foresee from this simple example what would happen if we had more spins. The states Ψ_n, with all spins aligned, would yield the diagonal terms H_{nn} that would contain the maximum number of terms such as $(-J_{ab})$. So these states would have the lowest energy.

Spin Hamiltonian

Now, we want to ask ourselves the following question: can we construct a Hamiltonian that will provide the same energy levels E_t and E_s, using one-electron wave functions, without paying attention, so to speak, to the requirement of the Pauli principle?

We have found two energy levels corresponding to eigenstate of the Hamiltonian that are also eigenstates of $\left(\hat{S}_1 + \hat{S}_2\right)^2$. Since $\left(\hat{S}_1 + \hat{S}_2\right)^2 = \hat{S}_1^2 + \hat{S}_2^2 + 2\hat{S}_1 \cdot \hat{S}_2$, the eigenvalues of $\hat{S}_1 \cdot \hat{S}_2$ are either 1/4 for the triplet or −3/4 for the singlet. Therefore, we can write our effective Hamiltonian as,

$$\mathcal{H}_{eff} = E_s(\frac{1}{4} - \hat{S}_1 \cdot \hat{S}_2) + E_t(\frac{3}{4} + \hat{S}_1 \cdot \hat{S}_2)$$
$$= cst - (E_s - E_t)\hat{S}_1 \cdot \hat{S}_2$$

(2.11)

It is useful, whenever we speak of exchange interaction, to remember that it stems from the procedure describe above. Namely, we have applied the Pauli principle and written an effective one-particle Hamiltonian that contains a term that looks like an interaction between electron spins. This interaction is in fact determined by the electrostatic interaction J_{ab} defined in (2.6). Therefore, its order of magnitude is that of a Coulomb energy.

2.3   DIRECT INTER-ATOMIC EXCHANGE

In many situations, a magnetic ion remains magnetic when it is embedded into a solid. Ferromagnetism might arise if these magnetic ions are strongly coupled to one another. The simplest form of interaction is the direct exchange, a consequence of the Pauli principle. To develop familiarity with this coupling, let us look at the simplest possible case of two identical atoms in close proximity. The extension from exchange in molecules to exchange in solids is addressed, e.g., in White's book on magnetism.[156]

Let us consider the problem of the hydrogen molecule (Fig. 2.1), as Heitler and London did in 1927.[154, 156] We want to take a perturbation approach to the calculation of the effect of the interactions between the electrons and of each electron with the nucleus of the other atom as both atoms approach one another. We denote by $\mathcal{H}_\infty(\boldsymbol{r}_1 - \boldsymbol{R}_a)$ and $\mathcal{H}_\infty(\boldsymbol{r}_2 - \boldsymbol{R}_b)$ the Hamiltonians of one electron on one atom when the other atom is far away. Thus, we want to consider the Hamiltonian:

$$\mathcal{H} = \mathcal{H}_\infty(\boldsymbol{r}_1 - \boldsymbol{R}_a) + \mathcal{H}_\infty(\boldsymbol{r}_2 - \boldsymbol{R}_b) - \frac{e^2}{4\pi\varepsilon_0}\left(\frac{Z}{|\boldsymbol{r}_1 - \boldsymbol{R}_b|} - \frac{Z}{|\boldsymbol{r}_2 - \boldsymbol{R}_a|} + \frac{1}{|\boldsymbol{r}_1 - \boldsymbol{r}_2|}\right) \quad (2.12)$$

The wave function that corresponds to the ground state of $\mathcal{H}_\infty(\boldsymbol{r})$ is some function $\varphi(\boldsymbol{r})$. For this two-electron problem, we need to create a basis of two one-electron states. We do so starting from the ground states of $\mathcal{H}_\infty(a)$ or $\mathcal{H}_\infty(b)$. We denote these atomic state functions centered on nuclei a and b as,

$$\varphi_a(\boldsymbol{r}) = \varphi(\boldsymbol{r} - \boldsymbol{R}_a) \qquad \varphi_b(\boldsymbol{r}) = \varphi(\boldsymbol{r} - \boldsymbol{R}_b) \quad (2.13)$$

Now, we can build our set of two-electron wave functions, including spin, as in the case of the intra-atomic exchange that we considered earlier. However, we had assumed that the two wave functions were orthogonal. In general, this is no longer so in this case. In other words, the integral,

$$l = \int d\boldsymbol{r}\, \varphi_a{}^*(\boldsymbol{r})\varphi_b(\boldsymbol{r}) \quad (2.14)$$

is in general different from zero when the atoms approach each other. We say that the orbitals overlap and ℓ is called the **overlap integral**. As for the intra-atomic problem of the previous section, we have readily two of the wave functions, which, taking care of their normalization, are,

$$\Psi_1 = \frac{1}{N}\begin{vmatrix} \varphi_a(1)u(1) & \varphi_a(2)u(2) \\ \varphi_b(1)u(1) & \varphi_b(2)u(2) \end{vmatrix} = \frac{1}{N}\left(\varphi_a(1)\varphi_b(2) - \varphi_a(2)\varphi_b(1)\right)u(1)u(2)$$

$$\Psi_4 = \frac{1}{N}\begin{vmatrix} \varphi_a(1)d(1) & \varphi_a(2)d(2) \\ \varphi_b(1)d(1) & \varphi_b(2)d(2) \end{vmatrix} = \frac{1}{N}\left(\varphi_a(1)\varphi_b(2) - \varphi_a(2)\varphi_b(1)\right)d(1)d(2) \quad (2.15)$$

where $N = \sqrt{2 - 2l^2}$. For the other two states, we could take as before the two combinations:

$$\begin{vmatrix} \varphi_a(1)d(1) & \varphi_a(2)d(2) \\ \varphi_b(1)u(1) & \varphi_b(2)u(2) \end{vmatrix} \qquad \begin{vmatrix} \varphi_a(1)u(1) & \varphi_a(2)u(2) \\ \varphi_b(1)d(1) & \varphi_b(2)d(2) \end{vmatrix} \quad (2.16)$$

However, they would not be orthogonal, whereas the sum of these two states and their difference are orthognal. We saw that they are states of S=1 and S=0, respectively. So, we choose to take for our basis states (taking care of the normalization) the wave functions:

$$\Psi_2 = \frac{1}{\sqrt{2 - 2l^2}}\left(\varphi_a(1)\varphi_b(2) - \varphi_a(2)\varphi_b(1)\right)\frac{1}{\sqrt{2}}\left(d(1)u(2) + u(1)d(2)\right)$$

$$\Psi_3 = \frac{1}{\sqrt{2 + 2l^2}}\left(\varphi_a(1)\varphi_b(2) + \varphi_a(2)\varphi_b(1)\right)\frac{1}{\sqrt{2}}\left(d(1)u(2) - u(1)d(2)\right) \quad (2.17)$$

The spin parts of the wave functions are orthogonal. Since the Hamiltonian has no spin part, this choice for the basis states diagonalizes the matrix $H_{ij} = <\Psi_i|\mathcal{H}|\Psi_j> (i,j = 1, ..., 4)$. Furthermore, our basis states have three states with identical orbital parts and one with a distinct orbital part. Thus, the matrix $\mathcal{H}$ in this basis is expected to have the form,

$$\mathcal{H} = \begin{pmatrix} E_t & 0 & 0 & 0 \\ 0 & E_t & 0 & 0 \\ 0 & 0 & E_s & 0 \\ 0 & 0 & 0 & E_t \end{pmatrix} \tag{2.18}$$

For our purpose, there is no need to go any further and examine the expressions of E_t and E_s. We have obtained a Hamiltonian that can be expressed with spin operators as in (2.11) in order to get the correct eigenvalues without imposing antisymmetrization of the wave functions. In this very simple approach, taking as trial wave functions those of electrons in atoms very far away, the prediction about the sign of $E_t - E_s$ is not as straightforward as it was in the mono-atomic case and even leads to an erroneous sign at large distances.[156] It turns out that, experimentally, the singlet state is generally the ground state.

2.4   SUPEREXCHANGE

In the previous section, we considered direct exchange, i.e., cases when one-electron wave functions centered on two atoms have a finite overlap. Let us now turn to a situation where instead of a direct overlap, there is an effective overlap mediated by an intermediate atom. Many magnetic systems are characterized by ions that carry a localized magnetic moment. These magnetic ions form bonds with non-magnetic ions located in between the magnetic ions. This intermediate atom is often oxygen or fluorine. So, in this case, there is no direct overlap of orbitals between the magnetic ions. How do we understand the magnetic coupling then? H. A. Kramers raised this question very early on in the development of the quantum theory of magnetism.[157] Often, the coupling among these local moments is **antiferromagnetic**. The formal explanation for this coupling was provided by Nobel Prize winner P. Anderson. [158–160]

Using a similar approach to the one used to infer the existence of intra-atomic and direct inter-atomic exchange, we start with a derivation based on molecular physics.[161] We consider once again two electrons and two nuclei, with the Hamiltonian (2.12) and the wave functions (2.13) centered on each atomic site. We assume that the overlap l given by (2.14) is negligible. As stated above, under the Heitler-London scheme, the singlet state is the lowest in energy, i.e., the state with a bonding molecular orbital (as opposed to an antibonding one). We write this state as:

$$\Psi_1 = \frac{1}{\sqrt{2}} \left(\varphi_a(1)\varphi_b(2) + \varphi_a(2)\varphi_b(1) \right) \frac{1}{\sqrt{2}} \left(d(1)u(2) - u(1)d(2) \right) \tag{2.19}$$

In order to refine our description, we allow for admixtures of the singlet state with the two ionic states in which both electrons reside on the same ion and in the same state. This is possible since we consider spin singlets only. The Hamiltonian contains no spin part, so it cannot mix singlet and triplet states. Evidently, the ionic state suffers a strong on-site repulsion, as we will find explicitly in the derivation that follows. We will see, however, that this admixture lowers the overall energy of the singlet state.

Let us determine the ground state for the Hilbert sub-space that is restricted to the space spanned by three states, which are Ψ_1 given by (2.19), and Ψ_2 and Ψ_3 defined by,

$$\Psi_2 = \Psi_{ionic,a} = \varphi_a(1)\varphi_a(2)\frac{1}{\sqrt{2}}\left(d(1)u(2) - u(1)d(2)\right)$$

$$\Psi_3 = \Psi_{ionic,b} = \varphi_b(1)\varphi_b(2)\frac{1}{\sqrt{2}}\left(d(1)u(2) - u(1)d(2)\right)$$

(2.20)

These three states are antisymmetric with respect to electron exchange and could be written as Slater determinants, as we have done before. Now, we want to find the eigenstates of the matrix of which the elements are $H_{ij} = < \Psi_i|\mathcal{H}|\Psi_j >$. Two assumptions greatly simplify the expression of the matrix elements : 1) the orbitals on site a and b are the same, and 2) the overlap of the localized orbitals is negligible. Let us start with H_{11}. There are terms that appear just as in the monoatomic case treated above, plus electrostatic terms due to the interaction of the electrons with the nuclei.

If we had considered antibonding states, i.e., a triplet spin state instead of the singlet, the matrix element H_{11} would be almost the same as for the singlet, except that the exchange integrals would appear twice, but with opposite signs. This means that J_{ab} would drop out in the triplet state energy. The same happened in the calculation for the intra-atomic exchange. So we write:[156]

$$H_{11} = E_T + J_{ab} \tag{2.21}$$

Likewise, we find,

$$H_{33} = H_{22} = E_T + U \tag{2.22}$$

where,

$$U = \langle \varphi_a(1)\varphi_a(2)| \frac{e^2}{4\pi\varepsilon_0 \left\| \mathbf{r}_1 - \mathbf{r}_2 \right\|} |\varphi_a(1)\varphi_a(2)\rangle \tag{2.23}$$

is called the **on-site repulsion**.

There is an off-diagonal matrix element between the ionic states. To the extent that the states can be taken as real, we can write:

$$H_{23} = \langle \varphi_a(1)\varphi_a(2)| \frac{e^2}{4\pi\varepsilon_0 \left\| \mathbf{r}_1 - \mathbf{r}_2 \right\|} |\varphi_b(1)\varphi_b(2)\rangle = J_{ab} \tag{2.24}$$

The remaining matrix elements involve a new type of term. Again, treating the overlap terms as negligible removes a lot of terms from the matrix elements and the remainder is given by,

$$H_{12} = \frac{2}{\sqrt{2}} \langle \varphi_a(1)\varphi_b(2)| \frac{e^2}{4\pi\varepsilon_0 \left\| \mathbf{r}_1 - \mathbf{r}_2 \right\|} |\varphi_a(1)\varphi_a(2)\rangle$$

$$H_{13} = \frac{2}{\sqrt{2}} \langle \varphi_a(1)\varphi_b(2)| \frac{e^2}{4\pi\varepsilon_0 \left\| \mathbf{r}_1 - \mathbf{r}_2 \right\|} |\varphi_b(1)\varphi_b(2)\rangle$$

(2.25)

These terms are equal, since the molecule is symmetric. We will write these terms as,

$$H_{12} = H_{13} = t \tag{2.26}$$

The double-center integral t is called a **transfer integral**.

Therefore, in the end, the matrix we want to diagonalize can be written as,

$$\mathcal{H} = \begin{pmatrix} E_T + J_{ab} & t & t \\ t & E_T + U & J_{ab} \\ t & J_{ab} & E_T + U \end{pmatrix} \tag{2.27}$$

Since the orbitals are centered around a and b, the integrand of t is rather small, whereas the integrand of U is large, so $t << U$. Finding the eigenvalues of $\mathcal{H}$ in (2.27) using Mathematica, for example, it can be found that the lowest energy level is given by,

$$E_S = E_T + J_{ab} - \frac{2t^2}{U} \tag{2.28}$$

and the associated eigenvector is given by,

$$\left(1, -\frac{t}{U}, -\frac{t}{U}\right) \tag{2.29}$$

In other words, the refined estimate of the ground state that we obtain by considering admixtures of the ionic states favors the singlet state (antiferromagnetic order), and the ground state is composed of the singlet state and an admixture of the ionic states,

$$\Psi_s = \psi_{singlet} - \frac{t}{U}\left(\Psi_{ionic,a} + \Psi_{ionic,b}\right) \tag{2.30}$$

This finding is the essence of the superexchange model. The singlet state is stabilized, or favored, by the ability of electrons to extend farther than the magnetic ion. The effect on energy depends on the transfer integrals t and the on-site repulsion U.

2.5 ZEEMAN INTERACTION

The non-relativistic limit of the Dirac equation for the electron contains an important term, called the **Pauli coupling**,[162]

$$\mathcal{H}_{\mathrm{P}} = \frac{e\hbar}{2m}\boldsymbol{\sigma} \cdot \boldsymbol{B} \tag{2.31}$$

It is the coupling to the magnetic induction field $\boldsymbol{B}$ of the intrinsic electron magnetic moment,([162] p. 105)

$$\boldsymbol{\mu}_B = \frac{-e\hbar}{2m}\boldsymbol{\sigma} = -g\frac{\mu_B}{\hbar}\boldsymbol{S} = -\mu_B\boldsymbol{\sigma} \tag{2.32}$$

where $g = 2$ and $\boldsymbol{S} = (1/2)\hbar\boldsymbol{\sigma}$ is the spin angular momentum of the electron.[3] The quantity

$$\mu_B = \frac{|e|\hbar}{2m} \tag{2.33}$$

is called the **Bohr magneton**.[4] Thus, the Pauli coupling term can be written as,

$$\mathcal{H}_{\mathrm{P}} = -\boldsymbol{\mu}_B \cdot \boldsymbol{B} = g\frac{\mu_B}{\hbar}\boldsymbol{S} \cdot \boldsymbol{B} \tag{2.34}$$

The orbital moment of an electron also gives rise to a magnetic dipole moment which couples to the magnetic induction field. This coupling term stems from the non-relativistic equation,

$$\mathcal{H}_{\mathrm{nr}} = \frac{1}{2m}\left(\boldsymbol{p} + e\boldsymbol{A}\right)^2 - e\Phi \qquad (e > 0)) \tag{2.35}$$

where Φ is the electrostatic potential and $\boldsymbol{A}$ is the potential vector, such that $\boldsymbol{B} = \boldsymbol{\nabla} \times \boldsymbol{A}$. Let us consider the case $\Phi = 0$, and the Coulomb gauge $div\boldsymbol{A} = 0$. For a uniform magnetic induction, we have,

$$\boldsymbol{A} = \frac{1}{2}\boldsymbol{B} \times \boldsymbol{r} \tag{2.36}$$

---

[3]Throughout this book, the charge of the electron is written $-e$.

[4]The Bohr magneton μ_B is defined with a positive sign, and it is approximatively $9.27{\cdot}10^{-24}$ J/T.

Then the Hamiltonian $\mathcal{H}_{\mathrm{nr}}$ reads,

$$\mathcal{H}_{\mathrm{nr}} = \frac{1}{2m}\boldsymbol{p}^2 + \frac{e}{2m}\boldsymbol{B} \cdot (\boldsymbol{r} \times \boldsymbol{p}) + \frac{e^2}{8m}\left(\boldsymbol{B}^2\boldsymbol{r}^2 - (\boldsymbol{B} \cdot \boldsymbol{r})^2\right) \tag{2.37}$$

Thus, the angular momentum $\boldsymbol{L} = \boldsymbol{r} \times \boldsymbol{p}$ appears in the second term, which expresses the coupling of an orbital dipolar moment to the magnetic induction. Combined with the Pauli term above, we have the following orbital and spin couplings, known as the **Zeeman interaction**, which is given by,([162] p. 121)

$$\mathcal{H}_Z = \frac{e}{2m}\left(\boldsymbol{L} + 2\boldsymbol{S}\right) \cdot \boldsymbol{B} = \frac{\mu_B}{\hbar}\left(\boldsymbol{L} + 2\boldsymbol{S}\right) \cdot \boldsymbol{B} \tag{2.38}$$

It is often convenient to define the unit-less operators $\hat{\boldsymbol{L}} = \boldsymbol{L}/\hbar$, $\hat{\boldsymbol{S}} = \boldsymbol{S}/\hbar$, and $\hat{\boldsymbol{J}} = \boldsymbol{J}/\hbar$ where $\boldsymbol{J} = \boldsymbol{L} + \boldsymbol{S}$. Then, we can write (2.38) as,

$$\mathcal{H}_Z = \mu_B \left(\hat{\boldsymbol{L}} + 2\hat{\boldsymbol{S}}\right) \cdot \boldsymbol{B} \tag{2.39}$$

2.6 SPIN-ORBIT INTERACTION

2.6.1 DEFINITION

Spin-orbit effects are ubiquitous in magnetism and spintronics. Notably, the so-called "Dzyaloshinskii-Moriya coupling" that is introduced at the end of this chapter can be understood as a second-order effect associated with spin-orbit effects. These effects account also for magnetocrystalline anisotropies (Chapter 4).

The existence of a spin-orbit coupling can be found by seeking the non-relativistic limit of the Dirac equation. Below, in §2.6.3, we give a non-relativistic derivation of this term. Where the electric field is static and there is no magnetic field, the spin-orbit term is:[162]

$$\mathcal{H}_{SO} = \frac{eh}{2m}\boldsymbol{\sigma} \cdot \frac{\mathbf{E} \wedge \mathbf{p}}{2mc^2} \tag{2.40}$$

Spin-orbit effects generate the fine structure of optical spectra. Basic reference materials on spin-orbit coupling in solids can be found in reviews and research articles.[163–165] In alkali metals, the electric field is negligible and there is practically no spin-orbit effect, in general. However, to the extent that phonons induce an electric field, the spin-orbit coupling can produce a Overhauser_1953, number [719] Recently, it has been shown that spin-orbit effects can also affect chemical reactivity. Spin non-conserving reactions are quasi forbidden when the electrons involved in the reactions belong to light elements. This selection rule is relaxed when heavy elements are involved.[167, 168]

2.6.2 ELECTRON IN A CENTRAL FIELD

In the case of one electron in a central field (e.g. hydrogen-like atom of nuclear charge Ze) the spin-orbit coupling becomes:

$$\mathcal{H}_{SO} = \frac{Ze^2}{4\pi\varepsilon_0}\frac{\hbar}{4m^2c^2}\left(\frac{\boldsymbol{r} \times \boldsymbol{p}}{r^3\hbar}\right) \cdot \boldsymbol{\sigma} \tag{2.41}$$

In this approximation, the orbital angular momentum $\boldsymbol{r} \times \boldsymbol{p}$ appears explicitly. The splitting of the energy levels of the hydrogen-like atom produced by the spin-orbit coupling can be found to be:[169]

$$E_{n,l+1/2} - E_{n,l-1/2} = \frac{mc^2}{2}\frac{Z^4\alpha^4}{n^3l(l+1)} \tag{2.42}$$

where α is the **fine structure constant**, $\alpha = e^2/(4\pi\varepsilon_0\hbar c)$. The fourth power dependence on Z in (2.42) indicates that strong spin-orbit coupling effects are expected in materials with heavy nuclei.

In a solid, a spherical approximation can be made. It consists in replacing the sum over all electrons, $\sum_i (Ze^2/r_i^3)(\boldsymbol{r} \times \boldsymbol{p}) \cdot \boldsymbol{\sigma}_i$ in an effective one-electron approach by the Hamiltonian,

$$\mathcal{H}_{\mathcal{SO}} = \lambda_{\text{SO}}\left(-\boldsymbol{\nabla}V(r) \times \boldsymbol{p}\right) \cdot \boldsymbol{\sigma} \tag{2.43}$$

where $V(r)$ is an effective potential. Electrostatic screening by the charges around the nuclei has two effects : $V(r)$ is large at small r only, and the wave function is also large at small r. Therefore, the spin-orbit coupling is dominated by the core of the atom. In estimating the spin-orbit coupling, one has to find wave functions wthat properly represent the core states. Hence, e.g., plane wave states would be a very poor approximation. This point is further discussed in §13.2.2 when examining the spin relaxation mechanisms that limit the spin diffusion length.

In a spherical symmetry approximation, the spin-orbit coupling has a significant value only near the nucleus, where $\boldsymbol{\nabla}V(r)$ is nearly radial, i.e., proportional to $\boldsymbol{r}$. Thus, (2.43) can be written in terms of the electron orbital angular momentum $\boldsymbol{l}$ as,

$$\mathcal{H}_{SO} = \xi(r)\boldsymbol{l} \cdot \boldsymbol{\sigma} \quad \text{with} \quad \xi(r) \propto \frac{1}{r}\frac{dV}{dr} \tag{2.44}$$

This is the reason why the interaction (2.40) is called the spin-orbit interaction.

2.6.3 NON-RELATIVISTIC DERIVATION

Let us consider an electron of charge $(-e)$ passing by a nucleus or an impurity of charge Q at a velocity $\boldsymbol{v}$ with $\|\boldsymbol{v}\| = v \ll c$, where c is the light velocity. In a reference frame where the electron is immobile, the charge Q has a velocity $-\boldsymbol{v}$. We expect that this moving charge will produce a magnetic induction that will act on the spin $\boldsymbol{S}$ according to the Pauli coupling (2.31). Let us apply the Biot-Savart law to calculate this magnetic induction field. We recall that when a current I passes down a segment $d\boldsymbol{y}$ of a conductor located at position $\boldsymbol{y}$, this segment contributes at the position $\boldsymbol{x}$ a contribution to the induction field of,[170]

$$d\boldsymbol{B}(\boldsymbol{x},\boldsymbol{y}) = \frac{\mu_0}{4\pi}\frac{Id\boldsymbol{y} \wedge (\boldsymbol{x} - \boldsymbol{y})}{\|\boldsymbol{x} - \boldsymbol{y}\|^3} \tag{2.45}$$

If the current I corresponds to the motion of a point charge, the current at any position $\boldsymbol{y}$ of the trajectory of this point charge will generate a current that is a delta function of time. To avoid this, let us assume that the charge Q is spread around a center position. Hence, we define a charge density $\rho(\boldsymbol{R},\delta r)$ for a point charge located at any $\boldsymbol{R}$, with the condition $\int \delta r^3 \rho(\boldsymbol{R},\delta r) = Q$. Since this integration will appear straightforwardly at the end of the derivation, we can simplify and assume a distribution over a cube of volume d^3 with two faces normal to the charge Q trajectory, such that a current density ρv threads through these faces.

Thus, we have a current $I = \rho v d^2$ that passes at the location $\boldsymbol{y}$ during a time $\delta t = d/v$. Then, in (2.45), $Id\boldsymbol{y}$ can be written as $d\boldsymbol{y} = -\boldsymbol{v}\delta t$ and,

$$Id\boldsymbol{y} = \left(\rho v d^2\right)(-\boldsymbol{v})\frac{d}{v} = Q\dot{s}\hat{e}_3 \tag{2.46}$$

In the last step, we have set the coordinate axis x_3 on the charge Q trajectory, so that we have $-\boldsymbol{v} = \dot{s}\hat{\boldsymbol{e}}_3$, where s designates the curvilinear coordinate for the charge Q, so $\boldsymbol{y} = s\hat{\boldsymbol{e}}_3$. We have thus found the magnetic induction,

$$\boldsymbol{B}(\boldsymbol{x}) = \frac{\mu_0}{4\pi} \frac{Q\dot{s}\hat{\boldsymbol{e}}_3 \wedge (\boldsymbol{x} - s\hat{\boldsymbol{e}}_3)}{||\boldsymbol{x} - s\hat{\boldsymbol{e}}_3||^3} \tag{2.47}$$

We can bring out in this last expression the electric field $\boldsymbol{E}(\boldsymbol{x})$ generated by Q,

$$\boldsymbol{E}(\boldsymbol{x}) = \frac{Q}{4\pi\varepsilon_0} \frac{(\boldsymbol{x} - s\hat{\boldsymbol{e}}_3)}{||\boldsymbol{x} - s\hat{\boldsymbol{e}}_3||^3} \tag{2.48}$$

Defining $\boldsymbol{p} = m\boldsymbol{v}$ as the linear momentum of the electron in the laboratory (e.g. nuclear or impurity rest frame), we have,

$$\boldsymbol{B}(\boldsymbol{x}) = \mu_0\varepsilon_0 \frac{\boldsymbol{E}(\boldsymbol{x}) \wedge \boldsymbol{p}}{m} \tag{2.49}$$

If we apply the relation $\mu_0\varepsilon_0 = 1/c^2$, we find a term mc^2 in the denominator.

We can calculate the Zeeman energy E' implied by (2.31) for the spin $\boldsymbol{S}$ in the magnetic induction field $\boldsymbol{B}(\boldsymbol{x})$,

$$E' = \left(\frac{e\hbar}{2m}\boldsymbol{\sigma}\right) \cdot \mu_0\varepsilon_0 \frac{\boldsymbol{E}(\boldsymbol{x}) \wedge \boldsymbol{p}}{m} \tag{2.50}$$

This energy E' is a factor of 2 larger than the spin-orbit energy implied by (2.40). This problem was identified by Thomas in his seminal paper and its follow-up.[171,172] Recently, Corben pointed out that "*relativistic corrections to non[-]relativistic theories are continuous functions of the velocities involved - relativity does not suddenly 'kick in' a factor of 2*".[173] Indeed, we can reason along the lines of the Thomas argument in a non-relativistic framework.

The orbital moment of the electron, $\boldsymbol{L} = \boldsymbol{r} \wedge \boldsymbol{v}$, which is defined with respect to the nucleus, has the same value in the electron rest frame, since it can be calculated as $(-\boldsymbol{r}) \wedge (-\boldsymbol{v}) = \boldsymbol{L}$. The Zeeman energy for the orbital moment, according to (2.38) reads,

$$E'_{\text{orb}} = \frac{e}{2m}\boldsymbol{L} \cdot \boldsymbol{B}(\boldsymbol{x}) \tag{2.51}$$

As we will learn in §7.2, the Hamiltonian corresponding to this energy implies a precession of the spin at the angular frequency,

$$\boldsymbol{\Omega}_{\text{orb}} = \frac{e}{2m}\boldsymbol{B}(\boldsymbol{x}) \tag{2.52}$$

We choose coordinate axes that follow the precession of the orbital angular momentum, in other words, the angular moment does not evolve in time in this frame. Let us now rewrite (2.50) in the form (2.34),

$$E' = 2\frac{\mu_B}{\hbar}\boldsymbol{S} \cdot \boldsymbol{B}(\boldsymbol{x}) \tag{2.53}$$

The Hamiltonian corresponding to this energy implies a precession of the spin at the angular frequency,

$$\boldsymbol{\Omega}' = \frac{e}{m}\boldsymbol{B}(\boldsymbol{x}) \tag{2.54}$$

When going back to the first reference frame, the change in angular velocity is simply,

$$\boldsymbol{\Omega} = \boldsymbol{\Omega}' - \boldsymbol{\Omega}_{\text{orb}} = \frac{1}{2}\boldsymbol{\Omega}' \tag{2.55}$$

The energy corresponding to this precession frequency is $E = E'/2$. This factor of 2 correction to (2.50) leads to the expression (2.40) for the Hamiltonian of the spin-orbit interaction. A formal derivation of the spin evolution of a moving particle in a homogeneous electromagnetic field was worked out by Bargmann, Michel, and Gelegdi.[174] The non-relativistic, many-body quantum theory of particles moving in an electromagnetic field was presented by Froehlich and Studer in 1993.[175]

2.7 THE DZYALOSHINSKII-MORIYA INTERACTION

2.7.1 SYMMETRY CONSIDERATIONS

The generalization to solids of the direct exchange that was introduced in §2.3 for the diatomic molecule leads to the **Heisenberg Hamiltonian**:[176]

$$\mathcal{H} = -\sum_{\alpha,\alpha'} \hat{S}(R_\alpha) \cdot J(R_\alpha, R_{\alpha'}) \, \hat{S}(R_{\alpha'}) \tag{2.56}$$

where R_α designates the location of the spins $\hat{S}(R_\alpha)$. From our derivation for the diatomic case, it is clear that, in the absence of spin dependence in the original Hamiltonian of the original many-electron problem, the tensors $J(R_\alpha, R_{\alpha'})$ are symmetric:

$$J(R_\alpha, R_{\alpha'}) = J(R_{\alpha'}, R_\alpha) \tag{2.57}$$

In this section, we will show that spin-orbit effects lead to an antisymmetric component in these $J(R_\alpha, R_{\alpha'})$ tensors. This type of antisymmetric exchange coupling is known as the **Dzyaloshinskii-Moriya (DM) interaction**.[177,178]

Assume that a part of the Heisenberg Hamiltonian (2.56) has the antisymmetric condition $J(R_\alpha, R_{\alpha'}) = -J(R_{\alpha'}, R_\alpha)$. Consider atomic sites 1 and 2. The coupling between the sites can be written in two ways that must give the same value for the coupling :

$$\hat{S}_1 \cdot J(1,2)\hat{S}_2 = \hat{S}_2 \cdot J(2,1)\hat{S}_1 = -\hat{S}_2 \cdot J(1,2)\hat{S}_1 \tag{2.58}$$

The equality between the first and last expressions for any value of $\hat{S}_1$ and $\hat{S}_2$ implies that the 3x3 matrix $J(1,2)$ must be antisymmetric. Therefore, the matrix $J(1,2)$ is defined by only three non-zero parameters. It is easy to show that the product $S_1 \cdot JS_2$ can be written in this case as,

$$\hat{S}_1 \cdot (J\hat{S}_2) = \hat{S}_1 \cdot (-D \times \hat{S}_2) = D \cdot (\hat{S}_1 \times \hat{S}_2) \tag{2.59}$$

This is the standard form used to express the DM interaction. The vector D can be treated as a set of three phenomenological parameters that can be inferred from the magnetic structure.

Significance

The DM interaction is important to explain the magnetic structure known as **spin canting** that is found in so-called **weak ferromagnets**. This term designates a family of compounds that are essentially antiferromagnetic, but with a slight deviation from the pure antiferromagnetic alignment. This misalignment results in a net magnetic moment. Spin canting occurs, for example, in La_2CuO_4, the parent compound of a family of inorganic high-T_c superconductors.[179]

The DM interaction plays a role in a great variety of other solids that have attracted much attention in recent years. This antisymmetric coupling is at the root of magnetic structures known as **skyrmions**. Surface spin canting due to DM interaction has been observed using near-field scanning techniques.[180]

Spin canting and skyrmions refer to the static properties of a magnetic material. The DM interaction can also influence the dynamic properties of a ferromagnet, namely, its spin wave dispersion (see §7.5), or create a long-range spin structure in an antiferromagnet, which has its own magnetic resonance modes (see §18.7).

2.7.2 SECOND-ORDER SPIN-ORBIT AND EXCHANGE COUPLINGS

Here, we consider a particular case to illustrate how an antisymmetric exchange term can arise. We follow a final comment Moriya (1960) made in his seminal paper.[178]

We consider the coupling between two ions, each one being described by states (n, m, ...) and (n',m', ...). The primed labels distinguish the states of one ion from those of the other. (n) and (n') designate the ground states of both ions. We consider now the second-order perturbation due to spin-orbit and Coulomb interactions, and we pick out those terms that are bilinear in each. We show that these terms are antisymmetric.

Thus, we evaluate the second-order correction to the energy given by:

$$\Delta E = \sum_{m,m'}$$
$$\frac{\langle nn'| \lambda \boldsymbol{L}_1 \cdot \boldsymbol{S}_1 + \frac{e^2}{4\pi\varepsilon_0 r_{12}} + \lambda \boldsymbol{L}_2 \cdot \boldsymbol{S}_2 |mm'\rangle \langle mm'| \lambda \boldsymbol{L}_1 \cdot \boldsymbol{S}_1 + \frac{e^2}{4\pi\varepsilon_0 r_{12}} + \lambda \boldsymbol{L}_2 \cdot \boldsymbol{S}_2 |nn'\rangle}{E_n + E_{n'} - E_m - E_{m'}}$$

(2.60)

Among the many terms that we obtain by developing this expression, we retain those that lead to the symmetry we are looking for. Namely, we retain from (2.60) the following terms:

$$\sum_m \frac{\langle n| \lambda \boldsymbol{L}_1 \cdot \boldsymbol{S}_1 |m\rangle \langle mn'| \frac{e^2}{4\pi\varepsilon_0 r_{12}} |nn'\rangle}{E_n - E_m} + \sum_m \frac{\langle nn'| \frac{e^2}{4\pi\varepsilon_0 r_{12}} |mn'\rangle \langle m| \lambda \boldsymbol{L}_1 \cdot \boldsymbol{S}_1 |n\rangle}{E_n - E_m}$$
$$+ \sum_{m'} \frac{\langle n'| \lambda \boldsymbol{L}_2 \cdot \boldsymbol{S}_2 |m'\rangle \langle nm'| \frac{e^2}{4\pi\varepsilon_0 r_{12}} |nn'\rangle}{E_{n'} - E_{m'}} + \sum_{m'} \frac{\langle nn'| \frac{e^2}{4\pi\varepsilon_0 r_{12}} |nm'\rangle \langle m'| \lambda \boldsymbol{L}_2 \cdot \boldsymbol{S}_2 |n'\rangle}{E_{n'} - E_{m'}}$$

(2.61)

In simplifying the sums, we use the fact that the spin-orbit terms concern one ion at a time, so only the ground state of the other ion can intervene. In other words, the spin-orbit effect of one ion is diagonal in the basis of the other ion.

As we are seeking exchange effects, we see their origins clearly in the expression above. We define exchange integrals as follows:

$$J(nn'mm') = \int d\boldsymbol{r}_1 \int d\boldsymbol{r}_2 \varphi_n^*(\boldsymbol{r}_1 - \boldsymbol{R})\varphi_{n'}^*(\boldsymbol{r}_2 - \boldsymbol{R}') \frac{e^2}{4\pi\varepsilon_0 r_{12}} \varphi_m(\boldsymbol{r}_2 - \boldsymbol{R})\varphi_{m'}(\boldsymbol{r}_1 - \boldsymbol{R}') \quad (2.62)$$

To complete the argument, we need to manipulate terms such as:

$$J(mn'n'n) = \int d\boldsymbol{r}_1 \int d\boldsymbol{r}_2 \varphi_m^*(\boldsymbol{r}_1 - \boldsymbol{R})\varphi_{n'}^*(\boldsymbol{r}_2 - \boldsymbol{R}') \frac{e^2}{4\pi\varepsilon_0 r_{12}} \varphi_{n'}(\boldsymbol{r}_2 - \boldsymbol{R}')\varphi_n(\boldsymbol{r}_1 - \boldsymbol{R})$$
$$= \int d\boldsymbol{r}_1 \int d\boldsymbol{r}_2 \varphi_m^*(\boldsymbol{r}_1 - \boldsymbol{R})\varphi_n(\boldsymbol{r}_1 - \boldsymbol{R})|\varphi_{n'}(\boldsymbol{r}_2 - \boldsymbol{R}')|^2 \frac{e^2}{4\pi\varepsilon_0 r_{12}}$$
$$= J^*(nn'n'm)$$

(2.63)

Therefore, the exchange terms in (2.61) are given by,

$$
\begin{aligned}
E_A = & \\
& \sum_m \frac{\langle n| \lambda \boldsymbol{L}_1 \cdot \boldsymbol{S}_1 |m\rangle \, J(mn'n'n)\boldsymbol{S}_1 \cdot \boldsymbol{S}_2}{E_n - E_m} + \sum_m \frac{J(nn'n'm)\boldsymbol{S}_1 \cdot \boldsymbol{S}_2 \, \langle m| \lambda \boldsymbol{L}_1 \cdot \boldsymbol{S}_1 |n\rangle}{E_n - E_m} \\
& + \sum_{m'} \frac{\langle n'| \lambda \boldsymbol{L}_2 \cdot \boldsymbol{S}_2 |m'\rangle \, J(m'nnn')\boldsymbol{S}_1 \cdot \boldsymbol{S}_2}{E_{n'} - E_{m'}} + \sum_{m'} \frac{J(n'nnm')\boldsymbol{S}_1 \cdot \boldsymbol{S}_2 \, \langle m'| \lambda \boldsymbol{L}_2 \cdot \boldsymbol{S}_2 |n'\rangle}{E_{n'} - E_{m'}}
\end{aligned}
\tag{2.64}
$$

To proceed with further simplifications, we assume that the orbital states are non-degenerate. In this case, the matrix elements of the orbital moment are pure imaginary numbers. Thus, we have,

$$
\langle n| \boldsymbol{L} |m\rangle = - \langle n| \boldsymbol{L} |m\rangle^*
\tag{2.65}
$$

This property leads to the quenching of the orbital moment, as we will see in section 3.3. The existence of the so-called "Kramer's doublet" is also linked to this symmetry; see Section 3.5. Symmetry under time reversal can be used to obtain this result. Further information about time reversal symmetry can be gained from L. Ballentine's book (Chapter 13).[181]

With the property (2.65), we can rewrite E_A as follows,

$$
\begin{aligned}
E_A = & \\
& \lambda \sum_m \frac{\langle n| \boldsymbol{L}_1 |m\rangle \, (J^*(nn'n'm)\boldsymbol{S}_1 \, (\boldsymbol{S}_1 \cdot \boldsymbol{S}_2) - J(nn'n'm) \, (\boldsymbol{S}_1 \cdot \boldsymbol{S}_2) \, \boldsymbol{S}_1)}{E_n - E_m} \\
& + \lambda \sum_{m'} \frac{\langle n'| \boldsymbol{L}_2 |m'\rangle \, (J^*(n'nnm')\boldsymbol{S}_2 \, (\boldsymbol{S}_1 \cdot \boldsymbol{S}_2) - J(n'nnm') \, (\boldsymbol{S}_1 \cdot \boldsymbol{S}_2) \, \boldsymbol{S}_2)}{E_{n'} - E_{m'}}
\end{aligned}
\tag{2.66}
$$

Now we extract from (2.66) the contributions that include the real part of $J(nn'n'm)$, noted $J'(nn'n'm)$:

$$
\begin{aligned}
E_A = & \lambda \sum_m \frac{\langle n| \boldsymbol{L}_1 |m\rangle \, J'(nn'n'm)}{E_n - E_m} \, (\boldsymbol{S}_1 \, (\boldsymbol{S}_1 \cdot \boldsymbol{S}_2) - (\boldsymbol{S}_1 \cdot \boldsymbol{S}_2) \, \boldsymbol{S}_1) \\
& + \lambda \sum_{m'} \frac{\langle n'| \boldsymbol{L}_2 |m'\rangle \, J'(n'nnm')}{E_{n'} - E_{m'}} \, (\boldsymbol{S}_2 \, (\boldsymbol{S}_1 \cdot \boldsymbol{S}_2) - (\boldsymbol{S}_1 \cdot \boldsymbol{S}_2) \, \boldsymbol{S}_2)
\end{aligned}
\tag{2.67}
$$

We point out that we have calculated matrix elements between orbital wave functions and left out the spin part as operators. Hence, the non-commutativity of these operators must be taken into account. We note that,

$$
S_{1x} \, (\boldsymbol{S}_1 \cdot \boldsymbol{S}_2) - (\boldsymbol{S}_1 \cdot \boldsymbol{S}_2) \, S_{1x} = i(\boldsymbol{S}_1 \times \boldsymbol{S}_2)_x
\tag{2.68}
$$

and likewise for the y and the z components. Therefore, we can write E_{DM} as,

$$
E_{DM} = i\lambda \left\{ \sum_m \frac{\langle n| \boldsymbol{L}_1 |m\rangle \, J'(nn'n'm)}{E_n - E_m} + \sum_{m'} \frac{\langle n'| \boldsymbol{L}_2 |m'\rangle \, J'(n'nnm')}{E_{n'} - E_{m'}} \right\} (\boldsymbol{S}_1 \times \boldsymbol{S}_2)
\tag{2.69}
$$

In summary, we have found that, because of spin-orbit coupling, there is an exchange term that adds a contribution to the coupling energy of the form (2.59).

2.8   FURTHER READINGS

Exchange

In this chapter, we presented the Heitler-London approximation. High resolution measurements find discrepancy from the prediction of this model for the dissociation of hydrogen molecules.[182]

The exchange energy relative relative to magnetostatic energy does not keep the same proportion when the size of a magnetic material is reduced,[183, 184] or when considering two-dimensional materials.[185]

Numerical simulations can shed some light on some features of magnetic properties. For example, the saturation magnetization M_s may change between its bulk value and the surface. Also, in Heusler alloys such as NiMnSb, it has been found that M_s depends on disorder.[186, 187] This may play a critical role in spintronics because NiMnSb is known as a half-metallic ferromagnet, i.e., one in which the conduction band of one spin is at the Fermi level.[50] So, it may be of great value when making tunnel magnetoresistive junctions. While the Heisenberg Hamiltonian (2.56) suggests, at first sight, either ferromagnetism or antiferromagnetism, *non-collinear magnetic structures* can also arise.[188] Ab-initio calculations of exchange interactions were carried out.[189]

Molecular magnets are solids where organometallics constitute the crystal units. The individual crystal units have an interest of their own,[190–192] and making materials out of them affords a special way of fine-tuning their magnetic properties.[193]

DM interaction

Terms that contain effective interactions similar to the DM interaction were derived long ago.[194–196] They are described in the textbook of Bethe and Salpeter (sec. 39)[197] and in Berestetskii et al. (sec. 83).[198] These results were cited and used by Stiles to calculate magnetic anisotropies in 3d ferromagnetic metals.[199]

The DM interaction can induce helical structures in magnetic materials called *helimagnets*. These structures were investigated for their spin transport properties in bulk materials, in particular, the formation of *magnetic solitons*.[200–206]

Non-collinear magnetic textures may be found in the bulk of chiral magnets, owing to bulk Dzyaloshinskii-Moriya interactions.[207–211] Such textures may also appear when interfaces induce DMI.[212–215] They were also observed in multilayers designed to promote *perpendicular magnetic anisotropy* (PMA).[216, 217]

The literature on *skyrmions* is extensive. Both topological and dynamical properties are of interest.[218] New aspects of DM interaction appear at interfaces,[219] giving rise to novel chiral magnetic structures.[220]

The DM interaction plays a role in the physics of organic charge-transfer salts. This is for example the case of the layered organic compound κ-$(ET)_2Cu[N(CN)_2]Cl$ which exhibits some unusual properties at low temperatures. ET stands for bis-ethelene-dithio-tetrathio-fulvalene (also abbreviated BEDT-TTF). At ambient pressure, this compound undergoes a transition to a magnetic state at 27 K.[221] When a pressure of 0.3 kbar is applied, however, no magnetic ordering is found and the system instead becomes an unconventional superconductor at $T_c = 12.8$ K.[222] The compound κ-$(ET)_2Cu[N(CN)_2]Cl$ belongs to a class of ambient-pressure organic superconductors, together with the isostructural compound κ-$(ET)_2Cu[N(CN)_2]Br$ ($T_c = 12.5$ K)[223] and κ-$(ET)_2Cu(NCS)_2$ ($T_c = 10.4$ K).[224] These are all unconventional superconductors for which spin fluctuations have been proposed as a pairing mechanism.[224, 225]

3 Crystal Field Effects

John Hasbrouck Van Vleck (1899–1980)

J.H. Van Vleck worked at the MIT Radiation Laboratory during World War II. Professor of physics at Harvard University, he developed a fundamental understanding of the physics of electrons in magnetic solids. He is regarded as a "father of modern magnetism" and was co-awarded the Nobel Prize in Physics in 1977, along with Philip W. Anderson and Sir Nevill Mott.

Spintronics: A primer

J.-Ph. Ansermet (https://orcid.org/0000-0002-1307-4864)

Chapter 3. Crystal Field Effects

Crystal field effects are calculated for a simple paramagnetic ion, leading to the notion of orbital moment quenching, g-factor anisotropy, and Van Vleck susceptibility. Likewise, simple models allow for straightforward calculations illustrating Kramer's degeneracy theorem and magnetic Jahn-Teller distortions.

DOI: 10.1201/9781003370017-3

3.1   HISTORICAL INTRODUCTION

William Gilbert (1544–1603) is known for having made the first carefully planned measurements on naturally magnetized minerals, notably lodestone (magnetite). In the preface to his 1600 book *"De Magnete,"*[226] Gilbert claims: *To you alone, true philosophers, ingenuous minds, who not only in books but in things themselves look for knowledge, have I dedicated these foundations of magnetic science – a new style of philosophising.* Galileo, who is famous for promoting the importance of experiments, mentioned Gilbert's book in his 1632 *Dialogue Concerning the Two Chief World Systems.*[227] Gilbert expresses – and rightfully so – his amazement that, when iron is smelted out from its ore, once cooled, this material has what he called "verticity" which we now call remanence ([226], chapter XIII). He tells us that, if a blacksmith hammers this iron to form elongated rods, they acquire magnetization, though they have not been magnetized by proximity to a a piece of lodestone. *This verticity [remanence] is acquired chiefly through the lengthening,* he tells us.

Nowadays, we would say that the origin of this effect is inverse magnetostriction. Magnetostriction refers to, e.g., the change in length of a ferromagnetic rod or wire when it is subjected to a magnetic field. The crystalline microstructure of the material is one parameter that determines the size of this effect.[228] The inverse effect is the influence of mechanical stress on magnetization. We can expect this effect to be noticeable only if magnetic anisotropy (see §4) is weak while magnetostrictive effects are large. These two conditions are rarely met. In general, when magnetostrictive effects are large, the magnetic anisotropy is also large. Hence, the search for such materials has gone on for decades.[229, 230] It has been shown that ultrasound can induce irreversible changes in the magnetization of carefully designed materials.[231, 232]

Quantum mechanics changed our understanding of magnetism in a profound way. The material presented in part 1 of this book attests to that. A short, yet insightful, history of magnetism was established by D. C. Mattis.[176] We mention here one founding father of quantum magnetism, Van Vleck, who received the Nobel prize with Philip W. Anderson and Sir Nevill Mott in 1977. In the 1930s, Van Vleck had developed quantum chemistry-type models to describe how electric fields in a crystal influence the properties of an impurity atom in that crystal, an effect now called ***crystal field***. Van Vleck also contributed to the understanding of the Jahn-Teller effect and was the first to point out the importance of electron correlations in the formation of local magnetic moments.

In this chapter, we examine how an atomic center that carries a magnetic moment interacts with its environment. We want to consider ferromagnets such that, to a fair approximation, we can consider them as composed of ions that retain at least some of the magnetic properties of the isolated ions. We want to know what changes occur as such a magnetic ion is introduced into a solid. This interaction with the environment of the magnetic ion is the crystal field. By considering how an atomic moment interacts with the lattice in which it resides, we can address a number of phenomena in magnetic solids, such as the susceptibility of paramagnetic centers, the quenching of orbital moment in 3d-metals, the g-factor anisotropy in magnetic resonance, Kramer's degeneracy, and the magnetic Jahn-Teller lattice distortions.

3.2   FREE PARAMAGNETIC IONS

The Lande factor

Here, we are concerned with the response of a local magnetic moment in a ferromagnet at temperatures above the Curie temperatureof that ferromagnet. A ***Curie-Weiss*** temperature dependence of the susceptibility T is often observed, characterized by,

$$\chi(T) = \frac{\chi_0}{T - T_c} \tag{3.1}$$

As we will see now, χ_0 depends directly on the magnetic moment per ion. To determine this, we focus on the magnitude of the moment $m = |\boldsymbol{m}|$ carried by each magnetic ion of the solid. Hund's rules applied to the ion imply that its ground state is a multiplet with given L, S, and J values.[1]

When a magnetic field is applied, the Zeeman interaction of the orbital and spin moment with the applied field implies an energy change, given to first order by:

$$\Delta E = \mu_B \boldsymbol{B} \langle LSJ, J_z | \left(\hat{\boldsymbol{L}} + 2\hat{\boldsymbol{S}} \right) | LSJJ_z \rangle \tag{3.2}$$

where μ_B is the **Bohr magneton** and the angular momenta are measured in units of $\hbar$. Let us set the z axis along the magnetic field $\boldsymbol{B}$.

The **Wigner-Eckart theorem**, which is discussed further at the end of this section, implies that:

$$\langle LSJ, J_z | (L_z + 2S_z) | LSJJ_z \rangle = g(LSJ) \langle LSJ, J_z | J_z | LSJJ_z \rangle = g(LSJ)J_z \tag{3.3}$$

Therefore, the magnetic moment can be written as,

$$\boldsymbol{\mu} = g(LSJ)\mu_B \hat{\boldsymbol{J}} \tag{3.4}$$

This expression must be applied with care because it is valid only to the extent that solely the ground-state multiplet is occupied. The coefficient g(LSJ) in (3.3) is called the **Landé factor**. It can be evaluated in the following manner.[2] From (3.3), we have,

$$\langle LSJ, J_z | (\boldsymbol{L} + 2\boldsymbol{S}) | LSJJ'_z \rangle = g(LSJ) \langle LSJ, J_z | \boldsymbol{J} | LSJJ'_z \rangle \tag{3.5}$$

Now, we consider,

$$\sum_{J'_z} \langle LSJ, J_z | (\boldsymbol{L} + 2\boldsymbol{S}) | LSJJ'_z \rangle \langle LSJ, J'_z | \boldsymbol{J} | LSJJ_z \rangle$$
$$= g(LSJ) \sum_{J'_z} \langle LSJ, J_z | \boldsymbol{J} | LSJJ'_z \rangle \langle LSJ, J'_z | \boldsymbol{J} | LSJJ_z \rangle \tag{3.6}$$

In this expression, we add sums over J'. This does not change anything since the matrix elements of $\boldsymbol{J}$ vanish when $J' \neq J$. Therefore, we can write,

$$\sum_{J', J'_z} \langle LSJ, J_z | (\boldsymbol{L} + 2\boldsymbol{S}) | LSJ'J'_z \rangle \langle LSJ'J'_z | \boldsymbol{J} | LSJJ_z \rangle$$
$$= g(LSJ) \sum_{J', J'_z} \langle LSJ, J_z | \boldsymbol{J} | LSJ'J'_z \rangle \langle LSJ'J'_z | \boldsymbol{J} | LSJJ_z \rangle \tag{3.7}$$

Now, we can make use of the completeness relation:

$$\sum_{J', J'_z} | LSJ'J'_z \rangle \langle LSJ'J'_z | = \mathbb{1} \tag{3.8}$$

---

[1] See, e.g., [72] chapter 31.

[2] See [72], appendix P.

Thus, we obtain,

$$\begin{aligned}
&\langle LSJ, J_z | \, (\boldsymbol{L} + 2\boldsymbol{S}) \cdot \boldsymbol{J} \, | LSJ J_z \rangle \\
&= g(LSJ) \langle LSJ, J_z | \, \boldsymbol{J}^2 \, | LSJ J_z \rangle = g(LSJ) J(J+1)
\end{aligned} \tag{3.9}$$

The matrix elements of the products $\boldsymbol{L} \cdot \boldsymbol{J}$ and $\boldsymbol{S} \cdot \boldsymbol{J}$ can be obtained from,

$$\begin{aligned}
\boldsymbol{S}^2 &= (\boldsymbol{J} - \boldsymbol{L})^2 = \boldsymbol{J}^2 + \boldsymbol{L}^2 - 2\boldsymbol{J} \cdot \boldsymbol{L} \\
\boldsymbol{L}^2 &= (\boldsymbol{J} - \boldsymbol{S})^2 = \boldsymbol{J}^2 + \boldsymbol{S}^2 - 2\boldsymbol{J} \cdot \boldsymbol{S}
\end{aligned} \tag{3.10}$$

using,

$$\begin{aligned}
\langle JLSJ_z | \, \boldsymbol{J}^2 \, | JLSJ_z \rangle &= J(J+1) \\
\langle JLSJ_z | \, \boldsymbol{S}^2 \, | JLSJ_z \rangle &= S(S+1) \\
\langle JLSJ_z | \, \boldsymbol{L}^2 \, | JLSJ_z \rangle &= L(L+1)
\end{aligned} \tag{3.11}$$

The final result is that the **_Landé factor_** is given by:

$$g(LSJ) = 1 + \frac{J(J+1) - L(L+1) + S(S+1)}{2J(J+1)} \tag{3.12}$$

Wigner-Eckart theorem

Within a multiplet of given total angular momentum, the expectation value of a vectorial operator is proportional to the expectation value of the total angular moment, with the coefficient of proportionality given by,[3]

$$\langle \tau', J, M' | \, V_m^{(1)} \, | \tau, J, M \rangle = \frac{\langle \tau', J, M' | \, \boldsymbol{J} \cdot \boldsymbol{V} \, | \tau, J, M \rangle}{\hbar^2 J(.J+1)} \langle J, M' | \, J_m^{(1)} \, | J, M \rangle \tag{3.13}$$

For any vector operator $\boldsymbol{V}$, its irreducible components are:

$$V_{+1} = -\sqrt{\frac{1}{2}} \, (V_x + iV_y) \quad V_0 = V_z \quad V_{-1} = -\sqrt{\frac{1}{2}} \, (V_x - iV_y) \tag{3.14}$$

Curie's law

Curie's law refers to a paramagnetic material, whereas Curie-Weiss's law refers to a system that turns magnetic below a critical temperature T_c. Here, we carry out the calculations without the added complication of a phase transition. If only the lowest multiplet of a paramagnetic ion is thermally excited, its free energy can be evaluated from,[4]

$$\exp\left(-\frac{F}{kT}\right) = \sum_{J_z = -J}^{J} \exp\left(-\frac{g(JLS)\mu_B B J_z}{kT}\right) \tag{3.15}$$

If there are N ions in a volume V, the magnetization M can be deduced from:

$$M = \mu_0 \frac{N}{V} \frac{\partial F}{\partial B} \tag{3.16}$$

---

[3] see [181], Eq. 7.125.

[4] see [72], Eq. 31.42.

The result is expressed in terms of **Brillouin functions** (see, e.g., [233]). In the limit of high temperatures, the expansion to first order in $g(JLS)B/kT$ yields Curie's law:

$$\chi = \frac{N}{V}\frac{(\mu_B p)^2}{kT} \qquad p = g(LSJ)\sqrt{J(J+1)} \qquad (3.17)$$

This law is fairly well verified for rare earth ions.[5] As Van Vleck showed in his 1932 book on susceptibilities,[234] the prediction (3.17) gives a fairly good estimate of the magnetic moment of 4f ions, except for Eu and Sm. This discrepancy is due to the fact that these ions have J-multiplets that are not accounted for in the present derivation of the susceptibility.

Things are quite different for 3d ions (see tables of values in, e.g., [72,234], Table 31.4). In that case, Curie's law is fairly well verified if, instead of J, the formula is used with only S, i.e., one must assume that $L = 0$. The phenomenon at the origin of this discrepancy between the free ion prediction and the experimental observation of vanishing orbital moment is referred to as the **quenching of the orbital moment**. This effect is also due to the crystal field, as explained in the following section.

3.3   QUENCHING OF THE ORBITAL MOMENT

This quenching has been inferred from data above the Curie temperature, i.e., in the paramagnetic state. Likewise, in the ferromagnetic state, it appears that the magnetic moment is determined by S, not J. The magnetic moment per atom in ferromagnets can be determined by measuring magnetization as a function of temperature below the Curie point. Thus, it was found that the magnetization as a function of temperature follows roughly a Brillouin behavior, as if the magnetization were due to the lining up of localized magnetic moments in an **exchange field**, or **Weiss molecular field**

$$\boldsymbol{B}_{ex} = \mu_0 \lambda \boldsymbol{M}(T) \qquad (3.18)$$

where $\boldsymbol{M}(T)$ is the magnetization. This approach constitutes a mean-field description of ferromagnetism. The temperature dependence of $M(T)$ can be found by solving:[79]

$$M(T) = g\mu_B \left\langle \hat{J}_z \right\rangle = B_J \left(\frac{g\mu_B J \lambda M(T)}{k_B T} \right) \qquad (3.19)$$

where B_J is the **Brillouin function** for an angular momentum $\hbar J$, for which there are $(2J+1)$ Zeeman energy levels. From fitting magnetization versus field data below the Curie temperature, magnets of 3d atoms have a magnetic moment proportional to $S(S+1)$, as if the orbital angular momentum did not intervene ($L = 0$), whereas ferromagnets containing 4f ions have a moment proportional to $J(J+1)$, with J deduced from Hund's rules.[72]

Van Vleck explained the quenching of orbital moment in the following manner. He took into account the interaction of the ion carrying a magnetic moment with the electrostatic charges of the surrounding atoms. These effects are referred to as the **crystal field**. When the crystal field dominates the spin-orbit coupling, then the wave functions that minimize the energy are functions that point to positive ions. Those are real wave functions (i.e., no imaginary parts). But the orbital moment operator is complex and its expectation value in the ground state must be a real quantity. Therefore, the expectation value must be zero.

Let us illustrate this argument with a concrete example, which was worked out in the framework of a study on color centers.[235] We consider one electron located at one ion that is surrounded by four positive charges located at the corners of a square centered on that ion. A displacement of the charges will be considered, as shown in Fig. 3.1. The ion is at the

---

[5]See tables in, e.g., [72,234], Table 31.3.

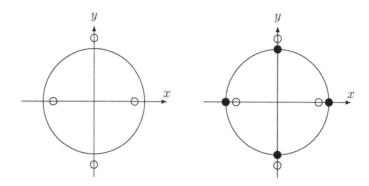

Figure 3.1 A negative ion (not shown) is located at the origin of the coordinate axes. Left: positive charges are displaced with respect to a configuration with a cubic symmetry. Right: in order to maintain charge neutrality, negative charges are added to simplify the calculation. Left and right circles are visual guides

center of the coordinate system. The generalized coordinate that defines the displacement of the positive charges (open circles) is noted $Q_{x^2-y^2}$. When the ion is subject to a crystal field, its Hamiltonian to first order is written as:

$$\mathcal{H}_{cf} = C(x^2 - y^2)Q_{x^2-y^2} \tag{3.20}$$

This can be derived by an expansion of the electrostatic interaction around the origin. With this system of charges, one would have an electrostatic contribution to the energy for any value of $Q_{x^2-y^2}$. In principle, there would be an electrostatic contribution even when the positive charges are symmetrically positioned. In order to avoid this extra term, we assume the presence of negative charges equidistant from the ion, hence having $Q_{x^2-y^2} = 0$ when the nearest neighbors are in a symmetric configuration. We assume that the ion at the origin has valence electrons in p states. We can choose either:

$$|x\rangle = xf(r) \qquad |y\rangle = yf(r) \qquad |z\rangle = zf(r) \tag{3.21}$$

or,

$$|+\rangle = \frac{1}{\sqrt{2}}\left(x + iy\right)f(r) \quad |z\rangle = zf(r) \quad |-\rangle = \frac{1}{\sqrt{2}}\left(x - iy\right)f(r) \tag{3.22}$$

The states in (3.22) are eigenstates of $\hat{L}_z$. When the ion is free, i.e., when $Q_{x^2-y^2} = 0$, the Zeeman orbital term lifts the degeneracy of the ground state.

Now, the set of p states (3.21), which are real functions, are the ones that diagonalize the perturbation Hamiltonian $\mathcal{H}_{cf}$. That is:

$$\langle x|\mathcal{H}_{cf}|y\rangle = \langle x|\mathcal{H}_{cf}|z\rangle = \langle y|\mathcal{H}_{cf}|z\rangle = 0 \tag{3.23}$$

whereas,

$$\langle x|\mathcal{H}_{cf}|x\rangle = \frac{1}{2}D \neq 0$$
$$\langle y|\mathcal{H}_{cf}|y\rangle = \frac{-1}{2}D \tag{3.24}$$
$$\langle z|\mathcal{H}_{cf}|z\rangle = 0$$

The first of these equations defines the constant D. In other words, the states (3.21) are the eigenstates of the Hamiltonian.

This being established, we now take into account the magnetic field. The Zeeman interaction of the orbital term, i.e., $\mathcal{H}_Z = \mu_B B \hat{L}_z$, is here assumed to be a small perturbation compared to the crystal field. It is easy to verify that:

$$\langle x| L_z |x\rangle = \langle y| L_z |y\rangle = \langle z| L_z |z\rangle = 0 \tag{3.25}$$

Therefore, this first order perturbation calculation predicts that there is no orbital contribution to the magnetic moment, i.e., the orbital momentum is quenched because of the crystal field. We can drive this point further by noticing that states $|x\rangle$ and $|y\rangle$ are made up (with equal probability) of eigenstates of L_z, which have opposite values of angular moment along z:

$$|x\rangle = \frac{|+\rangle + |-\rangle}{\sqrt{2}} \qquad |y\rangle = \frac{|+\rangle - |-\rangle}{i\sqrt{2}} \tag{3.26}$$

Thus, the contributions of these states to the orbital moment cancel out.

We can now examine how a magnetic field $\boldsymbol{B}$ mixes the ground state $|x\rangle$ with the other states. First order perturbation on the ground state gives:

$$|\psi_x\rangle = |x\rangle + \sum_n \frac{\langle n| \mu_B B(L_z/\hbar) |x\rangle}{E_x - E_n} |n\rangle = |x\rangle + i\frac{\mu_B B}{D} |y\rangle \tag{3.27}$$

Posing $\varepsilon = \mu_B B/D$, the perturbed ground state can be expressed as,

$$|\psi_x\rangle = \frac{|+\rangle (1 + \varepsilon) + |-\rangle (1 - \varepsilon)}{\sqrt{2}} \tag{3.28}$$

Now we see an imbalance building up under the action of the applied field between the L_z eigenstates of eigenvalue $+1$ and -1. This partial unquenching is at the origin of the chemical shifts in nuclear magnetic resonance (see [236], section 4.3).

We can exploit this model a little bit further. This will provide insight as to the effect of spin-orbit coupling, written written for this argument as $\mathcal{H}_{SO} = \boldsymbol{L} \cdot (\lambda \boldsymbol{S})$. In the approximation where the spin-orbit coupling is a perturbation, the average spin $\langle \boldsymbol{S} \rangle$ can be thought of as a magnetic field acting on the orbital moment. Therefore, we see that the spin-orbit interaction brings back a little contribution of the angular momentum to the magnetic moment, which would otherwise be quenched by the crystal field. This partial unquenching of the orbital moment also determines the magnetocrystalline anisotropies of 3d metals, as we will see in §4.5.

3.4   THE G-FACTOR ANISOTROPY AND VAN VLECK SUSCEPTIBILITY

In view of the experimental results on the susceptibility of 3d metal ions in solids, and in view of many other pieces of experimental evidence, we see that, to our initial description of an ion that contained intra-atomic interactions, we must add the effect of the crystal field. Furthermore, in 3d metals, the crystal field dominates the spin-orbit coupling and the Zeeman effect. This dominance of the crystal field has interesting consequences in electron paramagnetic resonance and helps interpret detailed studies of magnetic susceptibilities.

We consider the ground state to be a multiplet of given L and S, so the spin-orbit coupling has the form,

$$\mathcal{H}_{SO} = \lambda \boldsymbol{L} \cdot \boldsymbol{S} \tag{3.29}$$

and the Zeeman coupling is given by (2.39). We assume a non-degenerate ground state in which the angular momentum is quenched. The Zeeman and the spin-orbit interactions are treated as perturbations. The unperturbed eigenstates are eigenstates of the intra-atomic

Coulomb effects and the crystal field. Therefore, these states can be products of spin states and orbital wave functions $|\Gamma, \gamma\rangle \, |S, m_S\rangle$. To second order, we get an effective Hamiltonian with only spin operators remaining:[6]

$$E_{\text{eff}} = 2\mu_B \boldsymbol{B} \cdot \left\langle \hat{\boldsymbol{S}} \right\rangle - \sum_{\Gamma', \gamma'} \frac{\left| \langle \Gamma', \gamma' | \, \mu_B \boldsymbol{B} \cdot \hat{\boldsymbol{L}} + \lambda \hat{\boldsymbol{L}} \cdot \hat{\boldsymbol{S}} \, | \Gamma, \gamma \rangle \right|^2}{E_{\Gamma', \gamma'} - E_{\Gamma, \gamma}} \tag{3.30}$$

Expanding the square yields:

$$E_{\text{eff}} = 2\mu_B \sum_{\mu, \nu} (\delta_{\mu, \nu} - \lambda \Lambda_{\mu, \nu}) S_\mu B_\nu - \lambda^2 \sum_{\mu, \nu} \Lambda_{\mu, \nu} S_\mu S_\nu - {\mu_B}^2 \sum_{\mu, \nu} \Lambda_{\mu, \nu} B_\mu B_\nu \tag{3.31}$$

The first term in (3.31) shows that the g factor of 2 found in the Zeeman coupling (2.38) is now replaced by a tensorial relationship. Thus, the **g tensor** of a transition metal ion differs from the g factor of an isolated electron. We see that the difference is due to the spin-orbit coupling and is determined by the matrix $\Lambda_{\mu, \nu}$,

$$\Lambda_{\mu, \nu} = \sum_{\Gamma', \gamma'} \frac{\langle \Gamma, \gamma | \, L_\mu \, | \Gamma', \gamma' \rangle \, \langle \Gamma', \gamma' | \, L_\nu \, | \Gamma, \gamma \rangle}{E_{\Gamma', \gamma'} - E_{\Gamma, \gamma}} \tag{3.32}$$

This matrix may be non-diagonal since it depends on the crystal field. Thus, we have obtained a description of the so-called **g-factor anisotropy**. This g-tensor can be measured using electron paramagnetic resonance techniques (EPR). Thus, we can think of EPR spectroscopy as a means to probe the local environment of a paramagnetic ion.

The last term in (3.31) is diagonal in spin states. It gives an energy term that is quadratic in field. A double derivation with respect to the magnetic field would give a susceptibility tensor. This is a susceptibility that is independent of temperature. It arises from the admixture of multiplets of higher energy than the ground state. This contribution to the susceptibility is called the **Van Vleck susceptibility**. Here, we focused on localized electrons. A similar mechanism for delocalized electrons leads to the **orbital paramagnetism** of transition metals (see [237] and [156], section 3.2.1).

3.5   KRAMER'S DEGENERACY

A concrete example will illustrate the notions introduced in the above calculation. We consider a Co^{2+} ion in $CoCl_2$-$2H_2O$.[239] The ion Co^{2+} has 7 electrons on the 3d shell. As far as the spin is concerned, these 7 electrons behave like 3 holes in the 3d shell. The intra-atomic Coulomb interactions (i.e the first Hund rule, §2.2) lead to a ^4F ground state, i.e., a spin of 3/2 and an orbital moment L = 3. Under the effect of the crystal field, assumed to be cubic, the 3d states split, with a ground state having a 3-fold orbital degeneracy. Since the spin is 3/2, there are 12 degenerate states.[7]

We want to work out the effect of the spin-orbit coupling on this multiplet ground state defined by the crystal field. We will find that the ground state has a 2-fold degeneracy. This change in ground state due to the spin-orbit coupling is an important concept. It is this magnetic property of the ground state that determines the magneto-optical effect, not a magnetic interaction with light.

---

[6] See [156], Eq. 2.94ff.

[7] More on spin state depending on the relative strength of electron-electron and spin-orbit couplings can be found in [151].

Since we restrict our analysis to 3 degenerate orbital states, we can define inside this multiplet an effective orbital moment with $L = 1$. The spin-orbit coupling is given by (3.29). In order to find the ground state, we diagonalize the spin-orbit Hamiltonian in the basis of states:

$$|M_L, M_S\rangle \qquad M_L = 0, \pm 1 \quad M_S = \pm\frac{1}{2}, \pm\frac{3}{2} \tag{3.33}$$

It is convenient to write the spin-orbit coupling in the following form,

$$\frac{1}{\lambda}\mathcal{H}_{SO} = L_z S_z + \frac{1}{2}\left(L_+ S_- + L_- S_+\right) \tag{3.34}$$

with the usual notation $L_\pm = L_x \pm i L_y$. It is possible to find the eigenvectors and eigenvalues of the spin-orbit coupling in the basis of states (3.33) using Mathematica. In order to do so, it is necessary to define the tensorial product such as $L_z S_z$. Mathematica has a predefined function "Outer" which generates a matrix of matrices, which is not what we want. The following definition generates the desired matrix:

$$CrossP[x, y] := Table[Outer[Times, x, y][[$$
$$Floor[(i + Dimensions[y][[1]] - 1)/Dimensions[y][[1]]],$$
$$Floor[(j + Dimensions[y][[1]] - 1)/Dimensions[y][[1]]],$$
$$Mod[i + Dimensions[y][[1]] - 1, Dimensions[y][[1]]] + 1,$$
$$Mod[j + Dimensions[y][[1]] - 1, Dimensions[y][[1]]] + 1]],$$
$$i, 1, Dimensions[x][[1]] * Dimensions[y][[1]],$$
$$j, 1, Dimensions[x][[1]] * Dimensions[y][[1]]]$$

The spin-orbit Hamiltonian matrix restricted to the ground state multiplet is,

$$\frac{1}{2}\begin{pmatrix}
3 & 0 & 0 & 0 & 0 & 0 & 0 & 0 & 0 & 0 & 0 & 0 \\
0 & 1 & 0 & 0 & \alpha & 0 & 0 & 0 & 0 & 0 & 0 & 0 \\
0 & 0 & -1 & 0 & 0 & \alpha & 0 & 0 & 0 & 0 & 0 & 0 \\
0 & 0 & 0 & 3 & 0 & 0 & \alpha & 0 & 0 & 0 & 0 & 0 \\
0 & \alpha & 0 & 0 & 0 & 0 & 0 & 0 & 0 & 0 & 0 & 0 \\
0 & 0 & \alpha & 0 & 0 & 0 & 0 & 0 & \alpha & 0 & 0 & 0 \\
0 & 0 & 0 & \alpha & 0 & 0 & 0 & 0 & 0 & \alpha & 0 & 0 \\
0 & 0 & 0 & 0 & 0 & 0 & 0 & 0 & 0 & 0 & \alpha & 0 \\
0 & 0 & 0 & 0 & 0 & \alpha & 0 & 0 & -3 & 0 & 0 & 0 \\
0 & 0 & 0 & 0 & 0 & 0 & \alpha & 0 & 0 & -1 & 0 & 0 \\
0 & 0 & 0 & 0 & 0 & 0 & 0 & \alpha & 0 & 0 & 1 & 0 \\
0 & 0 & 0 & 0 & 0 & 0 & 0 & 0 & 0 & 0 & 0 & 3
\end{pmatrix} \tag{3.35}$$

where $\alpha = 2\sqrt{3/2}$. Mathematica can be used to find the eigenvalues and eigenvectors. It is thus found that the energy levels split. The energy of the lowest level is of about $-2.4\,\lambda$. It is a doublet, that is to say, there are two degenerate eigenstates, which are:

$$|\Psi^\pm\rangle = 2\left|m_L = \pm 1, m_S = \mp\frac{3}{2}\right\rangle$$
$$-1.5\left|m_L = 0, m_S = \mp\frac{1}{2}\right\rangle + 1.0\left|m_L = \mp 1, m_S = \pm\frac{1}{2}\right\rangle \tag{3.36}$$

Thus, we find that the spin-orbit coupling leads to a ground state that is a mixture of orbital states with different spin states. These two states are no longer expressible as a product of a spin state and an orbital wave function.

Kramer's theorem

The doublet found in the example above reflects the presence of a fundamental symmetry: the time reversal symmetry of the Hamiltonian. Kramer's theorem states that for an odd number of spins (in the example above, 3), no matter what the crystal field is like, there is at least a residual degeneracy. Only a magnetic field can lift this degeneracy.

Time reversal is not meant as the substitution $t \to -t$ by analogy to the space inversion $\boldsymbol{x} \to -\boldsymbol{x}$.[8] Instead, the time reversal operator $\mathcal{T}$ for operators is defined by:

$$\begin{aligned}
\mathcal{T}\boldsymbol{X}\mathcal{T}^{-1} &= \boldsymbol{X} \\
\mathcal{T}\boldsymbol{P}\mathcal{T}^{-1} &= -\boldsymbol{P} \\
\mathcal{T}\boldsymbol{L}\mathcal{T}^{-1} &= -\boldsymbol{L} \\
\mathcal{T}\boldsymbol{S}\mathcal{T}^{-1} &= -\boldsymbol{S}
\end{aligned} \tag{3.37}$$

$\mathcal{T}$ is an antilinear operator. This can be seen by working out the effect of $\mathcal{T}$ on the commutator of position and momentum, or components of the angular momentum, etc. An anti-linear operator A is one for which:

$$A(c_1 |\psi_1\rangle + c_2 |\psi_2\rangle) = c_1{}^* A |\psi_1\rangle + c_2{}^* A |\psi_2\rangle \tag{3.38}$$

The Dirac "brak-ket" notation does not work well for antilinear operators. It is best to consider that antilinear operators work only "to the right" in bra-ket notation. If we use coordinate representation for orbital variables and the standard representation for spin operators, i.e., one in which S_z is diagonal, then $\mathcal{T}$ can be expressed as:

$$\mathcal{T} = \exp\left(\frac{-i\pi S_y}{\hbar}\right) K_0 \tag{3.39}$$

where K_0 is the conjugation operator:

$$K_0 \Psi(\boldsymbol{x}, t) = \Psi(\boldsymbol{x}, t)^* \tag{3.40}$$

From this, it can be worked out that,

$$\mathcal{T}^2 = \exp\left(\frac{-i2\pi J_y}{\hbar}\right) \tag{3.41}$$

where J_y is the y component of the total angular momentum. In going from S to J, we used,

$$\exp\left(-i2\pi L_y/\hbar\right) = 1 \tag{3.42}$$

Thus, according to (3.41), if the state has an integer spin, the eigenvalue of $\mathcal{T}^2$ is +1. If the state is of angular momentum half odd-integer, then the eigenvalue of $\mathcal{T}^2$ is −1.

To conclude, we need to consider a Hamiltonian $\mathcal{H}$ such that $\mathcal{T}\mathcal{H} = \mathcal{H}\mathcal{T}$. Then, we have:

$$\mathcal{H}\mathcal{T} |\psi\rangle = \mathcal{T}\mathcal{H} |\psi\rangle = E |\psi\rangle \tag{3.43}$$

This says that $\mathcal{T} |\psi\rangle$ is an eigenstate of $\mathcal{H}$ of eigenvalue E. If $\mathcal{T}^2 |\psi\rangle = -|\psi\rangle$ (because $|\psi\rangle$ is a half odd-integer spin state), then $\mathcal{T} |\psi\rangle$ and $|\psi\rangle$ are linearly independent (as will be shown below), with the same energy. We have what is called a ***Kramers doublet***.

Let us show that $\mathcal{T} |\psi\rangle$ and $|\psi\rangle$ are linearly independent when $\mathcal{T}^2 |\psi\rangle = -|\psi\rangle$. *A contrario*, if $\mathcal{T} |\psi\rangle$ and $|\psi\rangle$ were linearly dependent, we could write $\mathcal{T} |\psi\rangle = a |\psi\rangle$. However, in that case we have,

$$-|\psi\rangle = \mathcal{T}\mathcal{T} |\psi\rangle = \mathcal{T}a |\psi\rangle = a^*\mathcal{T} |\psi\rangle = a^*a |\psi\rangle = |a|^2 |\psi\rangle \tag{3.44}$$

which is impossible. A more general demonstration was proposed by Roberts.[240]

---

[8]See, e.g., [181], Section 13.3.

Orbital moment quenching from a symmetry standpoint

One consequence of Kramer's theorem is the quenching of the orbital momentum. Suppose that we have $\mathcal{T}\mathcal{H} = \mathcal{H}\mathcal{T}$ and that an orbital state $|\psi\rangle$ is non-degenerate, presumably because the crystal field has lifted all degeneracies. Then, $\mathcal{T}^2 = 1$ and $\mathcal{T}|\psi\rangle = a|\psi\rangle$ with $|a|^2 = 1$. We can write $a = e^{i\varphi}$ and choose any value of φ. By the definition (3.37) of the action of $\mathcal{T}$ on angular momenta, we have:

$$\langle\psi|\boldsymbol{L}|\psi\rangle = -\langle\psi|\mathcal{T}\boldsymbol{L}\mathcal{T}^{-1}|\psi\rangle \tag{3.45}$$

The action of the inverse of $\mathcal{T}$ is obtained as follows:

$$\begin{aligned} \mathcal{T}|\psi\rangle &= e^{i\varphi}|\psi\rangle \\ \mathcal{T}^{-1}|\psi\rangle &= \mathcal{T}^{-1}\left(e^{-i\varphi}e^{i\varphi}|\psi\rangle\right) = e^{i\varphi}\mathcal{T}^{-1}\left(e^{i\varphi}|\psi\rangle\right) = e^{i\varphi}\mathcal{T}^{-1}\left(\mathcal{T}|\psi\rangle\right) = e^{i\varphi}|\psi\rangle \end{aligned} \tag{3.46}$$

Then,

$$\begin{aligned} \langle\psi|\boldsymbol{L}|\psi\rangle &= -\langle\psi|\mathcal{T}\boldsymbol{L}\mathcal{T}^{-1}|\psi\rangle \\ &= -e^{i\varphi}\langle\psi|\mathcal{T}\boldsymbol{L}|\psi\rangle = e^{i\varphi}\langle\psi|\boldsymbol{L}\mathcal{T}|\psi\rangle = e^{2i\varphi}\langle\psi|\boldsymbol{L}|\psi\rangle \end{aligned} \tag{3.47}$$

Taking $\varphi = \pi/2$ yields $\langle\psi|\boldsymbol{L}|\psi\rangle = -\langle\psi|\boldsymbol{L}|\psi\rangle$, so that $\langle\psi|\boldsymbol{L}|\psi\rangle = 0$, i.e., the orbital angular momentum is quenched.

3.6 MAGNETIC JAHN-TELLER IONS

Jahn-Teller effects come up in the description of manganates, for example. These materials exhibit colossal magnetoresistance and were studied extensively in the 1950s. Starting in the mid-90s, a renewed interest arose when perovskite manganates were found with T_c near room temperature.[39]

Consider that the crystal field lifts some degeneracy, but leaves a ground state as a degenerate multiplet. Consider then a distortion of the actual crystal that lowers the symmetry further, thus lifting this degeneracy. Suppose that this distortion lowers the energy of the ground state of the electrons. Call ε the relative distortion of the lattice. The electron energy is assumed to change linearly with ε:

$$\Delta E_{el} = -b\varepsilon \tag{3.48}$$

On the other hand, in the absence of this magnetic effect, the ions of the crystal were presumed to be in equilibrium, so this distortion will raise the energy of the ions, but it will only be quadratic in ε to the lowest order. Thus,

$$\Delta E_{ion} = a\varepsilon^2 \tag{3.49}$$

Now we look for a minimum, a compromise between these two effects. That is, we search for the minimum of the energy $\Delta E = a\varepsilon^2 - b\varepsilon$ as a function of ε:

$$\begin{aligned} \frac{\partial\Delta E}{\partial\varepsilon} &= 2a\varepsilon - b \quad \frac{\partial\Delta E}{\partial\varepsilon} = 0 \Rightarrow \quad \varepsilon = \frac{b}{2a} \\ \Delta E_{\min} &= a\left(\frac{b}{2a}\right)^2 - b\frac{b}{2a} = \frac{-b^2}{4a} < 0 \end{aligned} \tag{3.50}$$

Ions that carry a magnetic moment at a location where the crystal lowers its symmetry because of the electronic contribution are called magnetic Jahn-Teller ions.

3.7   FURTHER READINGS

Crystal field

Crystal field is ubiquitous in magnetism, as attests the *Crystal Field Handbook*.[241] As one might expect, the crystal field changes when the surface is reached. At the tip of a scanning tunneling microscope, the formation of molecular contacts was shown to modify strongly the local crystal field.[242]

At the surface of ferroelectrics, the electric field may be strong enough to switch the ordering of the energy level of the adsorbed molecules and thereby the spin state.[243] Molecules with this property are called ***spin-crossover molecules***.[244]

If the electrochemical potential of charges at the surface of a capacitor is spin-dependent, then applying a magnetic field to the capacitor will change the surface screening.[245] This has been observed at the surface of palladium.[246]

A partial unquenching of the angular moment can take place near surfaces. A numerical study reveals that this is the case for Co, Fe, and Ni surfaces.[247]

Jahn-Teller effects

The literature distinguishes three types of Jahn-Teller effects.[9] The intuitive description above corresponds to what is known as the static Jahn-Teller effect.[249] For example, Van Vleck examined the stability of a three-fold degenerate magnetic ion in an octahedral site.[250]

A situation might arise in which no static lifting of the degeneracy occurs, but there is a dynamic Jahn-Teller effect that causes a variation in the values of the coefficients of the effective spin Hamiltonian compared to their values when the system is static.[251, 252]

A third kind is a collective Jahn-Teller effect. It arises when many Jahn-Teller ions are coupled to one another.[253]

---

[9]See [248], Section 7.6.

4 Magnetic Anisotropies

Amikam Aharoni (1929–2002)

A. Aharoni earned a master's degree in physics from the Hebrew University and a PhD in physics from the Weizmann Institute, where he studied magnetoresistive memories. He continued his professional training with J. Bardeen at the University of Illinois Urbana-Champaign. In 1972, Aharoni joined the faculty of the Weizmann Institute where he became professor of theoretical magnetism.

Spintronics: A primer

J.-Ph. Ansermet (https://orcid.org/0000-0002-1307-4864)

Chapter 4. Magnetic Anisotropies

A phenomenological description of magnetic anisotropies can be very useful in predicting quasi-static magnetic behavior. As an example, the Stoner-Wohlfarth model is introduced. Research on magnetic memory bits calls for an empirical understanding of the physical mechanisms of these anisotropies. The interplay of crystal field and spin-orbit effects is calculated for a simple $3d$ magnetic ion with a predefined electronic structure. The case of a $4f$ ion is outlined.

DOI: 10.1201/9781003370017-4

4.1 HISTORICAL INTRODUCTION

Routinely, magnetic anisotropies are described empirically, using parameters deduced from measurements of hysteresis curves or ferromagnetic resonance. Theoretically, Van Vleck is known for having first proposed that these anisotropies arise from spin-orbit coupling effects in conjunction with crystal field effects, as we shall see in this chapter. In the late 1980s, Bruno used a tight-binding approach to show that, under certain circumstances, the anisotropy of the spin-orbit energy can be linked to the anisotropy of the orbital moment.[254]

In the transition metals Fe, Co, and Ni, the crystal symmetry is high. Consequently, the magnetocrystalline anisotropies are very low and their estimate using electronic structure calculations require special care. For example, Halilov et al. used over 300,000 points in k-space to get a value that matched experimental observations.[255] In multilayers, the case examined by Bruno, the anisotropies are two orders of magnitude larger; thus, a model based on chemical bonds is sufficient. Of particular interest for spintronics are multilayers that ensure perpendicular magnetic anisotropy (PMA).[256]

The study of magnetic nanostructures has shed new light on old issues, such as the strength of surface anisotropies. In 1954, Néel commented that surfaces ought to have strong anisotropies, owing to the dissymmetry of local bonds at the surface as compared to ions in the bulk.[257] Nowadays, it has been shown that nanometer-sized magnetic particles grown on well-defined substrates present enormous magnetic anisotropies.[258] In spintronics, research on the effect of spin-polarized currents on magnetization requires knowing about the magnetization configuration at interfaces. Thus, recent research activities point to the need for a thorough understanding of magnetic anisotropies at the microscopic level.

In this chapter, we define the Stoner-Wohlfarth model, which is a phenomenological description of magnetic anisotropies in a single-domain magnetic particle. We proceed with an estimate of the dipolar contribution to anisotropies. We then turn to a heuristic example illustrating magnetocrystalline anisotropies in a 3d magnet. Finally, we show how the description of anisotropies differs when treating 4f magnets and mention the contribution of exchange anisotropies.

4.2 PHENOMENOLOGY

Let us first express the notion that magnetic energy is a function of the orientation of the magnetization in a magnetic material. We will later examine the question of the physical origin of this anisotropy.

The magnetization orientation can be specified by the azimuthal (ϕ) and polar (θ) angles of the spherical coordinate system:

$$\boldsymbol{M} = M_s \left(\sin\theta \cos\phi \boldsymbol{e}_x + \sin\theta \sin\phi \boldsymbol{e}_y + \cos\theta \boldsymbol{e}_z \right) \tag{4.1}$$

The simplest case is a uniaxial magnetic anisotropy with an energy:

$$\frac{E}{V} = K_1 \sin^2\theta \tag{4.2}$$

In the case of tetragonal symmetry, the energy would be of the form:[238]

$$\frac{E}{V} = K_1 \sin^2\theta + K_2 \sin^4\theta + K_2^{(2)} \sin^4\theta \cos 4\phi + K_3 \sin^6\theta + K_3^{(3)} \sin^6\theta \cos 4\phi \tag{4.3}$$

Orders of magnitude

Typical values of magnetic energies are of the order of:

1 MJ/m^3 or 0.17 meV/atom for 3d magnets,
10 MJ/m^3 or 1.7 meV/atom for 4f magnets.

Here, the values expressed in the International Systems of Units (SI) were converted into meV/atom, assuming an atomic volume of $(0.3 \text{ nm})^3$. Of course, these values vary significantly depending on the material.

There are applications that require very low coercivity, such as in writing heads. To the contrary, some applications require very high coercivity, such as permanent magnets. Major efforts have been deployed in order to develop materials with a structure and composition that optimize these and other magnetic properties.

One quick estimate will give a feel for the order of magnitude these magnetic energies represent. The consideration proposed here stems from the challenge of using magnetic materials for recording. If magnetic particles are used as memory bits, there is a problem when the bit is so small that the magnetic anisotropy energy is comparable to the thermal energy. A value of 1 MJ/m^3 means that a spherical particle of 2 nm in diameter has the magnetic energy equal to $k_B T$ at room temperature. When $k_B T$ is much larger than the energy barrier for switching, the magnetic particle is said to be ***superparamagnetic***.[183] In recording applications, the magnetic anisotropy must be much larger than the thermal energy for the magnetization to remain stable over a time scale of several years.

4.3   THE STONER-WOHLFARTH MODEL

The model most frequently used to describe magnetization switching in a ferromagnetic particle is known as the ***Stoner-Wohlfarth model***.[259] This model assumes isolated magnetic particles, each one being single domain with a uniaxial magnetic anisotropy of the same strength. Let us put the z axis along the uniaxial anisotropy, and the applied field in the xz plane. Then, with the angles as defined in Fig. 4.1, we have:

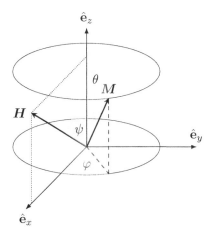

Figure 4.1   Magnetic field and magnetization orientation with respect to an axis of uniaxial anisotropy.

$$V(\theta, \varphi) = K \sin^2\theta - \mu_0 M_s H \left(\cos\theta \cos\psi + \sin\theta \cos\varphi \sin\psi \right) \qquad (4.4)$$

By assuming that the magnetic field is not aligned with the axis of magnetic anisotropy, a particular form of departure from axial symmetry occurs. [260] The situation occurs in experiments where the magnetization can be considered to remain in the plane defined by the magnetic field and the anisotropy axis ($\varphi = 0$). This simple case is sufficient to explain the notion of hysteresis. In the following, we show plots of the potential,

$$V\left(\theta, \psi\right) = \sin^2\theta - \bar{H}\left(\cos\theta \cos\psi + \sin\theta \sin\psi\right) \tag{4.5}$$

where $\bar{H} = \mu_0 M_s H / K$ is unitless. At zero field, using $\psi = \pi/4$, the potential as a function of θ has a double-well structure (Fig. 4.2, left). The field biases the double-well structure.

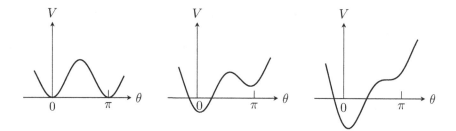

Figure 4.2 Stoner-Wohlfarth model (4.5) at zero applied field (left), at $\bar{H} = 0.5$ (middle) and at $H = 1.0$ (right). The middle figure is suggestive of the emergence of a meta-stable state at finite temperature when the field is sufficiently large. That state becomes unstable at any temperature when $H = 1.0$ (right).

At a magnetic field such that $\bar{H} = 0.5$, there is a metastable state (Fig. 4.2, middle). At $\bar{H} = 1.0$, this state becomes unstable, even at zero temperature (Fig. 4.2, right).

If the magnetic particle is prepared in a state where it is at zero field in the well on the right, then at $\bar{H} = 0.5$, it will move to the well on the left. This represents a jump in the magnetization, i.e., ***magnetization switching***. If the field is scanned further to a positive value, there is hardly any change in the orientation of the magnetization. If the field is ramped in the opposite direction, there is a switch to the well on the right.

4.4   ANISOTROPY ARISING FROM DIPOLAR INTERACTIONS

Let us now examine the contribution of dipole-dipole interactions to magnetic anisotropies. The dipolar coupling among local moments can be written as,[170]

$$E_{\mathrm{dip}} = \frac{\mu_0}{4\pi} \sum_{i<j} \frac{1}{r_{ij}^3} \left(\boldsymbol{m}_i \cdot \boldsymbol{m}_j - 3 \frac{(\boldsymbol{r}_{ij} \cdot \boldsymbol{m}_i)(\boldsymbol{r}_{ij} \cdot \boldsymbol{m}_j)}{r_{ij}^2} \right) \tag{4.6}$$

where $\boldsymbol{m}_i$ is the magnetic moment at site i, in units of Bohr magnetons. When the moments are all parallel (as they would be in a ferromagnetic domain), this expression can be reduced to:[261]

$$E_{\mathrm{dip}} = \frac{\mu_0}{4\pi} \sum_{i<j} \frac{m_i m_j}{r_{ij}^3} \left(1 - 3\cos^2\theta_{ij}\right) \tag{4.7}$$

where θ_{ij} is the angle between the magnetization and the bond between sites i and j. Here we consider localized moments, so we have a discrete sum. The case of delocalized electrons is beyond the scope of this book.[262]

Since the dipolar coupling (4.6) decreases with the cube of the distance, a sum over all the pairs in space converges so slowly that the dipoles at the surface of the sample play a significant role. Let us estimate the dipolar field $\boldsymbol{H}_{dip}(i)$ at site i produced by all the other dipole moments using the **Lorentz method**. In order to understand this approach, we first introduce the notion of **effective magnetic charges**.

In the absence of current, Ampère's law in Maxwell's equations reads $\boldsymbol{\nabla} \times \boldsymbol{H} = \boldsymbol{0}$. Therefore, $\boldsymbol{H}$ can be derived from a potential V_{mag}, i.e., $\boldsymbol{H} = -\boldsymbol{\nabla} V_{\text{mag}}$. V_{mag} is called the **magnetic potential**. Then the Gauss law $\boldsymbol{\nabla} B = 0$ implies $\nabla^2 V_{\text{mag}} = \boldsymbol{\nabla} \cdot \boldsymbol{M}$. Hence, the problem at hand resembles an electrostatic problem, where here the charge density ρ_M is given by,

$$\rho_M = -\boldsymbol{\nabla} \cdot \boldsymbol{M} \tag{4.8}$$

Thus, we have a Poisson-like equation $\nabla^2 V_{\text{mag}} = -\rho_M$. The boundary conditions are defined by the magnetic surface charges,

$$\sigma_M = \hat{n} \cdot \boldsymbol{M} \tag{4.9}$$

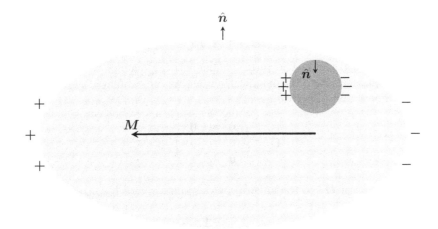

Figure 4.3 Illustration of the Lorentz method. The sample is supposed to be an ellipsoid with uniform magnetization. A cavity at site i is defined, outside of which the local magnetic moment $\mu_B \boldsymbol{m}_j$ can be taken as the macroscopic magnetization $\boldsymbol{M}(\boldsymbol{r})$ times a unit cell volume V. Plus and minus signs represent the surface effective magnetic charges defined in (4.8). Both unit vectors $\hat{n}$ point toward the outside of the matter confined between the ellipsoid and the cavity surfaces.

Let us now estimate the field $\boldsymbol{H}(i)$ at site i. Far away from this site, the details of how the magnetization might change from site to site are insignificant. So, at these far-away sites, we can replace the magnetization moment $\boldsymbol{m}_j$ by $\boldsymbol{M} V$, where V is the unit cell volume. Hence, we define a sphere around site i and apply this approximation outside this sphere (Fig. 4.3). Let us assume that the magnetization is uniform inside the sample volume. Then, out of the whole sample outside the sphere, only the magnetic charges at the surface of the sample give a contribution to $\boldsymbol{H}(i)$. This is called the **demagnetizing field**, $\boldsymbol{H}_d$.

The dipolar energy associated with the coupling of moment $\boldsymbol{m}_i$ to the other moments in the sphere is given by $E_{\text{dip}}(i) = \mu_0 \boldsymbol{m}_i \cdot \boldsymbol{H}_{\text{cav}}(i)$, where $\boldsymbol{H}_{\text{cav}}(i)$ is given by the sum,

$$\boldsymbol{H}_{\text{cav}}(i) = -\frac{1}{4\pi} \sum_{j \neq i}^{\boldsymbol{m}_j \in \text{cav}} \frac{1}{r_{ij}^3} \left(\boldsymbol{m}_j - 3 \frac{(r_{ij}\boldsymbol{m}_j)\, r_{ij}}{r_{ij}^2} \right) \tag{4.10}$$

which is called the **cavity field**. It can be shown that the cavity field vanished in a uniformly magnetized sample of cubic crystalline symmetry. Finally, the charges at the surface of the

Lorentz sphere give a uniform field:

$$H_L = \frac{1}{3}M \tag{4.11}$$

In summary, the dipolar field at site i is given by the sum of all these contributions:

$$H_{dip}(i) = H_{cav} + H_L + H_d \tag{4.12}$$

The magnetic energy associated with the demagnetizing field is called a ***shape anisotropy***. It is is calculated as:

$$E_{\text{shape}} = -\frac{\mu_0}{2} \int_V dV\, M(r) \cdot H_{dip}(r) \tag{4.13}$$

If the sample is an ellipsoid, the demagnetizing field H_d is uniform. It is expressed in terms of a demagnetizing tensor,

$$H_d = -\mathsf{D}\, M \tag{4.14}$$

where D is a tensor of trace 1. The term "demagnetizing" comes from the fact that this field opposes the magnetization. Demagnetizing factors are tabulated and can be found in various sources. For a thin plate, we have:

$$\mathsf{D} = \begin{pmatrix} 0 & 0 & 0 \\ 0 & 0 & 0 \\ 0 & 0 & 1 \end{pmatrix} \tag{4.15}$$

Thus, the contribution of H_d to the anisotropy per unit volume $K_{\text{shape}} \sin^2\theta$ of (4.4) in the case (4.15) is given by:

$$K_{\text{shape}} = \frac{\mu_0}{2} M_s^2 \tag{4.16}$$

The shape anisotropy energies of Fe, Co, and Ni are found to be of the order of 0.14, 0.09, and 0.01 meV/atom, respectively.

Let us now consider the contributions from the cavity and Lorentz fields to the magnetic anisotropy. It can be shown on the basis of symmetry arguments that a cubic symmetry implies the vanishing of the cavity field. However, if a cubic system is strained, then some magnetic anisotropy may arise. This contribution is generally small compared to the magnetocrystalline anisotropy described below, which arises from crystal field and spin-orbit effects. Crystals with lower symmetry may have a non-vanishing cavity field. This is the case of Co(hcp), for example. However, this dipolar contribution to the magnetocrystalline anisotropy is of about 4×10^{-7} eV/atom. Hence, it is quite small compared to the anisotropies observed experimentally.

In the case of a thin film, the Lorentz method does not apply. In particular, it is no longer possible to define a cavity.[263] Using another approach, Draaisma and De Jonge have shown that the magnetic anisotropy due to dipolar coupling in thin films turns out to be small, as well.[264] This contribution to the anisotropy was the biggest for Fe, and was estimated at 0.06 meV/atom. Again, this effect cannot be detected, because it is so small compared to surface magnetocrystalline anisotropies. In summary, it turns out that, for most cases, the dipolar contribution to the interface magnetocrystalline anisotropy is small.

4.5   FERROMAGNETS OF THE 3D ROW

Shape anisotropy is too small to account for the observed magnetic anisotropies. Therefore, there has to be some other mechanism by which the exchange-coupled electrons experience the presence of the underlying crystal. This mechanism is the result of the interplay of crystal field and the spin-orbit effects. The spin-orbit coupling is quite strong in magnets with 4f atoms and weaker in 3d-based magnets. Conversely, crystal field effects are weak in 4f atoms and quite strong in 3d-based systems. In 4f metals, the electronic shell that carries the magnetic moment is deep inside the atom; thus, the outer electrons shield the crystal field, and for this reason, the crystal field effect is much smaller than the spin-orbit coupling.

In this section, we focus on 3d atoms, for which the crystal field effect is larger than the spin-orbit coupling. The spin-orbit coupling can then be considered as a perturbation on the energy levels split by the crystal field. We encountered this idea in a general formula during our discussion of crystal field effects (Chapter 3). To illustrate the physics involved, we consider here a model involving two d orbitals,[238] namely $|xy\rangle$ and $|x^2 - y^2\rangle$ (Fig. 4.4).

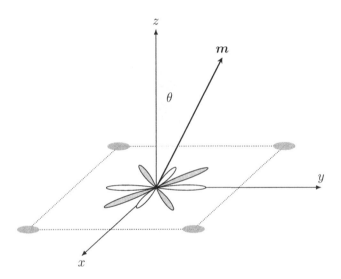

Figure 4.4   Wave-functions $|xy\rangle$ (pointing toward neighboring ions) and $|x^2 - y^2\rangle$ at an atomic site that carries a magnetic moment m oriented at an angle θ from the z axis in the (y, z) plane. Ellipses mark the location of ionic neighbors.

Let us consider a simple configuration of charges that determine the crystal field : 4 symmetrically positioned negative charges (Fig. 4.4). The d states we consider are degenerate until this crystal field effect is considered since the orbital that points toward the negative charges is energetically less favorable than the orbital that points along the axes instead. Therefore, the crystal field in this multiplet gives rise to a Hamiltonian of the form,

$$\mathcal{H} = \begin{pmatrix} A & 0 \\ 0 & -A \end{pmatrix} \tag{4.17}$$

If the magnetic moment m is that of a spin s=1/2, and it is inclined by an angle θ in the (y,z) plane, then the corresponding spin eigenstate $|\hat{m}\rangle$ can be written as, [1]

$$|\hat{m}\rangle = \cos\left(\frac{\theta}{2}\right)|+\rangle_z + i\sin\left(\frac{\theta}{2}\right)|-\rangle_z \tag{4.18}$$

where $|\pm\rangle_z$ are the eigenstates of S_z. In other words, the state (4.18) is the one with a probability of 1 to have the spin in the direction specified by $\hat{m}$.

To carry out the perturbation of the energy due to spin-orbit coupling (3.29), we need to work out matrix elements of the form:

$$\langle d_i|\langle\hat{m}|\,\mathcal{H}_{SO}\,|d_j\rangle\,|\hat{m}\rangle = \lambda\,\langle d_i|\langle\hat{m}|\,L_zS_z + \frac{1}{2}\left(L_+S_- + L_-S_+\right)|d_j\rangle\,|\hat{m}\rangle \tag{4.19}$$

An alternative approach to the calculation is to consider the spin state described using a quantization axis in a rotated system of axes (x',y',z'). In this reference frame, the spin state is simply $|+\rangle_{z'}$. Of the scalar product $L_{z'}S_{z'} + \frac{1}{2}\left(L_{+'}S_{-'} + L_{-'}S_{+'}\right)$, the only term that yields a matrix element of the spin operators different from zero is $L_{z'}\langle+|_{z'}S_{z'}|+\rangle_{z'} = \frac{1}{2}L_{z'}$. This operator can be expressed in terms of the orbital momentum operators in the (x,y,z) coordinates by effecting a simple rotation about the x-axis, thus giving,

$$L_{z'} = L_z\cos(\theta) + L_y\sin(\theta) \tag{4.20}$$

After some algebra, taking into account that our real wave functions are combinations of eigenstates of L_z, the matrix element of the Hamiltonian for the crystal field and the spin-orbit coupling is a 2 × 2 matrix given by:

$$\mathcal{H} = \begin{pmatrix} A & 0 \\ 0 & -A \end{pmatrix} + 2\lambda\cos\theta\begin{pmatrix} 0 & i \\ -i & 0 \end{pmatrix} \tag{4.21}$$

Hence, perturbation theory provides the energy eigenvalues:

$$E_\pm = \pm\sqrt{A^2 + 4\lambda^2\cos^2\theta} \tag{4.22}$$

That is to say, we find an angular dependence of the energy on the magnetization orientation with respect to the lattice. This is the concept of magnetocrystalline anisotropy we wished to illustrate with this simple example. The strength of this orientation dependence is determined by the spin-orbit coupling constant λ.

Based on this simple example, we can also make a remark on angular momentum quenching (§3.3). For $\theta = 0$, the ground state becomes:

$$|\Psi\rangle = \cos\frac{\chi}{2}\,|x^2 - y^2\rangle + i\sin\frac{\chi}{2}\,|xy\rangle \tag{4.23}$$

where χ is called the **mixing angle**, $\chi = \cot^{-1}(A/2\lambda)$. This state has the non-zero angular moment,

$$< L_z > = \hbar\frac{4\lambda}{\sqrt{A^2 + 4\lambda^2}} \tag{4.24}$$

In this example involving d states, we find as we did in §3.3 that the orbital moment is unquenched, owing to the spin-orbit coupling.

In §3.4, we carried out a similar perturbation calculation. The second term we obtained in (3.31) expresses a magnetic anisotropy. Its strength is determined by the strength of the spin-orbit coupling. It is an anisotropy that reflects the anisotropy of the crystal itself. For example, in magnetic oxides, the crystal field splitting may differ depending on the crystalline structure. The crystal field splitting of one-electron 3d states increases with decreasing symmetry of the unit cell.[238]

---

[1]see, e.g., Eq. 1.12 in [162] or Eq. 7.49 in [181].

4.6 FERROMAGNETS OF THE 4F ROW

In rare earth ferromagnets, the spin-orbit coupling dominates the crystal field effect, represented by the potential $V_{cf}(\boldsymbol{r})$ for an ion carrying a magnetic moment at $\boldsymbol{r} = \boldsymbol{0}$. As a consequence, the perturbation scheme used in this case consists in calculating the perturbation caused by the crystal field on states $|\Psi\rangle$ which are eigenstates of $\boldsymbol{J}^2$. The magnetostatic energy due to the crystal field effect is given by,

$$E = \langle \Psi | V_{cf}(\boldsymbol{r}) |\Psi\rangle \tag{4.25}$$

To the extent that the ions that contribute to the potential $V_{cf}(\boldsymbol{r})$ are far away from the ion carrying the magnetic moment, a point charge approximation can be used. Namely, the neighboring ions are considered as charges Q_i located at the crystalline sites $\boldsymbol{R}_i$. Then, $V_{cf}(\boldsymbol{r})$ reads:

$$V_{cf}(\boldsymbol{r}) = \frac{-e}{4\pi\varepsilon_0} \sum_{i=n.n.} \frac{Q_i}{|\boldsymbol{R}_i - \boldsymbol{r}|} \tag{4.26}$$

In case of equal charges ($Q_i = Q$) in cubic or tetragonal positions, the expansion of this potential to the lowest order, omitting constant terms, yields:

$$\begin{aligned}
V_{cubic}(x,y,z) &= \frac{35Qe}{8\pi\varepsilon_0 a^5}(x^2y^2 + y^2z^2 + z^2x^2) \\
V_{tetragonal}(x,y,z) &= \frac{Qe}{4\pi\varepsilon_0}\left(\frac{1}{a^3} - \frac{1}{c^3}\right)(3z^2 - r^2)
\end{aligned} \tag{4.27}$$

A systematic treatment of this problem for a ionic charge distribution $n_{4f}(\boldsymbol{r})$ would express the electrostatic energy as,

$$E = \int V_{cf}(\boldsymbol{r})n_{4f}(\boldsymbol{r})d\boldsymbol{r} \tag{4.28}$$

The electrostatic potential could be expanded in terms of a set of orthogonal functions. Likewise, the electron charge distribution $n_{4f}(\boldsymbol{r})$ could be expanded in the same set of basis functions. Then, the crystal symmetry properties could be invoked to simplify the expression of the integral (4.28). This was done by Abragam and Bleaney in their analysis of the magnetic resonance of paramagnetic ions.[2]

For the sake of clarity, let us consider one particular case for which the electrostatic energy V_{cf} is written as,[238]

$$V_{cf}(\boldsymbol{r}) = A_2^0(3z^2 - r^2) \tag{4.29}$$

The crystal field energy E given by (4.25) involves the integral $\langle \Psi | 3z^2 - r^2 |\Psi\rangle$. Since $|\Psi\rangle$ is an eigenstate of $\hat{\boldsymbol{J}}^2$, we can apply the Wigner-Eckart theorem, which yields that,

$$\langle \Psi | 3z^2 - r^2 |\Psi\rangle = Q_2 \langle J, M | 3\hat{J}_z^2 - \hat{J}^2 |J, M\rangle \tag{4.30}$$

and that Q_2 is independent of M.[3] Applying the last result for the state $|\Psi\rangle = |J, J\rangle$ gives,

$$\langle J, J | 3z^2 - r^2 |J, J\rangle = Q_2 J(2J - 1) \tag{4.31}$$

Thus, we find that Q_2 is positive for an elongated (prolate) charge distribution along the z axis, whereas Q_2 is negative for an oblate charge distribution.

---

[2]See [265], Section 3.3 and tables 14–16 of Appendix B.

[3]See, e.g., [236], Eq. (10.50).

In view of (4.30), we can write the crystal field Hamiltonian of your perturbation calculation as:

$$\mathcal{H}_{crystal} = A_2^0 Q_2 \left(3\hat{J}_z^2 - \hat{J}^2\right) \qquad (4.32)$$

Here, the z axis refers to the crystalline symmetry axis. We want to know the energy change when the magnetic moment (proportional to $\boldsymbol{J}$) rotates away from the z axis by an angle θ because a saturating magnetic field $\boldsymbol{B}$ is applied in a direction z' of the laboratory. In the presence of the Zeeman coupling, the total Hamiltonian reads:

$$\mathcal{H} = -g(LSJ)\mu_B B \hat{J}_{z'} + C\left(3\hat{J}_z^2 - \hat{J}^2\right) \qquad (4.33)$$

As the axes z and z' are an angle θ apart, we can write:

$$\hat{J}_z = \hat{J}_{z'}\cos(\theta) + \hat{J}_{x'}\sin(\theta) \qquad (4.34)$$

Finally, the first order perturbation yields for the energy as a function of angle,[4]

$$E = -g(LSJ)\mu_B B\, m + C\left(\frac{3\cos^2\theta - 1}{2}\right)\left(3m^2 - J(J+1)\right) \qquad (4.35)$$

for a state $|\Psi\rangle$ of angular momentum $J_z|\Psi\rangle = m|\Psi\rangle$.

If the ion is oblate, which is the case of Nd^{3+} for example, then A_2^0 must be positive in order for z ($\theta = 0$) to be an easy axis. If the ion is prolate, as is the case of Sm^{3+}, then A_2^0 must be negative in order for z to be an easy axis.

4.7   EXCHANGE ANISOTROPIES

The exchange given by the Dzialoshinskii-Moriya scheme includes the spin-orbit coupling, i.e., a coupling of spins to the lattice. For example, the expression (2.59) contains the vector $\boldsymbol{D}$, which is linked to the crystal structure.[266]

In §2.7, we showed that a tensorial Heisenberg exchange coupling can be decomposed into an antisymmetric tensor – such as the one responsible for the DM interaction – and a symmetric tensor. This symmetric tensor can be decomposed into an isotropic tensor and a traceless symmetric tensor. The latter tensor describes anisotropic exchange contributions to the magnetocrystalline anisotropies.

As the calculation in §2.7.2 suggests, there is a close relationship between the antisymmetric and the symmetric forms of exchange. It has been elegantly derived by Cepas.[267] We note in passing that skyrmion-like structures may arise that are not linked to the DM interaction, but to anisotropic exchange.[268]

4.8   FURTHER READINGS

Magnetocrystalline anisotropies can be characterized by **X-ray magnetic circular dichroism** (XMCD). This can be expected since there is a direct connection between magnetic anisotropies and the electronic structure.[269] A tight-binding description of transition metals can provide insight into the orbital contribution to the magnetic anisotropies.[254]

Magnetic anisotropies are modified by **mechanical stress**.[270] This phenomenon is associated with magnetostriction. Therefore, the magnetostrictive properties of a material have an influence on the switching field. Consequently, mechanical stress can affect the switching field.[271] Stress in a thin film can be induced by applying an electric field to a

---

[4]See, e.g., [236] Section 10.5

piezoelectric substrate onto which the film is deposited.[272] By growing multilayers in a specific way, it is possible to induce large strain in a deposited film that can, for example, induce a ***perpendicular magnetic anisotropy*** (PMA).[273]

Research is under way to determine whether organic molecules such as Mn-phthalocyanine, attached to a ferromagnetic metal, can change the metal magnetic anisotropy, owing to an exchange effect.[274,275] This phenomenon is referred to as ***molecular exchange biasing***. For example, an anisotropy was observed that followed Mn-porphyrin adsorption,[276] or other transition metal porphyrins. [274] One possibility is a superexchange mechanism linking the metal atom of a metal-organic chemisorbed molecule and an atom of the ferromagnet.[277,278] The key issue is to make sure that the anisotropy is not due to the spurious formation of a surface oxide.

One particularly intriguing magnetic hardening is the one observed when an organic chiral layer is deposited on a thin magnetic film (Fig. 4.5).[279,280] Chiral molecules appear to be able to induce a magnetic effect when they are subjected to an electric field.[281,282] Photoexcitation, leading to a charge transfer through chiral molecules, can also change the magnetic properties of the substrate where the chiral molecules are attached.[283] The peculiar electronic structure of chiral molecules is especially manifest in transport measurements, [281] but it has also been detected using spin-polarized photoemission.[284,285]

Figure 4.5   Magnetic force microscopy of a Co film with a 5 nm Au over-layer, onto which chiral molecules are chemisorbed in a square pattern. (Scale bar: 1 μm.) [280]

A great difficulty in making heterostructures of stacked magnetic metal layers and organic layers is to make sure that these multilayers are free of pinholes. When studying magnetic anisotropies of these structures, it is important to make sure one is studying intrinsic properties of the material, not the consequence of structural defects.[286,287] Pinhole-free layers can be produced, provided they are thick enough.[288]

Exchange biasing is the shift of the hysteresis loop generally obtained by depositing an antiferromagnetic layer on a ferromagnetic layer.[256,289,290] Alternatively, a ferromagnetic metal may form an antiferromagnetic oxide at its surface, like NiO or CoO. The exchange bias is normally determined simply by a hysteresis loop measurement. This is quite relevant when the exchange biasing is implemented in a magnetic sensor, e.g., based on GMR. However, when one seeks to characterize the physical origin of the phenomenon, one should keep in mind that the coercive field characterizes the weakest point of the structure. X-ray scattering can characterize the spin structure of the FM/AFM interface.[291] When applicable, nuclear magnetic resonance can provide a volume average of the exchange strength.[288]

Nanomagnetism refers to the study of magnetic nanostructures, i.e., magnets of sizes below 100 nm, roughly. For example, research on the magnetism of nanoscopic islands and atomic nanowires on surfaces revealed the critical role of perimeter atoms in determining the coercivity of the islands (Fig. 4.6).[292]

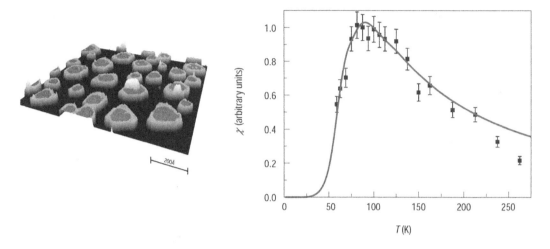

Figure 4.6   Left: STM image of atomically high islands of Pt with Co at the edges. Right: the peak of the susceptibility is indicative of the coercivity. The data was accounted for (continuous line) by assuming an anisotropy energy of about 0.9 meV/atom, whereas the bulk value is approximately 0.05 meV/atom.[292].

The ultimate size confinement, of course, is the *single atom magnetism*. When an atom is deposited at the surface of a substrate and carries a magnetic moment, the substrate-adatom interaction can be such that the atom moment experiences a PMA. This is the case for Fe on MgO.[293] Very large PMA is needed if this atomic moment is to be of practical use,[294] because the magnetic moment switching must be slow enough that operations on the spin can be envisaged.[295] The magnetic moment of holmium atoms on MgO has a coercive field of 8 T and switches on a time scale of minutes at 35K. [296]

5 Indirect Exchange

Tadao Kasuya and Kei Yosida

The so-called Rudermann-Kittel-Kasuya-Yosida (RKKY) interaction was originally proposed by M. Ruderman and C. Kittel of the University of California, Berkeley, as a means of explaining unusually broad nuclear spin resonance lines that were observed in natural metallic silver. The model was based on a second order perturbation of the Fermi contact interaction using the ground state of the free electron model. T. Kasuya of Nagoya University proposed the s-d and s-f exchange interactions in 1956 to explain the strongly correlated magnetic and magneto-transport properties of the iron family and the rare earth metals, showing a damped oscillating exchange interaction among localized magnetic moments that caused various types of magnetic orders. K. Yosida, at Berkeley in 1957, studied the coupling between nuclear spins and localized moments using the approach of Ruderman-Kittel and also found a spatially oscillating coupling.

Spintronics: A primer

J.-Ph. Ansermet (https://orcid.org/0000-0002-1307-4864)

Chapter 5. Indirect Exchange

When two entities of a system, such as atomic nuclei in a molecule or local moments in a magnet, are coupled to a bath, such as a cloud of conduction electrons, the two entities are coupled to one another via that bath. To make this notion clear, the hyperfine coupling between two nuclei and conduction electrons is shown to lead to a second-order perturbation that accounts for an indirect interaction. The generality of indirect couplings is made clear by characterizing the bath with a generalized susceptibility that characterizes the spatially distributed response to an inhomogeneous magnetic field.

DOI: 10.1201/9781003370017-5

5.1 HISTORICAL INTRODUCTION

This chapter is concerned with the coupling among spins that arises from their coupling to a bath. It is quite an important effect, as it is found in very different types of situations. In order to develop a feel for its importance, a brief historical survey is presented.

In 1953, an article reviewed experimental findings concerning a new coupling among spins that was detected by NMR.[297] The unusually long abstract summarized observations that could be systematized by assuming a spin-spin coupling of the form: $A_{12}\,\boldsymbol{\mu}_1\cdot\boldsymbol{\mu}_2$ between the nuclear moments $\boldsymbol{\mu}_1$ and $\boldsymbol{\mu}_2$, where the coupling constant A_{12} *"must involve the molecular electrons."* This coupling among nuclei in a molecule, which is indeed mediated by the electron cloud, is now referred to as the ***J-coupling*** or the ***scalar spin-spin coupling***.

This kind of indirect coupling among spins is also observed in solids. It plays a critical role in magnetism because it implies that localized moments in a magnet are coupled via their coupling to an electron bath. It is referred to as the ***RKKY coupling***. RKKY stands for the name of some of the scientists who in the 1950s showed the existence of an indirect coupling, in the case of nuclei coupled to electrons, Ruderman and Kittel, [298]; in the case of local moments coupled to conduction electrons, Kasuya and Yosida. [299] The response of the conduction electron around a local magnetic moment is a spatial oscillation of the spin polarization. If moments are close enough, equivalently, if the oscillation extends from one local moment to the next, the moments are coupled.

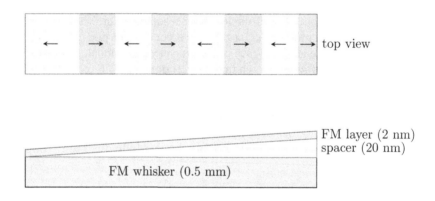

Figure 5.1 Bottom (not to scale) : A fantastically well-defined wedge-shaped spacer (from 0 to 20 nm of Cr over a length of 0.5 mm) is deposited on a ferromagnetic (FM) whisker (iron). A thin FM overlayer (2 nm of Fe) covers this wedge. Top: A view from the top of the capping layer magnetization. Alternating magnetic domains are observed that reflect the oscillatory behavior of the RKKY coupling across the spacer layer. Notice the great length of the whisker (0.5 mm) compared to the maximum height of the wedge spacer (20 nm).

In the 1980s, research was under way to detect and understand the magnetic coupling between two ferromagnetic layers separated by a non-magnetic metal. This idea was clearly an extension to layers of the notion of indirect coupling among localized spins via conduction electrons. Fig. 5.1 illustrates a beautiful experiment that clearly demonstrated the presence of a coupling between two magnetic layers, and that the sign of the coupling would oscillate as a function of the thickness. [81]

The magnetization data of Fig. 5.2 also reveal the presence of an ***interlayer exchange coupling*** between two ferromagnetic layers separated again by a non-magnetic layer.[81] The nanostructuring in this magnetic structure is often referred to as spin-engineering.

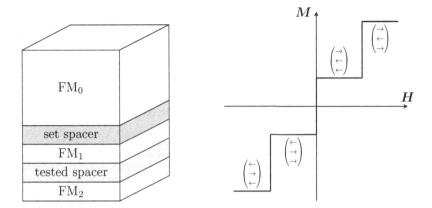

Figure 5.2 Left: Interlayer exchange coupling. A stack of three ferromagnetic (FM) layers is considered. The thick FM_0 layer is exchange-coupled to FM_0 by a fixed, already well-characterised spacer layer, giving rise to a strong antiferromagnetic coupling. Layers FM_1 and FM_2 are separated by a spacer layer which is responsible for an exchange coupling that one wants to characterize. Here, it is assume to be ferromagnetic. Right: The measurement of magnetization versus magnetic field shows the switching of FM_0 at nearly zero field. For the sake of clarity, the minor hysteresis loops are not shown. Owing to the antiferromagnetic coupling between FM_0 and FM_1, FM_1 switches whenever FM_0 does. Owing to the ferromagnetic coupling between FM_1 and FM_2, FM_2 follows FM_1 and also switches. Switching observed at high fields occurs when FM_2 aligns in the applied field. This switching field is a measure of the coupling between FM_1 and FM_2.

Here, a thick ferromagnetic layer FM_0 easily orients itself in the applied magnetic field. The thickness of the layer FM_0 is chosen such that the layer FM_1 is always antiferromagnetically coupled to FM_0. The spacer layer between FM_1 and FM_2 is a layer that one wishes to study, it may be gold or copper, for example. The field at which the layer F_2 switches is a measure of the coupling between F_2 and F_1. In that figure, the case where F_1 and F_2 are coupled ferromagnetically is shown.

This type of antiferromagnetic coupling between metallic ferromagnetic layers was important in early experiments on GMR, when the current was running parallel to the interface (CIP-GMR). Now, artificial antiferromagnets are created by using an interlayer, Ru generally, in structures where the current flows perpendicular to the interfaces, i.e., in CPP-GMR structures or in tunnel junctions (TMR). There has been a renewed interest in so-called "artificial antiferromagnets" because small skyrmions may be formed in such systems that can be used in memory applications.[300]

5.2 INDIRECT COUPLING OF NUCLEI VIA A BATH OF ELECTRONS

Here, we will see with a formal development that two spins that are coupled to a bath are coupled to one another through a second order process. Let us consider a system A composed of two nuclear spins that are coupled to a bath B of conduction electrons. Proceeding this way, we avoid some complications since nuclei and electrons are distinguishable particles. The interaction of conduction electrons with core electrons on atoms that carry a magnetic moment is the subject of §6.5.

The nuclear states are labelled $|a>$ and the electronic states $|b>$. The coupling of each nucleus to the electrons is described by a Hamiltonian $\mathcal{H}_{int}$. We consider the electrons in their ground state $|b = 0 >$. The second order perturbation on two states $|a >$ and $|a' >$ that are degenerate, i.e., $E_a = E_{a'}$, yields the following result:[301]

$$\mathcal{H}_{a' a} = \sum_{a''b'} \frac{\langle a'0| \mathcal{H}_{\text{int}} |a''b'\rangle \, \langle a''b'| \mathcal{H}_{\text{int}} |a0\rangle}{E_a + E_0 - E_{a''} - E_{b'}} \tag{5.1}$$

The two nuclei have spins $\boldsymbol{I}_1 = \hbar\hat{\boldsymbol{I}}_1$ and $\boldsymbol{I}_2 = \hbar\hat{\boldsymbol{I}}_2$. The Hamiltonian describing the interaction of each nucleus with the electrons is taken to be the **contact hyperfine interaction**:

$$\mathcal{H}_{\text{int}} = \hat{\boldsymbol{I}}_1 \cdot \sum_l J\hat{\boldsymbol{S}}_l \delta \left(\boldsymbol{R}_1 - \boldsymbol{r}_l\right) + \hat{\boldsymbol{I}}_2 \cdot \sum_l J\hat{\boldsymbol{S}}_l \delta \left(\boldsymbol{R}_2 - \hat{\boldsymbol{r}}_l\right) \tag{5.2}$$

where $\hbar\hat{\boldsymbol{S}}_l$ is the spin of an electron at location $\boldsymbol{r}_l$. We can write this interaction as,

$$\mathcal{H}_{\text{int}} = \hat{\boldsymbol{I}}_1 \cdot \boldsymbol{G}_1 + \hat{\boldsymbol{I}}_2 \cdot \boldsymbol{G}_2 = \mathcal{H}_1 + \mathcal{H}_2 \tag{5.3}$$

We omit the nuclear energies E_a, $E_{a''}$ which are negligible compared to the electronic energies in (5.1). Thus, we obtain,

$$\mathcal{H}_{a'a} = \sum_{a''b'} \frac{\langle a'0| \mathcal{H}_1 + \mathcal{H}_2 |a''b'\rangle \, \langle b'a''| \mathcal{H}_1 + \mathcal{H}_2 |a0\rangle}{E_0 - E_{b'}} \tag{5.4}$$

Each numerator in this sum is composed of four terms. Writing them explicitly yields,

$$\mathcal{H}_{a'a} = \sum_{a''b'} \frac{1}{E_0 - E_{b'}} \left\{ \begin{array}{l} \langle a'0| \mathcal{H}_1 |a''b'\rangle \, \langle b'a''| \mathcal{H}_1 |a0\rangle \\ + \langle a'0| \mathcal{H}_2 |a''b'\rangle \, \langle b'a''| \mathcal{H}_2 |a0\rangle \\ + \langle a'0| \mathcal{H}_1 |a''b'\rangle \, \langle b'a''| \mathcal{H}_2 |a0\rangle \\ + \langle a'0| \mathcal{H}_2 |a''b'\rangle \, \langle b'a''| \mathcal{H}_1 |a0\rangle \end{array} \right\} \tag{5.5}$$

Each of the first two terms would be present if we had only two independent nuclei. These two terms describe the second order hyperfine structure of each nuclear resonance. The last two terms in (5.5) express an effective interaction between the nuclear spins, which arises from their respective interaction with the bath of electrons. In (5.5), we select therefore,

$$\langle a'| \hat{\boldsymbol{I}}_1 |a''\rangle \cdot \langle 0| \boldsymbol{G}_1 |b'\rangle \, \langle a''| \hat{\boldsymbol{I}}_2 |a\rangle \cdot \langle b'| \boldsymbol{G}_2 |0\rangle$$
$$+ \langle a'| \hat{\boldsymbol{I}}_2 |a''\rangle \cdot \langle 0| \boldsymbol{G}_2 |b'\rangle \, \langle a''| \hat{\boldsymbol{I}}_1 |a\rangle \cdot \langle b'| \boldsymbol{G}_1 |0\rangle \tag{5.6}$$

We can write the scalar products with implicit summation over spatial indices $\alpha, \beta = 1 : 1 \ldots 3$. This yields,

$$\langle a'| \hat{I}_{1,\alpha} |a''\rangle \, \langle 0| G_{1\alpha} |b'\rangle \, \langle a''| \hat{I}_{2\beta} |a\rangle \, \langle b'| G_{2\beta} |0\rangle$$
$$+ \langle a'| \hat{I}_{2\beta} |a''\rangle \, \langle 0| G_{2\beta} |b'\rangle \, \langle a''| \hat{I}_{1\alpha} |a\rangle \, \langle b'| G_{1\alpha} |0\rangle \tag{5.7}$$

This expression once put in (5.5), is summed over all nuclear states $|a'' >$. We can use in (5.5) the closure relation $\sum_{a''} |a'' >< a''| = \mathbb{1}$. Then the sum in (5.5) contains only terms of the form,

$$\langle a'| \hat{I}_{1\alpha}\hat{I}_{2\beta} |a\rangle \, \langle 0| G_{1\alpha} |b'\rangle \cdot \langle b'| G_{2\beta} |0\rangle + \langle a'| \hat{I}_{2\beta}\hat{I}_{1\alpha} |a\rangle \, \langle 0| G_{2\beta} |b'\rangle \, \langle b'| G_{1\alpha} |0\rangle \tag{5.8}$$

Hence, we find a coupling between the two nuclear spins that can be represented by a Hamiltonian $\mathcal{H}_{\text{eff}}$ that operates on the nuclear Hilbert space and is written as,

$$\mathcal{H}_{\text{eff}} = \hat{I}_{1\alpha} \sum_{b'} \frac{\langle 0| G_{1\alpha} |b'\rangle \, \langle b'| G_{2\beta} |0\rangle + \langle 0| G_{2\beta} |b'\rangle \, \langle b'| G_{1\alpha} |0\rangle}{E_0 - E_{b'}} \hat{I}_{2\beta} \tag{5.9}$$

Using the definition of $G_{1\alpha}$ and $G_{2\beta}$, which are implicitly given in (5.2) and (5.3), we find,

$$\mathcal{H}_{\text{eff}} = J^2 \hat{I}_{1\alpha} \sum_{b'} \left(\frac{\langle 0| \sum_l \hat{S}_{l\alpha}\delta(r_l - \boldsymbol{R}_1)|b'\rangle \langle b'| \sum_l \hat{S}_{l\beta}\delta(r_l - \boldsymbol{R}_2)|0\rangle + \text{c.c.}}{E_0 - E_{b'}} \right) \hat{I}_{2\beta} \quad (5.10)$$

where c.c. stands for "complex conjugate". Notice that the coupling (5.10) has the form,

$$\mathcal{H}_{\text{eff}} = \hat{\boldsymbol{I}}_1 \cdot \chi(\boldsymbol{R}_1, \boldsymbol{R}_2)\hat{\boldsymbol{I}}_2 \quad (5.11)$$

where the matrix $\chi(\boldsymbol{R}_1, \boldsymbol{R}_2)$ can be understood as a generalized susceptibility, as will be shown below in §5.3.

5.3   RKKY INTERACTION IN TERMS OF GENERALIZED SUSCEPTIBILITY

What we have analyzed for nuclear spins we now expand to the case of local magnetic moments.[156] We assume that these local moments are each individually coupled to the conduction electron spins. We arrive at the notion of an indirect coupling by introducing the response properties of the electron bath.

Let us first consider one local magnetic moment of spin $\boldsymbol{S}_\alpha$ located at the origin, $\boldsymbol{R}_\alpha = \boldsymbol{0}$, coupled to the conduction electron spins $\boldsymbol{s}_i$ at the positions $\boldsymbol{r}_i$, owing to exchange interactions of the form:

$$\mathcal{H}_{int}(\alpha) = -J \sum_i \hat{\boldsymbol{S}}_\alpha \cdot \hat{\boldsymbol{s}}_i \, \delta(\boldsymbol{r}_i) \quad (5.12)$$

We now introduce the conduction electrons susceptibility $\chi(\boldsymbol{r} - \boldsymbol{r}')$ that links the magnetization $\boldsymbol{M}(\boldsymbol{r})$ associated with the electron spins to a spatially variable magnetic field $\boldsymbol{H}(\boldsymbol{r})$ by,[1]

$$\boldsymbol{M}(\boldsymbol{r}) = \int d\boldsymbol{r}' \chi(\boldsymbol{r} - \boldsymbol{r}')\boldsymbol{H}(\boldsymbol{r}') \quad (5.13)$$

The important notion introduced in (5.13) is that a field, present at location $\boldsymbol{r}'$, can have an effect on the magnetization at another location $\boldsymbol{r}$. It is expected that a bath has such a response, or else, it would be difficult to imagine how the electron spins could act as a bath if what happens at one location remains at that location. The bath would loose its identity as a system near equilibrium.

To use the generalized susceptibility $\chi(\boldsymbol{r} - \boldsymbol{r}')$, we interpret (5.12) as the interaction of the magnetic moments $\hbar\gamma_e \hat{\boldsymbol{s}}_i$ with the field,

$$\boldsymbol{H}_{\text{eff}}(\boldsymbol{r}) = \frac{-J}{\mu_0 \hbar\gamma_e} \hat{\boldsymbol{S}}_\alpha \delta(\boldsymbol{r}) \quad (5.14)$$

Thus, the magnetization $\boldsymbol{M}(\boldsymbol{r})$ of the conduction electron spins in response to this field, is given by,

$$\boldsymbol{M}(\boldsymbol{r}) = \chi(\boldsymbol{r}) \left(\frac{-J}{\mu_0 \hbar\gamma_e} \right) \hat{\boldsymbol{S}}_\alpha \quad (5.15)$$

The integral of $\boldsymbol{M}(\boldsymbol{r})$ over the entire volume V of the sample gives the resulting magnetic moments of all the conduction electrons. Hence, we have,

$$\sum_i \hbar\gamma_e \hat{\boldsymbol{s}}_i = \int d^3 r \boldsymbol{M}(\boldsymbol{r}) \quad (5.16)$$

---

[1]This definition, which conforms to the notation of [156], implies that $\chi(\boldsymbol{r} - \boldsymbol{r}')$ has units of 1/volume.

Therefore, $\boldsymbol{M}(\boldsymbol{r})$ can be written in the form of a discrete sum as,

$$M(\boldsymbol{r}) = \sum_i \hbar\gamma_e \hat{\boldsymbol{s}}_i \delta(\boldsymbol{r} - \boldsymbol{r}_i) \tag{5.17}$$

Let us suppose that there is a local magnetic moment of spin $\boldsymbol{S}_\beta$ located at $\boldsymbol{R}_\beta$. At the location, the electron spin magnetization (5.15) is,

$$\sum_i \hbar\gamma_e \hat{\boldsymbol{s}}_i \delta(\boldsymbol{R}_\beta - \boldsymbol{r}_i) = \chi(\boldsymbol{R}_\beta)\left(\frac{-J}{\mu_0 \hbar\gamma_e}\right)\hat{\boldsymbol{S}}_\alpha \tag{5.18}$$

At the location of $\boldsymbol{S}_\beta$, the exchange interactions (5.12) are given by,

$$\mathcal{H}_{\text{eff}}(\alpha,\beta) = -J\hat{\boldsymbol{S}}_\beta \cdot \sum_i \hat{\boldsymbol{s}}_i \delta(\boldsymbol{R}_\beta - \boldsymbol{r}_i) = \left(\frac{J^2}{\mu_0\hbar^2\gamma_e^2}\right)\hat{\boldsymbol{S}}_\beta \cdot \chi(\boldsymbol{R}_\beta - \boldsymbol{R}_\alpha)\hat{\boldsymbol{S}}_\alpha \tag{5.19}$$

where we have explicitly designated the location of $\hat{\boldsymbol{S}}_\alpha$ by $\boldsymbol{R}_\alpha$. Thus, we have obtained the form (5.11) by a description that clearly identifies the matrix of (5.10) with a generalized susceptibility.

In summary, at one site, the local magnetic moment provokes a response of the bath of electrons. This response has a spatial extension. Hence, this first local moment has an influence on the bath spin polarization at the location of a second local moment (Fig. 5.3). And the bath at this location is coupled to another local moment. Hence, both local moments are coupled, via the bath response. H. J. Ziegler and G. W. Pratt [302] gave an intuitive explanation of the RKKY coupling by saying that the electrons that collided with the moment at site 1 (via the exchange interaction) carry "information" about the polarization of site 1 until they collide with the moment at site 2. This is an interpretation of Fig. 5.3 based on a microscopic description of scattering processes.

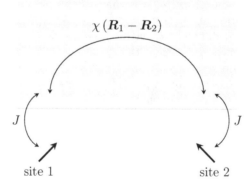

Figure 5.3 Schematics of the indirect coupling such as the RKKY coupling.

Remarks

1. In the formalism of §12.6 describing a subsystem A coupled to a bath, we will identify in the second order perturbation a Hamiltonian term with the matrix elements:

$$\Delta\mathcal{H}_{aa'} = \frac{-1}{2}\sum_{a''}\sum_{\alpha,\beta}\langle a|\,A_\alpha\,|a''\rangle\,\bar{\chi}_{\alpha\beta}(\omega_{a'a''})\,\langle a''|\,A_\beta\,|a'\rangle \tag{5.20}$$

Here, the indices α and β label the operators describing the coupling (12.125) of a bathed system A and a bath B. Let us consider the contribution to $\Delta\mathcal{H}_{aa'}$ of the coupling that involved the spin operators $I_{1\alpha}$ and $I_{2\beta}$ where now α and β refer to the coordinates only. The operators $I_{1\alpha}$ and $I_{2\beta}$ refer to angular momenta located at positions $\boldsymbol{R}_1$ and $\boldsymbol{R}_2$. Thus, the susceptibility must now be written as $\chi_{\alpha\beta}(\boldsymbol{R}_1,\boldsymbol{R}_2,\omega_{a'a''})$. We can assume that the susceptibility $\bar{\chi}_{\alpha\beta}(\omega_{a'a''})$, which describes an electronic response, does not vary much over the range of frequencies that are typical for the local moments. Therefore, we can take the argument of $\omega_{a'a''}$ to be zero. This allows us to apply the closure relationship in (5.20) and obtain:

$$\Delta\mathcal{H}_{a\,a'} = \frac{1}{\hbar}\sum_{\alpha\beta}\langle a|\,A_\alpha A_\beta\,|a'\rangle\,\bar{\chi}_{\alpha\beta}(\boldsymbol{R}_1,\boldsymbol{R}_2) \qquad (5.21)$$

Thus, the formal developments that will be presented in §12.6 also contain the notion of indirect couplings.

2. The susceptibility of the free electron gas is isotropic. Therefore, it depends only on the distance $r = \|\boldsymbol{R}_1 - \boldsymbol{R}_2\|$ between sites. It is given by,[2]

$$\chi(r) = \chi_{\text{Pauli}}\frac{k_F^3}{16\pi^2}\left(\frac{\sin(2k_F r) - k_F r\cos(2k_F r)}{(k_F r)^4}\right) \qquad (5.22)$$

Here, χ_{Pauli} is the Pauli susceptibility, characterizing the spin response $\boldsymbol{M} = \chi_{\text{Pauli}}\boldsymbol{H}$ of the free electron gas to a uniform field $\boldsymbol{H}$. The Fermi momentum k_F is defined by the expression $E_F = \hbar^2 k_F^2/(2m)$ for the Fermi energy, where m is the electron mass. Thus, we see in $\chi(r)$ an oscillatory behavior that accounts for the coupling observed between ferromagnetic layers separated by a non-magnetic metal (Fig. 5.1). This oscillatory behavior was also observed in the spin polarization of a non-magnetic metal, in the vicinity of a magnetic impurity (Fig. 15.8).

5.4   FURTHER READINGS

Indirect couplings

Pseudo-dipolar coupling, typically between nuclei, is a dipolar coupling mediated by the dipolar coupling to the electron spins in the bath.[236, 303] In some circumstances, the indirect coupling takes the form of an exchange coupling. It was first called a pseudo-exchange coupling, but is now referred to as a J-coupling.[297, 304, 305] The possibilities for indirect couplings are many. For examples, the coupling of magnetic impurities via a one-dimensional helical state was investigated theoretically.[306]

Artificial antiferromagnet

Two ferromagnetic layers separated by a spacer layer can have their respective magnetization in an antiferromagnetic configuration, provided the spacer layer thickness is such that the RKKY coupling due to the spacer layer is positive. This type of try-layer structure is called an artificial, or synthetic, antiferromagnet.[300, 307] Of course the dimensions are in this case on a nanometer scale, instead of being on an atomic scale as it is in an antiferromagnetic material. This type of magnetic nanostructures finds applications in spintronics, in particular in MRAMs.[308] Such structures have also been used when making spin torque oscillators, i.e., structures in which the magnetization oscillates when a large spin current is

---

[2]See, e.g., [156] Eq. (9.18).

driven through them.[309] The magnetic coupling between the layers can be modulated in situ with an electric field.[310] It's been predicted that a ferroelectric contact to the tri-layer structure may enhance the strength of this switching effect.[311] Electric-field-controlled RKKY has also been obtained in ferrimagnets.[312]

Spin-spin coupling for qubits

The electron spin can be thought of as a two-level system that can be used to form a qubit. The spin is localized in a *quantum dot*. We expect spin-spin coupling to be a necessary ingredient if spins are to be used in quantum computing. However, the quantum dots must be far enough apart in order to apply a magnetic field specifically to a specific site. In 2012, Yacoby, Loss et al. proposed a possible architecture based on a metallic gate that ensures strong enough coupling among several spins.[313]

6 Moments In Metals

Edmund Clifton Stoner (1899–1968)

E.C. Stoner is most famous for his model of itinerant ferromagnetism, which he developed as a professor at the University of Leeds. His prior theoretical work was completed at the Cavendish Laboratory of the University of Cambridge in the field of astrophysics and x-ray absorption. Thus, Stoner dealt with questions pertaining to the population of energy levels in the early days of quantum mechanics.

Spintronics: A primer

J.-Ph. Ansermet (https://orcid.org/0000-0002-1307-4864)

Chapter 6. Moments in Metals

Does a magnetic ion remain magnetic when it is embedded in a non-magnetic metal? That is one of the questions of the Kondo problem. When can band electrons turn ferromagnetic ? The Stoner model offers a first approximation to solve this question. The so-called Stoner enhancement factor accounts for why some metals have a much larger susceptibility than the Pauli susceptibility predicted for a free electron gas. Adding fluctuations to the Stoner model leads to the critical behavior of the Curie-Weiss susceptibility. The exchange interaction of conduction electrons with local magnetic moments is accounted for with the *sd* model. In conductors without spatial inversion symmetry, spin-orbit coupling can give rise to the Rashba effect. Conduction electrons can be in orbital angular momentum (OAM) states. A tight-binding example illustrates this possibility.

DOI: 10.1201/9781003370017-6

6.1 HISTORICAL INTRODUCTION

What happens when a magnetic ion is embedded into a metal? So far, we have considered atoms or ions that were presumed to carry a magnetic moment. This approach is acceptable for a large class of compounds. We have paired such ions or atoms and found some mechanisms that may couple them ferromagnetically or anti-ferromagnetically. We also considered the crystal field effect on isolated magnetic moments. However, we have not addressed the question of how the Fermi sea of a metal may interact with this moment. Does the moment remain? If so, how does the Fermi sea respond to the presence of this moment?

These questions were addressed with great experimental and theoretical efforts in the 1960s and 1970s. The problem is complex because of the fermionic character of the electron system. The non-magnetic case is already tough: determining the electronic structure of impurities in metals is quite challenging. Recent developments in the electronic structure of adsorbates on metal catalysts revisit this type of problem. Of course, further complications arise in considering what happens to the magnetic moment of an impurity. *Anderson's model* accounts for these effects (see §6.2). Anderson's model is closely related to the *Hubbard model* and the *Kondo effect*. The polarization of the electron cloud at the vicinity of a magnetic impurity gives rise to a resistivity minimum, observed at low temperatures in metals doped with magnetic impurities and in ternary alloys such as CsPtSn.[72] This resistivity minimum was first observed in the 1930s, and Kondo later gave an explanation in 1963. In the 1990s, the Kondo effect was examined again to see how it changed when sample dimensions were reduced to the nanometer scale, or the magnetic impurities were at a surface. There was a theoretical difficulty in the low temperature limit of the standard description of the Kondo effect until the 1980s, when P.B. Wiegman et al. found the exact solution to the Kondo problem at zero temperature. The Kondo effect was also investigated in exotic materials such as heavy fermions and so-called Kondo lattices were also studied.

Why are some metals ferromagnetic? This question was addressed in a symposium organized by C. Kittel in 1953. The invited speakers discussed various models to account for ferromagnetism. They were key players in the fields of theoretical solid state physics and magnetism, including: Zener, Slater, Stoner, and Van Vleck. Van Vleck's paper gave a remarkable synopsis of the various models.[314] Models for *itinerant ferromagnetism* are based on almost free electrons and have had limited success in describing ferromagnetism (see §6.3). Nonetheless, they reveal some key ingredients and are worth keeping in mind as a backdrop against which one can appreciate the newly discovered magnetism in *alkaline-earth hexaborides*, such as La-doped CaB6.[315] The origin of ferromagnetism in this system has been a longstanding controversy. It may be a case of ferromagnetism of stable excitons.[316]

EuB$_6$ is ferromagnetic below about 12 K. It has been known for decades and its colossal magnetoresistance has generated a lot of controversy.[317] It is a ferromagnetic material with a very low concentration of charge carriers. One model to explain ferromagnetism in such a case contends that it may arise from a lowering of the effective mass of the charge carriers associated with the polarization of their spins. Thus, ferromagnetism here would be a band structure effect.

The *sd-model* (see §6.5) has been applied by Berger to explain the interaction of a charge current with a domain wall, for example.[56] The sd-model is also useful to understand the physics of manganites, which present colossal magnetoresistance.

6.2 MAGNETIC IMPURITIES IN METALS

Some ions carry a magnetic moment, like Mn or Fe ions. What happens if they are diluted in a metal like Cu or Ag? Do these impurities keep their magnetic moment? In this section, we refer to Anderson's model of the electronic structure of an impurity in a non-magnetic metal. We report below the insight obtained by a mean field solution to Anderson's model.[318] To appreciate the conclusions of this approach, suffice it to say that the Anderson Hamiltonian includes:

- a description of the electronic structure of the host,
- the energy of an electron on the impurity with spin up or down,
- a repulsion term expressing the Coulomb repulsion of two electrons of opposite spins on the impurity,
- an interaction of the electron at the impurity with the electrons of the Fermi sea of the metal.

We assume that the impurity has only one d orbital available. The on-site repulsion U is of the order of 20 eV.[319] In comparison, the Coulomb interaction of two electrons at neighboring sites is of a few eV, while the exchange interaction is typically 1/40 eV. Clearly, the **on-site Coulomb repulsion** is by far the greatest term. Three energy parameters play a central role in Anderson's model. One is U, the on-site repulsion. Another is the energy of the d state of the impurity. We set the zero of the energy scale at the Fermi level, and call ε_d the energy level of the impurity d state. Typically, $\varepsilon_d \ll U$. So, from this alone, we would guess that the impurity has one electron and that the impurity would carry a magnetic moment. What makes the matter complex is that this impurity electron may interact with the conduction electrons at the Fermi level. Because of this interaction, the electron on the impurity has a finite **lifetime** at this location. If the energy scale of this interaction is Δ, the lifetime is of the order of $\tau = \hbar/\Delta$.

Let us now introduce the notion of a local density of states associated with the impurity. To do this, we consider $|d\sigma>$ the impurity state with spin σ and energy ε_d, and $|n\sigma\rangle$ the eigenstates of the Anderson Hamiltonian, of energy $\varepsilon_{n\sigma}$. The probability that the state $|n\sigma\rangle$ overlaps with $|d\sigma\rangle$ is $|\langle n\sigma |d\sigma\rangle|^2$. By calculating the average occupancy of the impurity with spin σ, we can define a density of states for the impurity site according to:

$$\langle n_{d\sigma}\rangle = \sum_{n|\varepsilon_{n,\sigma}<E_F} |\langle n\sigma |d\sigma\rangle|^2 = \int_{-\infty}^{E_F} d\varepsilon\, \rho_{d\sigma}(\varepsilon) \tag{6.1}$$

The energy levels of the impurity $E_{d\sigma}$ are shifted due to the interaction with the Fermi sea, with $E_{d\sigma} = \varepsilon_d + U \langle n_{d(-\sigma)}\rangle$ in the mean field approximation. The local density of states $\rho_{d\sigma}(\varepsilon)$ has a Lorentzian form,

$$\rho_{d\sigma}(\varepsilon) = \frac{1}{\pi} \frac{\Delta}{(\varepsilon - E_{d\sigma})^2 + \Delta^2} \tag{6.2}$$

Schematically, the energy levels of the impurity are as shown in Fig. 6.1. Their position on the energy scale is relevant because we consider Δ to be small compared to U. In Fig. 6.1, the schematic on the left represents a case where the impurity is likely to carry a moment. To the contrary, the schematic on the right represents a case where $\Delta > \varepsilon_d$. This second case is characterized by a **mixed valence**, in which the hybridization of the impurity state with the conduction electrons is critical. It is likely that the magnetic moment is wiped out due to the rapid fluctuation of the charge state of the impurity. The situation where U is small compared to Δ is not shown in Fig. 6.1, but corresponds to a situation with rapid spin fluctuation at the impurity site. Anderson's 1961 paper shows which region of the parameter space spanned by $-\varepsilon_d/U$ and Δ/U admits magnetic solutions.[320] Outside

Figure 6.1 Schematics for an impurity with one unpaired spin embedded in a Fermi sea. Left: the occupied impurity level is well below the Fermi level E_F. Right: the occupied impurity level is in near resonance with the Fermi level. The Lorentzian curve suggests a broadening of the impurity level due to the short residence lifetime of the spin on the impurity.

this region, the net moment at the impurity site vanishes. As P. Phillips shows in, [318] the limit between the two regions is determined by the condition,

$$\rho_d(E_F)\,U > 1 \qquad\qquad\qquad (6.3)$$

This says that, in order to ensure a moment, the on-site Coulomb repulsion and the density of states must be large. Both criteria together imply that the hybridization between the impurity state and the Fermi sea is small.

6.3 THE FERROMAGNETISM OF BAND ELECTRONS

We derive here the Stoner model of ferromagnetism.[238] In a first step, we assume that the electrons are non-interacting. A magnetic field H is applied. The densities of states (DOS) $D_s(E)$ of the two spin bands are split, as shown schematically in Fig. 6.2. We assume that $D_\uparrow(E)$ and $D_\downarrow(E)$ are simply translated by an amount $\pm\Delta E/2$ with respect to their position when no field is applied. The spin-dependent densities of states of Fe and Ni are presented in Fig. 6.3 to illustrate the grossly simplified character of the DOS functions shown in Fig. 6.2.

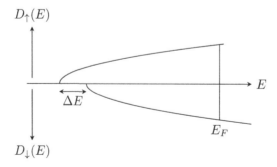

Figure 6.2 Very simple model of density of states $D_\uparrow$ and $D_\downarrow$ for exchange-split bands.

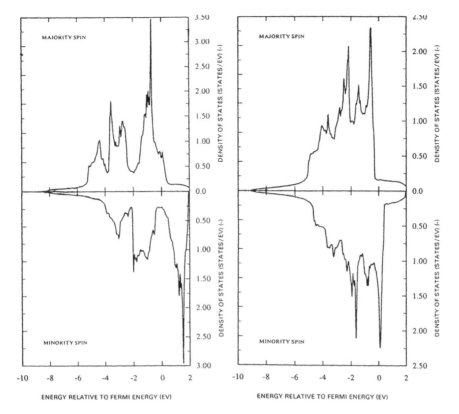

Figure 6.3 Density of states for minority and majority spins of Fe and Ni. From [81], citing [321]

Nowadays, the spin polarization of the conduction band is studied with synchrotron-based photomemission experiments. Thus, the question of the role of sp bands and d bands in 3d metals can be addressed experimentally. In Fig. 6.4, the band structure of Ni is compared to that of Cu. Careful analysis of these photoemission data yielded the value of the spin polarization of conduction electrons in Ni.[322]

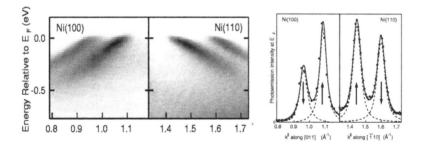

Figure 6.4 Left: energy versus k for different orientations of the k vector parallel to the surface. Right: photoemission intensity at the Fermi level E_F with spin assignment.[322]

We want to use the simple model illustrated in 6.2 and calculate the occupation number of each spin band. To do so, it is easier to work out the integrals using the picture of Fig. 6.5. This translation on the energy axis is no more than the graphical equivalent of a change of variables of integration.

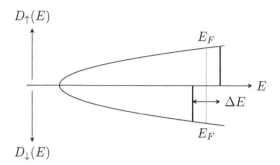

Figure 6.5 Schematics of spin-split bands of Fig. 6.2 with equalized DOS, thus presenting artificially unbalanced chemical potentials (or spin-dependent chemical potential).

Using Fig. 6.5, we find that the energy of the band E_B is given by:

$$E_B = \int_0^{E_F+\frac{\Delta E}{2}} ED(E)dE + \int_0^{E_F-\frac{\Delta E}{2}} ED(E)dE \qquad (6.4)$$

We assume $\Delta E \ll E_F$ and that the density of states near the Fermi level is the same for both sub-bands. Then:

$$E_B = 2\int_0^{E_F} ED(E)dE + \int_{E_F}^{E_F+\frac{\Delta E}{2}} ED(E)dE - \int_{E_F-\frac{\Delta E}{2}}^{E_F} E'D(E')dE'$$

$$\approx 2E_0 + D(E_F)\int_{E_F}^{E_F+\frac{\Delta E}{2}} EdE - \int_{E_F-\frac{\Delta E}{2}}^{E_F} EdE = 2E_0 + D(E_F)\frac{\Delta E^2}{4} \qquad (6.5)$$

Within the same approximation, the spin populations are:

$$N_\uparrow - N_\downarrow = \int_0^{E_F+\frac{\Delta E}{2}} D(E)dE - \int_0^{E_F-\frac{\Delta E}{2}} D(E)dE \approx \int_{E_F-\frac{\Delta E}{2}}^{E_F+\frac{\Delta E}{2}} D(E)dE = \Delta E\, D(E_F) \qquad (6.6)$$

The total number of electrons is $N = N_\uparrow + N_\downarrow$ and the spin polarization is defined as,

$$s = \frac{N_\uparrow - N_\downarrow}{N} \qquad (6.7)$$

Then, (6.7) and (6.6) can be combined to express s in terms of ΔE. The energy of the band E_B, including a magnetic energy term, using g $= 2$ for electrons and $\sigma = 1/2$, is given by:

$$E_B = 2E_0 + \frac{N^2}{4D_s(E_F)}s^2 - \mu_0\mu_B N H s \qquad (6.8)$$

Let us drop the constant term from here on. The equilibrium polarization is the one that minimizes E_B. The magnetization is given by:

$$M = \frac{\mu_B N s}{\Omega_T} \qquad (6.9)$$

where Ω_T is the volume per metal atom. This model gives the Pauli susceptibility:

$$\chi_P = \frac{2\mu_0\mu_B{}^2 D_s(E_F)}{\Omega_T} \tag{6.10}$$

As early as 1938, Stoner proposed that the exchange interaction among these electrons be parameterized phenomenologically by:

$$E_{ex} = -\frac{I}{4}(N_\uparrow - N_\downarrow)^2 \tag{6.11}$$

It turns out that the exchange term takes on a value of about 1 eV for most materials. The fact that this exchange energy must have this form can be seen from the following argument.

Let us assume that the interaction among electrons is simply a large Coulomb repulsion when they are on the same site:

$$U\hat{n}_\uparrow \hat{n}_\downarrow \tag{6.12}$$

Here, the $\hat{n}_\sigma$ would be the particle-number operators acting on the many-electron wave function. U is the energy cost for putting a second electron of opposite spin on a localized orbital. Using the operator $\hat{n} = \hat{n}_\uparrow + \hat{n}_\downarrow$, this Hamiltonian can be written in the form:

$$U\hat{n}_\uparrow \hat{n}_\downarrow = U\left(\frac{1}{4}\hat{n}^2 - \frac{1}{4}(\hat{n}_\uparrow - \hat{n}_\uparrow)^2\right) \tag{6.13}$$

Thus, we find that the spin-dependent part of this interaction has the form announced in (6.11).

By adding the Stoner term E_{ex} to the band energy E_B, the susceptibility becomes:

$$\chi = \frac{\chi_P}{1 - ID_s(E_F)} \tag{6.14}$$

The presence of the denominator in (6.14) is known as the **Stoner enhancement factor** of the susceptibility. Pd, for example, is a metal with a strongly exchange-enhanced Pauli susceptibility. This enhanced susceptibility leads to the so-called Stoner criterion for ferromagnetism. The onset of ferromagnetism is estimated to occur when the susceptibility diverges at a critical value of the density of states. Examination of the energy of the band,

$$E_B = \frac{N^2}{4}\left(\frac{1}{D_s(E_F)} - I\right)s^2 - \mu_0\mu_B NHs \tag{6.15}$$

shows that one can have spontaneous magnetization (s non vanishing at $H = 0$) if:

$$D_s(E_F)I \geq 1 \tag{6.16}$$

This is called the **Stoner criterion**. It is satisfied in the case of Fe, Co and Ni.[238] The paramagnets Sc, Y, and Pd are just below the Stoner criterion. Coleman in [323] rearranged the rows of the periodic table according to two axes: the tendency toward formation of a local magnetic moment, and the tendency toward superconductivity. In Coleman's table, Pt and Au appear at the border of the region where magnetic atoms are found, meaning that they are susceptible to turning magnetic. Both metals are often used as contacts at the surface of a ferromagnet. Hence, a controversy sometimes arises regarding whether such contacts become magnetic by a **proximity** effect.

The Stoner model that leads to (6.15) fails to predict anything other than the criterion (6.16). In particular, it yields a Pauli-type paramagnetism above the transition temperature, whereas ferromagnets in their paramagnetic phase display a Curie-Weiss susceptibility.

Furthermore, this model fails to predict a reasonable estimate of the Curie temperature. The present model predicts the destruction of ferromagnetism at the Stoner temperature T_S, at which thermal fluctuations cause spin flips that equalize the spin populations. If the DOS presents a smooth maximum, then this temperature is such that:

$$k_B T_S \approx D(E) \frac{\Delta E^2}{4} = \frac{N s}{4} \Delta E = \frac{m}{\mu_B} \frac{1}{4} I \qquad (6.17)$$

This predicts a Stoner temperature T_S of 6000 K for Fe, 4000 K for Co, and 2900 K for Ni. These values are well above the actual values of the Curie temperature T_C for these metals. In fact, spin fluctuations occur at much lower temperatures and the Curie temperature is that at which the orientation of the moments fluctuates critically.

With increasing temperatures, transition metals look more and more like they have localized moments rather than an itinerant character.[319] For example, the magnetization curves of Fe, Co, and Ni, when plotted as a function of temperature, closely resemble the Brillouin functions one would expect from localized moments acted on by a Weiss molecular field. Furthermore, the change in specific heat near the Curie temperature of Fe is just what one would predict for localized electrons. Reflecting on our understanding of magnetism, Fulde wrote in 1995: *"We leave open for the time being the question of whether (...) at temperatures considerably higher than T_C clusters with partial ferromagnetic alignment of, say, 10-50 atoms are still present. There are some arguments in favor of this latter picture. (...) even in the ferromagnetic case, it is not known at present how local the correlations are, i.e., whether they are predominantly restricted to an atom or to a larger cluster."*[1]

6.4   SPIN FLUCTUATIONS

The itinerant electron picture of the previous section can be considerably improved if spin fluctuations are taken into account, as shown in this section. The exchange interaction is represented in a mean field picture as an exchange field acting on each electron. Thus, in a mean field or molecular field approximation, we write:

$$h_{\mathrm{mf}} = \alpha \, m \qquad (6.18)$$

If a field h is applied, then the magnetization m is given by $m = \chi h = \chi_P (h + h_{\mathrm{mf}}) = \chi_P h + \alpha \chi_P m$, therefore,

$$m = \frac{\chi_P}{1 - \alpha \chi_P} h \qquad (6.19)$$

By comparison with (6.14), we see that $\alpha \chi_P = I D(E_F)$. Thus, we see that the introduction of a Weiss molecular field also leads to a Stoner enhanced susceptibility.

Let us introduce thermal fluctuations by stating that the exchange field has the following temperature dependence:

$$h_{\mathrm{eff}} = h_{\mathrm{mf}} (1 - A \, T) \qquad (6.20)$$

We could get this temeprature dependence by showing that the exchange field is reduced by a coefficient that depends on the correlation function $\langle s_i \cdot s_j \rangle$.[2] Applying the fluctuation-dissipation theorem (see §12.5), we would then infer that, in the high-temperature limit, this correlation function is proportional to temperature. Here, we will simply accept (6.20) as a reasonable approximation.

---

[1]See [319], pp. 267–268.

[2]see Fulde,[319], p. 245.

Based on the expression (6.20), it follows that the susceptibility is given by:

$$\chi(T) = \frac{\chi_P}{1 - D(E_F)I + AT} \tag{6.21}$$

The denominator vanishes at the Curie temperature given by:

$$1 - D(E_F)I + AT_C = 0 \quad \Rightarrow \quad T_C = \frac{D(E_F)I - 1}{A} \tag{6.22}$$

Therefore, the susceptibility takes on the Curie-Weiss form:

$$\chi(T) = \frac{\chi_P}{A} \frac{1}{T - T_C} \tag{6.23}$$

We must keep in mind that the origin of this Curie-Weiss behavior is different from the Curie-Weiss behavior of ferromagnets like Fe and Co, for which the localized moment picture applies. The susceptibility $\chi(T)$ in (6.21) does a fair job of describing nearly ferromagnetic materials, i.e., paramagnetic materials with a giant susceptibility. This model of itinerant electrons with fluctuations also describes **weak ferromagnets** such as ZrZn$_2$ and Ni$_3$Al. In particular, (6.22) predicts T_C of the correct order of magnitude. Bulk crystalline Ni$_3$Al is a prototypical itinerant ferromagnet. However, it has been claimed that Ni$_3$Al in nanocrystalline form is paramagnetic with Stoner-enhanced paramagnetism.[324]

6.5   THE SD-MODEL

The following description is appropriate for the 3d ferromagnets Fe, Co, and Ni, considered as having magnetic moments carried by atomic sites and conduction electrons.[325] A simple picture is used here, according to which conduction electrons form a sea of s-state electrons, while 3d electrons are responsible for the localized moments. A different approach leading to an s-d interaction was developed by Larsen.[326]

Our main concern is to obtain the form of the interaction between the s and the d electrons. We consider the scattering of a single conduction electron spin with N electrons forming a ferromagnetic moment. The localized electrons are described by **Wannier functions**, that are defined in terms of Bloch wave functions $b_{\boldsymbol{k}}^{\pm}(\boldsymbol{r})$ by, [162]

$$w_s^{\pm}(\boldsymbol{r} - \boldsymbol{R}_s) = \frac{1}{\sqrt{N}} \sum_{\boldsymbol{k}} e^{-i\boldsymbol{k}\cdot\boldsymbol{R}_s} b_{\boldsymbol{k}}^{\pm}(\boldsymbol{r}) \tag{6.24}$$

where $\boldsymbol{R}_s$ is a lattice vector. The state of the conduction electron is specified by a Bloch function $\psi_{\boldsymbol{k}}^{\pm}(\boldsymbol{r})$, which can be thought of as composed of Wannier functions $\phi_s(\boldsymbol{r})$:

$$\psi_{\boldsymbol{k}}^{\pm}(\boldsymbol{r}) = \frac{1}{\sqrt{N}} \sum_{s} e^{i\boldsymbol{k}\cdot\boldsymbol{R}_s} \phi_s(\boldsymbol{r}) \tag{6.25}$$

The state of the system composed of these $N + 1$ electrons must be an anti-symmetric linear combination of these states. There are many possible states of the system, corresponding to all of the spin configurations. The interaction of the conduction electron with the ferromagnetic electrons is given by the Hamiltonian:

$$\mathcal{H}_{ee} = \sum_{i<j}^{N+1} \frac{e^2}{4\pi\varepsilon_0 r_{ij}} \tag{6.26}$$

A perturbation calculation carried out to estimate the effect of the Coulomb interaction needs to work out the matrix elements of this Hamiltonian between these anti-symmetric states of N+1 electrons. When this is done,[325] it is possible to construct a Hamiltonian that would give the same matrix elements, were we to consider simple products of one-electron wave functions instead of the fully anti-symmetric ones. This Hamiltonian has the form:

$$\mathcal{H}(\boldsymbol{k}', \boldsymbol{k}) = -2 \sum_{j=1}^{N} J_j(\boldsymbol{k}', \boldsymbol{k}) \, \boldsymbol{S}_j \cdot \boldsymbol{s} \tag{6.27}$$

where,

$$J_j(\boldsymbol{k}', \boldsymbol{k}) = \int \int d\tau_{12} \frac{e^2}{4\pi\varepsilon_0 r_{12}} \psi_{\boldsymbol{k}'}^*(1) w_j(1) w_j^*(2) \psi_{\boldsymbol{k}}(2) \tag{6.28}$$

is an exchange integral. Now, by expressing the expansion of the Bloch wave function of the conduction electron in terms of Wannier functions, we have:

$$J_j(\boldsymbol{k}', \boldsymbol{k}) = \frac{1}{N} e^{i(\boldsymbol{k}-\boldsymbol{k}')\boldsymbol{R}_j} \int \int d\tau_1 d\tau_2 \frac{e^2}{4\pi\varepsilon_0 r_{12}} \phi_s^*(\boldsymbol{r}_1) w_j(\boldsymbol{r}_1) w_j^*(\boldsymbol{r}_2) \phi_s(\boldsymbol{r}_2) \tag{6.29}$$

Let us recall that, in this expression, the w_j are the localized wave functions of those electrons giving rise to the ferromagnetic moments; they are presumed to be 3d-electrons, while the ϕ_s represent a 4s-type electron, so the exchange integral is an s-d exchange integral. Atomic physics provides tabulated values for such integrals. In a solid, one must expect these integrals to take on different values because of screening effects.

Thus, we write (6.29) as:

$$J_j(\boldsymbol{k}', \boldsymbol{k}) = \frac{1}{N} e^{i(\boldsymbol{k}-\boldsymbol{k}')\boldsymbol{R}_j} J_{sd} \tag{6.30}$$

and treat J_{sd} as a phenomenological parameter. Therefore, the end result has the following form:[325]

$$\mathcal{H}(\boldsymbol{k}', \boldsymbol{k}) = -2J_{sd} \left(\frac{1}{N} \sum_j e^{i(\boldsymbol{k}-\boldsymbol{k}')\boldsymbol{R}_j} \boldsymbol{S}_j \right) \cdot \boldsymbol{s} \tag{6.31}$$

The case $\boldsymbol{k}' = \boldsymbol{k}$ corresponds to spin scattering without momentum scattering. This situation occurs when electrons pass through domain walls.[327, 328] In this case, $\sum_{i,j} \boldsymbol{S}_i \cdot \boldsymbol{S}_j$ commutes with $\mathcal{H}(\boldsymbol{k}', \boldsymbol{k})$, meaning that the magnitude of the d moments $\left(\sum_i \boldsymbol{S}_i \right)^2$ is constant. This means that the d moments undergo a rotation under the effect of this spin-only scattering (no momentum scattering). But even when $\boldsymbol{k}' = \boldsymbol{k}$, $\sum_i S_{i,z}$ does not commute with $\mathcal{H}(\boldsymbol{k}', \boldsymbol{k})$, meaning that spin flips take place. This will be discussed in §13.3 when considering magnetic contributions to the spin lifetime.

6.6   RASHBA EFFECT

In this section, we introduce the Rashba effect, which can occur in metals and semiconductors.[329] In this effect, electron momentum and spin are locked, i.e., they have a specific orientation with respect to one another. In other words, under the Rashba effect, spins present a special structure in $\boldsymbol{k}$-space. Specific symmetry conditions are needed for this to occur, which we analyze before discussing the Rashba Hamiltonian that accounts for the effect.

6.6.1 TIME REVERSAL AND SPATIAL INVERSION SYMMETRY

In full generality, if A is an operator expressing a symmetry of the physical system described by the Hamiltonian $\mathcal{H}$, then $A\mathcal{H}A^{-1} = \mathcal{H}$ or,

$$\cdot \ A\mathcal{H} = \mathcal{H}A \tag{6.32}$$

Consider now a Bloch state with its spin "up,"

$$\mathcal{H}\left|\boldsymbol{k},\uparrow\right\rangle = E(\boldsymbol{k},\uparrow)\left|\boldsymbol{k},\uparrow\right\rangle \tag{6.33}$$

Then:

$$A\mathcal{H}\left|\boldsymbol{k},\uparrow\right\rangle = \mathcal{H}A\left|\boldsymbol{k},\uparrow\right\rangle = E(\boldsymbol{k},\uparrow)A\left|\boldsymbol{k},\uparrow\right\rangle \tag{6.34}$$

The last equality means that $A\left|\boldsymbol{k},\uparrow\right\rangle$ is also an eigenstate with the same energy.

We now apply this reasoning for the case of a Hamiltonian with **_time inversion symmetry_**. This is the case in a band structure calculation, even if the Hamiltonian contains a spin orbit term, because the spin and angular momenta both change signs in time inversion. Now $T\left|\boldsymbol{k},\uparrow\right\rangle = \left|-\boldsymbol{k},\downarrow\right\rangle$, which is an eigenstate of $\mathcal{H}$ of energy $E(-\boldsymbol{k},\downarrow)$. Therefore, we have that,

$$E(\boldsymbol{k},\uparrow) = E(-\boldsymbol{k},\downarrow) \tag{6.35}$$

We now apply the same reasoning to the case of **_space inversion symmetry_**, i.e., we assume that $\mathcal{H}$ has inversion symmetry. If R is the inversion operator, $R\left|\boldsymbol{k},\uparrow\right\rangle = \left|-\boldsymbol{k},\uparrow\right\rangle$ is an eigenstate of $\mathcal{H}$ of energy $E(-\boldsymbol{k},\uparrow)$. Therefore, we have,

$$E(\boldsymbol{k},\uparrow) = E(-\boldsymbol{k},\uparrow) \tag{6.36}$$

When both time reversal and inversion symmetry are valid, we have,

$$E(\boldsymbol{k},\uparrow) = E(\boldsymbol{k},\downarrow) \tag{6.37}$$

This says that the bands are spin degenerate.

6.6.2 RASHBA HAMILTONIAN

At a surface, this degeneracy disappears and the bands can split according to their spin. We can assume that there is a gradient of the electrostatic potential that is normal to the surface. We note $\hat{\boldsymbol{e}}_z$ the unit vector normal to the surface. We expect a potential gradient $\nabla V_\perp = \hat{\boldsymbol{e}}_z dV_\perp/dz$. Hence, the spin-orbit coupling in this case must have the form:

$$\mathcal{H}_R = \alpha_R\left(\hat{\mathbf{e}}_z \wedge \boldsymbol{p}\right)\cdot\boldsymbol{\sigma} \tag{6.38}$$

As $\boldsymbol{\sigma} = (\sigma_x, \sigma_y, \sigma_z)$, we have,

$$\mathcal{H}_R = \alpha_R\left(p_x\sigma_y - p_y\sigma_x\right) \tag{6.39}$$

This Hamiltonian is a perturbation with respect to the electronic Hamiltonian, whose eigenstates are Bloch states, of momentum $\boldsymbol{p} = \hbar\boldsymbol{k}$. We assume that the electrons have an effective mass m^* and an unperturbed energy is given by $(k_x^2 + k_y^2)/2m^*$. Expressing the Pauli matrices in the basis of spins quantized along the z-axis, this gives:[3]

$$\mathcal{H} = \begin{pmatrix} \dfrac{1}{2m^*}\left(k_x^2 + k_y^2\right) & \alpha_R\hbar(-k_y - ik_x) \\[2ex] \alpha_R\hbar(-k_y + ik_x) & \dfrac{1}{2m^*}\left(k_x^2 + k_y^2\right) \end{pmatrix} \tag{6.40}$$

---

[3]These matrices can be found in (19.80).

This can be readily solved by an algebraic software package. Thus, we obtain the eigenvalues:

$$E(\boldsymbol{k}, \pm) = \frac{1}{2m^*} k^2 \pm \alpha_R \hbar k = \frac{1}{2m^*} (k \pm \alpha_R \hbar 2m^*)^2 - \alpha_R^2 \hbar^2 2m^* \qquad (6.41)$$

with $k = \sqrt{k_x^2 + k_y^2} = |\mathbf{k}|$. The minus sign gives an energy of lower value than the plus sign for any given k. The alternative expression on the right-hand side of (6.41) shows that the energy surfaces can be thought of as generated by the following geometric construction. First, the free-electron parabola is shifted in, say, the k_x direction by the amount $\alpha_R \hbar 2m^*$, and second, it is rotated about the z-axis (Fig. 6.6). As a consequence, in a cut at constant energy through this surface, the dispersion looks like two circles, or one, depending on the energy value (Fig. 6.7). The corresponding eigenvectors can be written as,

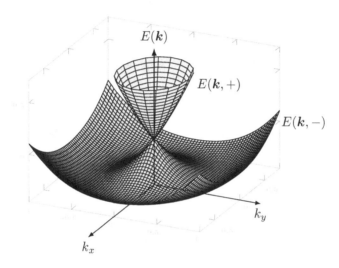

Figure 6.6 Rashba-split bands according to (6.41) (normalized as $z = x^2 + y^2 \pm 1.2\sqrt{x^2 + y^2}$). The top surface is drawn within a smaller domain for clarity.

$$\pm \frac{ik}{\mathrm{k}_x + i k_y} |\uparrow\rangle + |\downarrow\rangle = \pm e^{i\frac{\pi}{2}} e^{-i\alpha(\mathbf{k})} |\uparrow\rangle + |\downarrow\rangle \qquad (6.42)$$

where $\alpha(\mathbf{k})$ is the phase of the complex number $k_x + i k_y$. We note $\phi(\mathbf{k}) = \frac{\pi}{2} - \alpha(\mathbf{k})$ the angle between $\boldsymbol{k}$ and the y direction. We can normalize these eigenvectors and write them as :

$$\frac{1}{\sqrt{2}} \left(|\uparrow\rangle \mp e^{-i\phi(\mathbf{k})} |\downarrow\rangle \right) \qquad (6.43)$$

The off-diagonal Rashba terms give eigenstates that mix the up and down spin states. We can calculate the expectation value of the spin $\boldsymbol{\sigma}$. Respecting the sign stacking order, we calculate $\frac{1}{2} \left(\langle\uparrow| \mp e^{i\phi(\mathbf{k})} \langle\downarrow| \right) \left(|\uparrow\rangle \mp e^{-i\phi(\mathbf{k})} |\downarrow\rangle \right)$. This yields the following expectation values for $\sigma_x, \sigma_y,$ and σ_z,[330]

$$\frac{1}{2} \begin{pmatrix} 1 & \mp e^{i\phi(\mathbf{k})} \end{pmatrix} \begin{pmatrix} 0 & 1 \\ 1 & 0 \end{pmatrix} \begin{pmatrix} 1 \\ \mp e^{-i\phi(\mathbf{k})} \end{pmatrix} = \mp \cos\phi(\mathbf{k})$$

$$\frac{1}{2} \begin{pmatrix} 1 & \mp e^{i\phi(\mathbf{k})} \end{pmatrix} \begin{pmatrix} 0 & -i \\ i & 0 \end{pmatrix} \begin{pmatrix} 1 \\ \mp e^{-i\phi(\mathbf{k})} \end{pmatrix} = \pm \sin\phi(\mathbf{k}) \qquad (6.44)$$

$$\frac{1}{2} \begin{pmatrix} 1 & \mp e^{i\phi(\mathbf{k})} \end{pmatrix} \begin{pmatrix} 1 & 0 \\ 0 & -1 \end{pmatrix} \begin{pmatrix} 1 \\ \mp e^{-i\phi(\mathbf{k})} \end{pmatrix} = 0$$

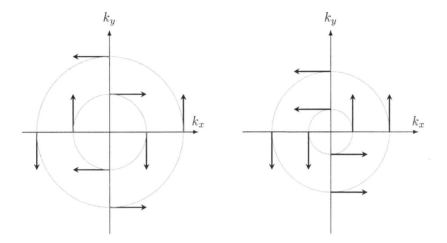

Figure 6.7 Spin-momentum locking due to the Rashba effect. Notice how the spin orientation depends on whether the energy is below (left) or above (right) the degeneracy point at $\boldsymbol{k} = 0$.

In other words, we have $\langle \boldsymbol{\sigma} \rangle = (\mp \cos \phi(\mathbf{k}), \pm \sin \phi(\mathbf{k}), 0)$. Our definition of $\phi(\boldsymbol{k})$ implies that the unit vector parallel to $\boldsymbol{k}$ is $\hat{\boldsymbol{k}} = (\sin \phi, \cos \phi, 0)$. We conclude that the spin $\boldsymbol{\sigma}$ is perpendicular to the vector $\boldsymbol{k}$ in these Rashba split bands. This circumstance, by which the spin direction is determined by the $\boldsymbol{k}$, is often referred to as ***spin locking***. Below the degeneracy point in Fig. 6.6, a cut at constant energy gives two circles belonging to one energy surface, whereas above this degeneracy point, one circle belongs to $E(\boldsymbol{k}, -)$ and the other to $E(\boldsymbol{k}, +)$ (Fig. 6.7).

6.7 ORBITAL ANGULAR MOMENTUM STATES

6.7.1 PHENOMENOLOGY

Giant spin splitting of the electron band structure is observed in some solids, including noble metal surface states,[331–333], semi-metals and their alloys,[334–336], semiconductors,[337,338] and atomically thin transition-metal dichalcogenides.[339]

It has been shown that this phenomenon can be linked to the presence of Bloch states that carry angular moment, which are called for this reason ***orbital angular momentum states*** (OAM states).[340] As in the Rashba effect, this can only happen if inversion symmetry is broken, which is of course the case with surfaces,[341] interfaces (Fig 6.8),[342,343] and non-inversion symmetric crystals.[344,345] Orbital angular momentum states also play a central role in the phenomenon known as current-induced spin orbit torque.[346–348]

There are many transport phenomena such as the anomalous Hall and Nernst effects for which the picture of charge carriers as point charges is too naive to account for what is observed.[349] Instead, we have to think that we are dealing with the transport of ***wave packets***. The ***side jump*** mechanism, for example, refers to a collision of a wave packet with a point defect that leads to a jump in the trajectory of the wave packet center of mass. This mechanism contributes to the Hall effect. The wave packet in real space may include a rotation about its center of mass (Fig. 6.9). It has a counterpart in k-space, i.e., it is composed of $\boldsymbol{k}$ vectors with a finite spread in $\boldsymbol{k}$-space.

We recall that a form of coarse graining is used when treating properly electromagnetism on a scale larger than atoms and molecules. This procedure leads to a distinction between

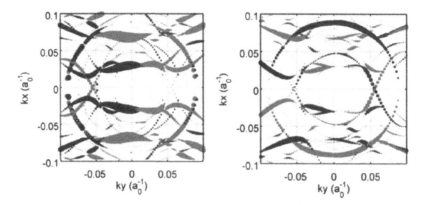

Figure 6.8 Band structure of a Co/Pt multilayer, with x-projection (left) and y-projection (right) of the spin in one Co layer. The grey scale indicates the sign (light grey, negative, dark grey, positive), the size of the dot the magnitude of the spin polarization. After [343].

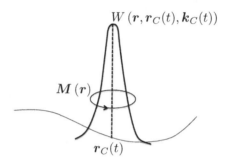

Figure 6.9 Wave packet $W(\boldsymbol{r},\boldsymbol{r}_C,\boldsymbol{k}_C)$ centered at $\boldsymbol{r}_C$ in real space; trajectory $\boldsymbol{r}_C$ of the wave packet in real space; orbital moment (represented as a circular motion) $\boldsymbol{M}$.

transport currents and bound currents, which are of the form $\boldsymbol{\nabla} \times \boldsymbol{M}$.[4] Likewise, when a wave packet has a rotational structure as illustrated in Fig. 6.9, we also have to distinguish a charge current associated with the motion of the center of mass of the wave packet, and a bound current. This distinction has proven to be important in analyzing the Nernst effect in a ferromagnetic metal.[349]

In conformity with the classical definition of angular momentum, it is natural to define an orbital magnetic moment in terms of the expectation value of the angular moment operator, $\boldsymbol{L} = (\boldsymbol{r} - \boldsymbol{r}_C) \times m\boldsymbol{v}$, where $\boldsymbol{v}$ is meant here as the dynamic variable $\partial \mathcal{H}/\partial(\hbar \boldsymbol{k})$ conjugate to the momentum $\hbar \boldsymbol{k}$ (see §11.2). Thus, the expectation value for the wave packet W is given by,

$$\boldsymbol{m_k} = -\frac{e}{2}\langle W|\,(\hat{\boldsymbol{r}} - \boldsymbol{r}_C) \times \hat{\boldsymbol{v}}\,|W\rangle \tag{6.45}$$

since the orbital magnetic moment $\boldsymbol{m} = (-e/2m)\boldsymbol{L}$ (see §2.5).

In the following, we take a simpler approach and construct a tight-binding state that has non-vanishing orbital angular momentum.

---

[4]Compare, for example, with (5.32) and (5.103) in Jackson's textbook.[350]

6.7.2 QUASI ONE-DIMENSIONAL HELICAL STATES

Let us construct a state that has an angular moment in some specific direction of space. We do this starting from a localized orbital $\Psi_a(\boldsymbol{x})$. First, we construct a Bloch state using this orbital, as one would for a line of atoms at periodic positions $\boldsymbol{R}$. The z-axis is set along this line of atoms and the x- and y-axes are normal to it (Fig. 6.10). As in the tight-binding method ([72] chapter 10), we write,

$$\Psi_{\boldsymbol{k}}(\boldsymbol{r}) = \sum_{\boldsymbol{R}} e^{i\boldsymbol{k}\cdot\boldsymbol{R}}\, \Psi_a(\boldsymbol{r} - \boldsymbol{R}) \tag{6.46}$$

Now, we introduce a twist in the wave function (6.46) by assuming that, at each location $\boldsymbol{R}$, the local wave function is rotated about the y-axis with respect to $\Psi_a(\boldsymbol{x})$ by an angle κR, where $R = \boldsymbol{R}\cdot\hat{\boldsymbol{e}}_z$ and κ defines the extent of the twisting (Fig. 6.10). We will see that the overlap provoked by this rotation plays a crucial role.

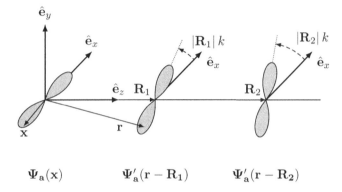

$$\Psi_a(\mathbf{x}) \qquad\qquad \Psi_a'(\mathbf{r} - \mathbf{R_1}) \qquad\qquad \Psi_a'(\mathbf{r} - \mathbf{R_2})$$

Figure 6.10 Construction of an OAM state. Each localized orbital $\Psi_a\,(\boldsymbol{r} - \boldsymbol{R}_i)$ is obtained from an atomic state $|\Psi_a(\boldsymbol{x})>$ by the combination of a rotation about the y-axis by an the angle $|\boldsymbol{R}_i|\kappa$ for every translation given by $\boldsymbol{R}_i$, resulting in an atomic orbital $|\Psi_a(\boldsymbol{r} - \boldsymbol{R}_i)>$ centered at the location defined by $\boldsymbol{R}_i$. The state $|\Psi_a(\boldsymbol{x})>$ is represented with a one-lobe function and the site-to-site distance is exaggerated, for the sake of clarity. Due to the relative rotation of the wave functions from site to site, the lobes from neighboring sites overlap.

We recall some basic results to represent this rotated-translated wave function in a convenient way for the upcoming calculation. For every spatial transformation $\tau(s)$ that depends on a continuous parameter s, there is a unitary transformation of the wave function such that,[5]

$$\Psi(\tau(s)\boldsymbol{x}) = U(s)\Psi(\boldsymbol{x}) \qquad U(s) = e^{iKs} \quad (K = K^\dagger) \tag{6.47}$$

K is called the **generator** of $\tau(s)$. The generator of translations is the linear momentum $\boldsymbol{p}$ and the generator of rotations is the angular momentum $\boldsymbol{L}$. Namely, we have for a translation by a vector $\boldsymbol{a}$,

$$|\boldsymbol{x} + \boldsymbol{a}\rangle = e^{-i\boldsymbol{a}\cdot\boldsymbol{p}/\hbar}\,|\boldsymbol{x}\rangle \tag{6.48}$$

and for a rotation by an angle α about an axis of unit vector $\hat{\boldsymbol{n}}$,

$$|\boldsymbol{x} + \alpha\hat{\boldsymbol{n}} \times \boldsymbol{x}\rangle = e^{-i\alpha\hat{\boldsymbol{n}}\cdot\boldsymbol{L}/\hbar}\,|\boldsymbol{x}\rangle \tag{6.49}$$

---

[5]see [181], chapt. 3.

Hence, the newly defined, rotated, local state $\Psi_a(\boldsymbol{r} - \boldsymbol{R})$ can be written as,

$$\Psi'_a(\boldsymbol{r} - \boldsymbol{R}) = e^{-iR\kappa L_y/\hbar}\Psi_a(\boldsymbol{r} - \boldsymbol{R}) \tag{6.50}$$

The newly defined Bloch state we have thus created is given by,

$$\Psi_{\boldsymbol{k}}(\boldsymbol{r}) = \sum_R e^{ikR}\, e^{-iR\kappa L_y/\hbar}\Psi_a(\boldsymbol{r} - \boldsymbol{R}) \tag{6.51}$$

Let us now calculate the expectation value of L_y in this state,

$$< L_y > = \frac{\sum_{R,R'} e^{ik(R-R')}\, \langle\Psi_a(\boldsymbol{r} - \boldsymbol{R}')|\, e^{iR'\kappa L_y/\hbar}\, L_y\, e^{-iR\kappa L_y/\hbar}\, |\Psi_a(\boldsymbol{r} - \boldsymbol{R})\rangle}{\sum_{R,R'} e^{ik(R-R')}\, \langle\Psi_a(\boldsymbol{r} - \boldsymbol{R}')|\, e^{i(R'-R)\kappa L_y/\hbar}\, |\Psi_a(\boldsymbol{r} - \boldsymbol{R})\rangle} \tag{6.52}$$

The denominator ensures the normalization of the Bloch state, which was not expressed in (6.51). We approximate that the overlaps are small in comparison to the norm $|\Psi_a(\boldsymbol{r}-\boldsymbol{R})|^2$. In the numerator, we add the identity operator written as $e^{-iR\kappa L_y/\hbar}e^{iR\kappa L_y/\hbar}$. This yields in,

$$< L_y > = \frac{\sum_{R,R'} e^{ik(R-R')}\, \langle\Psi_a(\boldsymbol{r} - \boldsymbol{R}')|\, e^{i(R'-R)\kappa L_y/\hbar}e^{iR\kappa L_y/\hbar}L_y\, e^{-iR\kappa L_y/\hbar}\, |\Psi_a(\boldsymbol{r} - \boldsymbol{R})\rangle}{\sum_R \langle\Psi_a(\boldsymbol{r} - \boldsymbol{R})|\Psi_a(\boldsymbol{r} - \boldsymbol{R})\rangle} \tag{6.53}$$

The type of operator conjugation that appears in the numerator will become very familiar when we analyze spin echo experiments in §19.4.1 (in particular (19.28)). It can be thought of as a rotation by an angle $-R\kappa$ of the operator L_y. Hence, it does not change.

Let us take for the local orbitals $\Psi_a(\boldsymbol{r} - \boldsymbol{R})$ eigenfunctions of L_y of eigenvalue $\hbar$. What is left in the numerator of (6.53) is the overlap of a local orbital with the orbital located at a neighboring site. It is quite reasonable to assume that only nearest-neighbors have non-vanishing overlaps. So, now, we can change summation variables in (6.53), using $\boldsymbol{R}' = \boldsymbol{R} + d\hat{\boldsymbol{e}}_z$. We take only the nearest-neighbor d values, namely $R' = R + d$ and $R' = R - d$, because overlaps will be negligible beyond the first neighbor. This yields for (6.53),

$$< L_y > = \frac{\sum_R \hbar \left(e^{-ikd}\, \langle\Psi_a^{(d\kappa)}(\boldsymbol{x} - \boldsymbol{d})|\Psi_a(\boldsymbol{x})\rangle + e^{ikd}\, \langle\Psi_a^{(-d\kappa)}(\boldsymbol{x} + \boldsymbol{d})|\Psi_a(\boldsymbol{x})\rangle\right)}{\sum_R \langle\Psi_a(\boldsymbol{r} - \boldsymbol{R})|\Psi_a(\boldsymbol{r} - \boldsymbol{R})\rangle} \tag{6.54}$$

where $\langle\Psi_a^{(\pm d\kappa)}(\boldsymbol{r} - \boldsymbol{d})|$ is $\langle\Psi_a(\boldsymbol{r} - \boldsymbol{d})|$ rotated by $\pm d\kappa$. The overlaps of the wave function localized at $\boldsymbol{R}$ with the wave functions centered at the neighboring sites are equal and independent of $\boldsymbol{R}$. Thus, we have,

$$< L_y > = \hbar\frac{2\, \langle\Psi_a^{(d\kappa)}(\boldsymbol{x} - \boldsymbol{d})|\Psi_a(\boldsymbol{x})\rangle}{\sum_R \langle\Psi_a(\boldsymbol{r} - \boldsymbol{R})|\Psi_a(\boldsymbol{r} - \boldsymbol{R})\rangle} \sum_R \cos(kR) \tag{6.55}$$

Let us assume that the localized orbitals are normalized, $\langle\Psi_a(\boldsymbol{r} - \boldsymbol{R})|\Psi_a(\boldsymbol{r} - \boldsymbol{R})\rangle = 1$. Then, we have at the Γ point of the reciprocal lattice, i.e., at $k = 0$,

$$< L_y > = \hbar\, 2\, \langle\Psi_a^{(d\kappa)}(\boldsymbol{r} - \boldsymbol{d})|\Psi_a(\boldsymbol{x})\rangle \tag{6.56}$$

Thus, we find that $< L_y > \neq 0$ because of the non-vanishing overlap of wave functions centered on neighboring sites.

6.8 FURTHER READINGS

Spin texture in carbon nanostructures

When a graphene ribbon has zigzagged edges, it may host unpaired spins.[351] The interface between ribbons of different topological properties seems to host spin-locked states.[352] Building on this notion, R. Fasel and collaborators explored the properties of planar carbon molecules. These non-Kekulé structures are topologically "frustrated."[353] When deposited on gold, they host Kondo resonances.[354, 355].

Chiral graphene nanoribbons could be produced,[356] as well as doping graphene structures with nitrogen.[357] Earlier, in 2000, Ando pointed out the importance of the overlap between neighboring p orbitals in his analysis of spin-dependent transport in carbon nanotubes.[358]

Spin-momentum locking

The spin versus momentum locking was observed, for example, in SmB_6 by spin-resolved photoemission.[359] In *topological insulators*, the surface states have a metallic character with spin-momentum locking.[360]

A Rashba effect in cylindrical geometry was shown to give rise to strong spin-orbit effects. Thus, helical states were observed in Ge/Si core-shell nanowires.[361] Enhanced spin-orbit coupling was also observed in core-shell nanowires based on GaAs.[362]

The Dresselhaus effect is a close cousin to the Rashba effect.[329] It was shown to arise in bulk wurtzite materials.[363]

Electric-field-induced magnetism

Pt is not far from satisfying the Stoner criterion. Therefore, it is conceivable that doping the surface of a Pt electrode using *ionic liquids* may induce a perpendicular magnetic anisotropy.[364–366] An *anomalous Hall effect* (AHE) (see Chapter 11) was found when Pt was subject to an electric field applied by means of an ionic liquid.[367]

s-d exchange interaction

In view of the discussion on s-d exchange, let us mention here that *itinerant antiferromagnetism* was identified in RuO_2.[368] For the same reason, we point to the observation of magnetic ordering in a paramagnet, YCO_2.[369] The magnetic moment and electronic structure of impurities in 3d metals were investigated using first-principles numerical calculations.[370]

Direct experimental evidence for vortex states

Generation in vacuum of an electron beam characterized by a vortex state has been accomplished by the following method. A nano-hole was produced in an Ag-thin film. A chiral plasmonic excitation was obtained by shining circularly polarized light on the Ag film (Fig. 6.11). It was a surface plasmon polariton. In the context of magnetic resonance, the notion of polariton is explained in §18.5. When an electron beam impinges on the chiral structure thus formed, the state of the beam develops a vortex structure, as observed with an ultrafast transmission electron microscope.[371]

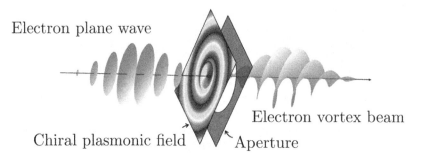

Figure 6.11 Conceptual rendering of an experimental method used to produce a vortex state with an electron beam. A plasmon polariton excited in an Ag film comprising a hole acts as a phase plate with a pronounced chiral structure. The chiral structure and the vortex nature of the beam could be verified with an ultrafast transmission electron microscope. (Figure adapted from [371].)

7 Magnetization Dynamics

Lev Davidovitch Landau (1908–1968)

L.D. Landau graduated from the Leningrad State University in 1927. Following his studies in theoretical physics, he visited several major centers for theoretical physics in Europe. Between 1932 and 1937, he became head of the theoretical physics department at Karkhiv, Ukraine. This is where he and and his friend E. Lifshitz began writing their world-famous Course of Theoretical Physics in ten volumes, one of which is dedicated to electromagnetic fields in continuous media. Landau received the Nobel Prize in Physics for his study of superfluidity.

Spintronics: A primer

J.-Ph. Ansermet (https://orcid.org/0000-0002-1307-4864)

Chapter 7. Magnetization Dynamics

The Larmor precession of an angular momentum in a magnetic field is derived semi-classically. When treating the case of local moments, a thermodynamic argument leads to an effective magnetic induction field defined as a functional derivative of a magnetic potential closely related to magnetic enthalpy. The Landau-Lifshitz-Gilbert equation describes the magnetization dynamics in this effect field, including damping. Spin waves are derived in the case where exchange contributions play an important role. One-magnon eigenstates are defined as an introduction to the Holstein-Primakoff description of spin waves. This description can account for spin waves where both dipolar and exchange interactions play a role.

7.1 HISTORICAL INTRODUCTION

It is generally assumed that Faraday invented a balance that could measure magnetization. Its principle is based on the force exerted on a magnetic material by an inhomogeneous magnetic field. As for the torque acting on a magnetic material when it is subjected to a uniform field, we have to turn to J. Larmor. Professor at Cambridge University since 1885, Larmor was interested in matter interactions with electromagnetic fields. In 1896, Larmor established that magnetic dipoles subjected to a magnetic induction field undergo a rotational motion, now known as the Larmor precession.

Landau and Lifshitz considered that magnetization must precess about the magnetic induction field that determines the equilibrium magnetization. This field, now referred to as "effective field", was obtained by a variational principle according to which a magnetic energy, function of magnetization and its spatial variation, is minimized under the condition that its modulus is at all times equal to the saturation magnetization.[372] Landau and Lifshitz introduced a phenomenological damping term as early as 1935. In 1955, T.L. Gilbert replaced that damping term by a Langevin-type damping term. Both forms of the evolution equation for magnetization will be introduced in this chapter.

The chapter starts out with a description of spin precession in a magnetic field. From this simple consideration, we can introduce the Landau-Lifshitz equation. Expanding on this result, exchange spin waves are inferred from a semi-classical argument consisting in using the Heisenberg exchange model for a ferromagnet with localized moments and working out a continuous medium approximation for spin waves in a classical picture. The quantum mechanical one-magnon eigenstate is presented as a way to introduce the concept of a spin wave represented as the collective result of correlated local spin precessions. The chapter ends with a presentation of the Holstein-Primakoff representation of spin waves. This method is applied to a Heisenberg-like Hamiltonian with dipolar couplings.

The notions of magnetization dynamics that are introduced here will be used throughout this book. For example, spin waves will be alluded to in Chapter 9 on Boltzmann's theory of transport, in connection with the magnon drag effects, and in Chapter 13 regarding the spin wave contribution to conduction electron spin relaxation. The Landau-Lifshitz equation, with an added phenomenological damping term, will be the basis on which we introduce magnetic relaxation in Chapter 14.

7.2 SPIN PRECESSION

Let us consider a magnetic moment $\boldsymbol{m}$ in a semi-classical picture according to which there is an angular momentum $\boldsymbol{J}$ associated with the magnetic moment $\boldsymbol{m}$, which is given by,

$$\boldsymbol{m} = \gamma \boldsymbol{J} \tag{7.1}$$

where γ is the called the **gyromagnetic ratio**. The relation (7.2) is a consequence of the Wigner-Eckart theorem of quantum mechanics.[373] We will also use (7.2) when considering $\boldsymbol{m}$ and $\boldsymbol{J}$ as classical properties. For the free electron, for example, (2.32) implies,

$$\gamma = \frac{-g|e|}{2m} = -\frac{g\mu_B}{\hbar} \tag{7.2}$$

Therefore, we have,

$$\boldsymbol{m} = -g\mu_B \hat{\boldsymbol{J}} \tag{7.3}$$

where $\hat{\boldsymbol{J}}$ is a unitless angular momentum, i.e., $\boldsymbol{J} = \hbar\hat{\boldsymbol{J}}$.

The evolution of the angular momentum in a magnetic induction field $\boldsymbol{B}$ is then given, according to classical mechanics, by the torque exerted on the moment,[170]

$$\frac{d\boldsymbol{J}}{dt} = \boldsymbol{m} \times \boldsymbol{B} \tag{7.4}$$

We could also arrive at this result in a quantum mechanical framework, applying the von Neumann equation of evolution for $\boldsymbol{J}$, when the dynamics is determined by the Zeeman Hamiltonian,

$$\mathcal{H} = -\boldsymbol{m} \cdot \boldsymbol{B} = -\gamma \hbar \boldsymbol{B} \cdot \hat{\boldsymbol{J}} = \mu_B \boldsymbol{B} \cdot \hat{\boldsymbol{J}} \tag{7.5}$$

Using the definition (7.2) of γ, equation (7.4) can be rewritten as:

$$\frac{d\boldsymbol{m}}{dt} = \boldsymbol{m} \times \gamma \boldsymbol{B} = \boldsymbol{\Omega} \times \boldsymbol{m} \tag{7.6}$$

where

$$\boldsymbol{\Omega} = -\gamma \boldsymbol{B} \tag{7.7}$$

It should be noted that, sometimes, in the magnetism literature, the gyromagnetic ratio is defined as a positive quantity. In other words, some authors refer to the ratio $ge/(2m)$ when they speak of the gyromagnetic ratio. As a way of bridging with that practice, but always keeping our definition (7.2), the evolution equation (7.6) for $\boldsymbol{m}$ can be written as,

$$\frac{d\boldsymbol{m}}{dt} = -\left(\frac{ge}{2m}\right) \boldsymbol{m} \times \boldsymbol{B} \qquad (e > 0) \tag{7.8}$$

Equation (7.6) describes the rotation of the vector $\boldsymbol{m}$ at the angular velocity $\boldsymbol{\Omega}$. $\|\boldsymbol{\Omega}\|/2\pi$ is called the **Larmor frequency** (Fig. 7.1). It may be useful to remember that the sense of precession is given by a vector of angular velocity opposite $\boldsymbol{B}_0$ if γ is positive.

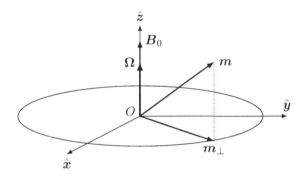

Figure 7.1 A magnetic moment $\boldsymbol{m}$ precesses in a magnetic induction field $\boldsymbol{B}_0$ with an angular velocity $\boldsymbol{\Omega}$.

7.3 EFFECTIVE MAGNETIC FIELD

Landau and Lifshitz derived an equation of evolution for the magnetization by applying a variational principle.[372] Their approach allows a determination of the effective magnetic induction field about which the magnetization precesses. It can be extended to the case of an inhomogeneous magnetization.[374] The effective magnetic induction field is derived from this variational calculation as the derivative of an energy function with respect to the magnetization,

$$\text{magnetic induction field} = \frac{\partial \,(\text{energy function})}{\partial \,(\text{magnetization})} \tag{7.9}$$

Confusion may arise as to which energy function this is. In all experiments involving magnetization evolution, it is evident that a magnetic field, defined as the applied magnetic field, is applied somewhat at a distance from the sample. As H. Callen puts it,[375] *the system is in contact with a magnetic field reservoir.* Therefore, it is the **magnetic enthalpy** that is at a minimum when the system is at an equilibrium with a magnetic field reservoir.

Let us now assume that the system is first described by the state variables entropy density s, number of moles of substances $\{n_A\}$, and magnetization $\boldsymbol{M}$. The internal energy density is the function $u(s, \{n_A\}, \boldsymbol{M})$.[376, 377] The magnetic enthalpy is defined by the Legendre transformation,

$$h_m(s, \{n_A\}, \boldsymbol{B}) = u - \boldsymbol{M} \cdot \boldsymbol{B} \tag{7.10}$$

As per the Legendre transform, $h_m(s, \{n_A\}, \boldsymbol{B})$ is a function of $\boldsymbol{B}$ and its differential reads,

$$dh_m(s, \{n_A\}, \boldsymbol{B}) = Tds + \sum_A \mu_A dn_A - \boldsymbol{M} \cdot d\boldsymbol{B} \tag{7.11}$$

By mere inspection of the differential, we draw the definition of magnetization as,

$$\boldsymbol{M} = -\frac{\partial h_m}{\partial \boldsymbol{B}} \tag{7.12}$$

According to the consideration that led to (7.9), we would like to calculate the derivative of $h_m(s, \{n_A\}, \boldsymbol{B})$ with respect to $\boldsymbol{M}$, which is the conjugate of the state variable $\boldsymbol{B}$.

A similar problem occurs when working with the state function enthalpy $H(S, p)$, which is a function of the entropy S and the pressure p of the system, and we wish to calculate the derivative of the enthalpy with respect to the temperature T, which is the conjugate of S. To do this clearly, we need to consider S as a function of T and p, thus writing $H(S(T, p), p)$. Then,

$$\frac{\partial H(S(T, p), p)}{\partial T} = \frac{\partial H(S, p)}{\partial S} \frac{\partial S(T, V)}{\partial T} = T K_p(T, p) \tag{7.13}$$

where $K_p(T, p)$ is the entropy capacity at constant pressure.[1]

Let us now proceed in a similar fashion for the magnetic enthalpy, which we write as,

$$h_m(s, n_a, \boldsymbol{B}) = h_m(s, n_a, \boldsymbol{B}(s, n_a, \boldsymbol{M})) \tag{7.14}$$

Then,

$$\frac{\partial h_m}{\partial \boldsymbol{M}} = \frac{\partial h_m(s, \{n_a\}, \boldsymbol{B}(s, \{n_a\}, \boldsymbol{M}))}{\partial \boldsymbol{M}} = \frac{\partial h_m}{\partial \boldsymbol{B}} \frac{\partial \boldsymbol{B}}{\partial \boldsymbol{M}} \tag{7.15}$$

Let us consider a linear medium and the tensorial susceptibility χ defined as,[170]

$$\boldsymbol{M} = \chi \boldsymbol{B} \Leftrightarrow \chi = \frac{\partial \boldsymbol{M}}{\partial \boldsymbol{B}} \tag{7.16}$$

Given this definition and equation (7.12). the equation (7.15) gives us,

$$\frac{\partial h_m}{\partial \boldsymbol{M}} = -\chi^{-1} \boldsymbol{M} = -\boldsymbol{B} \tag{7.17}$$

Therefore, the magnetic induction $\boldsymbol{B}$ can be calculated as,

$$\boldsymbol{B} = -\frac{\partial h_m}{\partial \boldsymbol{M}} \tag{7.18}$$

---

[1] see [376], §5.4.

In general, when studying magnetization dynamics, we do not need to consider the variables $s, \{n_a\}$. Then, we will write the magnetic enthalpy h_m, considered itself as a function of $\boldsymbol{B}(s, \{n_a\}, \boldsymbol{M})$ as simply $V(\boldsymbol{M})$, i.e.,

$$h_m\left(s, \{n_a\}, \boldsymbol{B}(s, \{n_a\}, \boldsymbol{M})\right) \to V(\boldsymbol{M}) \tag{7.19}$$

Thus, $V(\boldsymbol{M})$ is a short-hand notation that represents the magnetic enthalpy, a state function, in which the state field $\boldsymbol{B}$ is considered as a function of the magnetization $\boldsymbol{M}$. Then, (7.18) in this notation reads,

$$\boldsymbol{B} = -\frac{\partial V(\boldsymbol{M})}{\partial \boldsymbol{M}} \tag{7.20}$$

The definition (7.20) of $\boldsymbol{B}$ is the one that we need when calculating an effective field to be used in the Landau-Lifshitz equation.

In §7.2, we examined the semi-classical picture according to which the (classical) magnetization is proportional to the angular momentum. Both the angular momentum and the magnetization precess about the effective magnetic induction field $\boldsymbol{B}_{\text{eff}}$ due to the torque $\boldsymbol{M} \times \boldsymbol{B}_{\text{eff}}$. Thus, we have,

$$\frac{d\boldsymbol{M}}{dt} = \gamma\,\boldsymbol{M} \times \boldsymbol{B}_{\text{eff}} \tag{7.21}$$

7.4   THE LANDAU-LIFSHITZ-GILBERT EQUATION

Equation (7.21) does not account for any relaxation mechanism. It can be modified to include a viscosity term, called a Gilbert damping term. The magnetization evolution is then given by the the **Landau-Lifshitz-Gilbert equation** (LLG) for the magnetization evolution:

$$\frac{d\boldsymbol{M}}{dt} = \gamma\boldsymbol{M} \times \boldsymbol{B}_{\text{eff}} + \gamma\boldsymbol{M} \times \left(-\eta\frac{d\boldsymbol{M}}{dt}\right) \tag{7.22}$$

where γ is the gyromagnetic ratio of the electron, and η is a form of viscosity coefficient. We can introduce a dimensionless coefficient, known as the α Gilbert damping coefficient, by writing the equation in the form,

$$\frac{d\boldsymbol{M}}{dt} = \gamma\boldsymbol{M} \times \boldsymbol{B}_{\text{eff}} - \frac{\alpha}{M_s}\boldsymbol{M} \times \left(\frac{d\boldsymbol{M}}{dt}\right) \tag{7.23}$$

where α is the dimensionless parameter,

$$\alpha = \eta\gamma M_s \tag{7.24}$$

called the **Gilbert damping coefficient**. A typical value for α is 0.01, but experimental values vary greatly.[378,378] A value for α of 10^{-4} is considered very low. Derivation of the damping term from a variational approach can be found in [379].

The formula (7.20) for the effective field $\boldsymbol{B}_{\text{eff}}$ sometimes leads to confusion. In case of doubt as to how to operate such a derivative, it is best to return to the basic definition of this mathematical operation. In general, if we want the ith component of the effective field, we are looking for B_i, which obeys:

$$-B_i\delta M_i = V(\boldsymbol{M} + \delta M_i\hat{\boldsymbol{e}}_i) - V(\boldsymbol{M}) \tag{7.25}$$

We will consider the case in which the magnetic potential V is given in spherical coordinates: $V(\theta, \varphi)$. In this coordinate system, what we want is, for example:

$$-B_\theta\delta M_\theta = V(\theta + d\theta, \varphi) - V(\theta, \varphi) = \frac{\partial V}{\partial \theta}d\theta \tag{7.26}$$

The quantity $\delta M_\theta/M_s$ is an infinitesimal segment on the unit sphere, along the line coordinate on which φ is fixed. Its length is simply $d\theta$. Therefore, we have,

$$B_\theta = -\frac{1}{M_s}\frac{\partial V}{\partial \theta} \tag{7.27}$$

Likewise, we find,

$$-B_\varphi \delta M_\varphi = V(\theta, \varphi + d\varphi) - V(\theta, \varphi) = \frac{\partial V}{\partial \varphi}d\varphi \tag{7.28}$$

This time, the infinitesimal length on the unit sphere is $\delta M_\varphi/M_s = \sin\theta\, d\varphi$, so that,

$$B_\theta = -\frac{1}{M_s \sin\theta}\frac{\partial V}{\partial \varphi} \tag{7.29}$$

We recognize the expressions for the gradient in spherical coordinates, projected on the unit vectors tangent to the line of coordinates,

$$\boldsymbol{\nabla} = \left(\frac{\partial}{\partial \theta}, \frac{1}{\sin\theta}\frac{\partial}{\partial \varphi}\right) \tag{7.30}$$

Therefore, we can write the effective field as,

$$\boldsymbol{B}_{\text{eff}} = -\frac{\partial V}{\partial \boldsymbol{M}} = -\frac{1}{M_s}\boldsymbol{\nabla}V \tag{7.31}$$

The Gilbert form of the evolution equation can be recast into a ***Landau-Lifshitz equation*** by solving it for $d\boldsymbol{M}/dt$ as follows. First, we see that the equation of evolution (7.23) for $\boldsymbol{M}$ is written in a form that ensures that the modulus M_s is constant. This is often referred to as the ***micromagnetic hypothesis***. The constant-modulus hypothesis implies:

$$0 - \frac{dM^2}{dt} = 2\boldsymbol{M}\cdot\frac{d\boldsymbol{M}}{dt} \tag{7.32}$$

It indicates that $d\boldsymbol{M}/dt$ must be perpendicular to $\boldsymbol{M}$. This is ensured in the equation of evolution (7.23) by having all terms of the form $\boldsymbol{M}\times$.

By operating $\boldsymbol{M}\times$ to the left of the Gilbert equation (7.22), we obtain:

$$\boldsymbol{M}\times\frac{d\boldsymbol{M}}{dt} = \gamma\boldsymbol{M}\times(\boldsymbol{M}\times\boldsymbol{B}_{\text{eff}}) + \gamma(-\eta)\,\boldsymbol{M}\times(\boldsymbol{M}\times\frac{d\boldsymbol{M}}{dt}) \tag{7.33}$$

Substituting this expression in the Gilbert damping term of (7.22) and using the conversion of double cross-products into scalar products,[2] we find:

$$\frac{d\boldsymbol{M}}{dt} = \frac{\gamma}{(1+\eta^2\gamma^2 M_s^2)}\boldsymbol{M}\times\boldsymbol{B}_{\text{eff}} - \frac{\eta\gamma^2}{(1+\eta^2\gamma^2 M_s^2)}\boldsymbol{M}\times(\boldsymbol{M}\times\boldsymbol{B}_{\text{eff}}) \tag{7.34}$$

It is customary to define h' and g' by,

$$g' = \frac{\gamma}{(1+\alpha^2)M_s} \qquad h' = \frac{\alpha\gamma}{(1+\alpha^2)M_s} = \alpha g' \tag{7.35}$$

We can then write an equation of evolution for the unit vector $\boldsymbol{u} = \boldsymbol{M}/M_s$,

$$\frac{d\boldsymbol{u}}{dt} = -g'\boldsymbol{u}\times\boldsymbol{\nabla}V + h'\boldsymbol{u}\times(\mathbf{u}\times\boldsymbol{\nabla}V) \tag{7.36}$$

---

[2]$\boldsymbol{a}\times(\boldsymbol{b}\times\boldsymbol{c}) = (\boldsymbol{a}\cdot\boldsymbol{c})\boldsymbol{b} - (\boldsymbol{a}\cdot\boldsymbol{b})\boldsymbol{c}$

It may be helpful to remember that g' is essentially the precession term, whereas h' represents the dissipative term. In this equation, we can make use of the gradient expression for the effective field, as derived above. Therefore, the equation of motion in spherical coordinates is given by,

$$
\begin{aligned}
\frac{d\mathbf{u}}{dt} &= -g'\hat{\mathbf{r}} \times \left(\frac{\partial V}{\partial \theta}\hat{\boldsymbol{\theta}} + \frac{1}{\sin\theta}\frac{\partial V}{\partial \varphi}\hat{\boldsymbol{\varphi}} \right) + h'\hat{\mathbf{r}} \times \left(\hat{\mathbf{r}} \times \left(\frac{\partial V}{\partial \theta}\hat{\boldsymbol{\theta}} + \frac{1}{\sin\theta}\frac{\partial V}{\partial \varphi}\hat{\boldsymbol{\varphi}} \right) \right) \\
&= -g'\frac{\partial V}{\partial \theta}\hat{\boldsymbol{\varphi}} + g'\frac{1}{\sin\theta}\frac{\partial V}{\partial \varphi}\hat{\boldsymbol{\theta}} + h'\hat{\mathbf{r}} \times \left(\frac{\partial V}{\partial \theta}\hat{\boldsymbol{\varphi}} - \frac{1}{\sin\theta}\frac{\partial V}{\partial \varphi}\hat{\boldsymbol{\theta}} \right) \\
&= \left(-h'\frac{\partial V}{\partial \theta} + g'\frac{1}{\sin\theta}\frac{\partial V}{\partial \varphi} \right)\hat{\boldsymbol{\theta}} + \left(-g'\frac{\partial V}{\partial \theta} - h'\frac{1}{\sin\theta}\frac{\partial V}{\partial \varphi} \right)\hat{\boldsymbol{\varphi}}
\end{aligned}
\tag{7.37}
$$

We will use this last result in §14.2, when establishing the Fokker-Plank equation for the evolution of magnetization coupled to a thermal bath.

7.4.1 SUSCEPTIBILITY AND UNITS

We clarify here the definition of a magnetic susceptibility. As we have seen, the proper way to express the LLG equation is to write that the torque acting on the magnetization is of the form $\gamma \boldsymbol{M} \times \boldsymbol{B}$. Then, if a susceptibility is derived from the LLG equation, it naturally leads to a susceptibility $\chi^{(B)}$ defined as,

$$
\boldsymbol{M} = \boldsymbol{\chi}^{(\mathrm{B})}\boldsymbol{B}
\tag{7.38}
$$

where $\boldsymbol{B}$ is in this case the microwave perturbation applied to the magnetization, $\boldsymbol{\chi}^{(\mathrm{B})}$ designates a tensor, and we write $\chi^{(\mathrm{B})}$ when we want to designate a scalar susceptibility. When the LLG equation is used to determine a susceptibility (see §17.3 and §18.3), only $\mu_0\boldsymbol{H}$ contributes because $\boldsymbol{B}$ appears in the LLG equation (7.34) in the form $\boldsymbol{M} \times \boldsymbol{B}$, which is equal to $\boldsymbol{M} \times \mu_0\boldsymbol{H}$. Hence, we have, $\boldsymbol{M} = \boldsymbol{\chi}^{(\mathrm{B})}\mu_0\boldsymbol{H}$. Therefore, we have the equality,

$$
\boldsymbol{\chi} = \mu_0\boldsymbol{\chi}^{(\mathrm{B})} \qquad (\mathrm{LLG})
\tag{7.39}
$$

In general, because $\boldsymbol{B} = \mu_0(\boldsymbol{H} + \boldsymbol{M})$, we have,

$$
\boldsymbol{M} = \boldsymbol{\chi}\boldsymbol{H} = \boldsymbol{\chi}^{(\mathrm{B})}\boldsymbol{B} = \boldsymbol{\chi}^{(\mathrm{B})}\mu_0\boldsymbol{H}(1 + \boldsymbol{\chi})
\tag{7.40}
$$

Thus, when the susceptibility is very small, we have approximately,

$$
\boldsymbol{\chi} = \mu_0\boldsymbol{\chi}^{(\mathrm{B})} \qquad (\chi \ll 1)
\tag{7.41}
$$

7.5 EXCHANGE SPIN WAVES

Let us now refine the description of excitations in a ferromagnet by adding a term to the magnetization evolution equation (7.21) that accounts for exchange couplings. It is possible to define a magnetic induction field, in the sense of (7.20), that accounts for changes in exchange energy when the magnetization field is distorted.[183] In order to define this field, we need to work out a continuum description of exchange energy. We start with the Heisenberg exchange Hamiltonian (2.56), written as,

$$
E_{ex} = -\sum_{i,j} J_{ij}\hat{\boldsymbol{S}}_i \cdot \hat{\boldsymbol{S}}_j = -JS^2 \sum_{\mathrm{neighbors}} \cos\phi_{ij}
\tag{7.42}
$$

where J (>0 for a ferromagnet) has units of energy and $\hat{S}_z|m\rangle = m|m\rangle$. The angles ϕ_{ij} are

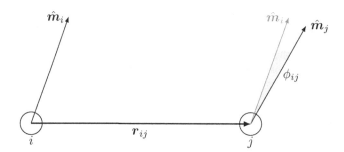

Figure 7.2 Nearest neighbor atomic sites, unit vectors $\hat{\boldsymbol{m}}_i$ parallel to the local magnetization $\boldsymbol{M}(\boldsymbol{r}_i)$ at sites $\boldsymbol{r}_i$.

defined using the unit vector $\hat{\boldsymbol{m}}_i$ on each site i, where the unit vector $\hat{\boldsymbol{m}}_i$ is parallel to the local magnetization $\boldsymbol{M}(\boldsymbol{r}_i)$ (Fig. 7.2).

In going to the continuous limit, it is assumed that the angles ϕ_{ij} are small. Then, we can write,

$$\phi_{ij}^2 = |\boldsymbol{\delta m}|^2 \tag{7.43}$$

where

$$\boldsymbol{\delta m} = \begin{pmatrix} m_x(\boldsymbol{r}_i + \boldsymbol{r}_{ij}) - m_x(\boldsymbol{r}_i) \\ m_y(\boldsymbol{r}_i + \boldsymbol{r}_{ij}) - m_y(\boldsymbol{r}_i) \\ m_z(\boldsymbol{r}_i + \boldsymbol{r}_{ij}) - m_z(\boldsymbol{r}_i) \end{pmatrix} = \begin{pmatrix} \frac{\partial m_x}{\partial x}\Delta x + \frac{\partial m_x}{\partial y}\Delta y + \frac{\partial m_x}{\partial z}\Delta z \\ \frac{\partial m_y}{\partial x}\Delta x + \frac{\partial m_y}{\partial y}\Delta y + \frac{\partial m_y}{\partial z}\Delta z \\ \frac{\partial m_z}{\partial x}\Delta x + \frac{\partial m_z}{\partial y}\Delta y + \frac{\partial m_z}{\partial z}\Delta z \end{pmatrix} \tag{7.44}$$

Thus, $\boldsymbol{\delta m}$ can be written as,

$$\boldsymbol{\delta m} = \boldsymbol{r}_{ij} \cdot \begin{pmatrix} \boldsymbol{\nabla} m_x \\ \boldsymbol{\nabla} m_y \\ \boldsymbol{\nabla} m_z \end{pmatrix} \tag{7.45}$$

When we consider the particular case $\boldsymbol{r}_{ij} = a\hat{\boldsymbol{x}}$, we have,

$$\phi_{ij}^2 = \boldsymbol{\delta m}^2 = a^2[(\partial_x m_x)^2 + (\partial_x m_y)^2 + (\partial_x m_z)^2] \tag{7.46}$$

Let us now consider a cubic lattice, with nearest neighbors at a distance a from one another. There are 6 neighbors and the sum in (7.42) gives,

$$\sum_{\text{neighbours}} \cos \phi_{ij} \approx 6 - \frac{1}{2} \sum_{\text{neighbours}} \phi_{ij}^2$$

$$
\begin{aligned}
\sum_{\text{neighbours}} \phi_{ij}^2 &= a^2[(\partial_x m_x)^2 + (\partial_x m_y)^2 + (\partial_x m_z)^2] \\
&+ a^2[(\partial_y m_x)^2 + (\partial_y m_y)^2 + (\partial_y m_z)^2] \\
&+ a^2[(\partial_z m_x)^2 + (\partial_z m_y)^2 + (\partial_z m_z)^2] \\
&= a^2[(\boldsymbol{\nabla} m_x)^2 + (\boldsymbol{\nabla} m_y)^2 + (\boldsymbol{\nabla} m_z)^2]
\end{aligned} \tag{7.47}
$$

The expansion of the cosine leaves a constant self energy term and a variation of energy due to the distortion of the magnetization field, which is now treated in the continuum limit. This is the contribution of the exchange energy that we need to consider in order to apply (7.20). The energy density contribution due to exchange is therefore,

$$V(\hat{\boldsymbol{m}}(\boldsymbol{r})) = \frac{JS^2}{2a}\left((\boldsymbol{\nabla} m_x)^2 + (\boldsymbol{\nabla} m_y)^2 + (\boldsymbol{\nabla} m_z)^2\right) \tag{7.48}$$

Deriving this term with respect to M as in (7.20) yields the exchange field. Here, we have an example where a mere derivation with respect to the variable M is not possible. We need to think in terms of functional derivatives, as in (7.25). Hence, let us calculate, for example, the change of energy over a finite volume of the energy density when the magnetization varies by δM in the x direction corresponding to a reorientation of $M/M_s = \delta m \hat{x}$:

$$
\begin{aligned}
-\int_V d^3 r\, B_x\, \delta M &= \int_V d^3 r\, \delta V = \int_V d^3 r\, \Big(V(m + \delta m \hat{x}) - V(m) \Big) \\
&= \frac{JS^2}{2a} \int_V d^3 r\, \Big([\boldsymbol{\nabla}(m_x + \delta m)]^2 - [\boldsymbol{\nabla}(m_x)]^2 \Big) \\
&= \frac{JS^2}{a} \int_V d^3 r\, \boldsymbol{\nabla}(m_x) \cdot \boldsymbol{\nabla}(\delta m) \\
&= \frac{JS^2}{a} \Big[\boldsymbol{\nabla}(m_x)\, \delta m \Big]_{\partial V} - \frac{JS^2}{a} \int_V d^3 r\, \boldsymbol{\nabla}^2(m_x)\, \delta m
\end{aligned}
\tag{7.49}
$$

The integration by parts is done for the x coordinate.[3] It produces two terms: an integral over the surface ∂V of the volume V, which is zero because δM is assumed to be zero at the surface; and a volume integral, which we can use to infer the volume quantity $B_x \delta M$. We can proceed likewise in the y and z directions. Therefore, the exchange field vector is of the form,

$$
\boldsymbol{B}_{ex} = D \boldsymbol{\nabla}^2 \boldsymbol{M}
\tag{7.50}
$$

Therefore, the Landau-Lifshitz equation of evolution for the magnetization in case of distortion of the magnetization field has the following form:

$$
\frac{\partial \boldsymbol{M}}{\partial t} = \gamma \boldsymbol{M} \times \boldsymbol{B} + D \boldsymbol{M} \times \boldsymbol{\nabla}^2 \boldsymbol{M}
\tag{7.51}
$$

Eigenmodes

We now assume that we have an infinite slab with the direction of the magnetic anisotropy axis and of the applied field normal to the slab. We set the z-axis normal to the slab and consider small deviations from equilibrium, as we did for the magnetostatic modes. We expect eigenmodes of the form:

$$
\begin{pmatrix} m_x e^{i\omega t} e^{\pm ikz} \\ m_y e^{i\omega t} e^{\pm ikz} \\ M_0 \end{pmatrix}
\tag{7.52}
$$

Substituting in the equation of motion (7.51), keeping linear terms only, and assuming an internal effective field:

$$
\begin{pmatrix} 0 \\ 0 \\ B_i \end{pmatrix}
\tag{7.53}
$$

we find the eigenvalue equation:

$$
\omega^2 = \big(\gamma B_i + D k^2 M_s \big)^2
\tag{7.54}
$$

---

[3]The integration by parts can be thought of as a generalized Gauss theorem ([380], vol. 1, section IV.3.4 after Eq. 23).

When B_i is small compared to the exchange field, the dispersion relation reads $\omega = Ak^2$. $A = DM_s$ is of the order of magnitude of 5 meV/nm^{2}or 2 10^{-40} J/m^2. In practice, it is very difficult to excite short wave length (large k) spin waves.[381, 382]

7.6 MAGNONS

It can be shown that a classical Hamiltonian description of waves as small excitations from equilibrium can be reformulated in terms of harmonic oscillators.[383] Likewise, the Holstein-Primakoff method, which is the subject of this section, transforms a Heisenberg exchange Hamiltonian into a Hamiltonian of harmonic oscillators. Before engaging in a formal development, we note that a general theorem states that the smallest excitations of a continuous system are waves of long wave length.

7.6.1 GOLDSTONE MODES

In his famous 1972 essay on the scientific merit of condensed matter physics,[384] Anderson invited us to be surprised that, associated with a Heisenberg Hamiltonian that has full rotational symmetry,[385] we are comfortable with the habit of stating that the ground state is ferromagnetic. This is an example of a ***broken symmetry***. The symmetry operation here is discrete in nature: it is a time-reversal symmetry. Above the Curie temperature, the system (taken as a whole) has randomly oriented magnets, so reversing time changes nothing to the macroscopic picture of its state. A simpler example of a broken symmetry is the translational symmetry of a gas condensing into a crystal upon cooling to a low enough temperature. The high-temperature phase has a continuous symmetry; the same is not true for the low-temperature phase.

Goldstone theorem states that when a continuous symmetry is broken, it is possible to produce long-wavelength excitations at an infinitely small cost in energy.[386] For example, acoustic phonons are the Goldstone modes of solids. In an isotropic ferromagnet, they are the spin waves that may have vanishingly small wave numbers and no energy gap between the ground state and these first excitations.[233, 387, 388]

7.6.2 ONE-MAGNON EIGENSTATES

Let us now find the low-lying excited state of a Heisenberg ferromagnet. Following Ashcroft and Mermin's famous textbook,[72] we consider a ferromagnetic crystal that can be described by an exchange Hamiltonian of the form,

$$\mathcal{H} = -\sum_{n.n.} J(\boldsymbol{R} - \boldsymbol{R}')\hat{S}(\boldsymbol{R}) \cdot \hat{S}(\boldsymbol{R}') - g\mu_B B \sum_i S_{z\,i} \qquad (7.55)$$

for a field B along the $+z$ axis. This Hamiltonian could describe a ferromagnetic insulator with localized moments.[176] The ground state of this ferromagnet has all spins pointing in the $+z$ direction. We denote this ground state $|0\rangle$. The dynamics of spin waves of short wavelengths is dominated by the exchange interaction and we can expect the Hamiltonian (7.55) to account reasonably well for the short wave length limit. In spintronics however, the electrical excitation of spin waves typically leads to long wave length magnons, for which dipolar effects play an important role. The magnon dynamics that includes dipolar coupling is explained in §7.6.3.

Let us now consider states where the angular momentum at position $\boldsymbol{R}$ has its z component decreased by one unit, that is:

$$|\boldsymbol{R}\rangle = \frac{1}{\sqrt{2S}}\hat{S}_-(\boldsymbol{R})\,|0\rangle \qquad (7.56)$$

where $\hat{S}_-(\boldsymbol{R})$ is the lowering operator for the spin at $\boldsymbol{R}$. These $|\boldsymbol{R}\rangle$ states are not eigenstates of $\mathcal{H}$ since,

$$\mathcal{H}\,|\boldsymbol{R}\rangle = E_0\,|\boldsymbol{R}\rangle + g\mu_B B\,|\boldsymbol{R}\rangle + S\sum_{\boldsymbol{R}'} J(\boldsymbol{R}-\boldsymbol{R}')\left(|\boldsymbol{R}\rangle - |\boldsymbol{R}'\rangle\right) \qquad (7.57)$$

Here, $E_0 = -Ng\mu_B BS - \frac{1}{2}NZJS^2$ and Z is the number of nearest neighbors. This result is found using,

$$\begin{aligned}
\hat{S}_x(\boldsymbol{R})\hat{S}_x(\boldsymbol{R}) + \hat{S}_y(\boldsymbol{R})\hat{S}_y(\boldsymbol{R}) &= \frac{1}{2}\left(\hat{S}_+(\boldsymbol{R})\hat{S}_-(\boldsymbol{R}) + \hat{S}_-(\boldsymbol{R})\hat{S}_+(\boldsymbol{R})\right)\\
\hat{S}_z(\boldsymbol{R}')\,|\boldsymbol{R}\rangle &= S\,|\boldsymbol{R}\rangle \qquad (\boldsymbol{R}\neq\boldsymbol{R}')\\
\hat{S}_+(\boldsymbol{R}')\,|\boldsymbol{R}\rangle &= 0 \qquad (\boldsymbol{R}\neq\boldsymbol{R}')\\
\hat{S}_-(\boldsymbol{R}')\hat{S}_+(\boldsymbol{R})\,|\boldsymbol{R}\rangle &= 2S\,|\boldsymbol{R}'\rangle
\end{aligned} \qquad (7.58)$$

The Hamiltonian $\mathcal{H}$ has the translational invariance of the lattice of the ferromagnet. In analogy with the more familiar case of phonons, let us consider the following linear combinations of $|\boldsymbol{R}'\rangle$ states:

$$|\boldsymbol{k}\rangle = \frac{1}{\sqrt{N}}\sum_{\boldsymbol{R}}\exp\left(i\boldsymbol{k}\cdot\boldsymbol{R}\right)|\boldsymbol{R}\rangle \qquad (7.59)$$

These states are eigenstates of $\mathcal{H}$ and $\boldsymbol{k}$ are vectors of the reciprocal lattice. We can verify that, indeed,

$$\mathcal{H}\,|\boldsymbol{k}\rangle = \left(E_0 + \hbar\omega(\boldsymbol{k})\right)|\boldsymbol{k}\rangle \qquad (7.60)$$

with

$$\hbar\omega(\boldsymbol{k}) = g\mu_B B + JS\sum_{\boldsymbol{\delta}}\left(1 - \cos\boldsymbol{k}\cdot\boldsymbol{\delta}\right) \qquad (7.61)$$

The vectors $\boldsymbol{\delta}$ connect one spin to its nearest neighbors. The order of magnitude of $\hbar\omega(\boldsymbol{k})$ is:

$$\hbar\omega(\boldsymbol{k}) = g\mu_B B + JSa^2 k^2 \qquad (7.62)$$

where $a = |\boldsymbol{\delta}|$.

The spin wave states $|\boldsymbol{k}\rangle$ have a simple physical interpretation. Since the $|\boldsymbol{k}\rangle$ states are superpositions of $|\boldsymbol{R}\rangle$ states that are fully polarized except for one spin, which has its magnetic number m_z reduced by one, the $|\boldsymbol{k}\rangle$ states have a magnetization $NS - 1$. Furthermore, the transverse spin correlation function defined by,

$$\hat{S}_\perp(\boldsymbol{R})\hat{S}_\perp(\boldsymbol{R}') = \left(\hat{S}_x(\boldsymbol{R})\hat{S}_x(\boldsymbol{R}') + \hat{S}_y(\boldsymbol{R})\hat{S}_y(\boldsymbol{R}')\right) \qquad (7.63)$$

has a simple expectation value in the $|\boldsymbol{k}\rangle$ states, namely:

$$\left\langle\boldsymbol{k}\left|\hat{S}_\perp(\boldsymbol{R})\hat{S}_\perp(\boldsymbol{R}')\right|\boldsymbol{k}\right\rangle = \frac{S}{N}\cos\left(\boldsymbol{k}\cdot(\boldsymbol{R}-\boldsymbol{R}')\right) \qquad (\boldsymbol{R}\neq\boldsymbol{R}') \qquad (7.64)$$

Hence, the transverse component of each spin $\hat{S}_\perp(\boldsymbol{R})$ has a non-vanishing expectation value. The expectation value of the angle between two such transverse spin components is given by $\boldsymbol{k}\cdot(\boldsymbol{R}-\boldsymbol{R}')$. The physical picture of these spin-wave states is consequently of the kind shown in Fig. 7.3.

This description can be extended to higher excitations by using coherent spin states.[389] The most commonly used method, however, is that of Holstein-Primakoff, described in section 7.6.3.

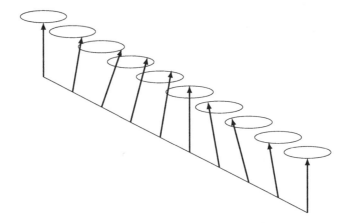

Figure 7.3 Schematics of a spin wave for a one-dimensional chain of spins, as deduced from (7.64). At each atomic site, the local spin precesses, forming the same cone angle, but the phase changes progressively from site to site, thus defining a wave for the spins.

7.6.3 HOLSTEIN-PRIMAKOFF METHOD

Let us consider a ferromagnet described by a Heisenberg exchange Hamiltonian:

$$\mathcal{H} = -J \sum_{\substack{i>j= \\ \text{n.n.}}} \hat{\boldsymbol{S}}_i \cdot \hat{\boldsymbol{S}}_j - g\mu_B B \sum_i \hat{S}_j^z \quad (J > 0) \tag{7.65}$$

Here, the exchange interaction is isotropic. We can write the scalar product as,

$$\hat{\boldsymbol{S}}_i \cdot \hat{\boldsymbol{S}}_j \equiv \hat{S}_i^x \hat{S}_j^x + \hat{S}_i^y \hat{S}_j^y + \hat{S}_i^z \hat{S}_j^z = \frac{1}{2}(\hat{S}_i^+ \hat{S}_j^- + \hat{S}_j^+ \hat{S}_i^-) + \hat{S}_i^z \hat{S}_j^z \tag{7.66}$$

In addition to this exchange interaction, we want to take into account long-range interactions, because we would like to be able to describe the long-wavelength limit of spin waves. The lowest order long-range interactions are dipolar in nature, given by:[170]

$$\mathcal{H}_d = \frac{1}{2} \sum_{i,j} \frac{\mu_0}{4\pi} \frac{\gamma^2}{r_{ij}^3} \left(\boldsymbol{S}_i \cdot \boldsymbol{S}_j - 3 \left(\boldsymbol{S}_i \cdot \hat{\boldsymbol{r}}_{ij} \right) \left(\boldsymbol{S}_j \cdot \hat{\boldsymbol{r}}_{ij} \right) \right) \tag{7.67}$$

Each term of the sum is the dipolar coupling energy of magnetic dipole moments $\gamma \boldsymbol{J}_i$ and $\gamma \boldsymbol{J}_i$, located at positions characterized by the separation given by the vector $\boldsymbol{r}_{ij}$, the distance r_{ij} and the unit vector $\hat{\boldsymbol{r}}_{ij} = \boldsymbol{r}_{ij}/r_{ij}$.

We note in passing that the dipolar sum (7.67) is a bilinear form that can be written as,

$$\mathcal{H}_d = \sum_{i<j} \boldsymbol{S}_i \cdot \mathsf{D}_{ij} \boldsymbol{S}_j \tag{7.68}$$

The matrices D_{ij} are symmetric, as can be expected readily by looking at (7.67). Indirect couplings, such as (5.10), can also be expressed using a symmetric matrix. Instead, the DM coupling (2.69) is expressed using D_{ij} matrices that are antisymmetric.

Let us now define the ***Holstein-Primakoff*** representation of spin operators. This will allow us to describe spin-waves as harmonic oscillators. The following development is a brief account of Mattis' *Quantum Theory of Magnetism*,[176]. Mattis himself followed the review on spin waves by Van Vleck and Van Kranendonk.[390] We designate the eigenstate of $S_{z,i}$ by,

$$|n_i\rangle \quad n_i = 0,...,2s+1 \tag{7.69}$$

We can think of n_i as numbering the eigenstates of $S_{z,i}$ starting from $-s$, then $-s+1$, etc. until $+s$. There are 2s+1 such states. The spin operators have the following matrix elements in this basis:

$$< n_i|\hat{S}_{x,j}|n_i+1 > =< n_i+1|\hat{S}_{x,i}|n_i>^* = \frac{1}{2}\sqrt{(n_i+1)(2s-n_i)}$$

$$< n_i|\hat{S}_{y,i}|n_i+1 > =< n_i+1|\hat{S}_{y,i}|n_i>^* = -\frac{1}{2}i\sqrt{(n_i+1)(2s-n_i)} \tag{7.70}$$

$$< n_i|\hat{S}_{z,i}|n_i > = s - n_i \quad (0 \le n_i \le 2s, \hbar = 1)$$

By way of comparison, consider now the eigenstates $|n_i\rangle$ of the Hamiltonian of a harmonic oscillator, for which,

$$\left(\frac{p_i^2}{2m} + \frac{1}{2}m\omega^2 x_i^2 - \frac{1}{2}\hbar\omega\right)|n_i> = n_i\hbar\omega|n_i> \tag{7.71}$$

The matrix elements of the position and momentum operators are:

$$< n_i|x_i|n_i+1 > =< n_i+1|x_i|n>^* = \sqrt{\frac{\hbar}{2m\omega}(n_i+1)}$$

$$< n_i|p_i|n_i+1 > =< n_i+1|p_i|n_i>^* = -\sqrt{\frac{\hbar m\omega}{2}(n_i+1)} \tag{7.72}$$

Clearly, these matrix elements will approach those of the angular momentum if we rescale the position and momentum operators according to:

$$Q_i = x_i\sqrt{\frac{m\omega}{\hbar}} \quad P_i = p_i\sqrt{\frac{1}{\hbar m\omega}} \quad [P_i, Q_j] = \delta_{ij}\frac{1}{i} \tag{7.73}$$

Indeed, we find,

$$< n_i|Q_i\sqrt{s}|n_i+1 >=< n_i+1|Q_i\sqrt{s}|n_i>^* = \frac{1}{2}\sqrt{(n_i+1)2s}$$

$$< n_i|P_i\sqrt{s}|n_i+1 >=< n_i+1|P_i\sqrt{s}|n_i>^* = -\frac{1}{2}i\sqrt{(n_i+1)2s} \tag{7.74}$$

$$< n_i|s - \frac{1}{2}(P_i^2 + Q_i^2 - 1)|n_i >= s - n_i$$

We will limit ourselves, from now on, to small values of n_i. In view of the theory of spin waves of §7.6.2, when taking small values n_i only, we are limiting ourselves to small deviations at any spin site from the ground state value $n_i = 0$. In this limit, one has,

$$\hat{S}_i^x = Q_i\sqrt{s}, \quad \hat{S}_i^y = P_i\sqrt{s}, \quad \hat{S}_i^z = s - \frac{1}{2}(P_i^2 + Q_i^2 - 1) \tag{7.75}$$

We proceed by expressing the exchange Hamiltonian in this representation, which is the Holstein-Primakoff representation of angular momentum. We substitute in the Hamiltonian these expressions for the angular momentum. As we want to look at the lowest order effect, we disregard in this substitution terms that are cubic and quartic in these operators. Hence, the exchange Hamiltonian becomes:

$$\mathcal{H}_{lin} = E_0 + g\mu_B(B + B_0)\sum_i \frac{1}{2}(P_i^2 + Q_i^2 - 1) - Js\sum_{n.n.}(P_iP_j + Q_iQ_j) \tag{7.76}$$

where,

$$B_0 = \frac{Jsz}{g\mu_B} \quad E_0 = -Ng\mu_B Bs - \frac{1}{2}NzJs^2 \tag{7.77}$$

We can identify B_0 with the **Weiss molecular field**.[79] E_0 is the energy of the system when all the spins are aligned in the applied field. We have thus obtained a Hamiltonian that bears some resemblance to that of a harmonic oscillator, except for the last terms. We make the expansion in plane waves in analogy to what we did in §7.6.2:

$$Q_i = \frac{1}{\sqrt{N}} \sum_k e^{ik \cdot R_i} Q_k \quad P_i = \frac{1}{\sqrt{N}} \sum_k e^{ik \cdot R_i} P_k \tag{7.78}$$

Note that,

$$Q_k^* = Q_{-k} \quad [Q_{k'}, P_k^*] = i\delta_{kk'} \tag{7.79}$$

The vector k belongs to the reciprocal lattice. This plane-wave expansion produces an exchange Hamiltonian of the form:

$$\mathcal{H}_{lin} = \sum_k \frac{1}{2}(P_k^* P_k + Q_k^* Q_k - 1)\hbar\omega(k) + E_0 = \sum_k n_k \hbar\omega(k) + E_0 \tag{7.80}$$

where,

$$n_k \equiv \frac{1}{2}(P_k^* P_k + Q_k^* Q_k - 1)$$
$$\hbar\omega(k) = g\mu_B B + Js\sum_\delta (1 - \cos k \cdot \delta) \cong g\mu_B B + Jsa^2 k^2 + 0(k^4) \tag{7.81}$$

Thus, we have decomposed our Heisenberg Hamiltonian into a sum of uncoupled harmonic oscillators. As V. L. Safonov puts it, "magnons are harmonic oscillators in k-space."[391]

Long-wavelength limit

We now address the issue of the effect on the spin-wave spectrum of the long range couplings characterized in (7.67) by $\mathcal{H}_d$. The direction cosines of a spin i coupled to a spin j are given by:

$$\frac{r_{ij}}{r_{ij}} = (\alpha_{ij}, \beta_{ii}, \gamma_{ij}) \tag{7.82}$$

The dipolar Hamiltonian can be decomposed into three parts:[392]

$$\mathcal{H}_{d,0} = \frac{1}{2}\frac{\mu_0}{4\pi}\sum_i \sum_j \frac{g^2\mu_B^2 s^2}{r_{ij}^3}(1 - 3\gamma_{ij}^2)$$

$$\mathcal{H}_{d,1} = -3\frac{\mu_0}{4\pi}\sum_i \sum_j \frac{g^2\mu_B^2 s^2}{r_{ij}^3}(\alpha_{ij}\gamma_{ij}Q_j + \beta_{ij}\gamma_{ij}P_j)(s^{-1/2}) \tag{7.83}$$

$$\mathcal{H}_{d,2} = \frac{1}{2}\frac{\mu_0}{4\pi}\sum_i \sum_j \frac{g^2\mu_B^2 s^2}{r_{ij}^3}\big[(1 - 3\alpha_{ij}^2)Q_iQ_j - 3\alpha_{ij}\beta_{ij}(Q_iP_j + P_iQ_j)$$
$$+(1 - 3\beta_{ij}^2)P_iP_j - (1 - 3\gamma_{ij}^2)(P_i^2 + Q_i^2 - 1)\big]$$

The first term is associated with the **demagnetizing field**. For an ellipsoid with a field in the z direction, the demagnetizing coefficient is given by:

$$N_z = \frac{a^3}{4\pi}\sum_j \frac{1}{r_{ij}^3}(1 - 3\gamma_i^2) + \frac{1}{3} \tag{7.84}$$

where a^3 is the volume containing one local moment. The magnetization of the ellipsoid is uniform. It is given by:

$$M_0 = \frac{g\mu_B s}{a^3} \tag{7.85}$$

Thus, $\mathcal{H}_{d,0}$ can be written as,

$$\mathcal{H}_{d,0} = -\frac{1}{2}VM_0\mu_0\left(\frac{1}{3}M_0 - N_z M_0\right) \tag{7.86}$$

where the parenthesis is consistent with the dipolar field (4.12).

The second term, $\mathcal{H}_{d,1}$, cancels out for any point of symmetry, such as in a cubic crystal. It would be small anyway, as shown by Mattis.[176] We are left with determining in the long-wavelength limit $\mathcal{H}_{lin}$ and $\mathcal{H}_{d,2}$. For $\mathcal{H}_{d,2}$, we also develop in plane waves using (7.78). After several algebraic manipulations, the result can be expressed as,

$$\mathcal{H} \to \frac{1}{2}\sum_k [A(k)Q_k^*Q_k + B(k)P_k^*P_k + 2C(k)Q_k^*P_k] + \text{const} \tag{7.87}$$

where,

$$
\begin{aligned}
A(\boldsymbol{k}) &= \hbar\omega(\boldsymbol{k}) + A_{xx}(\boldsymbol{k}) - A_{zz}(0) \\
B(\boldsymbol{k}) &= \hbar\omega(\boldsymbol{k}) + A_{yy}(\boldsymbol{k}) - A_{zz}(0) \\
C(\boldsymbol{k}) &= A_{xy}(\boldsymbol{k})
\end{aligned} \tag{7.88}
$$

The coefficients $A_{ij}(\boldsymbol{k})$ are the dipolar sums,

$$
\begin{aligned}
A_{xx}(\boldsymbol{k}) &= \frac{s}{N}\sum_{i,j} D_{ij}(1-3\alpha_{ij}^2)e^{i\boldsymbol{k}\cdot\boldsymbol{R}_{ij}} \quad A_{yy}(\boldsymbol{k}) = \frac{s}{N}\sum_{i,j} D_{ij}(1-3\beta_{ij}^2)e^{i\boldsymbol{k}\cdot\boldsymbol{R}_{ij}} \\
A_{zz}(\boldsymbol{k}) &= \frac{s}{N}\sum_{i,j} D_{ij}(1-3\gamma_{ij}^2)e^{i\boldsymbol{k}\cdot\boldsymbol{R}_{ij}} \quad A_{xy}(\boldsymbol{k}) = \frac{-3s}{N}\sum_{i,j} D_{ij}\alpha_{ij}\beta_{ij}e^{i\boldsymbol{k}\cdot\boldsymbol{R}_{ij}}
\end{aligned} \tag{7.89}
$$

Here, $\omega(\boldsymbol{k})$ is the dispersion relation (7.81), and α_{ij} are the direction cosines of the i–j-segments with respect to the x, y, and z axes.

Size effects are to be expected in these dipolar sums.[393] For example, for the case of a sphere of radius R, at a position near the centre of the crystal, we have:

$$A_{xx}(k) \cong \left(1 - 3\frac{k_x^2}{k^2}\right)\left[1 - \frac{3j_1(kR)}{kR}\right] \tag{7.90}$$

We can operate once again a canonical transformation to transform the Hamiltonian (7.87) into that of a sum of harmonic oscillators. The canonical transformation is of the form:

$$
\begin{aligned}
P_k' &= a_k P_k + b_k Q_k \qquad Q_k' = c_k Q_k + d_k P_k \\
[P_{k_1}', Q_{k_2}'] &= \frac{\delta_{k_1,k_2}}{i}
\end{aligned} \tag{7.91}
$$

If the dipolar sums (7.89) are known, the coefficients c_k and d_k can be determined and the Hamiltonian can be written as,

$$\mathcal{H} = \sum \frac{1}{2}(P_k'^*P_k' + Q_k'^*Q_k' - 1)\hbar\omega'(k) + \text{const} \tag{7.92}$$

After this transformation, the spin wave dispersion is given by,

$$\hbar\omega'(\boldsymbol{k}) = \sqrt{A(\boldsymbol{k})B(\boldsymbol{k}) - C^2(\boldsymbol{k})} \tag{7.93}$$

In a typical ferromagnetic resonance experiment, the rf field is fairly homogeneous and it can excite only the spin wave modes near k = 0. The resonance frequency is given by (7.93). In this limit, the $A_{xy}(\boldsymbol{k})$ terms can be neglected for the same reasons as $\mathcal{H}_{d,1}$. The other sums can be identified with demagnetizing factors, and the final result is the Kittel formula (17.8).

In an infinite medium, the dispersion relation for exchange spin waves takes the following form, [372,394]

$$\omega = \sqrt{(\omega_H + A\omega_M k^2)(\omega_H + A\omega_M k^2 + \omega_M \sin^2\theta_k)} \tag{7.94}$$

where A is the exchange constant and θ_k the angle between $\boldsymbol{M}_0$ and the spin wave momentum vector $\boldsymbol{k}$. This formula is modified when treating thin films, because boundary conditions play a critical role in spin waves.[395]

7.7 FURTHER READINGS

Damping

The LLG equation (7.23) includes a phenomenological model of damping that is also found in the thermodynamics of irreversible processes applied to magnetic systems.[396,397] Microscopic models for the relaxation of spin waves have been developed, based on magnon-phonon collisions.[398,399]. Relaxation by phonons is suppressed when the solid is so small that phonons are not excited, opening up the possibility of very low damping.[400]

Several authors have considered a generalized LLG equation (7.23) in which the Gilbert coefficient α in the damping term $\boldsymbol{M} \times (\alpha\, d\boldsymbol{M}/dt)$ is replaced by a tensor.[401,402] This tensor can account for a rotational anisotropy, which depends on the direction of $d\boldsymbol{M}/dt$, or an orientational anisotropy, which depends on the orientation of $\boldsymbol{M}$.[403] Furthermore, the damping mechanism can be non-local.[404] Typically, damping inside a ferromagnet may be due in part to the presence of a metal, such as Pt, at the surface of the ferromagnet. We will refer to this type of mechanism when introducing the concept of spin pumping in §17.6.

Spin waves

Some authors write the magnon dispersion relation in terms of an effective field:[99]

$$E(\boldsymbol{k}) = Dk^2 + g\mu_B B_{\text{eff}} \tag{7.95}$$

Dipolar terms can be cast into the formula for B_{eff}. For example, if the magnetization $\boldsymbol{M}_s$ is saturated parallel to the axis, the sample that has the form of a rod, and when the applied field $\boldsymbol{B}_E$ is also parallel to the rod, then $\boldsymbol{B}_{\text{eff}} = \boldsymbol{B}_E + \mu_0\frac{7}{15}\boldsymbol{M}_s$ was found to be a good approximation when considering the heat conductivity due to magnons.[405] If θ is the angle between the magnon wave vector $\boldsymbol{k}$ and the magnetization $\boldsymbol{M}_s$, then to first order in M_s/B_E:

$$E(\boldsymbol{k}) = Dq^2 + g\mu_B\left(B_E + \frac{1}{2}\mu_0 M_s\sin^2\theta_k\right) \tag{7.96}$$

The angular dependence of magnon dispersion was clearly observed by ferromagnetic resonance spectroscopy (Chapter 15) in patterned ferrite thin films.[406]

Dipolar spin waves were observed in artificial magnetic superlattices made of equally spaced ferromagnetic stripes in which the coupling between stripes was exclusively dipolar in nature.[407]

While the Holstein-Primakoff method is based on localized magnetic moments, let us point out that a local moment description of itinerant magnets was demonstrated, suggesting that the approach presented here may apply over a wide range of magnetic systems.[408,409] Spin wave dynamics can also be described using the Berry phase formalism which will be introduced in §11.[410]

Anisotropic spin wave propagation was observed in nanostructured thin films.[411] The damping anisotropy can be detected by measuring spin-wave spectra as a function of angle. In thin films of lanthanum strontium manganese oxide, an angular variation of the relaxation time by a factor of two was observed. The strong in-plane anisotropy was induced by the lattice mismatch with respect to the substrate.[412]

Spin waves and the DM interaction

Chiral interactions of various origins can influence spin wave propagation properties.[413] In §2.7, we saw that the **Dzyaloshinskii-Moriya interaction** (DMI) can give rise to spin canting. It also modifies the spin wave dispersion relation. In a bulk sample, when the wavevector is parallel to the equilibrium magnetization, the effect of the DMI on spin wave dispersion depends on the extent to which the spin waves are pure exchange spin waves, or spin waves with both dipolar and exchange interactions (Fig. 7.4).[414]

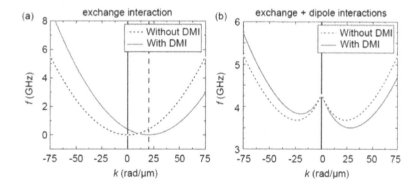

Figure 7.4   Magnon dispersion relation for a thin film with or without DM interaction, for the case of exchange spin waves (left) or spin waves where both exchange and dipolar interactions play a significant role. From [414].

Interfacial DMI leads to a spin wave dispersion which contains a term linear in k, when the wavevector is perpendicular to the equilibrium magnetization.[415] Since DMI can be linked to a second order effect of the spin-orbit interaction, the presence of a Rashba effect (§6.6) is an indicator of strong interfacial DMI.[416] In this case, we expect the electronic structure of the non-magnetic high-Z metal to be critical.[417] The **linear term** in the dispersion was verified experimentally, using Brillouin light scattering to measure dispersion,[418] or launching spin waves of a well-defined wave vector using structured antennas and electrical spin wave detection.[419]

It was possible to estimate the strength of an interfacial DMI by studying the spin wave propagation in thin ferromagnetic films.[420] DMI was found responsible for a magnon drift current observed in magnetic insulators capped with platinum.[421]

While the interfacial DMI can also be characterized by quasi-static measurements, [219] it also impacts domain wall velocity.[422] This is to be expected, since any inhomogeneity of the exchange interaction can affect domain wall velocity.[423]

Numerical simulations

In the absence of an external field, there might be a gap to the excitation of spin waves due to spin orbit interactions.[424,425] By way of *ab initio* calculations, the contribution of spin-orbit and magnetic anisotropy can be worked out.[199] Figure 7.5 presents the results of a numerical simulation [426] based on the spin-density functional method.[426] This yielded the magnons dispersion of iron in a very large range of wavevector values.

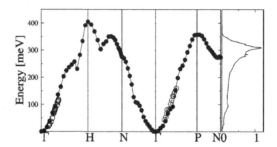

Figure 7.5 Magnon dispersion relation on high-symmetry lines and magnon density of states (in states/(meV*cell) $\times$ 10^{-2}) of bcc Fe. Solid circles mark calculated frequencies; lines are guide to the eye. For comparison, experimental data are also shown for pure Fe at 10K (dotted circles) and for Fe doped with 12% Si (circles) at room temperature, that were obtained by neutron scattering. From [426].

Section III

SPIN TRANSPORT

8 Thermodynamics of Spin-Dependent Transport

Lars Onsager (1903–1976)

With a degree in chemical engineering from the Norwegian Institute of Technology at Trondheim, L. Onsager spent some time working with P. Debye at the Swiss Federal Institute of Technology on modelling electrochemistry. Onsager was professor at Brown University, then at Yale University. He received the Nobel Prize in Chemistry in 1968 for his discovery of the reciprocal relations that bear his name.

Spintronics: A primer
J.-Ph. Ansermet (https://orcid.org/0000-0002-1307-4864)
Chapter 8. Thermodynamics of Spin-Dependent Transport

The thermodynamics of irreversible processes can be applied to spin-dependent transport by considering two types of charge carriers, corresponding to majority and minority spins. Combining this approach for scalar and vectorial quantities leads to a spin diffusion equation and a definition of a spin-diffusion length in terms of transport coefficients, including spin-mixing effects. A spin-dependent Ohm's law accounts for giant magnetoresistance in thin multilayers. The spin-diffusion equation implies spin accumulation at the interface between two thick magnetic layers and an associated spin-interface resistance.

DOI: 10.1201/9781003370017-8

8.1   HISTORICAL INTRODUCTION

The theory of irreversible processes was established by C. Eckart, L. Onsager, and I. Prigogine.[376] It allows a unified description of transport phenomena such as Ohm's law for isothermal charge transport, Fourier's law for heat diffusion, Fick's laws for mass transport in concentration gradients, and many cross-effects such as the Seebeck and Nernst effects.[375, 386, 427] For the sake of convenience, we recall the following definitions:

- *Ohm's law* expresses the proportionality of the charge current density and the gradient of the electrochemical potential, which is the sum of the chemical potential and the energy of the charge carriers in the electrostatic potential. If the proportionality is expressed with a tensor of rank two, a transverse effect, the Hall effect, is predicted.

- The *Seebeck effect* refers to the gradient of electrostatic potential which is induced by a temperature gradient. Thus, it is a volume effect. It is common to measure the voltage at the ends of a wire when one end is heated up and the other is kept cold. This experiment may give the impression that the Seebeck effect is due to a physical phenomenon taking place at the junction, which is generally not the case.

- If the proportionality between the temperature gradient and the gradient of the electrostatic potential is expressed with a tensor of rank two, the off-diagonal elements account for the *Nernst effect*. When applying a temperature gradient in one direction in a thin film, an in-plane voltage is induced at the edges of the sample perpendicular to the heat current induced by the temperature gradient (and accounted for by Fourier's law).

- When a charge current is applied across the junction between two different conductors, and the temperature is assumed to be the same throughout the junction, then a heat exchange is induced at the junction. This is the *Peltier effect*.

- The *Soret effect* refers to the relative change in the electrochemical potential of two substances when a mixture comprising both substances is subjected to a temperature gradient.

In 1971, Dyakonov and Perel pointed out that a current may induce spin polarization in semiconductors, owing to spin-orbit effects.[428]

In 1985, M. Johnson and R.H. Silsbee reported low temperature measurements on the resistance of a pair of ferromagnetic electrodes laid at close distance on an aluminum block, as a function of the relative orientation of the magnetization in both electrodes.[429] They analyzed their results using the thermodynamics of irreversible processes.[430]

Van Son, van Kempen, and Wyder also used a thermodynamic approach to argue that the inter-conversion of spin up and spin down during transport through the junction between a ferromagnetic metal and a non-ferromagnetic metal induces a spin dependence of the electrochemical potential.[431]

The material in this section relies on the *thermodynamics of irreversible processes*. To apply this formalism, we make the assumption that we have two "types" of electrons, in conformity with Mott's "two-current model" alluded to in §1.2.1.[69] The electrons are majority and minority spins in a ferromagnets. In a non-magnetic material, we are going to refer to "up" and "down" electrons. Thermodynamics of irreversible processes can conveniently treat the transport of two types of charge carriers and an entropy current. Thus, it is possible to include thermal effects in spin-dependent transport by using a "three-current model" that is derived from thermodynamics.[432, 433]

8.2 THERMODYNAMICS OF SPIN-DEPENDENT TRANSPORT

8.2.1 CONTINUITY OF SPIN AND CHARGE

Many aspects of spin-dependent transport can be captured by a thermodynamic description of transport. It is the case of giant magnetoresistance in the geometry where the current is driven perpendicular to the plane of the layers (for short, CPP-GMR). Thermodynamics can also account for the spin-diffusion length, which will be defined in this chapter. This thermodynamic picture relies on expressing a linear relationship between generalized currents and generalized forces.[376, 386, 434]

In the following, spin-dependent transport is modeled in terms of two matter currents, in other words, currents of two charge carriers, labeled with $(+)$ or $(-)$, depending on their orientation with a fixed direction of space.[435] When considering transport through magnetic layers, we will specify whether an electron $(+$ or $-$ in orientation) is a majority spin ($\uparrow$) or a minority spin ($\downarrow$). These two ways of defining the spin orientation should not be confused. It is possible to extend the thermodynamic description by using Pauli matrices, so that the spin can have any orientation.[436] Here, we explore what can be done with the two-current model.

Therefore, let us consider a thermodynamic system with two substances: electrons with spin up and electrons with spin down. By this assumption, we assume that it is possible to define a chemical potential for each substance. This is consistent with Mott's two-current model and it is justified by the fact that collisions within a spin channel are much more frequent than collisions with spin flip.

It can be shown that, in the presence of an electrostatic potential φ, the quantities that play the role of chemical potentials $\mu_\pm$ are the ***electrochemical potential***,[1]

$$\bar{\mu}_\pm = \mu_\pm + q\varphi \tag{8.1}$$

where q is the charge of the electron. Electrochemical potentials are used extensively in electrochemistry, for example. Below, we will refer to j_+ and j_- as the (charge) current densities for up and down spins. The number of moles of electrons have the following ***continuity equations*** in the reference frame of the magnetic layers:

$$\frac{dn_+}{dt} + \frac{1}{q}\boldsymbol{\nabla}\cdot\boldsymbol{j}_+ = \rho_+$$
$$\frac{dn_-}{dt} + \frac{1}{q}\boldsymbol{\nabla}\cdot\boldsymbol{j}_- = \rho_- \tag{8.2}$$

The source terms $\rho_\pm$ express the occurrence of spin flips. We can model the spin flips as "chemical" reactions in which substance "up" becomes substance "down," and vice versa. Thus, we write two chemical reactions:

$$(a)\ (+) \rightarrow (-)$$
$$(b)\ (-) \rightarrow (+) \tag{8.3}$$

Let us write these reactions in terms of ***stoichiometric coefficients***, passing all elements to the right hand side:

$$(a)\ 0 \rightarrow \nu_{a+}(+) + \nu_{a-}(-)$$
$$(b)\ 0 \rightarrow \nu_{b-}(-) + \nu_{b+}(+) \tag{8.4}$$

---

[1]see, e.g., [376], chaps. 8 and 10.

Hence, the stoichiometric coefficients are:

$$\nu_{a+} = \nu_{b-} = -1 \qquad \nu_{a-} = \nu_{b+} = +1 \tag{8.5}$$

The *reaction rates* are given by,

$$\frac{\dot{n}_+}{\nu_{a+}} = \frac{\dot{n}_-}{\nu_{a-}} = \omega_a$$
$$\frac{\dot{n}_+}{\nu_{b+}} = \frac{\dot{n}_-}{\nu_{b-}} = \omega_b \tag{8.6}$$

Under the combined effect of both reactions (i.e., spin flips), the source terms are given by,

$$\rho_+ = \nu_{a+}\omega_a + \nu_{b+}\omega_b = -\omega_a + \omega_b$$
$$\rho_- = \nu_{a-}\omega_a + \nu_{b-}\omega_b = \omega_a - \omega_b \tag{8.7}$$

The *affinities* of the reactions are defined by,

$$\mathcal{A}_a = -\left(\mu_+\nu_{a+} + \mu_-\nu_{a-}\right)$$
$$\mathcal{A}_b = -\left(\mu_+\nu_{b+} + \mu_-\nu_{b-}\right) \tag{8.8}$$

where μ_+ and μ_- are the *spin-dependent chemical potentials*. For any chemical reaction, the *entropy production rate* is given by:[376]

$$\pi_S = \frac{1}{T}\sum_i \mathcal{A}_i\omega_i \tag{8.9}$$

Applying the affinities (8.8) to π_S yields:

$$\pi_S = \frac{1}{T}\left(\mu_+ - \mu_-\right)\left(\omega_a - \omega_b\right) \tag{8.10}$$

In conformity with the *thermodynamics of irreversible processes* in the linear approximation, we ensure that π_S is positive by writing,

$$\left(\omega_a - \omega_b\right) = W\left(\mu_+ - \mu_-\right) \qquad \left(W > 0\right) \tag{8.11}$$

Thus, we can express the difference in the spin flip rates in terms of the difference in the chemical potentials:

$$\rho_+ = -W\left(\mu_+ - \mu_-\right) \qquad\qquad \rho_- = W\left(\mu_+ - \mu_-\right) \tag{8.12}$$

So far, we have characterized the spin flips that may take place in a small volume of matter. In the presence of currents that may not be homogeneous, the continuity equations (8.2), given the conditions (8.12), yield:

$$\frac{d\left(n_+ + n_-\right)}{dt} + \frac{1}{q}\boldsymbol{\nabla}\cdot\left(\boldsymbol{j}_+ + \boldsymbol{j}_-\right) = 0 \tag{8.13}$$

$$\frac{d\left(n_+ - n_-\right)}{dt} + \frac{1}{q}\boldsymbol{\nabla}\cdot\left(\boldsymbol{j}_+ - \boldsymbol{j}_-\right) = -2W\left(\mu_+ - \mu_-\right) \tag{8.14}$$

Equation (8.13) is the continuity equation for the charge and $\boldsymbol{j}_+ + \boldsymbol{j}_- = \boldsymbol{j}_e$ is the charge current density. It expresses that the charge is conserved. Equation (8.14) is a continuity equation for the spin polarization. Therefore, the current density

$$\boldsymbol{j}_p = \boldsymbol{j}_+ - \boldsymbol{j}_- \tag{8.15}$$

is the **spin polarization current**, commonly called spin current. It is customary to write: $\mu_\pm = \mu_0 \pm \Delta\mu$. This is always possible. It simply amounts to defining:

$$\mu_0 = \frac{1}{2}\left(\mu_+ + \mu_-\right) \qquad \Delta\mu = \frac{1}{2}\left(\mu_+ - \mu_-\right) \tag{8.16}$$

The average chemical potential μ_0 can be thought of as the chemical potential of the electrons when spin effects are negligible. The populations $n_\pm$ and the chemical potentials $\Delta\mu_\pm$ are linked with one another. This point will be made clear in the next chapter.

8.2.2 DIFFUSION EQUATION AND SPIN DIFFUSION LENGTH

The thermodynamics of irreversible processes in the linear regime yields linear relationships between the currents $\boldsymbol{j}_+$ and $\boldsymbol{j}_-$, on the one hand, and the generalized forces that are the gradients of the electrochemical potentials, on the other hand. The relationships between the generalized currents and forces is given by an Onsager matrix.[376] Thus, for transport in one dimension, we write:

$$\begin{pmatrix} j_+ \\ j_- \end{pmatrix} = -\begin{pmatrix} L_{++} & L_{+-} \\ L_{-+} & L_{--} \end{pmatrix}\begin{pmatrix} \nabla\mu_+ - e\nabla\varphi \\ \nabla\mu_- - e\nabla\varphi \end{pmatrix} \tag{8.17}$$

Both terms L_{-+} and L_{+-} represent **spin mixing** mechanisms since they add or remove a contribution to the current from one spin channel to the other. The conceptual association of these terms to spin mixing will be confirmed at the end of the present development, when their role is compared to the spin mixing terms that are accounted for in section 9.4 using Boltzmann's transport equation.

In order to bring out the affinity $\mu_+ - \mu_-$ in (8.17), we use the notation of (8.16) and write,

$$\begin{aligned} \mu_+ &= \frac{\mu_+ + \mu_-}{2} + \frac{\mu_+ - \mu_-}{2} = \mu_0 + \Delta\mu \\ \mu_- &= \frac{\mu_+ + \mu_-}{2} - \frac{\mu_+ - \mu_-}{2} = \mu_0 - \Delta\mu \end{aligned} \tag{8.18}$$

Let us define $\nabla V = \nabla\mu_0 - e\nabla\varphi$ $(e > 0)$ and the electrical current $j_e = j_+ + j_-$. In terms of ∇V, the current j_e and the spin current j_p given by (8.15), we can write (8.17) as,

$$\begin{pmatrix} j_e \\ j_p \end{pmatrix} = -\mathsf{L}\begin{pmatrix} \nabla V \\ \nabla(\Delta\mu) \end{pmatrix} \tag{8.19}$$

where L is the 2×2 matrix representing the rank-2 tensor,

$$\mathsf{L} = \begin{pmatrix} (L_{++} + L_{--}) + (L_{+-} + L_{-+}) & (L_{++} - L_{--}) - (L_{+-} - L_{-+}) \\ (L_{++} - L_{--}) + (L_{+-} - L_{-+}) & (L_{++} + L_{--}) - (L_{+-} + L_{-+}) \end{pmatrix} \tag{8.20}$$

When these relations are substituted in the continuity equations (8.13) for j_e and (8.14) for j_p, we find for the stationary state, after a few arithmetic manipulations,

$$\nabla^2 V = -\frac{(L_{++} - L_{--}) - (L_{+-} - L_{-+})}{(L_{++} + L_{--}) + (L_{+-} + L_{-+})}\nabla^2\Delta\mu \tag{8.21}$$

and

$$qW\Delta\mu = \left\{\frac{L_{++}L_{--} - L_{+-}L_{-+}}{(L_{++} + L_{--}) + (L_{+-} + L_{-+})}\right\}\nabla^2\Delta\mu \tag{8.22}$$

Equation (8.22) implies that the chemical potential difference $\Delta\mu$ follows the **diffusion equation**,

$$\nabla^2 \Delta\mu = \frac{\Delta\mu}{l_{\text{sf}}^2} \qquad (8.23)$$

where the **spin-diffusion length** l_{sf} is given by,

$$l_{\text{sf}}^2 = \frac{1}{qW}\left\{\frac{L_{++}L_{--} - L_{+-}L_{-+}}{(L_{++} + L_{--}) + (L_{+-} + L_{-+})}\right\} \qquad (8.24)$$

Concerning the units, (8.23) imposes that l_{sf} has the unit of a length. From (8.19) we see that the matrix element L_{ij} has the units of $j_e/(q\nabla\varphi)$. In the following, we will use the fact that the units of L_{ij} are that of a conductivity divided by q. In view of, for example, (8.14), qW has units of $\nabla^2 j_p/(q\nabla\varphi)$ since $\Delta\mu$ has the same units as $q\varphi$. Hence, the expression on the right-hand side of (8.24) has units of $1/\nabla^2$, i.e., a length squared.

8.3 GMR OF THIN MAGNETIC LAYERS

Let us now see that the constitutive equations (8.23) for $\Delta\mu$ and (8.21) for ΔV as a function of $\Delta\mu$ can describe CPP-GMR. The most immediate application of these equations is the case of a multilayer. To keep the algebraic equations as simple as possible, we assume that all magnetic layers have an equal thickness d, and that they are juxtaposed with their magnetic orientations in a parallel or antiparallel configuration. We assume an infinite stack of layers and focus on one bilayer, to which we impose periodic boundary conditions (Fig. 8.1). The current is assumed to be perpendicular to the interfaces. We further assume that the spin-diffusion length is very large compared to the thickness of the layers. Then, the gradients are uniform in each layer and integration of the potential gradients across each layer is trivial. We introduce the notation:

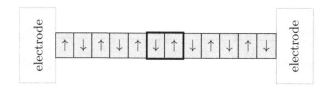

Figure 8.1 Schematics representing a quasi-infinite stack of magnetic layers that can be in either a parallel or an antiparallel configuration (the one shown here). Each layer has the same thickness d. Far away from the electrodes, the bilayers (as highlighted with a thick frame surrounding two adjacent layers) are the same.

$$\begin{aligned}
(1/q)\sigma_0 &= L_{++} + L_{--} & (1/q)\beta\sigma_0 &= L_{++} - L_{--} \\
(1/q)\sigma_{\rightleftarrows} &= L_{+-} + L_{-+} & (1/q)\eta\sigma_{\rightleftarrows} &= L_{+-} - L_{-+}
\end{aligned} \qquad (8.25)$$

The coefficient β characterizes the **spin asymmetry of the conductivity**. While the generality of the Onsager approach leads to a possible asymmetry in the spin mixing, i.e., $L_{+-} \neq L_{-+}$, in the following, we will neglect spin mixing altogether, i.e., $\sigma_{\rightleftarrows} = 0$. Using this notation, the spin diffusion length becomes:

$$l_{sf}^2 = \frac{\sigma_0(1 - \beta^2)}{q^2 W} \qquad (8.26)$$

Equation (8.21) for V can be written as:

$$\nabla^2 V = -\frac{\beta}{l_{sf}^2}\Delta\mu \tag{8.27}$$

The transport equations (8.19) become:

$$\begin{pmatrix} j_e \\ j_p \end{pmatrix} = -\sigma_0 \begin{pmatrix} 1 & \beta \\ \beta & 1 \end{pmatrix} \begin{pmatrix} \nabla V/q \\ \nabla\Delta\mu/q \end{pmatrix} \tag{8.28}$$

Let us assume that the two layers are made of identical materials, so that they have the same conductivity. Only the orientation of one layer with respect to the other may change. Let us say that, in the first layer, when the magnetization is in the up direction, the $(+)$ electrons are majority spins ($\uparrow$) and the $(-)$ are minority spins ($\downarrow$). In the second layer, the magnetization is either up (the P configuration) or down (the AP configuration). In the AP case, the $(+)$ and $(-)$ electrons are in reverse orientation with respect to the magnetization. Inspecting the definitions of β and η, we see that the reversal of the magnetization from one layer to the next amounts to changing the signs of β and η in the transport matrix.

Hence, in a layer of up magnetization, the transport matrix is:

$$\mathsf{L}_u = \sigma_0 \begin{pmatrix} 1 & \beta \\ \beta & 1 \end{pmatrix} \begin{pmatrix} \nabla V)/q \\ \nabla\Delta\mu/q \end{pmatrix} \tag{8.29}$$

In a layer with down magnetization, the transport matrix is:

$$\mathsf{L}_d = \sigma_0 \begin{pmatrix} 1 & -\beta \\ -\beta & 1 \end{pmatrix} \begin{pmatrix} \nabla V/q \\ \nabla\Delta\mu/q \end{pmatrix} \tag{8.30}$$

The continuity of the currents at the interface implies for the parallel configuration:

$$\begin{pmatrix} j_e \\ j_p \end{pmatrix} = -\mathsf{L}_u \begin{pmatrix} \nabla V_1{}^P/q \\ \nabla\Delta\mu_1{}^P/q \end{pmatrix} = -\mathsf{L}_u \begin{pmatrix} \nabla V_2{}^P/q \\ \nabla\Delta\mu_2{}^P/q \end{pmatrix} \tag{8.31}$$

In the antiparallel configuration, the continuity of the currents implies:

$$\begin{pmatrix} j_e \\ j_p \end{pmatrix} = -\mathsf{L}_u \begin{pmatrix} \nabla V_1{}^{AP}/q \\ \nabla\Delta\mu_1{}^{AP}/q \end{pmatrix} = -\mathsf{L}_d \begin{pmatrix} \nabla V_2{}^{AP}/q \\ \nabla\Delta\mu_2{}^{AP}/q \end{pmatrix} \tag{8.32}$$

Now assume that the gradients are uniform in each layer. Then for each configuration (P or AP), there are 4 equations with 6 parameters. One of the parameters, say, the current j_e, is determined by the experimental conditions. Furthermore, we imagine this bilayer as being deep inside a multilayer containing many bilayers. Then we can apply periodic boundary conditions for the chemical potential. This implies that the drop of chemical potential difference across the bilayer must vanish. This condition provides additional equations:

$$\begin{aligned} d\left(\nabla\Delta\mu_1{}^P + \nabla\Delta\mu_2{}^P\right) &= 0 \\ d\left(\nabla\Delta\mu_1{}^{AP} + \nabla\Delta\mu_2{}^{AP}\right) &= 0 \end{aligned} \tag{8.33}$$

Now there are 5 unknowns and 5 equations in each configuration. They can be solved in terms of j_e.

We seek to calculate the potential drop across the bilayer, i.e.:

$$\begin{aligned} \Delta V^P &= -d\left(\nabla V_1{}^P + \nabla V_2{}^P\right) \\ \Delta V^{AP} &= -d\left(\nabla V_1{}^{AP} + \nabla V_2{}^{AP}\right) \end{aligned} \tag{8.34}$$

In the parallel case, the gradients of $\Delta\mu$ must be equal, according to the continuity of the currents, and opposite, according to the condition of periodicity. So they must be zero. The potential drops must also be equal. Therefore, in the parallel configuration, we have simply:

$$\Delta V^P = -2d\nabla V_1{}^P = q2d\frac{j_e}{\sigma_0} \tag{8.35}$$

Recall that here, the V terms have units of q times an electrical potential. In other words, the effective conductivity is,

$$\sigma_{\text{eff}}^P = \sigma_0 \tag{8.36}$$

The antiparallel case requires some algebraic manipulations. To make things clearer, we use the vectorial notation,

$$\mathbf{J} = \begin{pmatrix} j_e \\ j_p \end{pmatrix} \qquad \mathbf{X} = \begin{pmatrix} \nabla V \\ \nabla\Delta\mu \end{pmatrix} \tag{8.37}$$

Hence, the transport equations (8.31) and (8.32) applied to layers 1 and 2 imply,

$$\begin{aligned}
\mathbf{J} &= -\mathsf{L}_u\mathbf{X_1} = -\mathsf{L}_d\mathbf{X_2} \\
\mathbf{X_1} &= -\mathsf{L}_u{}^{-1}\mathbf{J} \qquad \mathbf{X_2} = -\mathsf{L}_d{}^{-1}\mathbf{J} \\
\mathbf{X_1} + \mathbf{X_2} &= -\left(\mathsf{L}_u{}^{-1} + \mathsf{L}_d{}^{-1}\right)\mathbf{J} \\
\mathbf{J} &= -q\mathbf{Y}^{AP}\left(\mathbf{X_1} + \mathbf{X_2}\right)/q
\end{aligned} \tag{8.38}$$

where,

$$q\mathbf{Y}^{AP} = \left(\mathsf{L}_u{}^{-1} + \mathsf{L}_d{}^{-1}\right)^{-1} \tag{8.39}$$

is an admittance tensor. The integration over the layer thickness is immediate under the assumption of uniform gradients. With the boundary conditions that we have specified, what remains is,

$$\begin{pmatrix} j_e \\ j_p \end{pmatrix} = \mathbf{Y}^{AP}\begin{pmatrix} \Delta V^{AP} \\ 0 \end{pmatrix} \tag{8.40}$$

The effective conductivity in the AP configuration is given by,

$$\sigma_{\text{eff}}^{AP} = \frac{j_e}{\Delta V^{AP}/(2dq)} = 2\left(\mathbf{Y}^{AP}\right)_{11} \tag{8.41}$$

The voltage drop is given by,

$$\Delta V^{AP}/q = dj_e/\left(\mathbf{Y}^{AP}\right)_{11} \tag{8.42}$$

A little algebra gives:

$$\begin{aligned}
\mathsf{L}_u{}^{-1} &= \frac{q}{\sigma_0(1-\beta^2)}\begin{pmatrix} 1 & -\beta \\ -\beta & 1 \end{pmatrix} \qquad \mathsf{L}_d{}^{-1} = \frac{q}{\sigma_0(1-\beta^2)}\begin{pmatrix} 1 & \beta \\ \beta & 1 \end{pmatrix} \\
\left(\mathbf{Y}^{AP}\right)^{-1} &= \frac{2}{\sigma_0(1-\beta^2)}\begin{pmatrix} 1 & 0 \\ 0 & 1 \end{pmatrix} \Rightarrow \left(\mathbf{Y}^{AP}\right)_{11} = \frac{\sigma_0}{2}(1-\beta^2)
\end{aligned} \tag{8.43}$$

Finally, the conductivity in the antiparallel configuration is given by,

$$\sigma_{\text{eff}}^{AP} = \sigma_0(1-\beta^2) \tag{8.44}$$

It is customary to define the GMR ratio as:

$$GMR = \frac{\Delta V^{AP} - \Delta V^P}{\Delta V^{AP}} = 1 - \frac{\Delta V^P}{\Delta V^{AP}} \tag{8.45}$$

Thus, we find,

$$GMR = \beta^2 \tag{8.46}$$

In other words, in the simplest case, the spin-dependence of the conductivity, characterized in (8.25) by β, determines the strength of the GMR effect.

8.4 SPIN-INDUCED INTERFACE RESISTANCE

We now consider two ferromagnetic layers that are so thick that we can treat them in the limit where the outer edges of the layers are infinitely far from the interface between the ferromagnetic layers. Hence, the chemical potential difference $\Delta\mu$ vanishes at the outer face of the bilayer.[435]

In the parallel configuration, the bilayer is a homogeneous system. Therefore, the chemical potential difference vanishes everywhere. The voltage gradient is uniform and the conductivity is once again σ_{eff} given by (8.36). Neglecting once again spin mixing, the overall potential drop is given by,

$$\Delta V^P = \frac{2d}{\sigma_0} j_e q \tag{8.47}$$

In the antiparallel configuration, the diffusion equation must be solved. The general solution is a sum of exponentials. We use a coordinate axis Ox normal to the interface, with O at the interface. The solution that is finite on either side is of the form,

$$\Delta\mu(x) = \Delta\mu_0 \exp\left(-\frac{\varepsilon x}{l_{\text{sf}}}\right) \tag{8.48}$$

with

$$\varepsilon = +1(x > 0), \varepsilon = -1(x < 0) \tag{8.49}$$

The integration constant $\Delta\mu_0$ will result from imposing the boundary conditions. The electric field (voltage gradient) is obtained by integration of (8.27). Thus, we find,

$$-\nabla V_i = qE_i - \varepsilon \frac{\beta_i}{l_{sf}} \Delta\mu_0 \exp\left(-\frac{\varepsilon x}{l_{\text{sf}}}\right) \tag{8.50}$$

where ε is given by (8.49). The index i in (8.50) refers to the layer number. The continuity of the electric field at the interface implies:

$$-\nabla V|_{int} = qE_1 + \beta \frac{\Delta\mu_0}{l_{\text{sf}}} = E_2 + \beta \frac{\Delta\mu_0}{l_{\text{sf}}} \tag{8.51}$$

Thus, the continuity implies simply $E_1 = E_2 = E$ everywhere. The set current j_e is constant throughout the structure because $\nabla j_e = 0$ in the steady state.

Far from the interface, the electric field is just E. The linear transport equations provide $j_e = \sigma_0 E$. In other words, it is known to the extent that the conductivity of the material is known. In order to finalize the calculation, we need to integrate the electric field to get $\Delta\mu_0$ in terms of the potential drop imposed on the material,

$$\Delta V^{AP} = -\int_{-d}^{0} \nabla V_1{}^{AP} dx - \int_{0}^{d} \nabla V_2{}^{AP} dx \tag{8.52}$$

This integration is straightforward. Here, we used the assumption of thicknesses much greater than the spin diffusion length, implying $\exp(-d/l_{sf}) \approx 0$. We thus find an expression for $\Delta\mu_0$ in terms of the field at the interface:

$$\beta \frac{\Delta\mu_0}{l_{sf}} = \frac{\Delta V^{AP} - 2dqE}{2l_{\text{sf}}} \tag{8.53}$$

This result gives the voltage drop across the thick double layer in terms of $\Delta\mu_0$,

$$\Delta V^{AP} = 2dqE + 2\beta\Delta\mu_0 \tag{8.54}$$

The continuity of currents at the interface implies, as for the case of thin films, a relationship between the gradients at the interface:

$$j_e = (-)Y^{AP}_{11}\frac{\left(\nabla V_1^{AP} + \nabla V_2^{AP}\right)_{\text{int}}}{q} - Y^{AP}_{12}\frac{\left(\nabla\Delta\mu_1^{AP} + \nabla\Delta\mu_2^{AP}\right)_{\text{int}}}{q} \tag{8.55}$$

The second term vanishes because both Y^{AP}_{12} and the chemical potential sum vanish (the chemical potential is symmetric about the interface). So the continuity of the currents at the interface implies,

$$j_e = 2\,Y^{AP}_{11}\left(E + \beta\frac{\Delta\mu_0}{ql_{\text{sf}}}\right) \tag{8.56}$$

Applying (8.53), we get:

$$j_e = 2Y^{AP}_{11}\left(E + \frac{\Delta V^{AP} - 2dqE}{2ql_{\text{sf}}}\right) \approx Y^{AP}_{11}\left(\frac{\Delta V^{AP}}{ql_{\text{sf}}} - \frac{2dE}{l_{\text{sf}}}\right) \tag{8.57}$$

The approximation makes use of the assumption $d \gg l_{sf}$. Rearranging the terms yields,

$$\Delta V^{AP} = \left(\frac{l_{\text{sf}}}{Y^{AP}_{11}} + \frac{2d}{\sigma_0}\right)j_e q \tag{8.58}$$

The GMR ratio can now be evaluated:

$$GMR = 1 - \frac{\Delta V^P}{\Delta V^{AP}} = \frac{l_{\text{sf}}}{d}\frac{\sigma_0}{2Y^{AP}_{11}} = \left(\frac{l_{\text{sf}}}{d}\right)\frac{1}{(1-\beta^2)} \tag{8.59}$$

We see that in the limit of layers much thicker than the spin diffusion length, the GMR ratio is of the order of l_{sf}/d.

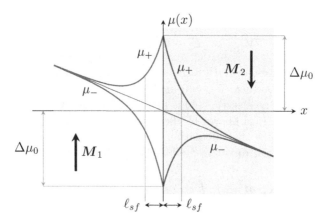

Figure 8.2 Spin-dependent electrochemical potential μ_+ and μ_- as a function of the position with respect to the interface between two magnetic metals of opposite magnetization orientations M_1 and M_2. At the interface, a spin accumulation builds up, characterized by $\Delta\mu_0$. At a distance much greater than ℓ_{sf} the usual ohmic behavior is recovered for both μ_+ and μ_-.

The magnitude of the spin accumulation effect, expressed in terms of the chemical potential difference, can be written as a function of the current using (8.56). We recall that we also have $j_e = (\sigma_0 + \sigma_{\rightleftarrows})\, E$. After a few algebraic manipulations, we find,

$$\Delta\mu_0 = j_e \frac{\beta l_{\text{sf}}}{\sigma_0 \left(1 - \beta^2\right)^2} \tag{8.60}$$

In the limit of a small β, we have:

$$\Delta\mu_0 = \frac{\beta}{\sigma_0} l_{\text{sf}} j_e \tag{8.61}$$

The spin effect that was shown in (8.54) can be expressed in terms of a resistance. If the junction has a cross section S and a current $I = j_e\, S$ runs through it, then the junction considered here, which is equivalent to an abrupt magnetic wall, presents a ***spin-interface resistance***,[437]

$$R_{\text{int}} = \frac{2\beta^2}{\sigma_0} \frac{l_{\text{sf}}}{S} \tag{8.62}$$

The chemical potential difference appears as the potential drop across this interface resistance, since we have,

$$\Delta\mu_0 = R_{\text{int}} I \tag{8.63}$$

8.5   FURTHER READINGS

Spin Caloritronics

In this chapter, we used the thermodynamics of irreversible processes to analyze GMR and account for a spin diffusion length. The temperature was presumed uniform throughout the sample. If a temperature gradient is present, we have a combination of heat, spin and charge currents. This field of research is now known as ***spin caloritronics***.[438, 439]

Let us consider, as a first example, a stack of ferromagnetic metal layers spaced with non-magnetic metal layers thinner than the spin diffusion length for that metal. Instead of driving an electrical current through it as we would to study magnetoresistance, we drive a heat current and measure the effective thermopower of this stack as a function of the magnetic field, i.e., the ***magneto-thermopower***. In the wake of magnetoresistance research, many researchers studied MTEP, first with current flowing in the plane (CIP) of the layers,[440, 441] and later with current perpendicular to this plane (CPP) or in granular systems.[433, 442, 443] In some nanostructures, it is not possible to simply assume that the temperature is uniform because the Peltier effect at each bilayer implies that a local temperature gradient builds up.[432, 444]

There are many cross effects that can be considered and thermodynamics is a good way to define them.[445, 446] In multilayered structures, there are boundary conditions that must be satisfied, such as the continuity of the temperature. But specifically interfacial effects can also occur and be described in a thermodynamic approach.[447, 448]

In many cases, we deal with a well-known phenomenon, and when it is studied in a magnetic system, we extend the description to include a spin-dependence to the transport coefficient, such as spin-dependent Seebeck and Nernst coefficients.[446] The spin-dependent Seebeck coefficients should not be confused with the ***spin Seebeck effect***. This effect is defined as the voltage generated on leads laid on a ferromagnet subjected to a temperature gradient. The leads are perpendicular to the temperature gradient. In view of the Nernst effect, the experimental conditions must be well under control in order to avoid confusing one effect for the other.[449] The spin Seebeck effect was first observed in a metallic magnet.[450,

451] It was also observed using magnetic semiconductors,[452] and insulators.[453] It was also reported in the case of antiferromagnetic materials.[454, 455]

In conductors, the extension of the two-current model with a heat current readily implies the existence of a **heat-driven spin current**.[456] This phenomenon is distinct from the spin current induced in spin valves with CPP currents, in the sense that the spin current in the non-magnetic spacer depends on the relative orientation of the magnetization in the magnetic layer. If the spacer layer is thicker than the spin diffusion length, the effective spin current vanishes. Instead, the heat-driven spin current is a bulk property.[456] In insulators, heat-driven spin current can be accounted for by the thermodynamics of irreversible processes if magnons are considered as quasi-particles, in which case a chemical potential can be attributed to magnons.[457]

Transverse spin currents

From a general thermodynamic point of view, magnetization is an extensive quantity per unit volume. As such, a continuity equation that includes the divergence of a tensorial spin current must apply.[458] Thus, the thermodynamics of irreversible processes can infer that there ought to be a **spin Hall current** induced by an electric field ∇V,[446]

$$\boldsymbol{j}_p = -\frac{\sigma_\perp}{q_A} \left(\hat{n} \times \boldsymbol{\nabla} V \right) \tag{8.64}$$

where $\hat{n}$ is a unit vector along the applied magnetic field $\boldsymbol{B}$ and the spins are either parallel or antiparallel to $\hat{n}$. The more general case where the spin current is a tensor can also be predicted from thermodynamics.[377] The **Edelstein** effect is an example of such a transverse spin current induced by an electric field. Its origin resides in the Rashba splitting of the conduction bands (Fig. 6.6).[459, 460]

Likewise, the thermodynamics of irreversible processes also predicts a **heat-induced spin Hall** current. In other words, it is a spin current induced by a temperature gradient ∇T,[446]

$$\boldsymbol{j}_p = -\frac{\sigma \varepsilon_\perp + \sigma_\perp \varepsilon}{q_A} \left(\hat{n} \times \boldsymbol{\nabla} T \right) \tag{8.65}$$

Here, $\varepsilon_\perp$ and $\sigma_\perp$ characterize the antisymmetric part of the conductivity and Seebeck tensors, respectively.

When electromagnetic fields are treated as state variables, the thermodynamics of irreversible processes shows that a temperature gradient induces a torque on the magnetization.[377] The Landau-Lifshitz equation that describes the evolution of magnetization (see §7.4) can also be written in a way that involves a spin current.[374, 455]

Spin injection

The injection of a spin current in a material can be tested by **non-local measurements**. Two magnetic electrodes are deposited on a material being tested. A current is driven through this structure. Next to this pair of electrodes, another magnetic electrode, or a pair of them, is laid out. The current implies a spin accumulation in the material under examination. A spin diffusion in all directions, not in the current direction only, is associated with this spin accumulation. Thus, a spin current reaches the other electrode (or a pair of them).[461] This method can be used to characterize spin transport in non-magnetic conductors.[462]

The spin diffusion length is quite long in pure Ag.[18, 463] Using a non-local geometry, it has been possible to drive spins over distances of several micrometers.[464] Not only that, but when a magnetic field was applied, the spins precessed coherently from the injection electrode to the non-local electrode. This spin rotation and its consequence for spin transport are called the *Hanle effect*.

Spin diffusion without bodily displacement of electrons or nuclei is a key ingredient of the family of techniques known as Dynamic Nuclear Polarization (see Section 16.3). Through magnetic resonance techniques, the high spin polarization of electrons is imparted on nuclear spins. Strategies can be designed to optimize spin diffusion from the surface of a nanoparticle to its interior.[465]

A heat current can induce a spin current in a metallic ferromagnet.[456] Likewise, *heat-driven spin injection* is also possible. This phenomenon has been observed using a non-local geometry with permalloy (NiFe alloy) as a spin source that was heated up by Joule heating.[466] A heat-driven spin current can act on the magnetization of a spin valve.[467]

By driving heat across a magnetic system, it is possible to inject a spin current into a free layer, i.e., to achieve *thermal spin injection*.[468, 469] A spin current can also be induced by heat in an insulating ferromagnet and affect spin wave propagation.[470]

9 Boltzmann Description of Transport

Ludwig Boltzmann, 1844–1906

L.E. Boltzmann studied physics at the University of Vienna. J. Loschmidt, J. Stefan, and A. von Ettingshausen were among his teachers. A bitter debate arose between those like Boltzmann who were developing physical theories based on an atomistic view and those who insisted on the preeminence of the concept of energy.

Spintronics: A primer

J.-Ph. Ansermet (https://orcid.org/0000-0002-1307-4864)

Chapter 9. Boltzmann Description of Transport

The Boltzmann transport equation is introduced, using the simplest form of the relaxation approximation. Derivations of Ohm's law and the Seebeck effect are shown for simple metals with parabolic bands, implying the Wiedemann-Franz law. This formalism can be used to derive Mott's formula for thermopower, as well as to describe phonon and magnon drag effects. The spin-dependent transport equation yields Mott's two-current model, a spin-diffusion equation, and a spin-diffusion length expressed in terms of electron and spin mean-free paths.

DOI: 10.1201/9781003370017-9

9.1 HISTORICAL INTRODUCTION

Boltzmann established in 1872 the equation that we introduce in this chapter and apply to describe spin-dependent transport. It should be noted that thermodynamics is a powerful tool to define a great variety of transport phenomena. It can be used to predict transport phenomena that involve various generalized forces and currents, in particular cross effects. However, thermodynamics by itself cannot be used to estimate quantitatively a transport coefficient, such as an electrical resistivity. The Boltzmann approach is used in this book to go beyond thermodynamics and introduce microscopic mechanisms to our understanding of transport.

Boltzmann was focused on developing a mechanical theory of heat. He introduced a statistical approach by considering that the state of a molecule with n degrees of freedom could be represented by a point in what we now call **states space**. He defined a statistical distribution of points in states space and wrote an equation for the time evolution of the distribution. Here, we will first derive this kinetic theory and show how it applies to charges without spin. We will then expand the states space by introducing spin as an additional dimension and apply it to spin-dependent transport, thus obtaining a result first derived by Fert and Campbell.[75, 76]

In 1877, after having developed his kinetic theory, Boltzmann obtained his famous formula for entropy, $S = k_B \ln(\Omega)$, where Ω is the number of microscopic configurations that correspond to the same macroscopic state, and k_B is now known as the Boltzmann constant.

Using the Boltzmann theory of transport, it is possible to derive some empirical laws that are often used to characterize the transport properties of materials. We shall see a few examples in this chapter. The **Wiedemann-Franz law** states that the ratio of the heat to charge conductivities is proportional to temperature, and the proportionality constant would be the same for all materials. The **Mott formula** links the Seebeck coefficient to the density of states at the Fermi level. **Phonon drag** refers to the effect of the electron-phonon coupling on electronic transport in a temperature gradient. It gives rise to an enhanced Seebeck effect in crystals at low temperature. In magnetic materials, a **magnon drag** effect can be expected due to the electron-magnon coupling. Baylin first discussed magnon drag using a variational principle.[471] Following these examples, we will expand the Boltzmann theory of transport to include spin and illustrate its application by analyzing the basic mechanisms of giant magnetoresistance.

9.2 BOLTZMANN DESCRIPTION OF TRANSPORT WITHOUT SPIN

9.2.1 AIM AND VALIDITY

We want to describe spin-dependent transport with Boltzmann theory of transport in the approximation of the two-current model (see §1.2.1). Compared to the thermodynamic approach, the Boltzmann equation has the advantage of providing means to add a microscopic vision to the description of transport phenomena. In order to do so, we first introduce the Boltzmann formalism and illustrate it with a couple of transport phenomena where spin is not involved.

Let us start with an introduction to the Boltzmann equation of transport for the case where no spin is involved. We want to consider electronic states as wave packets with energy defined to a greater level of accuracy than $k_B T$. This requirement defines a limiting size δr of the wave packet, and we should only consider situations where the electric field does not vary much over this distance. Let us use the relation,[71]

$$\delta r \, \delta p \approx \hbar \qquad (9.1)$$

to estimate δr. From $E = (1/2)mv^2$ we get $\delta E = mv\delta v = v\delta p$, which we want to be of the order of $k_B T$, implying $\delta p \cong k_B T/v$ and hence,

$$\delta r \cong \frac{\hbar}{\delta p} = \frac{\hbar v}{k_B T} \cong \frac{\hbar v}{\delta E} \qquad (9.2)$$

Assume, for an example, that we would wish to have $\delta r = 0.1$ nm, as if we wanted to describe collisions of conduction electrons with the field of an ion. Then, as $v \approx 10^6$ m/s in a metal, we would have,

$$\delta E \approx \frac{\hbar \mathrm{v}}{\delta r} = 10^{-18} J \approx 10 E_F \qquad (9.3)$$

This is too big compared to $k_B T$ by several orders of magnitude. Therefore, in order to satisfy the condition on the energy of the wave packets, we can consider calculating with this approach inhomogeneities that take place over macroscopic scales only. Collisions with ions, therefore, must be worked out in another framework and the result imported into this wave packet description.

9.2.2 BOLTZMANN TRANSPORT EQUATION

We define a probability distribution for states of a given momentum $\boldsymbol{p}$ at a position $\boldsymbol{r}$:

$$f(\boldsymbol{p}, \boldsymbol{r}, t) \qquad (9.4)$$

This means that the probability that an infinitesimal volume of the phase space is occupied is given by,

$$\frac{2}{\hbar^3} f(\boldsymbol{p}, \boldsymbol{r}, t) \, d^3 p \, d^3 r \qquad (9.5)$$

The factor of 2 accounts for the two spin states that can be found at each point $(\boldsymbol{p}, \boldsymbol{r})$ of phase space. We note that, since $\boldsymbol{p} = \hbar \boldsymbol{k}$ where $\boldsymbol{k}$ is a wave vector, $d^3 p \, d^3 r / \hbar^3$ is dimensionless.

Let us clarify the normalization of the Boltzmann distribution. We consider a small volume $\mathcal{V}$ in real space around a position $\boldsymbol{r}$, which contains on average just one electron. That is, $\mathcal{V}$ is a volume per electron, analogous to the molar volume used in thermodynamics. If there are N electrons in the volume V of a uniform sample, then, $\mathcal{V} = V/N$. the inverse quantity, $1/\mathcal{V}$, is the local density of electrons, i.e., the electron density $n(\boldsymbol{r}, t)$. Thus, we have the normalization,

$$\int d^3 r \int d^3 p \frac{2}{\hbar^3} f(\boldsymbol{p}, \boldsymbol{r}, t) = \mathcal{V} \int d^3 p \frac{2}{\hbar^3} f(\boldsymbol{p}, \boldsymbol{r}, t) = 1 \qquad (9.6)$$

Here, we have used the fact that over the small volume $\mathcal{V}$, $f(\boldsymbol{p}, \boldsymbol{r}, t)$ is independent of $\boldsymbol{r}$. Therefore, the integral over momentum space only gives the normalization condition,

$$\int d^3 p \frac{2}{\hbar^3} f(\boldsymbol{p}, \boldsymbol{r}, t) = n(\boldsymbol{r}, t) \qquad (9.7)$$

Let us now analyze how the distribution varies with time. To do so, we consider a fixed volume $\Delta p \, \Delta r$ of a phase space that has one dimension for momentum and one for space. Each particle of the system occupies a position $(\boldsymbol{r}, \boldsymbol{p})$ in the phase space. Over time, its position $\boldsymbol{r}$ changes because, presumably, the particle has a non-vanishing momentum, and its momentum $\boldsymbol{p}$ also changes because a force is applied to the particle. The time derivative of the content of the small phase space volume is given by,

$$\frac{\partial f}{\partial t} \Delta p \Delta r \qquad (9.8)$$

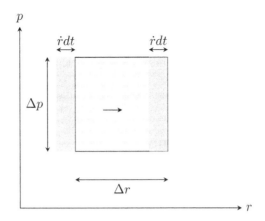

Figure 9.1 Schematic representation of phase space points entering a small volume of phase space, having dimensions Δr and Δp. These points are in the shaded "volumes" (area) $\Delta p\, v dt$. The arrow at the center of the small volume suggests the orientation of the velocity, so the phase space points enter at r and leave this small volume at $r + \Delta r$.

The rate (9.8) is due to the outflow and inflow of phase space positions into that volume, be it owing to a spatial or momentum change. The contribution to the occupation of this volume is represented in Fig. 9.1 for the spatial dimension r. There is an outflow at $\boldsymbol{r} + \Delta \boldsymbol{r}$ and an inflow at $\boldsymbol{r}$. We could reason likewise for the $\boldsymbol{p}$ direction. Within a time dt, the outflow of points has two contributions:

$$
\begin{aligned}
f\left(\boldsymbol{p}, \boldsymbol{r} + \Delta \boldsymbol{r}, t\right) \Delta p\, \dot{r} dt - f\left(\boldsymbol{p}, \boldsymbol{r}, t\right) \Delta p\, \dot{r}\, dt &= \frac{\partial f}{\partial \boldsymbol{r}} \frac{\Delta \boldsymbol{r}}{\Delta t} \Delta p \Delta r dt \\
f\left(\boldsymbol{p} + \Delta \boldsymbol{p}, \boldsymbol{r}, t\right) \dot{p} \Delta r\, dt - f\left(\boldsymbol{p}, \boldsymbol{r}, t\right) \dot{p} \Delta r\, dt &= \frac{\partial f}{\partial \boldsymbol{p}} \frac{\Delta \boldsymbol{p}}{\Delta t} \Delta p \Delta r dt
\end{aligned}
\tag{9.9}
$$

where we write $\dot{r}$ as $\Delta r/\Delta t$ and $\dot{p} = \Delta p/\Delta t$. As these two contributions correspond to phase space points leaving the small volume, from (9.8) and (9.9), we have,

$$
\frac{\partial f}{\partial t} = -\frac{\partial f}{\partial \boldsymbol{r}} \frac{\Delta \boldsymbol{r}}{\Delta t} - \frac{\partial f}{\partial \boldsymbol{p}} \frac{\Delta \boldsymbol{p}}{\Delta t}
\tag{9.10}
$$

These results hold if the phase space is multidimensional in position and momentum space. In a general case, we could recover the heuristic picture given here by choosing local r and p axes that are tangent to $\dot{\boldsymbol{r}}$ and $\dot{\boldsymbol{p}}$. Thus, taking the limit $\Delta t \to 0$, we find,

$$
\frac{\partial f}{\partial t} = -\frac{\partial f}{\partial \boldsymbol{r}} \frac{d\boldsymbol{r}}{dt} - \frac{\partial f}{\partial \boldsymbol{p}} \frac{d\boldsymbol{p}}{dt}
\tag{9.11}
$$

By analogy with the material derivative that is used in the physics of continuous media,[1] we define,

$$
\frac{df}{dt} = \frac{\partial f}{\partial t} + \frac{\partial f}{\partial \boldsymbol{r}} \frac{d\boldsymbol{r}}{dt} + \frac{\partial f}{\partial \boldsymbol{p}} \frac{d\boldsymbol{p}}{dt}
\tag{9.12}
$$

This is the time derivative of the distribution f that we would observe if we were, so to speak, following the flow of particles in phase space. The result (9.11) is known as the

---

[1] See, e.g., [376].

Liouville theorem, which states,

$$\frac{df}{dt} = 0 \tag{9.13}$$

Indeed, since so far no mechanism has been considered that would suddenly generate the presence of a particle, or its disappearance, within a small volume following the flow of phase space points, the number of points remains constant in that volume. We will introduce this type of mechanism in section 9.2.3.

9.2.3 RELAXATION TIME APPROXIMATION

Collisions are processes that alter (9.13). A collision in Fig. 9.1 would bring a point away from the volume we considered or suddenly generate the occupation of a point in it. In order to take into account collisions, we write:

$$\frac{df}{dt} = \frac{df}{dt}\bigg|_{coll} \tag{9.14}$$

Let $f_0\left(\boldsymbol{p},\boldsymbol{r}\right)$ be the distribution at equilibrium. At equilibrium, the collisions produce no change of $f\left(\boldsymbol{p},\boldsymbol{r},t\right)$. Now we assume that any departure of f from f_0 is supposed to decay exponentially under the effect of collisions. Then, f evolves according to,

$$\frac{df\left(\boldsymbol{p},\boldsymbol{r},t\right)}{dt} = -\frac{f\left(\boldsymbol{p},\boldsymbol{r},t\right) - f_0\left(\boldsymbol{p},\boldsymbol{r}\right)}{\tau\left(\boldsymbol{p}\right)} \tag{9.15}$$

In the following, there are no cases where the relaxation time $\tau\left(\boldsymbol{p}\right)$ needs to be considered as a function of position.

Let us now specify the charge carrier dynamics in (9.12). Thus, we write,

$$\frac{d\boldsymbol{p}}{dt} = \boldsymbol{F} \tag{9.16}$$

where $\boldsymbol{F}$ is the force applied to the charge carrier, represented by a point charge or a wave packet at $\left(\boldsymbol{p},\boldsymbol{r}\right)$, depending on the representation we use of the state of the charge carrier (classical or semi-classical). The ***Boltzmann equation in the relaxation time approximation*** can be written as,

$$\frac{\partial f\left(\boldsymbol{p},\boldsymbol{r},t\right)}{\partial t} + \frac{\partial f\left(\boldsymbol{p},\boldsymbol{r},t\right)}{\partial \boldsymbol{r}}\dot{\boldsymbol{r}} + \frac{\partial f\left(\boldsymbol{p},\boldsymbol{r},t\right)}{\partial \boldsymbol{p}}\boldsymbol{F} = -\frac{f\left(\boldsymbol{p},\boldsymbol{r},t\right) - f_0\left(\boldsymbol{p},\boldsymbol{r}\right)}{\tau\left(\boldsymbol{p}\right)} \tag{9.17}$$

In the upcoming developments, we will make the simplifying assumption that τ does not depend on momentum. We will have to improve on this description when we describe phonon or magnon drag (§9.3) effects, spin-dependent collisions (§9.4.2), or spin relaxation in metals (§13.2). In our developments, we will always consider the steady state of the system, for which the Boltzmann equation in the momentum-independent relaxation time approximation reads,

$$\frac{\partial f\left(\boldsymbol{p},\boldsymbol{r},t\right)}{\partial \boldsymbol{r}}\dot{\boldsymbol{r}} + \frac{\partial f\left(\boldsymbol{p},\boldsymbol{r},t\right)}{\partial \boldsymbol{p}}\boldsymbol{F} = -\frac{f\left(\boldsymbol{p},\boldsymbol{r},t\right) - f_0\left(\boldsymbol{p},\boldsymbol{r}\right)}{\tau} \tag{9.18}$$

9.2.4 STEADY STATE LINEARIZED BOLTZMANN EQUATION

Let us determine the linear response of the system when it is in a steady state. We assume that the equilibrium distribution is a function $f_0(\boldsymbol{p},\boldsymbol{r})$. We continue to keep track of a possible spatial variation in order to account later for thermoelectric effects; see §9.2.6. In view of the constraints discussed in §9.2.1, the spatial dependence in $f_0(\boldsymbol{p}, \boldsymbol{r})$ must take place over an extended length scale. In the example of §9.2.6, this spatial dependence is due to the fact that the chemical potential μ and the temperature depend on position.

We write the distribution function $f(\boldsymbol{p}, \boldsymbol{r})$ as the sum of two terms, the equilibrium distribution and the response to the force field $\boldsymbol{F}$,

$$f(\boldsymbol{p}, \boldsymbol{r}) = f_0(\boldsymbol{p}, \boldsymbol{r}) + f_1(\boldsymbol{p}, \boldsymbol{r}) \tag{9.19}$$

We assume that the function f_1 is a small deviation from the equilibrium $f_0(n)$. By substituting the decomposition (9.19) for f in the Boltzmann equation (9.18), we get,

$$\frac{\partial f_0(\boldsymbol{p}, \boldsymbol{r})}{\partial \boldsymbol{r}} \dot{\boldsymbol{r}} + \frac{\partial f_0(\boldsymbol{p}, \boldsymbol{r})}{\partial \boldsymbol{p}} \boldsymbol{F} + \frac{\partial f_1(\boldsymbol{p}, \boldsymbol{r})}{\partial \boldsymbol{p}} \boldsymbol{F} = -\frac{f(\boldsymbol{p}, \boldsymbol{r}) - f_0(\boldsymbol{p}, \boldsymbol{r})}{\tau} \tag{9.20}$$

We have discarded the term containing $\partial f_1/\partial \boldsymbol{r}$ because it is always much smaller than $\partial f_0/\partial \boldsymbol{r}$. We cannot discard the term containing $\partial f_1/\partial \boldsymbol{p}$ because, in the case of a Lorentz force, which will be treated in chap. 10, the term $(\partial f_0/\partial \boldsymbol{p})\boldsymbol{F}$ vanishes.

In the absence of a Lorentz force, we have simply,

$$\frac{\partial f_0(\boldsymbol{p}, \boldsymbol{r})}{\partial \boldsymbol{r}} \dot{\boldsymbol{r}} + \frac{\partial f_0(\boldsymbol{p}, \boldsymbol{r})}{\partial \boldsymbol{p}} \boldsymbol{F} = -\frac{f_1(\boldsymbol{p}, \boldsymbol{r})}{\tau} \quad \text{(no Lorentz force)} \tag{9.21}$$

This is the result that we will exploit in the remainder of this chapter. Let us remark that a quantum mechanical version of the Boltzmann equation can be worked out.[472] These equations can be obtained by considering the quantum mechanical evolution of a many-electron system coupled to a bath (see §12.7).

9.2.5   OHM'S LAW

We will generalize (9.20) to the case of spin-dependent transport in §9.4. Here, in order to gain familiarity with the application of the Boltzmann description of transport, we derive Ohm's law. In the next section, we will consider thermoelectric and conductivity effects together. The system is assumed uniform; in particular, there is no temperature gradient as we wish to consider Ohm's law under isothermal conditions. Then (9.21) reduces to,

$$\frac{\partial f_0(\boldsymbol{p}, \boldsymbol{r})}{\partial \boldsymbol{p}} \boldsymbol{F} = -\frac{f_1(\boldsymbol{p}, \boldsymbol{r})}{\tau} \tag{9.22}$$

Let us assume that we have charge carriers of charge q with,

$$q = -e \quad (e > 0) \tag{9.23}$$

and that the charge carriers experience an electrostatic potential $\varphi(\boldsymbol{r})$. They are subject to an electric field $-\boldsymbol{\nabla}\varphi$ and the force is given by,

$$\boldsymbol{F} = -e(-\boldsymbol{\nabla}\varphi) = e\boldsymbol{\nabla}\varphi \tag{9.24}$$

We choose a notation that seeks to clarify what we mean by an electric field. In an elementary treatment of Ohm's law, it is common to think of the electric field as being applied. In fact, a voltage is applied at the boundaries of the system and, in cases where no special contact effect takes place (Schottky barriers), this vision is good enough. In this chapter and the next, however, we want to treat thermoelectric effects. We will see that an electric field builds up in the system in response to a temperature gradient. Speaking of an "applied electric field" may bring some confusion in that case. Using $-\boldsymbol{\nabla}\varphi$ is a reminder that we are looking at properties in the bulk of the system.

In a metal, the equilibrium distribution f_0 is the Fermi-Dirac distribution,

$$f_0(\boldsymbol{p}) = f_{FD}\left(\frac{E(\boldsymbol{p}) - \mu}{k_B T}\right) \tag{9.25}$$

Since f_0 depends on $\boldsymbol{p}$ through the energy E only, we write,

$$\frac{\partial f_0}{\partial \boldsymbol{p}} = \frac{\partial f_0}{\partial E}\frac{\partial E}{\partial \boldsymbol{p}} = \frac{\partial f_0}{\partial E}\boldsymbol{v} \tag{9.26}$$

In the last equality, we have used,

$$\frac{\partial E}{\partial \boldsymbol{p}} = \boldsymbol{v} \tag{9.27}$$

This definition of the veloctiy $\boldsymbol{v}$ is obviously correct when assuming parabolic bands. The definition (9.27) is actually quite general, as it conforms to the thermodynamic notion of velocity as conjugate of the momentum $\boldsymbol{p}$. Solving (9.22) for f_1 gives,

$$f_1 = e\tau\left(-\frac{\partial f_0}{\partial E}\right)\boldsymbol{v}\cdot\boldsymbol{\nabla}\varphi \tag{9.28}$$

The current is determined by the velocity $\dot{\boldsymbol{r}}$ of the charge carriers and the distribution function f. Only f_1 contributes to the current since f_0 describes the equilibrium, and at equilibrium, there is no current. Thus, the current is given by summing charge times velocity for all possible values of the momentum $\boldsymbol{p}$. In order to carry out this sum, or the equivalent integral, we need to know the density of points in the momentum phase space, weighted by the statistical distribution $f_1(\boldsymbol{p})$. We determine this density by an incursion into the representation of electrons as Bloch states, such that $\boldsymbol{p} = \hbar\boldsymbol{k}$. Furthermore, we assuming period boundary conditions at the edge of a large volume compared to atomic dimensions. We assume a cube of dimension L. Hence, the momentum is given by multiples of $2\pi/L$ in each direction and the density of points in $\boldsymbol{k}$-space is,

$$\frac{L^3}{(2\pi)^3} \tag{9.29}$$

In $\boldsymbol{k}$, we will have integrals of the form,

$$\int d^3k \frac{L^3}{(2\pi)^3}\cdots \tag{9.30}$$

In general, we will seek physical properties per unit volume, i.e., we take this expression divided by L^3. Furthermore, we consider that there are two possible spin states per wave vector $\boldsymbol{k}$. In terms of $\boldsymbol{p}$ instead of the wave vector $\boldsymbol{k}$, we thus have integrals of the form,

$$\frac{2}{(2\pi\hbar)^3}\int d^3p\ldots \tag{9.31}$$

For example, the current density can be calculated as the integral,

$$\boldsymbol{j} = \frac{2}{(2\pi\hbar)^3}\int d^3p\,(-e\dot{\boldsymbol{r}})\,f_1 \tag{9.32}$$

We note that the units of $\boldsymbol{j}$ are charge per second and per meter^2, as it should be.

We have distinguished between $\boldsymbol{v}$ and $\dot{\boldsymbol{r}}$ in the definition of the current. This precaution is taken in preparation for the notion of anomalous velocity, which is the subject of chapter 11. Here, we simply have,

$$\dot{\boldsymbol{r}} = \boldsymbol{v} \tag{9.33}$$

Thus, the current is given by,

$$\boldsymbol{j} = -\frac{2e^2\tau}{(2\pi\hbar)^3} \int d^3p \left(-\frac{\partial f_0}{\partial E}\right) (\boldsymbol{v} \cdot \boldsymbol{\nabla}\varphi) \, \boldsymbol{v} \tag{9.34}$$

We now choose the x axis to be along the electric field, i.e., $\boldsymbol{E} = E_x \hat{\boldsymbol{x}} = -\nabla_x \varphi \, \hat{\boldsymbol{x}}$. Then,

$$\boldsymbol{j} = \frac{2e^2\tau}{(2\pi\hbar)^3} \int d^3p \left(-\frac{\partial f_0}{\partial E}\right) v_x E_x \boldsymbol{v} \tag{9.35}$$

Let us assume that $\partial f_0/\partial E$ is isotropic. Then, the integral (9.35) is non-vanishing in the x direction only. We further assume, as is customary at reasonable temperatures, that the Fermi-Dirac distribution is a fairly sharp step, so its derivative picks out only contributions near the Fermi level E_F. Thus, we can write,

$$\boldsymbol{j} = \frac{2e^2\tau}{(2\pi\hbar)^3} E_x \hat{\boldsymbol{x}} \int d^3p \left(-\frac{\partial f_0}{\partial E}\right) v_x{}^2 \tag{9.36}$$

The assumption that the Fermi surface is a sphere implies that,

$$\int d^3p \left(-\frac{\partial f_0}{\partial E}\right) v_x{}^2 = \frac{1}{3} \int d^3p \left(-\frac{\partial f_0}{\partial E}\right) v^2 \tag{9.37}$$

Therefore, we may write the current projection j_x as,

$$j_x = E_x \frac{2e^2\tau}{3(2\pi\hbar)^3} \int d^3p \left(-\frac{\partial f_0}{\partial E}\right) v^2 \tag{9.38}$$

Assuming that we have parabolic bands, we can rewrite this integration using,

$$\int d^3p \to 4\pi \int p^2 dp$$
$$v = p/m \tag{9.39}$$
$$\left(-\frac{\partial f_0}{\partial E}\right) v = \left(-\frac{\partial f_0}{\partial E}\right) \frac{\partial E}{\partial p} = -\frac{\partial f_0}{\partial p}$$

This gives,

$$j_x = E_x \frac{2e^2\tau}{3(2\pi\hbar)^3 m} \int 4\pi dp \, p^3 \left(-\frac{\partial f_0}{\partial p}\right) \tag{9.40}$$

Doing an integration by part produces two terms. The first term is proportional to $p^3 f_0 (p \to \infty)$, which vanishes because $f_0 \to 0$ when $p \to \infty$. The second term contains,

$$\frac{2}{(2\pi\hbar)^3} \int 4\pi dp \, p^2 \, f_0 = n \tag{9.41}$$

which is the electron density (9.7). Finally, we get the following expression of **Ohm's law**,

$$j_x = E_x \frac{e^2\tau}{m} n \tag{9.42}$$

Therefore, the conductivity is given by,

$$\sigma = \frac{e^2\tau}{m} n \tag{9.43}$$

This is known as the **Drude conductivity**.

9.2.6 SEEBECK EFFECT

Soon after the discovery of GMR, the so-called magneto-thermopower became a frequent subject of study.[441] Here, we use the Boltzmann equation (9.21) to predict the **Seebeck effect**. Thus, we want to show that a temperature gradient can induce a voltage across the sample in the direction of the temperature gradient. Since the temperature depends on position, the equilibrium distribution f_0 now reads,

$$f_0\left(E(\boldsymbol{p}), \mu(\boldsymbol{r}), T(\boldsymbol{r})\right) = \left(1 + \frac{E(\boldsymbol{p}) - \mu(\boldsymbol{r})}{k_B T(\boldsymbol{r})}\right)^{-1} \tag{9.44}$$

One point must be made clear in using the expression (9.44) for f_0. We speak here of the statistical distribution of microscopic states. The Fermi level μ is determined by filling states from the lowest energy until all electrons are placed on the energy levels. If the electrons are in a region where there is an electrostatic potential $\varphi(\boldsymbol{r})$, then all of the energy levels are shifted accordingly. As a consequence, the Fermi level is shifted by as much. Therefore, the difference $E(\boldsymbol{p}) - \mu(\boldsymbol{r})$ which appears in (9.44) does not change in the presence of an electrostatic potential $\varphi(\boldsymbol{r})$.

Let us now calculate the derivatives that appear in the linearized Boltzmann's equation (9.21). First, we consider $\partial f_0 / \partial \boldsymbol{r}$. By applying the standard rules for derivatives of fractions, we then find,

$$\frac{\partial f_0}{\partial \boldsymbol{r}} = -\frac{\partial f_0}{\partial E}\left(\boldsymbol{\nabla}\mu + \left(\frac{E - \mu}{T}\right)\boldsymbol{\nabla}T\right) \tag{9.45}$$

We now set the x coordinate axis along the temperature gradient, i.e., $\boldsymbol{\nabla}T = \nabla_x T \hat{\boldsymbol{x}}$. We show now that the system responds by having a non-vanishing gradient of electrostatic potential in the x direction, i.e., $\boldsymbol{\nabla}\varphi = \nabla_x \varphi\, \hat{\boldsymbol{x}}$. Hence, anticipating the presence of this gradient, we deduce from the Boltzmann equation (9.21),

$$\frac{\partial f_0}{\partial E}v_x\left(e\nabla_x\varphi\right) - \frac{\partial f_0}{\partial E}\left(\nabla_x\mu + \left(\frac{E - \mu}{T}\right)\nabla_x T\right)v_x + \frac{f_1}{\tau} = 0 \tag{9.46}$$

The chemical potential and the electrostatic potential add up in (9.46). Consequently, let us define the **electrochemical potential**,

$$\bar{\mu} = \mu - e\varphi \tag{9.47}$$

We can solve (9.46) for f_1 and find,

$$f_1 = v_x \tau \frac{\partial f_0}{\partial E}\left(\nabla_x\bar{\mu} + \left(\frac{E - \mu}{T}\right)\nabla_x T\right) \tag{9.48}$$

Since,

$$\frac{1}{e}\nabla_x\bar{\mu} = \frac{1}{e}\boldsymbol{\nabla}\bar{\mu}\cdot\hat{\boldsymbol{x}} = -\boldsymbol{\nabla}\left(\varphi + \frac{\mu}{(-e)}\right)\cdot\hat{\boldsymbol{x}} \tag{9.49}$$

we can consider $(1/e)\nabla_x\bar{\mu}$ as a generalized electric field. In other words, we may define an effective electric field given by,

$$\boldsymbol{E} = -\boldsymbol{\nabla}\left(\varphi + \frac{\mu}{(-e)}\right) \tag{9.50}$$

where $(-e)$ is the charge of the electron (9.23).

The voltage probed by an experimentalist is in effect a difference in the electrochemical potentials of the two contacts to the sample. Now, at each contact, there is an equilibrium between the contact and the material at that contact, characterized by the equality of the electrochemical potentials in the material under test and in the conductor used for the measurement. It is clear that if μ depends on position (because of a temperature gradient for example), then a voltage measurement will be observed that is due to this dependence of the chemical potential. In other words, it makes more physical sense to continue, as in thermodynamics, to work with $\boldsymbol{\nabla}\bar{\mu}$ as a generalized force. We can make one more argument in favor of this approach. The use of the electric field defined in (9.50) would, later in this chapter, lead to an apparent paradox when analyzing the two-current model. We will have two charge carrier types, each with its own chemical potential. If we were to insist on using (9.50) in that context, we would have an electric field that depends on the charge carrier type. This is a rather confusing notion. Hence, we will refrain from thinking in terms of electric fields. A possible exception might be that, at the very end of a calculation, identifying $(1/e)\nabla_x\bar{\mu}$ with an electric field clarifies what we measure practically. To think of $-\boldsymbol{\nabla}\bar{\mu}$ as a generalized force, in the sense of thermodynamics, is less prone to confusion. If we deal with one charge carrier, then at the end of our transport analysis using the Boltzmann formalism, we may choose to write

$$\boldsymbol{\nabla}\bar{\mu} = e\boldsymbol{E} \qquad (e > 0) \tag{9.51}$$

in order to identify our result with the conventional notation of books that are not concerned with spin-dependent transport. When considering spin-dependent transport, the contacts are regions of the sample where the spin-dependence vanishes. If this condition were not fulfilled, the measurement would be ill-defined.

We can also arrive at the out-of-equilibrium distribution (9.48) if we consider a situation where the potential φ is imposed and the temperature gradient is the response to this gradient. Therefore, it makes sense to use the solution (9.48) for f_1 and calculate the current in the x direction, with the current given according to the definition (9.32) as,

$$j_x = \frac{2}{\hbar^3}\int d^3p(-e)v_x f_1 \tag{9.52}$$

Thus we have,

$$j_x = (-e^2\tau)\frac{2}{3\hbar^3}\int(d^3p)v^2\frac{\partial f_0}{\partial E}\left(E_x + \left(\frac{E-\mu}{eT}\right)\nabla_x T\right) \tag{9.53}$$

Isotropy allows us to replace v_x^2 in the integral by $(1/3)v^2$. Then, applying the parabolic band assumption, in particular $\boldsymbol{v} = \boldsymbol{p}/m$ as in (9.39), we obtain,

$$j_x = \frac{e^2\tau}{m}\frac{2}{3\hbar^3}\int(d^3p)p\left(-\frac{\partial f_0}{\partial p}\right)\left(E_x + \left(\frac{E-\mu}{eT}\right)\nabla_x T\right) \tag{9.54}$$

In this equation, there are two integrals that must be worked out, namely,

$$\begin{aligned}
I_1 &= \frac{2}{3\hbar^3}\int(d^3p)\left(-\frac{\partial f_0}{\partial p}\right)p \\
I_2 &= \frac{2}{3\hbar^3}\int(d^3p)\left(-\frac{\partial f_0}{\partial p}\right)p(E-\mu)
\end{aligned} \tag{9.55}$$

In terms of these integrals over momentum space, the current reads,

$$j_x = \frac{e^2\tau}{m}\left(I_1 E_x + \left(I_2\frac{1}{eT}\right)\nabla_x T\right) \tag{9.56}$$

Let us now work out the two integrals I_1 and I_2. We use integration by part to find I_1:

$$I_1 = \frac{2}{3\hbar^3} \int dp 4\pi p^3 \left(-\frac{\partial f_0}{\partial p} \right) = \frac{2}{\hbar^3} \int dp 4\pi p^2 f_0 = n \qquad (9.57)$$

where the last step uses the definition (9.41) of the electron density. The other term of the integration by part vanishes because the integrand vanishes at both bounds. Thus, we have found $I_1 = n$.

To calculate I_2, we write,

$$I_2 = \frac{2}{\hbar^3} \int dp \left(-\frac{\partial f_0}{\partial p} \right) \frac{4\pi p^3}{3} (E - \mu) = \left(\frac{2}{\hbar^3} \int dp \left(-\frac{\partial f_0}{\partial p} \right) \frac{4\pi p^3}{3} \frac{p^2}{2m} \right) - n\mu \qquad (9.58)$$

Integration by part yields,

$$I_2 = \left(\frac{2}{\hbar^3} \int dp f_0 \frac{10\pi p^4}{3m} \right) - n\mu = \left(\frac{5}{3} \int d^3 p f_0 \frac{p^2}{2m} \right) - n\mu = n\Delta E_{th} \qquad (9.59)$$

where the last equality defines a statistical thermal energy ΔE_{th}. Since the last parenthesis represents an average energy, it is an extensive quantity and its value must be proportional to the density n. The integral as we have it is strongly dependent on the assumption that we have free electrons. However, for any other band structure, we could also write $I_2 = n\Delta E_{th}$, with a different definition of ΔE_{th}. More insight into the calculation of I_2 is given in §9.2.8.

We are now ready to conclude on the thermopower. The Seebeck effect is measured under the condition of zero current, i.e., $j_x = 0$. Thus, we have,

$$E_x = - \left(\frac{\Delta E_{th}}{eT} \right) \nabla_x T \qquad (9.60)$$

From the point of view of an experimentalist, this electric field is the gradient of an electro-static potential V, i.e., $\boldsymbol{E} = -\boldsymbol{\nabla} V$. If the sample is a rod, the integral over the length of the rod gives a voltage ΔV for an imposed ΔT. It is customary to define the Seebeck coefficient ε with the sign convention $\Delta V = -\varepsilon \Delta T$. Thus, we have found the Seebeck coefficient to be,

$$\varepsilon = \frac{-1}{e} \left(\frac{\Delta E_{th}}{T} \right) \qquad (9.61)$$

9.2.7 FOURIER AND WIEDEMANN-FRANZ LAW

The thermodynamic theory of irreversible processes shows that the heat current j_Q is the energy current minus the convective current $\mu(j/(-e))$.[2] Thus, the heat current can be written as:

$$j_Q = \frac{2}{\hbar^3} \int d^3 p \, (E - \mu) \, v_x f_1 \qquad (9.62)$$

Let us clarify the analogy with the thermodynamic consideration. Here, the energy should be counted as $E - e\varphi$ and the chemical potential in the presence of electrostatic potential should be $\mu - e\varphi$. It is the difference in currents of these two energies that represents a heat current. This difference is independent of φ. This is quite understandable since $j_Q = T j_s$, where j_s is an entropy current, and the entropy is not modified by the presence of an electrostatic potential.

---

[2]See,[376], Eq. 10.87.

We now apply the solution (9.48) for f_1 to calculate the heat current j_Q, again using the parabolic band assumption and referring to the electric field as in (9.51),

$$j_Q = \frac{e\tau}{m}\frac{2}{3\hbar^3}\int d^3p\,(E-\mu)\,p\frac{\partial f_0}{\partial p}\left(E_x + \left(\frac{1}{e}\frac{E-\mu}{T}\right)\nabla_x T\right) \qquad (9.63)$$

A new integral over momentum space appears in this expression. Let us define, by analogy with I_1 and I_2,

$$I_3 = \frac{2}{3\hbar^3}\int d^3p\left(-\frac{\partial f_0}{\partial p}\right)p(E-\mu)^2 \qquad (9.64)$$

If the function $(E-\mu)^2$ were not in the integrand of I_3, we would simply have I_1, the density

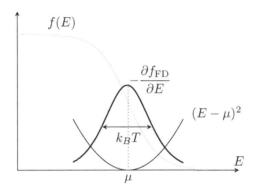

Figure 9.2 The Fermi-Dirac distribution and its derivative. The function $(E-\mu)^2$ is also represented to illustrate the argument concerning the value of I_3.

of electrons, n. We can estimate I_3 by replacing $(E-\mu)^2$ with some average value of this function, which, as shown in Fig. 9.2, is of the order of $(k_B T)^2$ in the interval where $\partial f_0/\partial E$ is non-vanishing. Hence, from this qualitative argument, we have that $I_3 \approx n(k_B T)^2$. Using the Sommerfeld expansion described in §9.2.8, it can be shown that,

$$I_3 = \frac{\pi^2}{3}n(k_B T)^2 \qquad (9.65)$$

The heat current j_Q can be written in terms of the integrals I_2 and I_3 as,

$$j_Q = \frac{e\tau}{m}\left(-I_2 E_x - I_3\frac{1}{eT}\nabla_x T\right) \qquad (9.66)$$

The linear relationships between currents and generalized forces (9.56) and (9.66), can be written as,

$$\begin{pmatrix} j_x \\ j_Q \end{pmatrix} = \begin{pmatrix} L_{11} & L_{12} \\ L_{21} & L_{22} \end{pmatrix}\begin{pmatrix} E_x \\ -\nabla_x T \end{pmatrix} \qquad (9.67)$$

These equations are equivalent to the Onsager relations deduced from the thermodynamics of irreversible processes. The Boltzmann formalism has provided us with the following expressions for the Onsager coefficients,

$$L_{11} = \frac{e^2\tau}{m}I_1 = \frac{e^2\tau}{m}n$$

$$L_{12} = \frac{e^2\tau}{m}\left(\frac{-1}{eT}\right)I_2 \qquad L_{21} = TL_{12} \qquad (9.68)$$

$$L_{22} = \frac{e^2\tau}{m}\left(\frac{1}{e^2T}\right)I_3$$

The electrical conductivity σ is measured in the absence of temperature gradient. From (9.67), we find Ohm's law, as we should, with $L_{11} = \sigma$. This σ is the **isothermal electrical conductivity** that we established in (9.43).

In the absence of electrical current, i.e., $j_x = 0$, the Onsager relations (9.67) imply a coupling between electrochemical potential gradient and temperature gradient. This is the thermoelectric effect or **Seebeck effect**. The Seebeck coefficient ε is defined as,

$$\frac{1}{e}\nabla_x\bar{\mu} = E_x = \varepsilon\nabla_x T \tag{9.69}$$

From the first line of (9.67), we find that the Seebeck coefficient is given by,

$$\varepsilon = \frac{L_{12}}{L_{11}} = \frac{-1}{e}\frac{\Delta E_{th}}{T} \tag{9.70}$$

Once again, let us reflect on the meaning of E_x and $(1/e)\nabla_x\bar{\mu}$. If we think in terms of electric fields, then integrating (9.69) over the length of a sample gives a voltage drop $-\Delta V$ and a temperature rise ΔT, so,

$$\Delta V = -\varepsilon\Delta T \tag{9.71}$$

This is the voltage measured by a high-impedance voltmeter. Another way to think of the result (9.69) is to integrate it over the length of the sample and obtain a rise in electrochemical potential $\Delta\bar{\mu}$. From this point of view, we need to realize that our voltmeter generates a potential difference $\Delta V = \Delta\bar{\mu}/(-e)$, such that no current flows. The leads at the points of contact have the same electrochemical potential as the contacted parts of the sample because the leads and points of contact can be assumed to be at equilibrium with one other.

The equations (9.67) express another transport property. Under the condition of zero charge current, i.e., $j_x = 0$ the second line of (9.67) can be expressed as a linear relationship between heat current and temperature gradient, which is **Fourier's law** of heat conduction. Hence, the thermal conductivity is found and expressed in terms of the Onsager coefficients as,

$$\kappa = \frac{L_{22}L_{11} - L_{12}L_{21}}{L_{11}} \tag{9.72}$$

To the extent that we can neglect the correction terms associated with the thermoelectric effect, i.e., the term $L_{12}L_{21}/L_{11}$ and anticipating on the result (9.81) which is derived by the method described in the next section, the heat conductivity reads,

$$\kappa = \frac{\pi^2}{3}\frac{k_B^2}{e^2}T\sigma \tag{9.73}$$

Thus, the **heat conductivity** is proportional to the electrical conductivity. The result (9.73) is known as the **Wiedemann-Franz law** law and the coefficient $\kappa/(\sigma T)$ is called the **Lorentz number**.

$$L_0 = \frac{\kappa}{\sigma T} = 2.44 \cdot 10^{-8} \text{ W}\Omega\text{K}^{-2} \tag{9.74}$$

Remark on the temperature dependence of the chemical potential

The expression (9.61) for the Seebeck coefficient, which we wrote here in terms of the ratio $\Delta E_{th}/T$ is the result of transport phenomena. This ratio represents an entropy, in conformity with the interpretation of the Seebeck coefficient as an entropy per unit amount of charge carrier, as the theory of irreversible processes shows [376], Eq. 11.95). In recent literature, researchers have observed very large effective Seebeck coefficients, which they attribute to the temperature dependence of the chemical potential. This contribution is the solid-state equivalent to what is observed in thermo-galvanic cells.[376, 473] From thermodynamics, we know that $\partial\mu/\partial T = -s$, the molar entropy ([376], Eq. 8.22). Thus, if a system undergoes a change in entropy that involves the charge carriers, this term will contribute to the Seebeck effect. Some call this solid state contribution the **presence term**. Such a contribution has been relevant in the study of order-disorder phase transitions, which can give rise to very large Seebeck coefficients.[474–477] When the presence term is considered, the conventional contribution is called the **transport term**.

9.2.8 MOTT'S FORMULA

We have found that the transport coefficients in (9.67) can all be determined from three integrals, of which the definitions (9.55) and (9.64) can be cast as,

$$I_\alpha = \frac{2}{3\hbar^3} \int d^3p \left(-\frac{\partial f_0}{\partial p} \right) p(E - \mu)^{(\alpha-1)} \tag{9.75}$$

We can simplify the determination of these integrals over momentum space by introducing an energy dependent conductivity,

$$\sigma(E) = e^2 \tau(E) \int \frac{d\boldsymbol{k}}{4\pi^3} \delta(E - E(\boldsymbol{k})) v_x^2(\boldsymbol{k}) \tag{9.76}$$

Using the energy-dependent conductivity (9.76), we can write,

$$I_\alpha = \int dE \left(-\frac{\partial f_0}{\partial E} \right) (E - \mu)^{(\alpha-1)} \sigma(E) \quad (\alpha = 1, 2, 3) \tag{9.77}$$

At this point, we make use of the fact that, at reasonable temperatures, the Fermi-Dirac distribution is roughly a rectangle. As shown in [72], appendix C, for any function $H(E)$, we can write:

$$\int_{-\infty}^{\infty} H(E) f_0(E, \mu) dE = \int_{-\infty}^{\mu} H(E) dE + \frac{\pi^2}{3} \left(k_B T \right)^2 H'(\mu) \tag{9.78}$$

where $H'(\mu)$ is the derivative of $H(\mu)$. By means of this development, known as a **Sommerfeld expansion**, we find :

$$L_{11} = \sigma \tag{9.79}$$

$$L_{21} = TL_{12} = -\frac{\pi^2}{3e} \left(k_B T \right)^2 \sigma' \tag{9.80}$$

$$L_{22} = \frac{\pi^2}{3} \frac{k_B^2 T}{e^2} \sigma \tag{9.81}$$

Therefore, we find for the Seebeck coefficient (9.70) an expression called the **Mott's formula** :

$$\varepsilon = -\frac{\pi^2}{3} \frac{k_B^2 T}{e} \frac{\sigma'}{\sigma} \tag{9.82}$$

In metals, it turns out that the Seebeck coefficient is roughly proportional to the temperature, as suggested by (9.82). In crystalline samples at low temperatures (typically 5–10 K), there is an additional contribution attributed to the coupling of electrons and phonons called phonon-drag; see §9.3.3. There can also be a magnon drag effect.[478, 479]

It can be shown that the energy derivative σ' of σ depends strongly on the density of states at the Fermi level. Indeed, σ' can be written as,[3]

$$\sigma' = \frac{\tau'}{\tau}\sigma + e^2\tau \int \frac{d^3k}{4\pi^3}\delta(\mu - E(\boldsymbol{k}))M^{-1}(\boldsymbol{k}) \qquad (9.83)$$

where $M(\boldsymbol{k})$ is the effective mass. To the extent that $M(\boldsymbol{k}) = m$ is independent of $\boldsymbol{k}$,(9.83) contains a term proportional to the density of states,[4]

$$DOS(\mu) = \int \frac{d^3k}{4\pi^3}\delta(\mu - E(\boldsymbol{k})) \qquad (9.84)$$

The practical implication of (9.82) is that a large Seebeck coefficient can be expected when a sharp feature appears in the density of states at the Fermi level.

9.3 PHONON DRAG AND MAGNON DRAG

9.3.1 PHENOMENOLOGY

While the Seebeck coefficient of a typical metal is roughly proportional to temperature, an enhanced value can be observed in very pure crystals at low temperatures. For example, the Seebeck coefficient of germanium, which is about 1 mV/K in the temperature range of 100–250 K, shoots up when the temperature is lowered and reaches a maximum of 18 mV/K at about 20 K.[480] It is common to find a broad maximum in the temperature dependence of the Seebeck coefficient of metals at low temperatures. This effect, called *phonon drag*, has to do with electron-phonon collisions, as we shall see in this and the following sections. The temperature cannot be too low, or else the phonon population vanishes. Phonon drag has been observed in a two-dimensional electron gas, which can occur at the interface between two materials.

Likewise, collisions of electrons with magnons can give rise to an enhanced thermoelectric effect called the *magnon drag*.[103, 481] This phenomenon is thought to play a role in the spin Seebeck effect.[482] In a 1967 study on the Seebeck effect in nickel and iron, Blatt et al. concluded that, in nickel (Fig. 9.3, left), the low temperature deviation from a linear temperature dependence was a phonon-drag effect. Notice that annealing, which is presumed to improve on the crystalline microstructure of the sample, enhances this effect. To the contrary, the magnon-drag effect dominated in iron (Fig. 9.3, right). A very large, positive peak was observed near 200 K. Annealing did not have much effect in this case. Adding platinum impurities reduced the magnon-drag effect, presumably because of spin wave scattering at defects.

A heuristic argument can clarify the concept of drag when electron transport is considered in the presence of a temperature gradient.[480] Let us consider phonons (or magnons) as quasi-particles that constitute an ideal gas. This gas exerts a pressure on the electrons. If $U_{ph}(T)$ is the internal energy (in the thermodynamic sense) of this gas, its pressure p is given by,

$$p = \frac{1}{3}U(T) \qquad (9.85)$$

as can be shown by applying the kinetic theory of gases to massless particles.[483] Let us write $U(T) = c_V T$, and assume that the heat capacity per unit volume of the solid c_V

---

[3]See,[72] Eqs. 13.63–13.65.

[4]See,[72], Eq. 8.57.

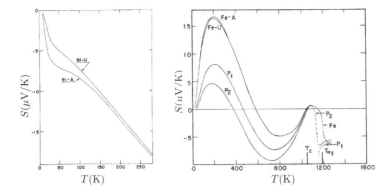

Figure 9.3 Left: Seebeck coefficient S for a nickel sample before (Ni-U) and after annealing (Ni-A). Right: S for an Iron sample before (Fe-U) and after annealing (Fe-A), and for an Iron alloy containing 1 at.% (P_1) or 2 at.% (P_2) platinum. At the Curie temperature (T_C) and the α to γ phase transition, S changes drastically. Figure adapted from [103]

is dominated by either the lattice vibration (in case we are interested in phonon drag) or the spin waves (for the case of magnon drag). The force per unit volume exerted on the electron is $-\boldsymbol{\nabla}p$. In the condition under which the thermoelectric effect is detected, there is no current, so there must be an electric field $\boldsymbol{E}_{\mathrm{drag}}$ acting on the electrons that compensates this force,

$$N(-e)\boldsymbol{E}_{\mathrm{drag}} - \boldsymbol{\nabla}p = 0 \qquad (9.86)$$

where N is the number of electrons per unit volume. Since $\boldsymbol{\nabla}p = c_V\boldsymbol{\nabla}T$, we obtain,

$$\boldsymbol{E}_{\mathrm{drag}} = \frac{c_V}{N(-e)}\boldsymbol{\nabla}T \qquad (9.87)$$

In Iron (Fig. 9.3, right), the deviation from a linear dependence of the temperature was found to follow a $T^{3/2}$ behavior from 15 to 70 K. This is the typical temperature dependence of the magnetic contribution to the heat capacity c_V.

In view of this heuristic argument, we see that, in order to account for the phonon (or magnon) drag effect in the Boltzmann description of transport, we need to consider phonons (or magnons) that are locally at equilibrium. However, they are not at equilibrium on a global scale, since a temperature gradient is forced on the system. The electrons are slightly out of a local equilibrium because of the effective field due to phonon or magnon drag, and because of the Seebeck effect, as described in §9.2.6. In order to account for electron-phonon (or electron-magnon) collisions, we need to refine the relaxation time approximation that we have used so far. We will proceed in two steps, first considering collisions at point defects, then at collisions with wave-like excitations.

In the following, we describe electrons as Bloch states, thus assuming that the material is a crystal. These states would propagate freely in a perfect crystal. Vibrations of the crystal, or spin waves, constitute defects of this ideal crystal and cause the Bloch states to scatter. In order to gain familiarity with a more detailed description of the mechanisms responsible for the Boltzmann distribution relaxation than the relaxation time approximation, we first describe collisions with point defects.

Notice that, in this chapter, we focus on transport properties and momentum relaxation is our main concern. In Part IV, we will study spin relaxation. In §13.2, we will describe the spin relaxation of conduction electrons due to collisions at point defects. In §13.3.2, we will evaluate the electron spin flip rate due to electron-magnon collisions.

9.3.2   SCATTERING AT POINT DEFECTS

In order to describe the interaction of electrons with magnons or phonons, we need to specify the relaxation mechanism leading to the relaxation introduced in §9.2.3. We proceed in two steps. First, we work out the relaxation of the Boltzmann distribution function due to scattering at defects. Then, we examine scattering by phonons or magnons.

To mark the difference between the Boltzmann approach used thus far and a more detailed description of relaxation mechanisms based on Bloch states, we write $p = \hbar k$. A region of phase space around a momentum k is scattered into a state of momentum k' due to a collision of the electron with a point defect. We represent the point defect by a scattering potential V. According to the Born formula, the scattering rate is given by,

$$W(k, k') = \frac{2\pi}{\hbar} |\langle k| V |k'\rangle|^2 \delta(E_k - E_{k'}) \tag{9.88}$$

This scattering rate has the symmetry,

$$W(k, k') = W(k', k) \tag{9.89}$$

Since we are dealing with electrons, we must take into account the exclusion principle. Therefore, a scattering event from state $|k\rangle$ to a state $|k'\rangle$ occurs with a probability proportional to the probability that the state $|k\rangle$ is occupied, and the final state $|k'\rangle$ is not yet occupied. Thus, the transition rate must be proportional to the product,

$$f_0(E_k)(1 - f_0(E_{k'})) \tag{9.90}$$

Therefore, the rate of electrons leaving the region of phase space around k and arriving at k' depletes the density of occupied points at k at the rate,

$$P_- = \sum_{k'} W(k, k') f_0(E_k)(1 - f_0(E_{k'})) \tag{9.91}$$

Likewise, the rate of increase of points fed into the region of phase space near k is,

$$P_+ = \sum_{k'} W(k', k) f_0(E_{k'})(1 - f_0(E_k)) \tag{9.92}$$

Owing to the symmetry (9.89) we obtain,

$$\left.\frac{df}{dt}\right|_{\text{pt defect}} = P_+ - P_- = \sum_{k'} W(k, k')(f_0(E_{k'}) - f_0(E_k)) \tag{9.93}$$

Therefore, the linearized Boltzmann equation, taking into account collisions at point defects, becomes,

$$v_k \cdot \frac{\partial f_0(E_k)}{\partial r} - \frac{e}{\hbar} E \cdot \frac{\partial f_0(E_k)}{\partial k} = \sum_{k'} W(k, k')(f_0(E_k) - f_0(E_{k'})) \quad \text{(no Lorentz force)} \tag{9.94}$$

where the electric field E was defined in (9.50) and $v_k = \partial E_k / \partial(\hbar k)$.

9.3.3   SCATTERING WITH PHONONS OR MAGNONS

Since we have a temperature gradient and we want to consider a local equilibrium, we should consider electron collisions as localized events. This can be done by describing the electron as a wave packet, i.e., a superposition of Bloch states. Likewise, the local vibration of the region of the lattice where the collision takes place can be described in terms of a superposition of phonons (magnons). This approach is customary when treating, e.g., neutron scattering.[156] Here, it is not necessary to go into such details. However, when we refer to a local equilibrium, we are implying a calculation in terms of wave packets.

Let us treat the case of phonons. The calculations would proceed in the same way with magnons. The magnitude of the matrix elements would differ, but the energy and momentum conservation would have exactly the same form. When an electron in a Bloch state $|k\rangle$ scatters into a state $|k'\rangle$ because of a collision with a bath of phonons, a phonon of momentum Q must be created or annihilated, and the total momentum is conserved,

$$k' = k + Q \tag{9.95}$$

This conservation is a consequence of the translational symmetry of the crystal. The statement that a phonon of momentum Q is created or annihilated means that the mode of momentum Q is in a state of excitation that is increased or decreased by the amount of energy $\hbar\omega_Q$.

In order to describe the contribution of electron-phonon collisions to the relaxation of the Boltzmann distribution function f_k of Bloch electrons of momentum k, we take into account the events that bring electrons into state $|k\rangle$ and those that cause the population of state $|k\rangle$ to decrease. Thus, we have,

$$\left.\frac{df_k}{dt}\right|_{\text{ph}} = \sum_{k'} \left[f_{k'}\left(1 - f_k\right) P_{k'k} - f_k \left(1 - f_{k'}\right) P_{kk'} \right] \tag{9.96}$$

where $P_{kk'}$ are scattering rates, and the Fermi-Dirac statistics of electrons is taken into account.

The scattering rates $P_{kk'}$ are proportional to the square of the matrix element of the interaction, $|M_{kk'}(Q)|^2$. The electron-phonon matrix element has the symmetry property :

$$M_{kk'}(Q) = M_{k'k}^*(-Q) \tag{9.97}$$

There are two ways to go from $|k\rangle$ to $|k'\rangle$. One way is to annihilate a phonon of momentum Q. The number of phonons of momentum Q is N_Q, given by the Bose-Einstein distribution. As a consequence of this annihilation, the electron gains this momentum, i.e., $k' = k + Q$. The energy after the collision is the electron energy decreased by the energy $\hbar\omega_Q$ of the annihilated phonon. The phonon annihilation is described by an annihilation operator a_Q, for which ([484], Eq. 19-1),

$$a_Q |N_Q\rangle = \sqrt{N_Q} |N_Q - 1\rangle \tag{9.98}$$

Therefore, the collision rate associated with this process is proportional to N_Q. This annihilation process is accounted for in the first term of (9.100).

The other way to arrive at the same state $|k'\rangle$ by a collision process that creates a phonon of momentum $-Q$, i.e., $k' - Q = k$. The energy of the final state is the sum of the electron and phonon energies. The phonon creation is described by a creation operator $a_{-Q}^\dagger$ for which ([484], Eq. 19-1),

$$a_{-Q}^\dagger |N_{-Q}\rangle = \sqrt{N_{-Q} + 1} |N_{-Q} + 1\rangle \tag{9.99}$$

Thus, we find that the electron-phonon collision rate is given by,

$$
\begin{aligned}
P_{kk'} =& \frac{2\pi}{\hbar} \sum_Q |M_{kk'}(Q)|^2 N_Q \delta\left(\varepsilon_{k'} - \varepsilon_k - \hbar\omega_Q\right) \\
&+ \frac{2\pi}{\hbar} \sum_Q |M_{kk'}(-Q)|^2 \left(N_{-Q} + 1\right) \delta\left(\varepsilon_{k'} - \varepsilon_k + \hbar\omega_{-Q}\right)
\end{aligned}
\tag{9.100}
$$

We assume crystal symmetries such that $\hbar\omega_{-\boldsymbol{Q}} = \hbar\omega_{\boldsymbol{Q}}$. Consequently, we also have $N_{-\boldsymbol{Q}} = N_{\boldsymbol{Q}}$. Using the symmetry property (9.97) and changing the summation variables, it is possible to bring out in $P_{\boldsymbol{k},\boldsymbol{k}'}$ terms that are symmetric in $\boldsymbol{k}$ and $\boldsymbol{k}'$, and to distinguish them for those that are not. Thus, we find,

$$
\begin{aligned}
P_{\boldsymbol{k}\boldsymbol{k}'} &= \frac{2\pi}{\hbar} \sum_{\boldsymbol{Q}} |M_{\boldsymbol{k}\boldsymbol{k}'}(\boldsymbol{Q})|^2 N_{\boldsymbol{Q}} \left[\delta\left(\varepsilon_{\boldsymbol{k}'} - \varepsilon_{\boldsymbol{k}} - \hbar\omega_{\boldsymbol{Q}}\right) + \delta\left(\varepsilon_{\boldsymbol{k}'} - \varepsilon_{\boldsymbol{k}} + \hbar\omega_{-\boldsymbol{Q}}\right) \right] \\
&\quad + \frac{2\pi}{\hbar} \sum_{\boldsymbol{Q}} |M_{\boldsymbol{k}\boldsymbol{k}'}(\boldsymbol{Q})|^2 \delta\left(\varepsilon_{\boldsymbol{k}'} - \varepsilon_{\boldsymbol{k}} + \hbar\omega_{\boldsymbol{Q}}\right) \\
P_{\boldsymbol{k}'\boldsymbol{k}} &= \frac{2\pi}{\hbar} \sum_{\boldsymbol{Q}} |M_{\boldsymbol{k}\boldsymbol{k}'}(\boldsymbol{Q})|^2 N_{\boldsymbol{Q}} \left[\delta\left(\varepsilon_{\boldsymbol{k}'} - \varepsilon_{\boldsymbol{k}} - \hbar\omega_{\boldsymbol{Q}}\right) + \delta\left(\varepsilon_{\boldsymbol{k}'} - \varepsilon_{\boldsymbol{k}} + \hbar\omega_{-\boldsymbol{Q}}\right) \right] \\
&\quad + \frac{2\pi}{\hbar} \sum_{\boldsymbol{Q}} |M_{\boldsymbol{k}\boldsymbol{k}'}(\boldsymbol{Q})|^2 \delta\left(\varepsilon_{\boldsymbol{k}'} - \varepsilon_{\boldsymbol{k}} - \hbar\omega_{\boldsymbol{Q}}\right)
\end{aligned}
\tag{9.101}
$$

Indeed, we see that the first lines of each of these expressions for the rates $P_{\boldsymbol{k}\boldsymbol{k}'}$ and $P_{\boldsymbol{k}'\boldsymbol{k}}$ are identical, whereas the second lines are not, because of the asymmetry (9.97). To clarify this point, we define the symmetric part as,

$$
P_{\boldsymbol{k}\boldsymbol{k}'}^{s} = \frac{2\pi}{\hbar} \sum_{\boldsymbol{Q}} |M_{\boldsymbol{k}\boldsymbol{k}'}(\boldsymbol{Q})|^2 N_{\boldsymbol{Q}} \left[\delta\left(\varepsilon_{\boldsymbol{k}'} - \varepsilon_{\boldsymbol{k}} - \hbar\omega_{\boldsymbol{Q}}\right) + \delta\left(\varepsilon_{\boldsymbol{k}'} - \varepsilon_{\boldsymbol{k}} + \hbar\omega_{-\boldsymbol{Q}}\right) \right]
\tag{9.102}
$$

The symmetric terms $P_{\boldsymbol{k},\boldsymbol{k}'}^{s}$ give rise to a contribution to the collision rate that is just like the impurity term (9.88). Thus, we obtain one contribution to the relaxation due to these symmetric terms,

$$
\left. \frac{\partial f_{\boldsymbol{k}}}{\partial t} \right|_{\text{ph,s}} = \sum_{\boldsymbol{k},\boldsymbol{k}'} P_{\boldsymbol{k}\boldsymbol{k}'}^{s} \left(f_0(E_{\boldsymbol{k}}) - f_0(E_{\boldsymbol{k}'}) \right)
\tag{9.103}
$$

Typically, phonon drag and magnon drag occur at low temperatures. From experiments, we known that impurity scattering dominates over phonon (or magnon) scattering at low temperatures (see, e.g., Fig. 1.11). Thus, we neglect the contribution (9.103) to the collision rate when considering phonon (or magnon) drag.

9.3.4 BOLTZMANN APPROACH

The non-symmetric terms in the expressions (9.101) of $P_{\boldsymbol{k},\boldsymbol{k}'}^{s}$ and $P_{\boldsymbol{k}',\boldsymbol{k}}^{s}$ give rise to the phonon (magnon) drag effect. The phonons (or magnons) constitute a local bath, even though the temperature is not constant across the system. To account for the phonon (magnon) drag effect, we have to assume that the phonons (magnons) are brought slightly out of their local equilibrium. Following Miele et al. ,[485] we write the local phonon (magnon) distribution $N_{\boldsymbol{Q}}$ as,

$$
N_{\boldsymbol{Q}} = N_{0\boldsymbol{Q}} - \frac{\partial N_0}{\partial(\hbar\omega_{\boldsymbol{Q}})} G(\boldsymbol{Q})
\tag{9.104}
$$

where $N_{0\boldsymbol{Q}}$ is the local equilibrium distribution, which is the Bose-Einstein distribution,

$$
N_{0\boldsymbol{Q}}(T(\boldsymbol{r})) = \left[\exp\left(\frac{\hbar\omega_{\boldsymbol{Q}}}{k_B T(\boldsymbol{r})} \right) - 1 \right]^{-1}
\tag{9.105}
$$

and the second term is the response to the electrons, written in a way that is convenient for the rest of the calculations. The term $G(Q)$ characterizes the out-of-equilibrium phonon (magnon) distribution. In the limit of low temperatures, the derivative $-\partial N_0/\partial(\hbar\omega_{\boldsymbol{Q}})$ is approximately $1/(k_B T)$. Thus, the non-symmetric terms of $P_{\boldsymbol{k}'\boldsymbol{k}}$ in the sum (9.96) give rise to the relaxation term,

$$\left.\frac{\partial f_{\boldsymbol{k}}}{\partial t}\right|_{drag} = -\frac{1}{k_B T}\sum_{\boldsymbol{k}',\boldsymbol{Q}} G(\boldsymbol{Q})\left(\Gamma_{\boldsymbol{k}\boldsymbol{k}'}(\boldsymbol{Q}) - \Gamma_{\boldsymbol{k}'\boldsymbol{k}}(\boldsymbol{Q})\right) \qquad (9.106)$$

with

$$\Gamma_{\boldsymbol{k}\boldsymbol{k}'} = \frac{2\pi}{\hbar}|M_{\boldsymbol{k}\boldsymbol{k}'}(\boldsymbol{Q})|^2 f_{0\boldsymbol{k}}\left(1 - f_{0\boldsymbol{k}'}\right)N_{0\boldsymbol{Q}}\delta\left(\varepsilon_{\boldsymbol{k}'} - \varepsilon_{\boldsymbol{k}} - \hbar\omega_{\boldsymbol{Q}}\right) \qquad (9.107)$$

The out-of-equilibrium phonon distribution can be assumed to relax with a single relaxation time τ_{pm}, where the subscript pm stands for phonon or magnon, depending on the drag effect considered. When applying the stationary Boltzmann equation (9.18) to the distribution $N_{\boldsymbol{Q}}$ of phonons (or magnons), we have to replace the velocity $\dot{\boldsymbol{r}}$ by its thermodynamic definition (9.27) as the conjugate variable of momentum.[5] With the energy being $\hbar\omega_{\boldsymbol{Q}}$ and the momentum $\hbar\boldsymbol{Q}$, we thus have,

$$\frac{\partial N_{\boldsymbol{Q}}}{\partial \boldsymbol{r}} \cdot \frac{\partial \omega_{\boldsymbol{Q}}}{\partial \boldsymbol{Q}} = -\frac{1}{\tau_{pm}}\left(N_{\boldsymbol{Q}} - N_{0\boldsymbol{Q}}\right) = \frac{1}{\tau_{pm}}\frac{\partial N_{0\boldsymbol{Q}}}{\partial(\hbar\omega_{\boldsymbol{Q}})}G(\boldsymbol{Q}) \qquad (9.108)$$

We used the definition 9.104 to write the last equality. To the lowest order of approximation, we can replace $N_{\boldsymbol{Q}}$ by $N_{0\boldsymbol{Q}}$ in (9.108). Calculating both derivatives in (9.108) yields an explicit expression for the response coefficient $G(\boldsymbol{Q})$,

$$G(\boldsymbol{Q}) = -\frac{\hbar\omega_{\boldsymbol{Q}}}{T}\tau_{pm}\frac{\partial\omega_{\boldsymbol{Q}}}{\partial\boldsymbol{Q}} \cdot \boldsymbol{\nabla}T \qquad (9.109)$$

From this point on, we follow Miele et al. ,[485] in their analysis of phonon drag. Thus, the relaxation time is written as τ_{ph}. For a simple isotropic model of the material, the phonon (magnon) velocity has the form :

$$\frac{\partial\omega_{\boldsymbol{Q}}}{\partial\boldsymbol{Q}} = s_\lambda\hat{\boldsymbol{Q}} \qquad (9.110)$$

where λ labels the vibration mode (longitudinal or transversal). Given this isotropy, we have $G(\boldsymbol{Q}) = -G(-\boldsymbol{Q})$. The phonon velocity is given by

$$s_\lambda = \frac{\omega_{\boldsymbol{Q}}}{|\boldsymbol{Q}|} \qquad (9.111)$$

Hence, the phonon drag contribution to the collision term of the Boltzmann transport equation reads :

$$\left.\frac{\partial f_{\boldsymbol{k}}}{\partial t}\right|_{drag} = -\frac{1}{k_B T^2}\sum_{\boldsymbol{k}',Q}\hbar\tau_{ph}s_\lambda^2\boldsymbol{\nabla}T \cdot \boldsymbol{Q}\left(\Gamma_{\boldsymbol{k}\boldsymbol{k}'}(Q) + \Gamma_{\boldsymbol{k}'\boldsymbol{k}}(-Q)\right) \qquad (9.112)$$

The sharpness of the step in the Fermi-Dirac distribution implies the following approximations:

$$\begin{aligned}
f_0(E)\left(1 - f_0(E + \hbar\omega)\right)N_0(\omega) &= -k_B T\frac{\partial f_0}{\partial E}\left(N_0(\omega) + f_0(E + \hbar\omega)\right)\\
f_0(E - \hbar\omega)\left(1 - f_0(E)\right)N_0(\omega) &= -k_B T\frac{\partial f_0}{\partial E}\left(N_0(\omega) + 1 - f_0(E - \hbar\omega)\right)
\end{aligned} \qquad (9.113)$$

---

[5]see, e.g., [457], appendix A.

Making use of these approximations, and after some lengthy calculations,[485] the **phonon drag** contribution can be written as,

$$\left.\frac{\partial f_{\boldsymbol{k}}}{\partial t}\right|_{drag} = \frac{\partial f_0}{\partial E} \sum_\lambda \frac{m^* s_\lambda^2 \tau_{ph}}{\tau_{ep}^\lambda} \boldsymbol{v}_{\boldsymbol{k}} \cdot \frac{\boldsymbol{\nabla} T}{T} \tag{9.114}$$

with,

$$\frac{1}{\tau_{ep}^\lambda} = \frac{2\pi}{\hbar} \sum_{\boldsymbol{k}', \boldsymbol{Q}} \left(1 - \frac{k'}{k} \cos\theta_{\boldsymbol{k}\boldsymbol{k}'}\right) |M_{\boldsymbol{k}\boldsymbol{k}'}(Q)|^2$$
$$\left[\left(N_{0Q} + f_0(E')\right)\delta\left(E' - E - \hbar\omega_{\boldsymbol{Q}}\right) + \left(N_{0Q} + 1 - f_0(E')\right)\delta\left(E' - E + \hbar\omega_{\boldsymbol{Q}}\right)\right] \tag{9.115}$$

We note that the phonon drag contribution (9.114) to the relaxation term of the Boltzmann equation (9.94) can be recast as an effective electric field $\boldsymbol{E}_{ph}$,

$$\boldsymbol{E}_{ph} = \sum_\lambda \frac{m^* s_\lambda^2 \tau_{ph}}{\tau_{ep}^\lambda} \frac{\boldsymbol{\nabla} T}{T} \tag{9.116}$$

9.4   SPIN-DEPENDENT TRANSPORT

Let us now turn to a description of spin-dependent transport by expanding on the introductory treatment of §9.2. In this section, we assume that the conduction electron spins can be described as either up or down. In a ferromagnet, we refer to the spin orientation with respect to the magnetization and use the terms majority and minority spins. In a non-magnetic system, we have to decide on the quantization direction with which to apply this formalism. In the Boltzmann description of spin-dependent transport, we suppose that we have two Boltzmann distributions, one for spin up and one for spin down. In this formalism, we consider collisions with or without spin flips. We can also take into account events that can flip the spin without changing the momentum of the electrons, i.e., contributions to spin mixing, first introduced in §8.2.2.

9.4.1   SPIN-DEPENDENT STATISTICAL DISTRIBUTIONS

Let us assume that we can define a statistical distribution in the phase space of positions and momenta of one-electron states $(\boldsymbol{r}, \boldsymbol{p})$ for each spin orientation s (s is "up" or "down", $+$ or $-$, $\uparrow$ or $\downarrow$, depending on the context). This does not preclude transfers between the spin channels, but such transfers must be slow enough that an equilibrium per channel can be defined. Thus, we consider two statistical distributions,

$$f_s\left(\boldsymbol{r}, \boldsymbol{p}, t\right) \qquad (s = \uparrow, \downarrow) \tag{9.117}$$

This description is justified by experimental data: the mean distance between two collisions (which contribute to the momentum relaxation) is much less than the mean distance between two spin flips. So, we can assume that we have two sub-populations of charge carriers that have a well-defined statistical state. Thus, we assume that we can define two local equilibrium distributions, one for each spin:

$$f_{0s}\left(\boldsymbol{r}, \boldsymbol{p}\right) = f_{\text{FD}}\left(\frac{E_s(\boldsymbol{p}) - \mu_s(\boldsymbol{r})}{k_B T}\right) \tag{9.118}$$

Here, f_{FD} is the Fermi-Dirac distribution (9.44), $E_s(\boldsymbol{p})$ is the energy of electrons of momentum $\boldsymbol{p}$ and spin s, $\mu_s(\boldsymbol{r})$ is the position-dependent and spin-dependent chemical potential,

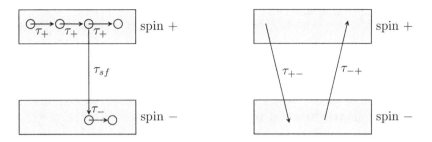

Figure 9.4 Sketch representing two types of collisions, with or without loss of momentum. Left: electrons are accelerated in between collisions and stopped fully at each inelastic collisions. Right: spins flip (i.e., changing channel) without any change in the momentum of the charge carrier.

k_B is the Boltzmann constant, and T is the temperature. In all of the following developments, we will analyze only isothermal systems. We learned from the thermodynamical approach of §8.4 that we ought to consider a position-dependent chemical potential nonetheless.

Let us distinguish two types of scattering events that modify the spin: those that happen without changing the momentum of the electron and those that do (Fig. 9.4). The effect of collisions is included as a distinct contribution to the time evolution of the probability distributions. Thus, we write the spin-dependent extension of the stationary, linearized Boltzmann equation (9.21) as,

$$0 = \frac{\partial f_s}{\partial \boldsymbol{r}} \frac{d\boldsymbol{r}}{dt} + \frac{\partial f_s}{\partial \boldsymbol{p}} \boldsymbol{F} + \frac{\partial f_s}{\partial t}\bigg|_{coll} \tag{9.119}$$

We seek the electronic response to an electric field. Thus, the electrons are subjected to the force,

$$\boldsymbol{F} = e\frac{\partial \varphi}{\partial \boldsymbol{r}} \tag{9.120}$$

This force is the same for both spin orientations. To find the linear response, we write,

$$f_s(\boldsymbol{r},\boldsymbol{p}) = f_{0s}(\boldsymbol{r},\boldsymbol{p}) + f_{1s}(\boldsymbol{r},\boldsymbol{p}) \tag{9.121}$$

where $f_{1s}(\boldsymbol{r},\boldsymbol{p})$ is a small perturbation. In (9.119), we have for the momentum term, to first order:

$$\frac{\partial f_s}{\partial \boldsymbol{p}} \boldsymbol{F} \approx \frac{\partial f_{0s}}{\partial E_s} \frac{\partial E_s}{\partial \boldsymbol{p}} \left(e\frac{\partial \varphi}{\partial \boldsymbol{r}}\right) \tag{9.122}$$

Since here we are concerned with the isothermal conductivity, the position term in (9.119) gives:

$$\frac{\partial f_s}{\partial \boldsymbol{r}} \frac{d\boldsymbol{r}}{dt} \approx -\boldsymbol{v}\frac{\partial f_{0s}}{\partial E_s} \frac{\partial \mu_s(\boldsymbol{r})}{\partial \boldsymbol{r}} \tag{9.123}$$

Here, as in the momentum term (9.122), we keep only the distribution at equilibrium f_{0s} because the spatial variation of the chemical potential comes from the application of the electric field. Hence, the expression (9.123) is of first order in the electric field.

Let us assume that we have parabolic bands that are the same for each spin orientation:

$$E_s(\boldsymbol{p}) = \frac{\boldsymbol{p}^2}{2m} \tag{9.124}$$

From (9.119), using (9.123) and (9.122), we find,

$$\frac{\boldsymbol{p}}{m} \frac{\partial \left(\mu_s\left(\boldsymbol{r}\right) - e\varphi\left(\boldsymbol{r}\right)\right)}{\partial \boldsymbol{r}} \frac{\partial f_{0s}}{\partial E_s} = \left.\frac{\partial f_s}{\partial t}\right|_{coll} \tag{9.125}$$

Just as in thermodynamics or in the spinless thermopower analysis presented in §9.2.6, the chemical potential and electrostatic potential come together. We combine them into a **spin-dependent electrochemical potential**:

$$\bar{\mu}_s\left(\boldsymbol{r}\right) = \mu_s\left(\boldsymbol{r}\right) - e\varphi\left(\boldsymbol{r}\right) \tag{9.126}$$

In summary, the linearized spin-dependent Boltzmann equations reads,

$$\frac{\boldsymbol{p}}{m} \frac{\partial \bar{\mu}_s\left(\boldsymbol{r}\right)}{\partial \boldsymbol{r}} \frac{\partial f_{0s}}{\partial E_s} = \left.\frac{\partial f_s}{\partial t}\right|_{coll} \tag{9.127}$$

9.4.2 COLLISIONS WITH SPIN FLIPS

In order to find an expression for the collision term in (9.127), we rely on a quantum mechanical calculation to provide us with scattering rates $P\left(\boldsymbol{k}, i; \boldsymbol{k}', j\right)$ with $(i = j)$ or without $(i \neq j)$ spin flips. We construct the collision term in (9.127) by counting events that bring spins to the s channel at $\boldsymbol{k}$ and events that remove spins from the s channel at the same $\boldsymbol{k}$:

$$\left.\frac{\partial f_s}{\partial t}\right|_{coll} = \sum_{i=\uparrow}^{\downarrow} \sum_{\boldsymbol{k}'} \left\{ \begin{array}{c} f_i\left(\boldsymbol{k}'\right)\left(1 - f_s\left(\boldsymbol{k}\right)\right) P\left(\boldsymbol{k}', i; \boldsymbol{k}, s\right) \\ -f_s\left(\boldsymbol{k}\right)\left(1 - f_i\left(\boldsymbol{k}'\right)\right) P\left(\boldsymbol{k}, s; \boldsymbol{k}', i\right) \end{array} \right\} \qquad (s = \uparrow, \downarrow) \tag{9.128}$$

We note that we could improve on this picture if we introduced the statistical weight of the initial state as in (12.175). Doing so would introduce a difference in transition rates from up to down spins and down to up. This approach would be necessary if we wanted to describe spin relaxation so as to predict the correct spin temperature.

The Born approximation (or Fermi golden rule) gives the symmetry $P\left(\boldsymbol{k}', i; \boldsymbol{k}, s\right) = P\left(\boldsymbol{k}, s; \boldsymbol{k}', i\right)$. Hence, the collision terms take on the same form as (9.93),

$$\left.\frac{\partial f_s}{\partial t}\right|_{coll} = \sum_{i=\uparrow}^{\downarrow} \sum_{\boldsymbol{k}'} \left\{\left[f_i\left(\boldsymbol{k}'\right) - f_s\left(\boldsymbol{k}\right)\right] P\left(\boldsymbol{k}', i; \boldsymbol{k}, s\right)\right\} \qquad (s = \uparrow, \downarrow) \tag{9.129}$$

In view of the concept of spin mixing, we pay special attention to those collisions that leave the momentum unchanged. As we will see below, these terms give rise to the off-diagonal elements in the the transport matrix that describes the linear relationships between the spin-dependent currents and the gradients of the electrochemical potentials $\bar{\mu}_s\left(\boldsymbol{r}\right)$. Furthermore, we decompose each sum over spin orientations to make explicit the collisions with spin flips and those without. Hence, we rewrite (9.129) as,

$$\begin{aligned} \left.\frac{\partial f_s}{\partial t}\right|_{coll} &= \left[f_{-s}\left(\boldsymbol{k}\right) - f_s\left(\boldsymbol{k}\right)\right] P\left(\boldsymbol{k}, -s; \boldsymbol{k}, s\right) \\ &+ \sum_{\boldsymbol{k}'} \left\{\left[f_s\left(\boldsymbol{k}'\right) - f_s\left(\boldsymbol{k}\right)\right] P\left(\boldsymbol{k}', s; \boldsymbol{k}, s\right)\right\} \\ &+ \sum_{\boldsymbol{k}'} \left\{\left[f_{-s}\left(\boldsymbol{k}'\right) - f_s\left(\boldsymbol{k}\right)\right] P\left(\boldsymbol{k}', -s; \boldsymbol{k}, s\right)\right\} \end{aligned} \tag{9.130}$$

In anticipation of the mathematical expressions that we will handle, we make an "educated guess" as to the form of the perturbation $f_{1s}(\boldsymbol{r},\boldsymbol{p})$ in (9.121). Thus, we write:

$$f_s(\boldsymbol{r},\boldsymbol{p},t) = f_{0s}(\boldsymbol{r},\boldsymbol{p}) + \alpha_s\,\boldsymbol{k}\,\frac{\partial\bar\mu_s(\boldsymbol{r})}{\partial\boldsymbol{r}}\frac{\partial f_{0s}(E_s(\boldsymbol{k}))}{\partial E} \tag{9.131}$$

So long as α_s is not specified, there is no loss of generality in writing the perturbation in this way.

We further assume that the collision centers do not have internal degrees of freedom. In other words, the value of their energies does not change after the collision and, therefore, the electrons undergo elastic collisions. When the collisions are elastic, we have $|\boldsymbol{k}| = |\boldsymbol{k}'|$, since we assume spherical bands, $E_s(\boldsymbol{k}) = E_s(|\boldsymbol{k}|)$. Hence, in (9.130), we have $f_s(\boldsymbol{r},\boldsymbol{k}) = f_s(\boldsymbol{r},\boldsymbol{k}')$. Thus, substituting (9.130) and (9.131) into (9.127), we find,

$$\begin{aligned}
\frac{\boldsymbol{p}}{m}\frac{\partial\bar\mu_s(\boldsymbol{r})}{\partial\boldsymbol{r}}\frac{\partial f_{0s}}{\partial E} &= [f_{-s}(\boldsymbol{k}) - f_s(\boldsymbol{k})]\,P(\boldsymbol{k},-s;\boldsymbol{k},s) \\
&+ \sum_{\boldsymbol{k}'}\alpha_s\,[\boldsymbol{k}' - \boldsymbol{k}]\,\frac{\partial\bar\mu_s(\boldsymbol{r})}{\partial\boldsymbol{r}}\frac{\partial f_{0s}(E_s(\boldsymbol{k}))}{\partial E}P(\boldsymbol{k}',s;\boldsymbol{k},s) \\
&+ \sum_{\boldsymbol{k}'}\left[\alpha_{-s}\,\boldsymbol{k}'\,\frac{\partial\bar\mu_{-s}(\boldsymbol{r})}{\partial\boldsymbol{r}}\frac{\partial f_{0-s}(E_{-s}(\boldsymbol{k}'))}{\partial E}\right. \\
&\qquad\left. -\alpha_s\,\boldsymbol{k}\,\frac{\partial\bar\mu_s(\boldsymbol{r})}{\partial\boldsymbol{r}}\frac{\partial f_{0s}(E_s(\boldsymbol{k}))}{\partial E}\right]P(\boldsymbol{k}',-s;\boldsymbol{k},s)
\end{aligned} \tag{9.132}$$

Let us assume that the scattering potential is spherically symmetric. Then, the scattering probability $P(\boldsymbol{k}',i;\boldsymbol{k},j)$ for any given i and j depends only on the angle θ between $\boldsymbol{k}'$ and $\boldsymbol{k}$. For every $\boldsymbol{k}'$ in (9.132) that has an angle θ with respect to $\boldsymbol{k}$, there is a $\boldsymbol{k}''$ at an angle $-\theta$. The projections normal to $\boldsymbol{k}$ of these $\boldsymbol{k}'$ and $\boldsymbol{k}''$ have contributions to the scattering that cancel each other out. So, in the summation, we can replace $\boldsymbol{k}'$ by $\boldsymbol{k}\cos\theta$:

$$\begin{aligned}
\frac{\boldsymbol{p}}{m}\frac{\partial\bar\mu_s(\boldsymbol{r})}{\partial\boldsymbol{r}}\frac{\partial f_{0s}}{\partial E} &= [f_{-s}(\boldsymbol{k}) - f_s(\boldsymbol{k})]\,P(\boldsymbol{k},-s;\boldsymbol{k},s) \\
&- \sum_{\boldsymbol{k}'}[1 - \cos\theta_{\boldsymbol{k}\boldsymbol{k}'}]\,\alpha_s\boldsymbol{k}\,\frac{\partial\bar\mu_s(\boldsymbol{r})}{\partial\boldsymbol{r}}\frac{\partial f_{0s}(E_s(\boldsymbol{k}))}{\partial E}P(\boldsymbol{k}',s;\boldsymbol{k},s) \\
&+ \sum_{\boldsymbol{k}'}\left[\cos\theta_{\boldsymbol{k}\boldsymbol{k}'}\,\alpha_{-s}\,\boldsymbol{k}\,\frac{\partial\bar\mu_{-s}(\boldsymbol{r})}{\partial\boldsymbol{r}}\frac{\partial f_{0-s}(E_{-s}(\boldsymbol{k}'))}{\partial E}\right. \\
&\qquad\left. -\alpha_s\,\boldsymbol{k}\,\frac{\partial\bar\mu_s(\boldsymbol{r})}{\partial\boldsymbol{r}}\frac{\partial f_{0s}(E_s(\boldsymbol{k}))}{\partial E}\right]P(\boldsymbol{k}',-s;\boldsymbol{k},s)
\end{aligned} \tag{9.133}$$

In this expression, the difference between μ_s and μ_{-s} is not taken into account. That is, in estimating the relaxation rates, the spin dependence of the chemical potential is of the second order. Noting that (9.131) implies,

$$\alpha_s\,\boldsymbol{k}\,\frac{\partial\bar\mu_s(\boldsymbol{r})}{\partial\boldsymbol{r}}\frac{\partial f_{0s}(E_s(\boldsymbol{k}))}{\partial E} = [f_s(\boldsymbol{r},\boldsymbol{p},t) - f_{0s}(\boldsymbol{r},\boldsymbol{p})] \tag{9.134}$$

we get from (9.133),

$$\begin{aligned}
\frac{\boldsymbol{p}}{m}\frac{\partial\bar\mu_s(\boldsymbol{r})}{\partial\boldsymbol{r}}\frac{\partial f_{0s}}{\partial E} &= [f_{-s}(\boldsymbol{k}) - f_s(\boldsymbol{k})]\,P(\boldsymbol{k},-s;\boldsymbol{k},s) \\
&- \sum_{\boldsymbol{k}'}\{[1 - \cos\theta_{\boldsymbol{k}\boldsymbol{k}'}]\,P(\boldsymbol{k}',s;\boldsymbol{k},s)\}[f_s(\boldsymbol{r},\boldsymbol{p},t) - f_{0s}(\boldsymbol{r},\boldsymbol{p})] \\
&- \sum_{\boldsymbol{k}'}\left\{\left[-\cos\theta_{\boldsymbol{k}\boldsymbol{k}'}\frac{\alpha_{-s}}{\alpha_s} + 1\right]P(\boldsymbol{k}',-s;\boldsymbol{k},s)\right\}[f_s(\boldsymbol{r},\boldsymbol{p},t) - f_{0s}(\boldsymbol{r},\boldsymbol{p})]
\end{aligned} \tag{9.135}$$

Given this result, we define the following relaxation rates:

$$\frac{1}{\tau_{\uparrow\downarrow}} = P(\boldsymbol{k}, -s; \boldsymbol{k}, s) \tag{9.136}$$

$$\frac{1}{\tau_s} = \sum_{\boldsymbol{k}'} \left\{ [1 - \cos\theta_{\boldsymbol{k}\boldsymbol{k}'}] P(\boldsymbol{k}', s; \boldsymbol{k}, s) \right\} \tag{9.137}$$

$$\frac{1}{\tau_{sf}} = \sum_{\boldsymbol{k}'} \left\{ \left[1 - \cos\theta_{\boldsymbol{k}\boldsymbol{k}'} \left(\frac{\alpha_{-s}}{\alpha_s} \right) \right] P(\boldsymbol{k}', -s; \boldsymbol{k}, s) \right\} \tag{9.138}$$

The definition (9.136) of $\tau_{\uparrow\downarrow}$ clearly states that the initial and final states before and after the collision have the same $\boldsymbol{k}$ vector. Thus, $1/\tau_{\uparrow\downarrow}$ characterizes the spin mixing rate. The relaxation times defined in (9.136)–(9.138) are implicitly functions of the momentum $\boldsymbol{k}$ and the spin s. With this analysis of the collision terms, the Boltzmann transport equation becomes,

$$\frac{\boldsymbol{p}}{m} \cdot \frac{\partial \bar{\mu}_s(\boldsymbol{r})}{\partial \boldsymbol{r}} \frac{\partial f_{0s}}{\partial E} = -\left(\frac{1}{\tau_s} + \frac{1}{\tau_{sf}} \right) [f_s - f_{0s}] + \frac{f_{-s} - f_s}{\tau_{\uparrow\downarrow}} \qquad (s = \uparrow, \downarrow) \tag{9.139}$$

This is the result established by Fert and Campbell in 1968, when they analyzed spin-dependent transport in nickel and iron.[486]

9.4.3 THE TWO-CURRENT MODEL

Let us apply (9.139) to calculate spin-dependent currents $\boldsymbol{j}_s(\boldsymbol{r})$. This leads to the **two-current model**, first introduced in §1.2.1. Thus, we define the current for each spin channel as,

$$\boldsymbol{j}_s(\boldsymbol{r}) = \sum_{\boldsymbol{k}} (-e) \frac{\hbar \boldsymbol{k}}{m} f_s(\boldsymbol{r}, \boldsymbol{p}) \tag{9.140}$$

We note that the equilibrium distribution does not contribute to the current (there is no net current at zero applied field). Let us multiply (9.139) by $(-e)\hbar\boldsymbol{k}/m$, apply $\boldsymbol{p} = \hbar\boldsymbol{k}$, and sum over all $\boldsymbol{k}$'s:

$$\begin{aligned}
\sum_{\boldsymbol{k}} (-e) \frac{\hbar \boldsymbol{k}}{m} \left(\frac{\hbar \boldsymbol{k}}{m} \cdot \frac{\partial \bar{\mu}_s(\boldsymbol{r})}{\partial \boldsymbol{r}} \right) \frac{\partial f_{0s}}{\partial E} & \\
= -\sum_{\boldsymbol{k}} \left(\frac{1}{\tau_s} + \frac{1}{\tau_{sf}} \right) (-e) \frac{\hbar \boldsymbol{k}}{m} f_s(\boldsymbol{k}) & \\
+ \sum_{\boldsymbol{k}} \frac{1}{\tau_{\uparrow\downarrow}} (-e) \frac{\hbar \boldsymbol{k}}{m} f_{-s}(\boldsymbol{k}) - \sum_{\boldsymbol{k}} \frac{1}{\tau_{\uparrow\downarrow}} (-e) \frac{\hbar \boldsymbol{k}}{m} f_s(\boldsymbol{k}) &
\end{aligned} \tag{9.141}$$

We recognize in the left-hand side of (9.141) the spin-dependent currents, $\boldsymbol{j}_s$ and $\boldsymbol{j}_{-s}$. In the approach of the relaxation time approximation, we take the values of the relaxation times at the Fermi level and write,

$$\sum_{\boldsymbol{k}} (-e) \frac{\hbar \boldsymbol{k}}{m} \left(\frac{\hbar \boldsymbol{k}}{m} \cdot \frac{\partial \bar{\mu}_s(\boldsymbol{r})}{\partial \boldsymbol{r}} \right) \frac{\partial f_{0s}}{\partial E} = -\left(\frac{1}{\tau_s} + \frac{1}{\tau_{sf}} + \frac{1}{\tau_{\uparrow\downarrow}} \right) \boldsymbol{j}_s(\boldsymbol{r}) + \frac{1}{\tau_{\uparrow\downarrow}} \boldsymbol{j}_{-s}(\boldsymbol{r}) \tag{9.142}$$

We treat the right-hand side of (9.142) in a manner similar to the calculation for the spin-less current in §9.2.5. That is, we take into account the fact that only the terms near the Fermi level (the chemical potential) contribute to the current (Fig. 9.5).[6] The equilibrium

---

[6]Mott and Jones, in their wonder book of 1936.

Fermi surface, by definition, yields no current. Thus, the only non-vanishing contributions to the sum in (9.142) come from the slight distortions of the Fermi surface that are due to the force $\boldsymbol{F}$ associated with the applied electric field.

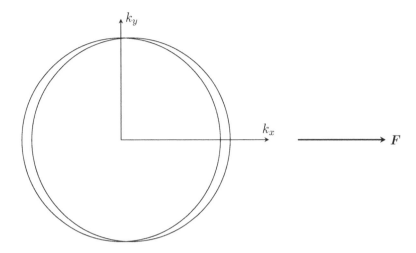

Figure 9.5   A Fermi sphere is shifted when the electrons are subjected to the force $\boldsymbol{F}$.

Let us work out the sum over $\boldsymbol{k}$ in (9.142) for an electric field in the z direction. Then, only the k_z terms contribute. As the other terms of this sum depend on $|\boldsymbol{k}|$ only, $k_z{}^2 = k^2 \cos^2 \theta$ provides the only non-vanishing contributions. Hence, the left-hand side of (9.142) can be written as,[7]

$$\frac{\partial \bar{\mu}_s (\boldsymbol{r})}{\partial z} \int\limits_0^{2\pi} d\varphi \int\limits_{-\pi/2}^{\pi/2} d\theta \sin\theta \int\limits_0^{k_F} \frac{k^2 dk}{(2\pi)^3} (-e) \frac{\hbar^2 k^2}{m^2} \cos^2 \theta \frac{\partial f_{0s}}{\partial E} \tag{9.143}$$

We recall that at low enough temperatures, we can make the approximation,

$$\partial f_{0s}/\partial E \approx -\delta \left(E - \mu_s \right) \tag{9.144}$$

Therefore, the k integral in (9.143) is readily calculated. The angular integral gives a factor of $4\pi/3$. In summary, we have,

$$-\frac{\partial \bar{\mu}_s (\boldsymbol{r})}{\partial z} \frac{4\pi}{3} \frac{k_F{}^4}{(2\pi)^3} (-e) \frac{\hbar^2}{m^2} = \left(\frac{1}{\tau_s} + \frac{1}{\tau_{sf}} + \frac{1}{\tau_{\uparrow\downarrow}} \right) j_{s,z} (\boldsymbol{r}) - \frac{1}{\tau_{\uparrow\downarrow}} j_{-s,z} (\boldsymbol{r}) \tag{9.145}$$

At this point, we want to transform this result in order to identify coefficients that can be understood as resistivity terms. Thus, we bring forth a term which can be thought of as the electric field, in the sense of (9.50). Hence, we write $\bar{\mu}_s (\boldsymbol{r}) = (-e) \left(\varphi (\boldsymbol{r}) - \frac{1}{e} \mu_s (\boldsymbol{r}) \right)$ in (9.145):

$$-\nabla_z \left(\varphi (\boldsymbol{r}) - \frac{1}{e} \mu_s (\boldsymbol{r}) \right) \frac{e^2 k_F{}^2 E_F}{3\pi^2 m} = \left(\frac{1}{\tau_s} + \frac{1}{\tau_{sf}} + \frac{1}{\tau_{\uparrow\downarrow}} \right) j_{s,z} (\boldsymbol{r}) - \frac{1}{\tau_{\uparrow\downarrow}} j_{-s,z} (\boldsymbol{r}) \tag{9.146}$$

---

[7]In §9.2.5, the density of points in k-space was $2/(2\pi)^3$. Here, treating spin dependence, we use $1/(2\pi)^3$.

Now we can identify three resistivities in this equation:

$$\rho_s = \frac{3\pi^2 m}{e^2 k_F{}^2 E_F}\left(\frac{1}{\tau_s} + \frac{1}{\tau_{sf}}\right)$$
$$\rho_{\uparrow\downarrow} = \frac{3\pi^2 m}{e^2 k_F{}^2 E_F}\frac{1}{\tau_{\uparrow\downarrow}} \tag{9.147}$$

The two equations implicit in (9.145) can be written as,

$$\begin{pmatrix} -\boldsymbol{\nabla}\left(\varphi\left(\boldsymbol{r}\right) - \frac{1}{e}\mu_\uparrow\left(\boldsymbol{r}\right)\right) \\ -\boldsymbol{\nabla}\left(\varphi\left(\boldsymbol{r}\right) - \frac{1}{e}\mu_\downarrow\left(\boldsymbol{r}\right)\right) \end{pmatrix} = \begin{pmatrix} \rho_\uparrow + \rho_{\uparrow\downarrow} & -\rho_{\uparrow\downarrow} \\ -\rho_{\uparrow\downarrow} & \rho_\downarrow + \rho_{\uparrow\downarrow} \end{pmatrix}\begin{pmatrix} j_\uparrow\left(\boldsymbol{r}\right) \\ j_\downarrow\left(\boldsymbol{r}\right) \end{pmatrix} \tag{9.148}$$

This matrix is readily inverted, giving:

$$\begin{pmatrix} j_\uparrow\left(\boldsymbol{r}\right) \\ j_\downarrow\left(\boldsymbol{r}\right) \end{pmatrix} = \frac{1}{D}\begin{pmatrix} \rho_\downarrow + \rho_{\uparrow\downarrow} & \rho_{\uparrow\downarrow} \\ \rho_{\uparrow\downarrow} & \rho_\uparrow + \rho_{\uparrow\downarrow} \end{pmatrix}\begin{pmatrix} -\boldsymbol{\nabla}\left(\varphi\left(\boldsymbol{r}\right) - \frac{1}{e}\mu_\uparrow\left(\boldsymbol{r}\right)\right) \\ -\boldsymbol{\nabla}\left(\varphi\left(\boldsymbol{r}\right) - \frac{1}{e}\mu_\downarrow\left(\boldsymbol{r}\right)\right) \end{pmatrix} \tag{9.149}$$

where

$$\frac{1}{D} = \frac{1}{\rho_\uparrow\rho_\downarrow + \rho_{\uparrow\downarrow}\left(\rho_\uparrow + \rho_\downarrow\right)} \tag{9.150}$$

As we know from the thermodynamic description of transport (§8.2.2), the spatial dependence of the chemical potential dies down over a characteristic distance, the spin diffusion length. If we are deep into a homogeneous material, then the terms containing the chemical potentials vanish and the total current $\boldsymbol{j} = \boldsymbol{j}_s + \boldsymbol{j}_{-s}$ is given by,

$$\boldsymbol{j} = \frac{\rho_\downarrow + \rho_\uparrow + 4\rho_{\uparrow\downarrow}}{\rho_\uparrow\rho_\downarrow + \rho_{\uparrow\downarrow}\left(\rho_\uparrow + \rho_\downarrow\right)}\left(-\boldsymbol{\nabla}\varphi\right) \tag{9.151}$$

This is equivalent to the result (1.10), which we derived by applying the principle of detailed balance to two spin channels. This two-current model was first proposed by Mott,[69,70]. It was refined in papers by Fert and Campbell.[75,76] The model can explain deviations of the resistivity from Matthiessen's law [489] in bulk ternary alloys (see §1.2.1).[78]

Spin flip scattering without momentum relaxation will be discussed in §13.3.2, when considering collisions of electron spins with magnons. In most transport experiments, however, this effect is negligible compared to the contribution of collisions to impurities. When spin mixing can be omitted, the off-diagonal elements of (9.149) can be neglected, and we then have,

$$\boldsymbol{j}_s\left(\boldsymbol{r}\right) = \sigma_s\left[-\boldsymbol{\nabla}\left(\varphi\left(\boldsymbol{r}\right) - \frac{1}{e}\mu_s\left(\boldsymbol{r}\right)\right)\right] = \sigma_s\left[\frac{-1}{(-e)}\boldsymbol{\nabla}\bar\mu_s\left(\boldsymbol{r}\right)\right] \tag{9.152}$$

where

$$\sigma_s = \frac{1}{\rho_s} = \frac{e^2 k_F{}^2 E_F}{3\pi^2 m}\left(\frac{1}{\tau_s} + \frac{1}{\tau_{sf}}\right)^{-1} \tag{9.153}$$

9.4.4 DIFFUSION EQUATION AND SPIN ACCUMULATION

Thus, we find Ohm's law for each spin channel. In a one-dimensional system, we have,

$$j_{s,z} = \sigma_s\frac{1}{e}\left(\frac{\partial\bar\mu_s}{\partial z}\right) \tag{9.154}$$

From the continuity equations (8.2) for each electron spin current, we have that, in the stationary regime, the divergence of each spin current is equal to the source of electrons in this channel. This source term is equal to the rate of spin flips causing electrons to come into the channel minus the rate of spin flips causing electrons to leave the channel:

$$div(\boldsymbol{j}_s) = \frac{\partial j_{s,z}}{\partial z} = \int \frac{1}{\hbar^3} d^3p \left(-\frac{f_s - f_{-s}}{\tau_{sf}} \right)(-e) \tag{9.155}$$

In this integrand, we need only consider $(f_{0+} - f_{0-})$ because the terms $f_{1,s}$ and $f_{1,-s}$ are proportional to the electric field, i.e., a perturbation. Graphically, we see that the integral of $f_{0+} - f_{0-}$ is the area between two Fermi-Dirac functions, displaced with respect to one another by $2\Delta\mu$ (Fig. 9.6). Thus, we approximate $f_{0+} - f_{0-}$ with a function proportional to a delta function. Let us work this out explicitly. Developing to first order, we have,

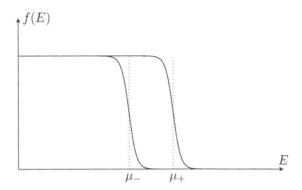

Figure 9.6 Fermi-Dirac distributions with the chemical potentials $\mu_+ = E_F + \Delta\mu$ and $\mu_- = E_F - \Delta\mu$.

$$f_0 \left(\frac{E + \varepsilon}{kT} \right) = f_0(u) = f_0(\varepsilon = 0) + \frac{\partial f_0}{\partial u}\, \varepsilon \frac{1}{kT}$$

$$= f_0(\varepsilon = 0) + \frac{\partial f_0}{\partial u}\, \varepsilon \frac{\partial u}{\partial E} = f_0(\varepsilon = 0) + \frac{\partial f_0}{\partial E}\, \varepsilon \tag{9.156}$$

Therefore,

$$f_{o+} - f_{o-} = f_0 \left(\frac{E - \mu_o - \Delta\mu}{kT} \right) - f_0 \left(\frac{E - \mu_o + \Delta\mu}{kT} \right) = -\frac{\partial f_0}{\partial E}2\Delta\mu \tag{9.157}$$

and (9.155) reads, using the definition (9.153) of σ_s,

$$\frac{\partial j_s}{\partial z} = \frac{\sigma_s}{\frac{1}{3}\left(\tau_s^{-1} + \tau_{sf}^{-1} \right)^{-1} v_F^2 e} \left(\frac{(\bar{\mu}_s - \bar{\mu}_{-s})}{\tau_{sf}} \right) \tag{9.158}$$

Let us distinguish in this expression the **electron mean free path**

$$\lambda_s = \frac{1}{3}\left(\tau_s^{-1} + \tau_{sf}^{-1} \right)^{-1} v_F \tag{9.159}$$

and the **electron spin-flip mean free path**,

$$\lambda_{sf} = v_F\, \tau_{sf} \tag{9.160}$$

Equation (9.158) can be written in terms of the coherence length,

$$l_s{}^2 = \frac{1}{3}\tau_{sf}\left(\tau_s^{-1} + \tau_{sf}^{-1} \right)^{-1} v_F^2 = \lambda_s \lambda_{sf} \tag{9.161}$$

as

$$\frac{e}{\sigma_s}\frac{\partial j_s}{\partial z} = \left(\frac{(\bar{\mu}_s - \bar{\mu}_{-s})}{l_s^2}\right) \tag{9.162}$$

On the basis of (9.162) and (9.154), we find that the chemical potential difference $2\Delta\mu = (\bar{\mu}_+ - \bar{\mu}_-) = (\mu_+ - \mu_-)$ follows a diffusion equation. Indeed, we have,

$$\mu_{\pm} = \mu_o \pm \Delta\mu$$

$$\frac{j_+}{\sigma_+} - \frac{j_-}{\sigma_-} = \frac{1}{e}\left(\frac{\partial\Delta\bar{\mu}}{\partial z}\right) \tag{9.163}$$

$$\frac{1}{e}\frac{\partial^2\Delta\bar{\mu}}{\partial z^2} = \frac{1}{\sigma_+}\frac{\partial j_+}{\partial z} - \frac{1}{\sigma_-}\frac{\partial j_-}{\partial z} = \frac{1}{e}\frac{\Delta\bar{\mu}}{l_+^2} + \frac{1}{e}\frac{\Delta\bar{\mu}}{l_-^2}$$

In short, we have thus obtained a diffusion equation for $\bar{\Delta}\mu$,

$$\frac{\partial^2\Delta\bar{\mu}}{\partial z^2} = \frac{\Delta\bar{\mu}}{l_{sf}^2} \tag{9.164}$$

where,

$$\frac{1}{l_{sf}^2} = \frac{1}{l_+^2} + \frac{1}{l_-^2} \tag{9.165}$$

If we take $\lambda_+ = \lambda_- = \lambda_e$, then :

$$l_{sf}^2 = \frac{1}{6}\lambda_{sf}\lambda_e \tag{9.166}$$

The practical significance of this result is that the spin-flip mean free path may be long, but if the electron mean free path is short, the spin diffusion length is short also.

When a chemical potential difference $\Delta\mu$ builds up at an interface in a transport experiment (Fig. 8.2), it is customary to say that there is **spin accumulation**. This term is used because the spin-dependent chemical potential difference $\Delta\bar{\mu}$ is closely linked to the spin polarization in the system. This can be seen easily by assuming a parabolic band, i.e., $E = \hbar^2 k^2/(2m)$. The relative spin polarization is defined as,

$$\frac{n_+ - n_-}{n_+ + n_-} = \frac{\sum_k (f_+ - f_-)}{\sum_k (f_+ + f_-)} \tag{9.167}$$

We calculate the spin densities n_+ and n_- using (9.41) without the factor 2 since now the densities are for a definite spin. When calculating the denominator in (9.167), we can use the approximation $\mu_1 \approx \mu_- \approx \mu_0$. Thus, we have,

$$n_+ \approx n_- \approx \frac{1}{(2\pi)^3}\int_0^\infty 4\pi k^2 dk \left(1 + \exp\left(\frac{\hbar^2 k^2/2m - \mu_0}{k_B T}\right)\right)^{-1} \tag{9.168}$$

We make the approximation (valid at low temperature) that the Fermi-Dirac distribution has the value 1 up to k_F and 0 beyond k_F, where k_F is defined by

$$\mu_0 = \frac{\hbar^2 k_F^2}{2m} \tag{9.169}$$

Thus, we find,

$$n_+ + n_- \approx 2\frac{4\pi}{(2\pi)^3}\frac{1}{3}k_F^3 \tag{9.170}$$

To calculate $n_+ - n_-$, we use the approximations (9.157) for $(f_+ - f_-)$ and write,

$$\frac{\partial f_0}{\partial E} = \frac{\partial f_0}{\partial k}\left(\frac{\partial E}{\partial k}\right)^{-1} \tag{9.171}$$

Thus, with an integration by parts, we find,

$$n_+ - n_- \approx -2\Delta\mu \frac{4\pi}{(2\pi)^3} \int_0^\infty \left(\frac{m}{\hbar^2}\right) k\,dk \frac{\partial f_0}{\partial k} = 2\Delta\mu \frac{4\pi}{(2\pi)^3} \left(\frac{m}{\hbar^2}\right) k_F \qquad (9.172)$$

So, we find the following relationship between spin polarization and spin-dependence $\Delta\mu$ of the chemical potential,

$$\frac{n_+ - n_-}{n_+ + n_-} = \frac{3\Delta\mu}{2\mu_0} \qquad (9.173)$$

9.5   FURTHER READINGS

Thermoelectric effects

The Seebeck effect (§9.2.6) is remarkably large in a class of layered crystal structures called *delafossite*.[490]

Deviations from the Lorentz number (9.74) would arise if inelastic scattering played a significant role, for example, because of the scattering of electrons with spin waves. It was found that the Wiedemann-Franz law holds in Co/Cu multilayers.[491] Likewise, the topological insulator $Bi_{1-x}Sb_x$, was found to have an enhanced thermal conductivity when driven into a Weyl semimetal state, but nonetheless the Wiedemann-Franz law also applied.[492] In a one-dimensional conductor, namely, lithium molybdenum bronze, the ration $\kappa/(\sigma T)$ diverges at low temperatures, a phenomenon attributed to the spin-charge separation described by the Luttinger liquid model.[493]

The validity limits of the Mott formula can be evaluated using a more elaborate representation of the electronic structure of a solid than the one used in this chapter, accounting for, in particular, electron-phonon interactions.[494]

In the framework of a detailed study of magnon drag in Fe, Co, and Ni, a simple model was proposed to explain magnon drag contribution to the Seebeck effect.[495] The model is based on momentum transfer between two subsystems, electrons and magnons. The approach is akin to the detailed balance (1.6) worked out to account for spin mixing effect. An alternative account of magnon drag was proposed in a study that showed how magnons can be considered as having a defined chemical potential.[457] A thermopile was developed to carry out detailed studies on electron-magnon collisions and the magnon drag effect.[496] A magnon-drag Peltier effect was observed in Ni-Cu alloy,[99] and was considered as a means of achieving cooling.[497]

Just like we defined the spin-dependent conductivities (9.153), it is possible to define *spin-dependent Seebeck coefficients*.[433, 446] In GMR measurements, the magnetic field dependence of the resistivity is measured. Likewise, measurements were carried out on the magnetic field dependence of the Seebeck coefficient of multilayers. The variation of the thermopower with magnetic field is often referred to as magneto-thermoelectric power (MTEP).[440, 441] MTEP has also been measured in spin valves,[498, 499] and tunnel junctions.[500]

Ultra-short laser pulses can give rise to a non-thermal spin-dependent Seebeck effect.[501] The magnon drag contribution to the spin dependence of the thermoelectric power was also investigated.[481]

Spin accumulation and magnetoresistance

Valet and Fert introduced the concept of GMR in a famous paper where they gave several predictions that experimentalists could test.[502] They analyzed the transport with currents perpendicular to the interfaces of a tri-layer, two ferromagnetic layers separated by a non-magnetic layer. This tri-layer was part of a periodic structure representing a multilayer.

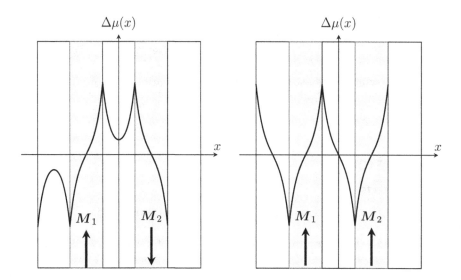

Figure 9.7 Chemical potential difference $\Delta\mu(x)$ as a function of position in an infinite stack of magnetic layers alternating with non-magnetic layers. Left: antiparallel magnetic configuration, presenting a non-vanishing average spin accumulation in the spacer layer. Right: parallel magnetic configuration. Adapted from [502].

On the left-hand side of Fig. 9.7, the magnetization vectors are anti-parallel. On the right-hand side, the magnetization vectors are parallel. Fig. 9.7 shows the spatial variation of the difference in chemical potential $\Delta\mu$ across each layer. Owing to the relation (9.173) between chemical potential and electron spin polarization, this graph represents the local spin polarization. For example, we see that, whether the magnetic configuration is in the parallel or antiparallel configuration, the spin polarization changes sign inside each magnetic layer.

Let us note that it is not necessary to have a spin valve or multilayer to induce spin accumulation. For example, Stiles et al. have calculated the spin accumulation in a single magnetic layer and in the non-magnetic leads to which this layer is electrically connected (Fig. 9.8).[503]

Several groups have studied CPP-GMR in detail. They determined, in particular, the spin asymmetry β that was introduced in §8.3. It should be noted that the definition (8.29) of β refers to a bulk transport property. It turns out that spin asymmetry is also found for the interface resistance and is an important contribution to the GMR effect. A concrete example is shown in Fig. 9.9. This figure brings out the importance of making nanostructures with sizes less than the spin diffusion length since the spin dependence of the conductivity can be manifest in a multilayer only when the typical sizes of the device are less than the spin diffusion length. In chapter 13, we will introduce models of spin-scattering processes responsible for spin relaxation and, ultimately, the spin diffusion length.

In Co/Cu multilayers, β values were found in the range of roughly 0.4–0.5, whereas γ values ranged from 0.7 to 0.8.[504–507] The spin diffusion length l_{sf} defined in (9.165) was deduced from measurements of GMR as a function of the thicknesses of the layers. In Co, it was estimated to be of 44 nm.[505, 506] In the Cu layers, it was shown to be dependent on doping.[507]. In [505, 506], the growth method was such that the spin diffusion length was found to be 140 nm, and in [504], 40 nm. Multilayers composed of Cu and permalloy showed that the permalloy β and γ values are larger than that of Co, with values ranging

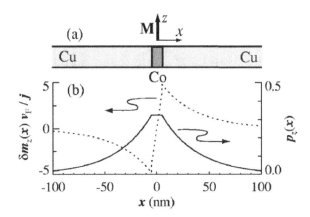

Figure 9.8 Spin polarization (left axis) and relative current polarization at a single cobalt layer contacted with semi-infinite copper leads.[503]

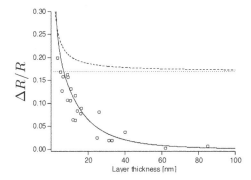

Figure 9.9 Magnetoresistance $\Delta R/R$ of Co/Cu multilayers of equal Co and Cu thicknesses. Dotted line: bulk spin-dependent scattering only ($\beta = 0.4$); dashed line: same with interface scattering ($\gamma = 0.7$); continuous line: same with spin diffusion effects (=40 nm).[504]

from 0.8 to 0.9,[508, 509] while the permalloy spin diffusion length was estimated at 4 nm in [508] and at 5.5 nm in [509].

While the existence of this spin accumulation can be inferred from transport measurements that are interpreted with, e.g., the Valet and Fert model, direct evidence of spin accumulation was also obtained. Positron annihilation was used to detect surface spin accumulation.[510] Kerr microscopy was used to detect spin injection from magnetic electrodes into a III–V semiconductor, GaAs.[511] Muon spin resonance was shown to be able to detect, as a function of depth beneath the injector, the spin polarization obtained by spin injection into an organic semiconductor.[512] There is a close correlation between the spin asymmetry β and the magnetic moment per atom as displayed in a Stoner-Pauling plot (Fig. 9.10).[514, 515] This dependence on composition has been accounted for with a numerical simulation of transport.[516–519]

The Boltzmann formalism can be used to infer that a ***spin-transfer torque*** is expected when a current is driven through a ***spin-valve***.[520] The literature on spin torque is abundant; see, e.g., [521]. Spin torques are introduced in §17.5.

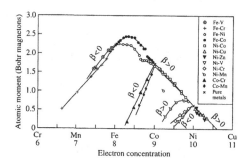

Figure 9.10 Sign of the bulk conductivity spin asymmetry β versus composition, displayed in a so-called Stoner-Pauling plot.[513]

10 Perpendicular Transport Phenomena

Edwin Herbert Hall (1855–1938)

E.H. Hall earned his PhD from Johns Hopkins University in 1880. He discovered the effect that bears his name in 1879 and became professor at Harvard University in 1895. He developed a set of experiments based on very simple equipments that secondary school students could use when preparing for admission at universities. In 1937, he received the award of the American Association of Physics Teachers for his contributions to teaching physics.

Spintronics: A primer

J.-Ph. Ansermet (https://orcid.org/0000-0002-1307-4864)

Chapter 10. Perpendicular Transport Phenomena

The Hall effect and the Nernst effect are described for simple metals as an illustration of the Boltzmann transport equation without spin. The Onsager reciprocity relations are derived using the Boltzmann transport equation. The Hall measurement of Van der Pauw is analyzed using a thermodynamic description of transport.

10.1 HISTORICAL INTRODUCTION

The year Hall discovered the eponymous Hall effect, he wrote an article disclosing his motivation and failed attempts in the American Journal of Mathematics.[522] Hall had been intrigued by a remark Maxwell made in his treatise on electricity and magnetism that the current path distribution in a conductor is the same, whether a magnetic field is applied or not, after transient induced currents have died out. (This point was revisited by van der Pauw more than fifty years later.[523, 524]) Hall found an article that seemed to contradict Maxwell's prediction and decided to test these ideas with an experiment. With the approval of his thesis advisor, Prof. H.A. Rowland, Hall engaged in detecting the magnetoresistance that would result from electrons drawn to the surface of a wire coiled within the gap of an electromagnet. In view of the scatter in his data, Hall concluded that there was no effect detected by his method. Nonetheless, he reasoned that an effect ought to happen and resolved to monitor the voltage point by point in a thin film subjected to a current and a magnetic field. On October 28, 1879, recounts Hall, he saw a clear deviation of his galvanometer needle that could not be attributed to any spurious effect.

There are transport phenomena other than the Hall effect in which the drive (be it a charge or a heat current) and the resulting effect (an electric field or a temperature gradient) are perpendicular, such as,[375, 376]

- the Nernst effect: a heat current giving rise to a voltage on contacts at the edge of the sample,
- the Ettingshausen effect: a charge current inducing a temperature gradient,
- the Righi-Leduc effect: a heat current driving a temperature gradient.

For each of these effects, there can be situations in which the two-current model applies, i.e., one can define spin-dependent Nernst, Ettingshausen or Rigghi-Leduc transport coefficients.

In this chapter, we apply the Boltzmann theory of transport to describe these effects in the most simple case of a metal with a single parabolic band. In doing so, we simply want to illustrate the calculation principles and prepare the reader for the next chapter, which will analyze the Hall effect in a magnetic metal using Berry phase formalism. The Hall effect in conductors with more than one band, like semiconductors, and the case of complex Fermi surfaces are not addressed in this book. Likewise, the quantum Hall effect and the quantum anomalous Hall effects are beyond the scope of this book.

In 1931, Onsager carried out a thermodynamic analysis of transport phenomena and established relationships between a generalized current j_α and a generalized force f_β, so that the overall effect of all the generalized forces on the α current is given by the sum $j_\alpha = \sum_\beta \mathsf{L}_{\alpha\beta} f_\beta$.[376, 525] The same year, a statistical analysis based on the assumption of reciprocity at the microscopic level allowed Onsager to show that the contribution of the force f_β on the current j_α was equal to that of the force f_α on the current j_β.[526] At the end of this chapter, we use the Boltzmann theory of transport to derive this result. The possibility of arriving at this result within the context of the thermodynamics of irreversible processes has been discussed.[527]

10.2 THERMODYNAMIC DEFINITIONS OF PERPENDICULAR TRANSPORT EFFECTS

Let us consider the vectorial relationships between generalized currents and forces that can be obtained in the framework of the thermodynamics of irreversible processes. When there is one substance A, we have,[376]

$$\boldsymbol{j}_s = \mathsf{L}_{ss}(-\boldsymbol{\nabla}T) + \mathsf{L}_{sA}\left(-\boldsymbol{\nabla}\mu_A - q_A\boldsymbol{\nabla}\varphi\right)$$
$$\boldsymbol{j}_A = \mathsf{L}_{As}(-\boldsymbol{\nabla}T) + \mathsf{L}_{AA}\left(-\boldsymbol{\nabla}\mu_A - q_A\boldsymbol{\nabla}\varphi\right) \tag{10.1}$$

where all $\mathsf{L}_{\alpha\beta}$ are tensors. Here, the generalized forces are time-reversal invariant. Therefore, the Onsager reciprocity relations read,[156, 525, 526, 528]

$$\mathsf{L}_{\alpha\beta}\left(\boldsymbol{B}\right) = \mathsf{L}_{\beta\alpha}\left(-\boldsymbol{B}\right) \tag{10.2}$$

This implies that the tensors $\mathsf{L}_{\alpha\alpha}\left(\boldsymbol{B}\right)$ are even functions of $\boldsymbol{B}$, i.e., they are functions of B^2, B^4 ... It also implies that, if $\mathsf{L}_{\alpha\beta}\left(\boldsymbol{B}\right)$ ($\alpha \neq \beta$) contains a term that is an odd function of $\boldsymbol{B}$, that term must be antisymmetric. We could have written (10.1) in cartesian coordinates. Then, this antisymmetry would also be imposed on the cartesian components of the tensors. In particular, the Lorentz force, as we shall see, leads to off-diagonal components that are linear in $\boldsymbol{B}$ field. Thus, we have,

$$\left[\mathsf{L}_{\alpha\beta}\left(\boldsymbol{B}\right)\right]_{ij} = -\left[\mathsf{L}_{\alpha\beta}\left(\boldsymbol{B}\right)\right]_{ji} \qquad i \neq j \ (i,j = x, y, z) \tag{10.3}$$

In §10.7, we will derive these symmetry properties within the Boltzmann description of transport.

Here, the substance A is the charge carrier, assumed to be electrons. Thus, the charge is $q_A = -e$ ($e > 0$). We once again use the electrochemical potential $\bar{\mu} = \mu - e\varphi$. We rewrite (10.1) in terms of the charge current, $\boldsymbol{j}_q = (-e)\boldsymbol{j}_e$ and the heat current $\boldsymbol{j}_Q = T\boldsymbol{j}_s$.

$$\boldsymbol{j}_Q = -T\mathsf{L}_{ss}\boldsymbol{\nabla}T - T\mathsf{L}_{es}(-\boldsymbol{B})\boldsymbol{\nabla}\bar{\mu}$$
$$\boldsymbol{j}_q = e\mathsf{L}_{es}(\boldsymbol{B})\boldsymbol{\nabla}T + e\mathsf{L}_{ee}\boldsymbol{\nabla}\bar{\mu} \tag{10.4}$$

The transverse transport phenomena are characterized by an excitation that is defined for a charge current or a temperature gradient. Therefore, it is convenient to rewrite (10.4) using $\boldsymbol{j}_q$ and $\boldsymbol{\nabla}T$ as if they were the independent variables. Thus, we get from the second equation of (10.4),

$$\boldsymbol{\nabla}\bar{\mu} = \frac{1}{e}\mathsf{L}_{ee}^{-1}\boldsymbol{j}_q - \mathsf{L}_{ee}^{-1}\mathsf{L}_{es}(\boldsymbol{B})\boldsymbol{\nabla}T \tag{10.5}$$

Substituting for $\boldsymbol{\nabla}\bar{\mu}$ in the heat current equation (10.4), we obtain,

$$\boldsymbol{j}_Q = \frac{-T}{e}\mathsf{L}_{es}(-\boldsymbol{B})\mathsf{L}_{ee}^{-1}\boldsymbol{j}_q - \left(T\mathsf{L}_{ss} - T\mathsf{L}_{es}(-\boldsymbol{B})\mathsf{L}_{ee}^{-1}\mathsf{L}_{es}(\boldsymbol{B})\right)\boldsymbol{\nabla}T \tag{10.6}$$

Now we proceed to an identification of the tensors L_{ee} and L_{es} with transport properties of known physical phenomena. We assume that the chemical potential does not vary with position. Therefore, $\boldsymbol{\nabla}\bar{\mu} = -e\boldsymbol{\nabla}\varphi$. The generalized Ohm's law under the isothermal condition reads, $\boldsymbol{\nabla}\varphi = -\boldsymbol{\rho}\boldsymbol{j}_q$.[376] By comparing this with (10.5) under the condition $\boldsymbol{\nabla}T = 0$, we can make the identification,

$$\boldsymbol{\rho}(\boldsymbol{B}) = \frac{1}{e^2}\mathsf{L}_{ee}^{-1} \tag{10.7}$$

In view of the discussion on the symmetry (10.2), we have that $\boldsymbol{\rho}(\boldsymbol{B})$ is an even function of $\boldsymbol{B}$, which is the case in magnetic systems without hysteresis, for example. The generalized Seebeck law reads $\boldsymbol{\nabla}\varphi = -\boldsymbol{\varepsilon}\boldsymbol{\nabla}T$ where $\boldsymbol{\varepsilon}$ is a tensor.[376] By comparing this expression of the thermopower law with (10.5) under the condition $\boldsymbol{j}_q = 0$, we can make the identification,

$$\boldsymbol{\varepsilon}(\boldsymbol{B}) = -e\boldsymbol{\rho}(\boldsymbol{B})\mathsf{L}_{es}(\boldsymbol{B}) \Rightarrow \mathsf{L}_{es}(\boldsymbol{B}) = \frac{-1}{e}\boldsymbol{\rho}^{-1}(\boldsymbol{B})\boldsymbol{\varepsilon}(\boldsymbol{B}) \tag{10.8}$$

In (10.6), we recognize in the second term of the right-hand side a generalized Fourier law $j_Q = -\kappa \nabla T$ under the condition $j_q = 0$. Hereafter, we will specify that the off-diagonal elements of κ may depend on the magnetic field B. Hence, we can now rewrite (10.5) and (10.6) in terms of the resistivity tensor ρ, the thermopower tensor ε and the heat conductivity tensor κ as follows,

$$\nabla \varphi = -\rho(B) j_q - \varepsilon(B) \nabla T$$
$$j_Q = T \rho^{-1}(-B) \varepsilon(-B) \rho(B) j_q - \kappa(B) \nabla T \tag{10.9}$$

Let us now specify the system further. We consider that transverse transport phenomena are measured in a thin film that can be considered as a two-dimensional system. The magnetic field is assumed normal to the thin film, which is isotropic. Finally, we posit that there is no mechanism that generates off-diagonal elements to the symmetric tensors $\mathsf{L}^s_{\alpha\beta}$. Put otherwise, only the magnetic field can cause off-diagonal terms and these terms are antisymmetric. Therefore, the tensors of resistivity, thermopower and heat conductivity have the form,[529]

$$\rho(B) = \rho \begin{pmatrix} 1 & \theta_H \\ -\theta_H & 1 \end{pmatrix}$$
$$\varepsilon(B) = \varepsilon \begin{pmatrix} 1 & \theta_N \\ -\theta_N & 1 \end{pmatrix} \tag{10.10}$$
$$\kappa(B) = \kappa \begin{pmatrix} 1 & \theta_{RL} \\ -\theta_{RL} & 1 \end{pmatrix}$$

where the Hall, Nernst and Righi-Leduc angles are defined, respectively, as,

$$\theta_H = \frac{\rho_{xy}}{\rho_{xx}} \qquad \theta_N = \frac{\varepsilon_{xy}}{\varepsilon_{xx}} \qquad \theta_{RL} = \frac{\kappa_{xy}}{\kappa_{xx}} \tag{10.11}$$

These angles are proportional to the magnetic field. In terms of components, the first line of (10.9) now reads,

$$\nabla_x \varphi = -\rho j_{qx} - \rho \theta_H j_{qy} - \varepsilon \nabla_x T - \varepsilon \theta_n \nabla_y T$$
$$\nabla_y \varphi = \rho \theta_H j_{qx} - \rho j_{qy} + \varepsilon \theta_N \nabla_x T - \varepsilon \nabla_y T \tag{10.12}$$

Let us proceed now with the prediction of transverse transport processes. We start with the *isothermal Hall effect*. We see from (10.12) that we have,

$$\nabla_y \varphi = \rho \theta_H j_{qx} \qquad (0 = j_{qy} = \nabla_x T = \nabla_y T) \tag{10.13}$$

Likewise, we see from (10.12) that we have the *isothermal Nernst effect*,

$$\nabla_y \varphi = \varepsilon \theta_N \nabla_x T \qquad (0 = j_{qx} = j_{qy} = \nabla_y T) \tag{10.14}$$

For the sake of completeness, we define two more transverse transport phenomena, the Ettingshausen and the Rigghi-Leduc effects, which can be deduced from the second equation (10.9). Expressing tensor components in cartesian coordinates, we calculate to first order,

$$\rho^{-1}(-B) \varepsilon(-B) \rho(B) = \varepsilon \begin{pmatrix} 1 & -\theta_N + 2\theta_H \\ \theta_N - 2\theta_H & 1 \end{pmatrix} \tag{10.15}$$

Therefore, by writing (10.9) in terms of tensor components, we get,

$$j_{Qx} = T\varepsilon j_{qx} + T\varepsilon(-\theta_N + 2\theta_H) j_{qy} - \kappa \nabla_x T - \kappa \theta_{RL} \nabla_y T$$
$$j_{Qy} = T\varepsilon(\theta_N - 2\theta_H) j_{qx} - T\varepsilon j_{qy} + \kappa \theta_{RL} \nabla_x T - \kappa \nabla_y T \tag{10.16}$$

The ***Ettingshausen effect*** is the temperature gradient induced by a current in the perpendicular direction under the adiabatic condition. That is,

$$\nabla_y T = -\frac{T\varepsilon(\theta_N - 2\theta_H)}{\kappa} j_{qx} \qquad (0 = j_{qy} = j_{Qy} = \nabla_x T) \qquad (10.17)$$

The ***Righi-Leduc effect*** is the temperature gradient induced in one direction when a temperature gradient is imposed in the perpendicular direction, under the adiabatic condition in the observing direction and zero current in both directions. That is,

$$\nabla_y T = \theta_{RL} \nabla_x T \qquad (0 = j_{Qy} = j_{qx} = j_{qy}) \qquad (10.18)$$

The Nernst effect arguably yields a more efficient thermoelectric generator than the Seebeck effect because, in the Nernst generator, the Ettingshausen effect tends to lower the thermal conductance of the device, whereas in a Seebeck-based device, the Peltier effect enhances the heat flow.[530, 531] Let us conclude this section by remarking that equations (10.12) and (10.9) can be used to define adiabatic Hall and Nernst effects. Experiments have been carried out to show the difference between the isothermal and adiabatic Nernst effects.[532, 533]

10.3   VAN DER PAUW ANALYSIS OF A HALL MEASUREMENT

We want to measure the Hall effect on a thin metallic layer deposited on a flat and insulating substrate. The layer is of a constant thickness, homogeneous and without holes (represented below by the gray area). On the perimeter of this thin layer, four contacts are established that can be considered as point contacts (their size is infinitely smaller than that of the sample). Perpendicular to the layer, a uniform magnetic induction field $\boldsymbol{B}$ is applied on the whole sample. The current I_{in} injected in the layer is set, being maintained as constant throughout. In particular, I_{in} is independent of the magnetic field applied. The current leaving, I_{out}, is also independent of the intensity of the induction field $\boldsymbol{B}$, owing to charge conservation. We make the following hypotheses:

- The gradient of the electrostatic potential φ is given by,

$$\boldsymbol{\nabla}\varphi = -\rho\,\boldsymbol{j}(\boldsymbol{x}) - \mathcal{H}\left(\boldsymbol{j}(\boldsymbol{x}) \times \boldsymbol{B}\right) \qquad (10.19)$$

  where ρ, the resistivity of the metal, and H, the Hall coefficient, are scalar parameters, independent of $\boldsymbol{B}$;
- Measurements are carried out in the stationary regime;
- The metal electron density n_e is independent of time and position;
- Temperature is the same everywhere and independent of time, i.e., there is no Joule effect, not even near the contacts;
- The metal film is so thin that we can admit that $j_z(\boldsymbol{x}) = 0$, where z is normal to the plane of the film; and
- The charge continuity equation reads,

$$\dot{n}_e + \boldsymbol{\nabla} \cdot \boldsymbol{j}_e = \boldsymbol{0} \qquad (10.20)$$

  where $\boldsymbol{j}_e$ is the electron current density. The electronic charge per mole is q_e.

From the phenomenological law (10.19), we have,

$$\begin{pmatrix} \nabla_x\varphi \\ \nabla_y\varphi \\ 0 \end{pmatrix} = \begin{pmatrix} -\rho & 0 & 0 \\ 0 & -\rho & 0 \\ 0 & 0 & 0 \end{pmatrix} \begin{pmatrix} j_x \\ j_y \\ j_z \end{pmatrix} + \mathcal{H}B \begin{pmatrix} j_y \\ -j_x \\ 0 \end{pmatrix} \qquad (10.21)$$

Hence, we can write,

$$\begin{pmatrix} \nabla_x \varphi \\ \nabla_y \varphi \end{pmatrix} = \begin{pmatrix} -\rho & \rho_{xy} \\ -\rho_{xy} & -\rho \end{pmatrix} \begin{pmatrix} j_x \\ j_y \end{pmatrix} \qquad \text{with } \rho_{xy} = \mathcal{H}B \qquad (10.22)$$

This phenomenological equation can be inverted to read,

$$\begin{pmatrix} j_x \\ j_y \end{pmatrix} = \begin{pmatrix} -\sigma & \sigma_{xy} \\ -\sigma_{xy} & -\sigma \end{pmatrix} \begin{pmatrix} \nabla_x \varphi \\ \nabla_y \varphi \end{pmatrix} \qquad (10.23)$$

From the continuity equation (10.20), given that we are working in the stationary regime, we have $\boldsymbol{\nabla} \cdot \boldsymbol{j}_e = \boldsymbol{0}$. As $\boldsymbol{j} = q_e \boldsymbol{j}_e$, we have $\boldsymbol{\nabla} \cdot \boldsymbol{j} = \boldsymbol{0}$, or,

$$\nabla_x j_x + \nabla_y j_y = 0 \qquad (10.24)$$

Using (10.23), we deduce from the continuity equation,

$$\nabla_x j_x = -\sigma \nabla_x^2 \varphi + \sigma_{xy} \nabla_x \nabla_y \varphi \qquad (10.25)$$
$$\nabla_y j_y = -\sigma_{xy} \nabla_x \nabla_y \varphi - \sigma \nabla_y^2 \varphi \qquad (10.26)$$

From (10.23) we can also write,

$$\nabla_y j_x = -\sigma \nabla_y \nabla_x \varphi + \sigma_{xy} \nabla_y^2 \varphi \qquad (10.27)$$
$$\nabla_x j_y = -\sigma_{xy} \nabla_x^2 \varphi - \sigma \nabla_x \nabla_y \varphi \qquad (10.28)$$

Thus, we conclude,

$$\nabla_x j_y = \nabla_y j_x \qquad (10.29)$$

We arrive at the important conclusion that the current $\boldsymbol{j}(\boldsymbol{x})$ is determined by the differential equations (10.24) and (10.29), which are independent of the applied magnetic field. Furthermore, the boundary conditions are also independent of the applied magnetic field, so the current density $\boldsymbol{j}(\boldsymbol{x})$ is independent of the magnetic field. In other words, if we do two measurements, one without a magnetic field and one with a magnetic field applied, we have

$$\begin{pmatrix} \nabla_x \varphi(\boldsymbol{B}) \\ \nabla_y \varphi(\boldsymbol{B}) \end{pmatrix} - \begin{pmatrix} \nabla_x \varphi(\boldsymbol{0}) \\ \nabla_y \varphi(\boldsymbol{0}) \end{pmatrix} = \rho_{xy} \begin{pmatrix} j_y \\ -j_x \end{pmatrix} \qquad (10.30)$$

Let us say that the contacts are at the positions $\boldsymbol{x}_1$ and $\boldsymbol{x}_2$. The voltage at the leads is given by $\varphi(\boldsymbol{x}_1) - \varphi(\boldsymbol{x}_2)$. The change of this voltage when a magnetic field is applied depends only on ρ_{xy} (which depends on $\mathcal{H}$ and $\boldsymbol{B}$) because the current density does not change when a magnetic field is applied. This is what Van der Pauw meant to clarify in his papers.[523,524]

10.4 BOLTZMANN EQUATION IN THE PRESENCE OF A MAGNETIC FIELD

Let us extend the analysis of transport phenomena based on the Boltzmann formalism and consider the isothermal Hall and Nernst effects. Both are measured under the condition of zero transverse current and a magnetic field applied in a direction normal to the sample plane. The sample is assumed to be a thin film (Fig. 10.1).

Let us apply the Boltzmann equation (9.20), in a case where the force on the charge carriers $\boldsymbol{F}$ is given by,

$$\boldsymbol{F} = e\boldsymbol{\nabla}\varphi - e\dot{\boldsymbol{r}} \times \boldsymbol{B} \qquad (10.31)$$

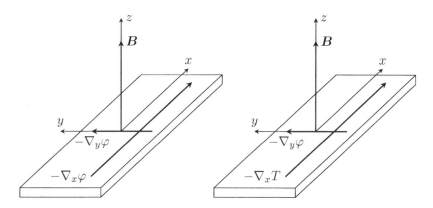

Figure 10.1 Geometry for analyzing perpendicular transport effects. Both are characterized by the transverse field $-\nabla_y\varphi$, where the driving force is in the x direction. It is either the electric field $-\nabla\varphi_x$ (Hall) or the temperature gradient $\nabla_x T$ (Nernst).

We apply (9.26) for $\partial f_0/\partial \boldsymbol{p}$, (9.45) for $\partial f_0/\partial \boldsymbol{r}$ and the definition (9.27) of $\boldsymbol{v}$ as conjugate to $\boldsymbol{p}$. Thus, we find,

$$
\begin{aligned}
&-\frac{\partial f_0}{\partial E}\left[\boldsymbol{\nabla}\mu + \left(\frac{E-\mu}{T}\right)\boldsymbol{\nabla}T\right]\cdot\dot{\boldsymbol{r}} \\
&+\frac{\partial f_0}{\partial E}\boldsymbol{v}\cdot[e\boldsymbol{\nabla}\varphi - e\dot{\boldsymbol{r}}\times\boldsymbol{B}] \\
&+\frac{\partial f_1}{\partial \boldsymbol{p}}\cdot[e\boldsymbol{\nabla}\varphi - e\dot{\boldsymbol{r}}\times\boldsymbol{B}] = -\frac{f_1}{\tau}
\end{aligned}
\tag{10.32}
$$

There are two terms proportional to $e\boldsymbol{\nabla}\varphi$. We consider the second as negligible since,

$$
\frac{\partial f_0}{\partial E}\,\boldsymbol{v}\cdot e\boldsymbol{\nabla}\varphi = \frac{\partial f_0}{\partial \boldsymbol{p}}\cdot e\boldsymbol{\nabla}\varphi \gg \frac{\partial f_1}{\partial \boldsymbol{p}}\cdot e\boldsymbol{\nabla}\varphi
\tag{10.33}
$$

In Chapter 11, we will introduce the notion of anomalous velocity. For this reason, we have been careful to distinguish $\dot{\boldsymbol{r}}$ and the velocity $\boldsymbol{v}$, which we have defined as the conjugate variable of the momentum $\boldsymbol{p}$. However, from now on in this chapter, we will not consider anomalous velocities. Hence, we write $\dot{\boldsymbol{r}} = \boldsymbol{v}$, as we had done in (9.33). Then, since $\boldsymbol{v}\cdot(\boldsymbol{v}\times\boldsymbol{B}) = \boldsymbol{0}$, there is only one non-vanishing term in (10.32) that contains $\boldsymbol{B}$. The possibility that $\boldsymbol{\nabla}\mu \neq 0$ leads to effects we do not address in this introductory derivation of the Hall effect. Hence, we will discard this term. Thus, after a rearrangement of the mixed product, we find,

$$
\left(-\frac{\partial f_0}{\partial E}\right)\left[-e\boldsymbol{\nabla}\varphi + \left(\frac{E-\mu}{T}\right)\boldsymbol{\nabla}T\right]\cdot\boldsymbol{v} + e\boldsymbol{B}\cdot\left[\boldsymbol{v}\times\frac{\partial f_1}{\partial \boldsymbol{p}}\right] = -\frac{f_1}{\tau}
\tag{10.34}
$$

Written in this way, it is evident that $\partial f_1/\partial \boldsymbol{p}$ is not parallel to $\boldsymbol{v}$; otherwise, there would be no magnetic field contributing to f_1.

10.5 HALL EFFECT FOR PARABOLIC BANDS

Let us analyze the electrical conductivity of a non-magnetic metal in parabolic bands,

$$
E(\boldsymbol{p}) = \frac{\boldsymbol{p}^2}{2m}
\tag{10.35}
$$

The metal is subjected to a constant, uniform magnetic induction field $\boldsymbol{B} = B\hat{\boldsymbol{z}}$ (Fig. 10.1) and the sample temperature is kept homogeneous (independent of position). Then (10.34) reduces to,

$$\left(-\frac{\partial f_0}{\partial E}\right) \boldsymbol{\nabla}\bar{\mu} \cdot \boldsymbol{v} + e\boldsymbol{B} \cdot \left[\boldsymbol{v} \times \frac{\partial f_1}{\partial \boldsymbol{p}}\right] = -\frac{f_1}{\tau} \tag{10.36}$$

Taking into account the definition (9.50) of the electric field, we have,

$$-\frac{\partial f_0}{\partial E} \boldsymbol{\nabla}\bar{\mu} = -e\frac{\partial f_0}{\partial E} \boldsymbol{E} \tag{10.37}$$

We want to show that $\boldsymbol{\nabla}_y\bar{\mu}$ is proportional to B. The experimental conditions in which the Hall effect is measured are such that $\boldsymbol{\nabla}_x\bar{\mu}$ must be considered as applied to the system. Then, in $\boldsymbol{\nabla}\bar{\mu}$, the contribution of the electrostatic potential dominates. We also assume that the experimental conditions are such that $\boldsymbol{\nabla}_z\bar{\mu} = 0$, which is the case when the sample is a thin film.

Thus, we obtain,

$$-e(E_x v_x + E_y v_y)\frac{\partial f_0}{\partial E} = -\frac{f_1}{\tau} - eB\left(v_x\frac{\partial f_1}{\partial p_y} - v_y\frac{\partial f_1}{\partial p_x}\right) \tag{10.38}$$

In comparison to our analysis of conductivity in §9.2.5, things are different here for the velocities in the (x, y) plane. Since we are considering parabolic bands, we can write the velocity components as,

$$v_x = \frac{p_x}{m} \qquad v_y = \frac{p_y}{m} \tag{10.39}$$

We are looking for a solution that is linear in p_x and p_y, i.e., of the form,

$$f_1 = a_x p_x + a_y p_y = \boldsymbol{a} \cdot \boldsymbol{p} \tag{10.40}$$

where $\boldsymbol{a} = (a_x, a_y, 0)^T$. Substituting for f_1 in (10.38) and grouping terms, we thus find,

$$\left(-\frac{e}{m}E_x\frac{\partial f_0}{\partial E} + \omega_c a_y + \frac{a_x}{\tau}\right)p_x = \left(-\frac{a_y}{\tau} + \omega_c a_x + \frac{e}{m}E_y\frac{\partial f_0}{\partial E}\right)p_y \tag{10.41}$$

where ω_c is the **cyclotron frequency**,

$$\omega_c = \frac{eB}{m} \tag{10.42}$$

Since p_x and p_y represent microscopic states, they can take on any value. Thus, the only way to satisfy this equation is to set the terms in parentheses to zero. Hence, we find the coupled equations for a and b,

$$\omega_c\tau a_y + a_x = \frac{e\tau}{m}E_x\frac{\partial f_0}{\partial E}$$
$$-\omega_c\tau a_x + a_y = \frac{e\tau}{m}E_y\frac{\partial f_0}{\partial E} \tag{10.43}$$

We can express a_x and a_y deduced from (10.43) in terms of E_x and E_y,

$$\begin{pmatrix} a_x \\ a_y \end{pmatrix} = \frac{e\tau}{m}\frac{\partial f_0}{\partial E}\frac{1}{1+\omega_c^2\tau^2}\begin{pmatrix} 1 & -\omega_c\tau \\ \omega_c\tau & 1 \end{pmatrix}\begin{pmatrix} E_x \\ E_y \end{pmatrix} \tag{10.44}$$

This result can be expressed in vectorial form as,

$$\boldsymbol{a} = \frac{e\tau}{m}\frac{\partial f_0}{\partial E}\frac{1}{1+\omega_c^2\tau^2}\left(\boldsymbol{E} + \omega_c\tau\hat{\boldsymbol{z}} \times \boldsymbol{E}\right) \tag{10.45}$$

We recall that here, E_y is the unknown that defines the strength of the Hall effect and E_x is set by the experiment.

Using for f_1 in (10.40) the value of $\boldsymbol{a}$ given by (10.45), we can calculate the current (9.32) with $\dot{\boldsymbol{r}} = \boldsymbol{v} = \boldsymbol{p}/m$. After an elementary manipulation of the mixed product, the current can be written as,

$$\boldsymbol{j} = \frac{e^2\tau/m^2}{1+\omega_c^2\tau^2} \frac{2}{\hbar^3} \int dp^3 \left(\frac{-\partial f_0}{\partial E}\right) [(\boldsymbol{p} + (\boldsymbol{p} \times \omega_c\tau\hat{\boldsymbol{z}})) \cdot \boldsymbol{E}] \, \boldsymbol{p} \tag{10.46}$$

Notice that the term in square brackets is a scalar. We can readily see in this result a tensorial relationship of the form $\boldsymbol{j} = \boldsymbol{\sigma} \cdot \boldsymbol{E}$, i.e., a generalized Ohm's law. In order to work out the integrals, we expand the scalar product in (10.46) and, because of the spherical symmetry of the dispersion of free electrons, we keep only terms in p_x^2 and p_y^2,

$$\boldsymbol{j} = \frac{e^2\tau/m^2}{1+\omega_c^2\tau^2} \frac{2}{\hbar^3} \int dp^3 \left(\frac{-\partial f_0}{\partial E}\right) \begin{pmatrix} p_x^2 [E_x - \omega_c\tau E_y] \\ p_y^2 [E_y + \omega_c\tau E_x] \end{pmatrix} \tag{10.47}$$

By applying relations (9.39), which are valid for parabolic bands, we get,

$$\int dp^3 \left(\frac{-\partial f_0}{\partial E}\right) p_i^2 = \frac{4\pi m}{3} \int p^2 dp \left(\frac{-\partial f_0}{\partial p}\right) p \quad (i = x, y, z) \tag{10.48}$$

Integration of (10.48) by part gives two terms, one of which vanishes as in (9.40). The other term is,

$$m \int 4\pi p^2 dp f_0 = n\frac{m\hbar^3}{2} \tag{10.49}$$

where n is the electron density. Using the Drude conductivity (9.43) in (10.47), we conclude that the conductivity tensor is given by,

$$\boldsymbol{\sigma} = \frac{ne^2\tau}{m} \begin{pmatrix} \dfrac{1}{1+\omega_c^2\tau^2} & \dfrac{-\omega_c\tau}{1+\omega_c^2\tau^2} & 0 \\ \dfrac{\omega_c\tau}{1+\omega_c^2\tau^2} & \dfrac{1}{1+\omega_c^2\tau^2} & 0 \\ 0 & 0 & 1 \end{pmatrix} \tag{10.50}$$

Here, we have extended the result to three dimensions by including the case of an electric field in the z direction, for which the $\boldsymbol{B}$ field does not change anything and the calculation is the same as in §9.2.5.

The experiment is conducted with the condition of zero current in the y direction. Therefore, the transverse field E_y is given by,

$$E_y = -\omega_c\tau E_x \tag{10.51}$$

Thus, the transverse electric field is proportional to B. This is the intended result.

We remark at this point that we have the antisymmetry,

$$\sigma_{xy} = -\sigma_{yx} \tag{10.52}$$

This antisymmetry is due to the fact that the force $\boldsymbol{F}$ of (9.20) contains the Lorentz force. We will generalize this result in §10.7.

The resistivity is given by $\boldsymbol{E} = \rho\boldsymbol{j}$. The inversion of σ is straightforward and gives,

$$\boldsymbol{\rho} = \frac{m}{ne^2\tau} \begin{pmatrix} 1 & \omega_c\tau & 0 \\ -\omega_c\tau & 1 & 0 \\ 0 & 0 & 1 \end{pmatrix} \tag{10.53}$$

The structure of $\boldsymbol{\rho}$ implies that the electric field in the direction of the current $\boldsymbol{j}$ is not dependent on the magnetic induction field. In other words, there is no longitudinal magnetoresistance in the plane perpendicular to the $\boldsymbol{B}$ field. The origin of this vanishing magnetoresistance can be traced back to the trial solution (10.40) for f_1, that relied on the velocity definitions (10.39), valid for isotropic parabolic band, i.e., $\varepsilon = \varepsilon(|\boldsymbol{k}|)$.

It is customary to characterize the Hall effect by the the coefficient R_H defined by,

$$E_y = R_H B j_x \tag{10.54}$$

The expression (10.53) of ρ implies that,

$$R_H = \frac{-1}{ne} \tag{10.55}$$

When this simple model applies, the Hall measurement can be used to determine the charge density n.

10.6 NERNST EFFECT FOR PARABOLIC BANDS

Let us now show that a temperature gradient imposed in the x direction (Fig. 10.1) induces an electric field in the y direction, under the experimental condition that there is no charge current flowing in either direction. This effect is called the **Nernst effect**. We assume that the experimental conditions are such that the temperature gradient is only in the x direction, i.e., $\nabla_y T = 0$. This will be the case if the substrate is a good thermal conductor. Thus, we are analyzing the Nernst effect under the isothermal condition in the y direction. Hence, we have,

$$\left(-\frac{\partial f_0}{\partial E}\right)\left[\boldsymbol{\nabla}\bar{\mu} + \left(\frac{E-\mu}{T}\right)\nabla_x T \hat{\boldsymbol{x}}\right]\cdot\boldsymbol{v} + e\boldsymbol{B}\cdot\left[\boldsymbol{v}\times\frac{\partial f_1}{\partial\boldsymbol{p}}\right] = -\frac{f_1}{\tau} \tag{10.56}$$

In (10.56), only the in-plane components of the velocity $\boldsymbol{v}$ contribute to the Nernst effect. Thus, we obtain,

$$\left(-\frac{\partial f_0}{\partial E}\right)\left[\boldsymbol{\nabla}\bar{\mu} + \left(\frac{E-\mu}{T}\right)\nabla_x T \hat{\boldsymbol{x}}\right]\cdot\boldsymbol{v} = -\frac{f_1}{\tau} - eB\left(v_x\frac{\partial f_1}{\partial p_y} - v_y\frac{\partial f_1}{\partial p_x}\right) \tag{10.57}$$

Let us again consider the free electron case (10.39). As with the Hall effect, we use a trial solution (10.40) for f_1 that is a linear function of momentum. This yields,

$$\frac{1}{m}\left(\frac{\partial f_0}{\partial E}\right)\left[\boldsymbol{\nabla}\bar{\mu} + \left(\frac{E-\mu}{T}\right)\nabla_x T \hat{\boldsymbol{x}}\right]\cdot\boldsymbol{p} = \frac{\boldsymbol{a}\cdot\boldsymbol{p}}{\tau} + \omega_c\left(p_x a_y - p_y a_x\right) \tag{10.58}$$

As before, this equation is satisfied for any p_x and p_y only if a_x and a_y are such that the coefficients of p_x and p_y vanish. This condition yields two equations for a_x and a_y,

$$\begin{aligned}\frac{\tau}{m}\frac{\partial f_0}{\partial E}\left[\nabla_x\bar{\mu} + \left(\frac{E-\mu}{T}\right)\nabla_x T\right] &= a_x + \omega_c\tau a_y \\ \frac{\tau}{m}\frac{\partial f_0}{\partial E}\left[\nabla_y\bar{\mu}\right] &= a_y - \omega_c\tau a_x\end{aligned} \tag{10.59}$$

In the y direction, there is no temperature gradient, so $\nabla_y\bar{\mu} = -e\nabla_y\varphi = eE_y$. In the x direction, to the extent that we are concerned with only transport thermopower (see §9.2.7), we can also write $\nabla_x \bar{\mu} = (-e)\nabla_x \varphi = eE_x$. Then, we have two equations to solve for a_x and a_y,

$$\frac{\tau}{m}\frac{\partial f_0}{\partial E}\left[eE_x + \left(\frac{E-\mu}{T}\right)\nabla_x T\right] = a_x + \omega_c \tau a_y$$

$$\frac{\tau}{m}\frac{\partial f_0}{\partial E}\left[eE_y\right] = -\omega_c \tau a_x + a_y \tag{10.60}$$

Therefore, we obtain,

$$a_x = \frac{\frac{\tau}{m}\frac{\partial f_0}{\partial E}}{1+\omega_c^2\tau^2}\left[eE_x - eE_y\omega_c\tau + \left(\frac{E-\mu}{T}\right)\nabla_x T\right] \tag{10.61}$$

$$a_y = \frac{\frac{\tau}{m}\frac{\partial f_0}{\partial E}}{1+\omega_c^2\tau^2}\left[eE_x\omega_c\tau + eE_y + \omega_c\tau\left(\frac{E-\mu}{T}\right)\nabla_x T\right] \tag{10.62}$$

Since we started out with the linearized Boltzmann equation, second order terms should be discarded in the solutions for a_x and a_y. For example, the term containing $E_y\omega_c$ is of the order of B^2, because E_y is of the order of B. Here, E_x is the linear response to $\nabla_x T$ and its product with $\omega_c\tau$ should also be discarded. Hence, we keep,

$$a_x = \frac{\frac{\tau}{m}\frac{\partial f_0}{\partial E}}{1+\omega_c^2\tau^2}\left[eE_x + \left(\frac{E-\mu}{T}\right)\nabla_x T\right]$$

$$a_y = \frac{\frac{\tau}{m}\frac{\partial f_0}{\partial E}}{1+\omega_c^2\tau^2}\left[eE_y + \omega_c\tau\left(\frac{E-\mu}{T}\right)\nabla_x T\right] \tag{10.63}$$

Thus, owing to (10.40), we have expressed f_1 in terms of E_x, E_y and $\nabla_x T$. We can calculate the current $\boldsymbol{j}$ according to (9.32), using $\dot{\boldsymbol{r}} = \boldsymbol{v} = \boldsymbol{p}/m$. Retaining only terms in p_x^2 and p_y^2 because of the parabolic band assumption, and applying $\partial f_0/\partial E = (m/p)\partial f_0/\partial p$, we get,

$$j_x = \frac{\frac{e^2\tau}{m}}{1+\omega_c^2\tau^2}\frac{2}{3(2\pi\hbar)^3}\int d^3 p \frac{-\partial f_0}{\partial p}p\left[E_x + \left(\frac{E-\mu}{eT}\right)\nabla_x T\right]$$

$$j_y = \frac{\frac{e^2\tau}{3m}}{1+\omega_c^2\tau^2}\frac{2}{(2\pi\hbar)^3}\int d^3 p \frac{-\partial f_0}{\partial p}p\left[E_y + \omega_c\tau\left(\frac{E-\mu}{eT}\right)\nabla_x T\right] \tag{10.64}$$

Using the definitions (9.55) of I_1 and I_2, we obtain,

$$j_x = \frac{\frac{e^2\tau}{m}}{1+\omega_c^2\tau^2}\left(I_1 E_x + I_2\left(\frac{1}{eT}\right)\nabla_x T\right)$$

$$j_y = \frac{\frac{e^2\tau}{3m}}{1+\omega_c^2\tau^2}\left(I_1 E_y + I_2\omega_c\tau\left(\frac{1}{eT}\right)\nabla_x T\right) \tag{10.65}$$

The Nernst measurement is done under the experimental condition that no charge current is flowing. Thus, under the conditions $j_x = 0 = j_y$ (10.65), yields two equations for the two unknowns E_x and E_y. The expression for E_x is identical to the one obtained in (9.60). The expression for E_y corresponds to the Nernst effect, with E_y given by,

$$E_y = \frac{I_2}{I_1}\omega_c\tau\left(\frac{1}{eT}\right)\nabla_x T \tag{10.66}$$

10.7 ONSAGER RECIPROCITY RELATIONS

Let us use the Boltzmann theory of transport to demonstrate (10.2) and (10.3). We again use the relaxation time approximation to represent the effect of collisions (see §9.2.3). However, we want to introduce here the notion that the transport properties fluctuate. In order to do this, we introduce the following distributions:[488]

- $f_0(t)$, the local equilibrium distribution at location $r(t)$ and momentum $p(t)$ when smoothly varying fields are applied; and
- $f(t)$, the statistical distribution when time-fluctuating fields are also applied.

The notation $f_0(t)$ stands for $f_0(r(t), p(t))$ where $r(t)$ and $p(t)$ are solutions to the semi-classical equations,

$$v = \dot{r} = \frac{\partial E}{\partial p}$$
$$\dot{p} = -eE - ev \times B \tag{10.67}$$

The notation $f(t)$ stands for the distribution $f(t, r, p)$ where (r, p) are any points in phase space.

We apply the relaxation condition (9.15) to the distribution $f(t)$. We thus have a first-order linear differential equation for $f(t)$ containing $f_0(t)$ as an inhomogeneous term,

$$\frac{df}{dt} + \frac{f}{\tau} = \frac{f_0}{\tau} \tag{10.68}$$

Without the inhomogeneous term, the solution to this equation would be $f(t) = c \exp\left(-\frac{t}{\tau}\right)$. Without loss of generality, we consider c to be a function of time, $c(t)$. Substituting for $c(t)$ in (10.68) yields,

$$\frac{dc}{dt} = \frac{f_0}{\tau} \exp\left(\frac{t}{\tau}\right) \tag{10.69}$$

We integrate from $-\infty$ to a time t. We set $c(-\infty) = 0$ so that $f(t)$ remains finite at all times. Thus, we obtain,

$$c(t) = \int_{-\infty}^{t} dt' \frac{f_0(t')}{\tau} \exp\left(\frac{t'}{\tau}\right) \tag{10.70}$$

and the Boltzmann distribution is now expressed in the integral representation,

$$f(t, r, p) = \int_{-\infty}^{t} dt' \frac{f_0(t')}{\tau} \exp\left(-\frac{t - t'}{\tau}\right) \tag{10.71}$$

We notice that what happens at a time t' in the distant past does not contribute significantly to $f(t)$ because of the exponential decay in the integrand. Integration by part of (10.71), i.e.,

$$f_0(t') = u \Rightarrow du = \frac{df_0(t')}{dt}$$
$$\frac{1}{\tau} \exp\left(-\frac{t - t'}{\tau}\right) dt' = dv \Rightarrow v = \exp\left(-\frac{t - t'}{\tau}\right) \tag{10.72}$$

This allows us to write $g(t)$ as follows:

$$f(t, r, p) = f_0(r(t), p(t)) - \int_{-\infty}^{t} dt' \frac{df_0(r(t'), p(t'))}{dt'} \exp\left(-\frac{t - t'}{\tau}\right) \tag{10.73}$$

Now, we develop the derivative of $f_0(t)$ with respect to time. First, we notice that,

$$\frac{df_0\left(\boldsymbol{r}(t),\boldsymbol{p}(t)\right)}{dt} = \frac{\partial f}{\partial \boldsymbol{r}}\boldsymbol{v} + \frac{\partial f}{\partial \boldsymbol{p}}\dot{\boldsymbol{p}} \tag{10.74}$$

Using (9.44) for $f_0(t)$, we can work out that,

$$\frac{\partial}{\partial \boldsymbol{r}}\left(1 + \exp\left(\frac{E-\mu}{k_B T}\right)\right)^{-1} = -\left(1 + \exp\left(\frac{E-\mu}{k_B T}\right)\right)^{-2}\exp\left(\frac{E-\mu}{k_B T}\right)$$
$$\left[\frac{1}{k_B T}\left(-\frac{\partial \mu}{\partial \boldsymbol{r}} - \frac{\mu}{k_B T^2}\frac{\partial T}{\partial \boldsymbol{r}}\right)\right] \tag{10.75}$$

and note that,

$$\frac{\partial}{\partial E}\left(1 + \exp\left(\frac{E-\mu}{k_B T}\right)\right)^{-1} = -\left(1 + \exp\left(\frac{E-\mu}{k_B T}\right)\right)^{-2}\exp\left(\frac{E-\mu}{k_B T}\right)\frac{1}{k_B T} \tag{10.76}$$

Therefore,

$$\frac{\partial f_0}{\partial \boldsymbol{r}} = \frac{\partial f_0}{\partial E}\left(-\frac{\partial \mu}{\partial \boldsymbol{r}} - \frac{E-\mu}{T}\frac{\partial T}{\partial \boldsymbol{r}}\right) \tag{10.77}$$

$$\frac{\partial f_0}{\partial \boldsymbol{p}} = \frac{\partial f_0}{\partial E}\frac{\partial E}{\partial \boldsymbol{p}} = \frac{\partial f_0}{\partial E}\boldsymbol{v} \tag{10.78}$$

We also note that $\partial f_0/\partial E = -\partial f_0/\partial \mu$. We can insert (10.77) and (10.78) into (10.73) to produce the following expression for the Boltzmann distribution f,

$$f(t,\boldsymbol{r},\boldsymbol{p}) = f_0(\boldsymbol{r}(t),\boldsymbol{p}(t))$$
$$- \int_{-\infty}^{t} dt' \exp\left(-\frac{t-t'}{\tau}\right)\boldsymbol{v}\left[-e\boldsymbol{\nabla}\varphi + \boldsymbol{\nabla}\mu + \left(\frac{E-\mu}{T}\right)\frac{\partial T}{\partial \boldsymbol{r}}\right]\frac{\partial f_0(t')}{\partial \mu} \tag{10.79}$$

The term $\boldsymbol{v}\times\boldsymbol{B}$ does not appear because $\boldsymbol{v}\cdot(\boldsymbol{v}\times\boldsymbol{B}) = 0$.

We remark that if we push the relaxation time approximation further, and assume that in the integrand of (10.79) only the exponential is strongly varying in time, then we have,

$$\int_{-\infty}^{t} dt' \exp\left(-\frac{t-t'}{\tau}\right) = \tau \tag{10.80}$$

As a consequence, we have,

$$f = f_0 - \boldsymbol{v}\tau\left[-e\boldsymbol{\nabla}\varphi + \boldsymbol{\nabla}\mu + \frac{\varepsilon-\mu}{T}\boldsymbol{\nabla}T\right]\frac{\partial f}{\partial \mu} \tag{10.81}$$

where f_0 is the Fermi-Dirac distribution (9.44). This result is equivalent to (9.48), as it should be. One word of caution: it looks like the magnetic induction field has disappeared from our analysis. This is not the case, because $\boldsymbol{v}$ should be understood as satisfying the dynamical equations (10.67), that depend on $\boldsymbol{B}$.

We will now make use of (10.79) to demonstrate the Onsager reciprocity relations. Onsager made the hypothesis that the linear transport relations that are defined for macroscopic properties are also valid when applied to the fluctuating quantities corresponding to these macroscopic transport properties. Thus, we need to recall here the key results of the thermodynamics of irreversible processes, which deal with macroscopic properties. In the thermodynamic approach, any small volume of the system is assumed to be in a local thermodynamic equilibrium. The energy density e obeys the continuity equation,[376]

$$\dot{e} + (\boldsymbol{\nabla} \cdot \boldsymbol{v}) + \boldsymbol{\nabla} j_e = \boldsymbol{v} \cdot \boldsymbol{f}^{ext} \tag{10.82}$$

where $\dot{e}$ designates the material derivative of e, and likewise for the other reduced extensive quantities. Here, the system is assume incompressible, i.e., $(\boldsymbol{\nabla} \cdot \boldsymbol{v}) = 0$, because no lattice deformations are considered. The developments in [376] show that $\boldsymbol{f}^{ext}$ refers to the overall force exerted on the volume element. We assume charge neutrality, so this overall force vanishes. Therefore, we have,

$$\dot{e} + \boldsymbol{\nabla} j_e = 0 \tag{10.83}$$

When we consider the energy density e as a function of momentum $\boldsymbol{p}$, density n and entropy s, its time derivative is given by,

$$\dot{e} = \boldsymbol{v} \cdot \dot{\boldsymbol{p}} + \mu \dot{n} + T \dot{s} \tag{10.84}$$

If an electron is subjected to the force $-e\boldsymbol{E}$, the force per unit volume is $\dot{\boldsymbol{p}} = \boldsymbol{F} = n(-e\boldsymbol{E})$. The Lorentz force does not contribute to the product $\boldsymbol{v} \cdot \dot{\boldsymbol{p}}$. Since there are no charge recombinations or charge traps, the continuity equation for the number density reads,

$$\dot{n} + \boldsymbol{\nabla} j_n = 0 \tag{10.85}$$

The density currents for the energy density e and the internal energy density u are equal when the medium is free of mechanical torque, which is the case in the present description ([376], Eq. 10.44). Thus, here, we have,

$$\boldsymbol{j}_u = \boldsymbol{j}_e \tag{10.86}$$

The thermodynamics of continuous media leads to the following expression for the entropy current:

$$\boldsymbol{j}_s = \frac{\boldsymbol{j}_u - \mu \boldsymbol{j}_n}{T} \tag{10.87}$$

When we insert the continuity equations (10.83) and (10.85) into the equation of evolution (10.84) for e, after a few elementary manipulations, we get:

$$\dot{s} = -\boldsymbol{\nabla} \cdot \boldsymbol{j}_s - \frac{1}{T} \boldsymbol{j}_n \cdot \boldsymbol{\nabla} \mu + (\boldsymbol{j}_e - \mu \boldsymbol{j}_n) \cdot \boldsymbol{\nabla} \left(\frac{1}{T} \right) + \frac{\boldsymbol{F} \cdot \boldsymbol{v}}{T} \tag{10.88}$$

Thus, we obtain a continuity equation for s with a source term given by,

$$\pi_s = -\frac{1}{T} \boldsymbol{j}_n \cdot \boldsymbol{\nabla} \mu + (\boldsymbol{j}_e - \mu \boldsymbol{j}_n) \boldsymbol{\nabla} \left(\frac{1}{T} \right) + \frac{\boldsymbol{F} \cdot \boldsymbol{v}}{T} \tag{10.89}$$

In the framework of the semi-classical Boltzmann description of transport, each charge carrier has an energy E and we have $\boldsymbol{j}_e = \varepsilon \boldsymbol{j}_n$. Therefore, the entropy source density can be written as,

$$\pi_s = \left(\frac{1}{T} \right) \left(e\boldsymbol{\nabla}\varphi - \boldsymbol{\nabla}\mu - \left(\frac{E - \mu}{T} \right) \boldsymbol{\nabla} T \right) \cdot \boldsymbol{j}_n \tag{10.90}$$

When Onsager obtained his reciprocity relations from statistical physics considerations, he assumed that the fluctuating generalized currents and forces obey the same linear relations as the macroscopic quantities. In order to apply this idea to the present Boltzmann description of transport, we define a microscopic equivalent of the source term (10.90) and consider the microscopic power dissipation,[488]

$$\dot{Q}(\boldsymbol{r}, \boldsymbol{k}) = \left(e\boldsymbol{\nabla}\varphi - \boldsymbol{\nabla}\mu - \left(\frac{E-\mu}{T}\right)\boldsymbol{\nabla}T \right)\cdot \boldsymbol{v}\, f(\boldsymbol{r}, \boldsymbol{k}) \tag{10.91}$$

The magnetic induction field $\boldsymbol{B}$ does not explicitly appear in this definition, but $\boldsymbol{v}$ in (10.91) does depend on $\boldsymbol{B}$.

We apply the definition (10.91) in the expression (10.79) for the distribution $f(t)$ and obtain,

$$f = f_0 + \int_{-\infty}^{t} dt' \exp\left(-\frac{t-t'}{\tau}\right) \dot{Q}(\boldsymbol{r}, \boldsymbol{k}) \frac{1}{f_0(t')} \frac{\partial f_0(t')}{\partial \mu} \tag{10.92}$$

At the root of the thermodynamics of irreversible processes is the observation that the entropy production, or the heat power $\dot{Q}$, can be written as a sum of products. Each product has one term identifiable as a generalized force x_α and the associated term is a generalized current X_α. That is,

$$\dot{Q} = \sum_{\alpha} X_\alpha x_\alpha \qquad X_\alpha = \frac{\partial \dot{Q}}{\partial x_\alpha} \tag{10.93}$$

Following the method of the thermodynamics of irreversible processes, we ensure that the the second law $\dot{Q} \geq 0$ is fulfilled by writing,

$$X_\alpha = \sum_{\alpha} \mathsf{L}_{\alpha\beta} x_\beta \tag{10.94}$$

The tensors $\{\mathsf{L}_{\alpha\beta}\}$ must be definite positive. In the linear regime, the tensors $\mathsf{L}_{\alpha\beta}$ do not depend on the forces x_α. From now on, we consider that α and β number both a type of generalized force (or current), and also the cartesian components of the vectorial quantities such as current densities or gradients.

Averages are calculated using (9.5) the Boltzmann distribution f given in (10.92). For any quantity $G(\boldsymbol{r}, \boldsymbol{p})$, its average is calculated as,

$$G = \int \frac{2d\boldsymbol{p}}{(\hbar)^3} d\boldsymbol{r}\, g(\boldsymbol{r}, \boldsymbol{p}) G(\boldsymbol{r}, \boldsymbol{p}) \tag{10.95}$$

We apply (10.95) to find the average current X_α. Thus, we have,

$$X_\alpha = \int \frac{2d\boldsymbol{k}}{(\hbar)^3} d\boldsymbol{r} \int_{-\infty}^{t} dt' \exp\left(-\frac{t-t'}{\tau}\right) \sum_{\beta} x_\beta \frac{d\dot{Q}(t')}{dx_\beta} \frac{1}{f_0(t')} \frac{\partial f_0(t')}{\partial \mu} \frac{d\dot{Q}(t)}{dx_\alpha} \tag{10.96}$$

Then, we can identify one by one the terms in this equation and in (10.94). This yields,

$$\mathsf{L}_{\alpha\beta} = \int \frac{2d\boldsymbol{k}}{(\hbar)^3} d\boldsymbol{r} \int_{-\infty}^{t} dt' \exp\left(-\frac{t-t'}{\tau}\right) \frac{d\dot{Q}(t')}{dx_\beta} \frac{1}{f_0(t')} \frac{\partial f_0(t')}{\partial \mu} \frac{d\dot{Q}(t)}{dx_\alpha} \tag{10.97}$$

Notice that $2d\boldsymbol{k}/(\hbar)^3$, $d\boldsymbol{r}$ and $d\dot{Q}(t)/dx_\alpha$ are evaluated at the same time t.

The term $(1/f_0)(\partial f_0/\partial \mu)$ is the main obstacle that keeps us from deducing the Onsager reciprocity relations from (10.97). Notice that the exponential in (10.97) suppresses the contributions to the integral of terms at times t' that are much more than τ away from t. We assume that, during a time interval of the order of τ, the statistical distribution does not change much. In other words, we can evaluate $(1/f_0)(\partial f_0/\partial \mu)$ at time t.

To clarify the argument that we need to carry out now, we choose the zero of the time scale at the time noted t so far. Furthermore, we make the change of variable $t'' = t - t'$, implying $dt'' = -dt'$. Then, the Onsager coefficient is expressed as,

$$L_{\alpha\beta} = \int \frac{2d\boldsymbol{k}}{(\hbar)^3} d\boldsymbol{r} \int_0^{\infty} dt'' \exp\left(-\frac{t''}{\tau}\right) \frac{d\dot{Q}(-t'')}{dx_\beta} \frac{1}{f_0(0)} \frac{\partial f_0(0)}{\partial \mu} \frac{d\dot{Q}(0)}{dx_\alpha} \quad (10.98)$$

The product $(d\dot{Q}(-t'')/dx_\alpha)(d\dot{Q}(0)/dx_\beta)$, which is a microscopic quantity, must be invariant under time reversal. We see whether this is the case by considering the definition (10.91) of this product. When a magnetic field is present, the time inversion symmetry is satisfied only if $\boldsymbol{B}$ is reversed, since the force $-e\boldsymbol{v} \times \boldsymbol{B}$, must remain unchanged. Thus, keeping in mind the requirement that we need to change $\boldsymbol{B}$ into $-\boldsymbol{B}$ when reversing time, we can write,

$$L_{\alpha\beta} = \int \frac{2d\boldsymbol{k}}{(\hbar)^3} d\boldsymbol{r} \int_0^{\infty} dt'' \exp\left(-\frac{t''}{\tau}\right) \frac{d\dot{Q}(t'')}{dx_\beta} \frac{1}{f_0(0)} \frac{\partial f_0(0)}{\partial \mu} \frac{d\dot{Q}(0)}{dx_\alpha} \quad (10.99)$$

We have written $L_{\alpha\beta}$ in the form of a correlation function. Its value depends only on the spacing between the times t and t', i.e., t''. Therefore, we can also write ([386] Eq. 14.27)

$$L_{\alpha\beta} = \int \frac{2d\boldsymbol{k}}{(\hbar)^3} d\boldsymbol{r} \int_0^{\infty} dt'' \exp\left(-\frac{t''}{\tau}\right) \frac{d\dot{Q}(0)}{dx_\beta} \frac{1}{f_0(0)} \frac{\partial f_0(0)}{\partial \mu} \frac{d\dot{Q}(-t'')}{dx_\alpha} \quad (10.100)$$

This expression corresponds to the definition of $L_{\beta\alpha}$ according to (10.98), evaluated with the reversed magnetic field. Thus, we have,

$$L_{\alpha\beta}(\boldsymbol{B}) = L_{\beta\alpha}(-\boldsymbol{B}) \quad (10.101)$$

This result covers both statements (10.2) and (10.3) (when the response to the $\boldsymbol{B}$ field is linear), because α and β in (10.94) number both the type of generalized forces and their spatial components.

10.8 FURTHER READINGS

Hall effect

The main purpose of this chapter was to introduce perpendicular transport phenomena described within the Boltzmann formalism. The simplest situation only was considered, in which the Fermi surface is a sphere. Measurements of the Hall effect in semiconductors can be used to characterize the Fermi surface topology.[534, 535]

The Hall voltage is commonly found by setting the transverse current to zero. It is presumed that this occurs because charges accumulate at the edges. By combining the Poisson equation for electrostatics with thermodynamic transport equation and the principle of least entropy production, it's been found that surface currents run in opposite directions at the edges of the Hall bar.[536]

Mott relation

In a two-dimensional electron gas (2DEG) with Rashba interaction, strong deviation from Mott's relation has been observed when the Fermi level reaches a band crossing point.[537]

It is possible to derive a generalized form of Mott's formula that applied for the Nernst effect as well as for the Seebeck effect.[538]

Nernst effect

Here we have examined perpendicular transport using the Boltzmann transport formalism. It is also possible to gain insight into perpendicular transport using thermodynamics. For example, thermodynamic arguments can show that a spin current can be associated with the Nernst effect.[446, 539]

In general, thermoelectric properties can give further insight into the electronic structure of materials,[540] notably in the case of strongly correlated systems.[541, 542] Nanoscale Nernst measurements have helped to characterize charge density waves in 2-dimensional chalcogenide materials.[543] Using non-local spin valves, an anomalous Nernst-Ettingshausen effect has been studied.[544]

A spin-dependent Nernst effect is expected in chalcogenide quantum wells, in which the Rashba effect plays an important role.[545] A spin Nernst effect has been observed in tungsten.[546]

Onsager reciprocity relations

The applicability of the *Onsager reciprocity relations* (10.2) has been debated in the context of research on chirality induced spin selectivity (CISS). It has been shown that this effect cannot be observed in a two-terminal device, if the response is linear and the Onsager reciprocity relations apply.[547] However, typical transport measurements through chiral molecules are intrinsically non-linear, because they involve either a tunnel effect or an electrochemical process that is typically non-linear.[548] It has been shown theoretically that the dephasing at an electrode that constitutes a bath causes non-symmetrical magnetic effects.[549]

11 Anomalous Transport

Michael Victor Berry (born 1941)

Sir M.V. Berry graduated with a PhD from the University of Saint Andrews. During his entire professorial career, he was based at the University of Bristol. He is known for his work on geometrical phases and his contributions to mathematical physics. He was elected Fellow of the Royal Society in 1982, knighted in 1996 and received many prizes such as the Dirac Medal of the International Center for Theoretical Physics and the Wolf Prize.

Spintronics: A primer

J.-Ph. Ansermet (https://orcid.org/0000-0002-1307-4864)

Chapter 11. Anomalous Transport

The Berry phase formalism is introduced by applying the adiabatic approximation to band electrons in the presence of a slowly varying electric field. This leads to constitutive equations for velocities in real and momentum spaces that can be used within the Boltzmann theory of transport. The Berry curvature gives a contribution of zeroth order of perturbation of the Boltzmann distribution, accounting for anomalous Hall and Nernst effects, and leading to an anomalous longitudinal magnetoresistance. The intrinsic spin Hall effect is described with an effective Hamiltonian. The extrinsic spin Hall effect is accounted for using a perturbation scheme that describes skew scattering.

DOI: 10.1201/9781003370017-11

11.1 HISTORICAL INTRODUCTION

In a ferromagnetic conductor, the Hall resistance R_{xy} is not proportional to the field B as found in (10.54). Instead, R_{xy} has a contribution proportional to the magnetization M. Often, the Hall effect in a ferromagnet is characterized by,

$$R_{xy} = r_0\, H + r_a\, M \qquad (11.1)$$

where H is the magnetic field. The coefficients r_0 and r_a characterize the conventional and the **anomalous Hall effect** (AHE), respectively. The AHE can be used to characterize the magnetization of very small samples, such as nanostructures.[550]

There is a long history of measurements and modelling of the AHE. For decades, the concepts of **side jumps** and **skew scattering** were the mechanisms invoked to account for this effect.[551] These concepts were challenged in particular by the discovery of very large Nernst effects in non-ferromagnetic materials, the **anomalous Nernst effect** (ANE). In some materials, even at reasonably low magnetic fields, the Nernst coefficient reaches values far beyond what can be observed in ferromagnets.[532, 552–555] These effects can be accounted for by the introduction of **Berry curvature** and **anomalous velocities**. Berry showed in his seminal paper that the now-called Berry curvature can be derived from considerations on adiabatic transformations. Similar ideas had been brought forth by Mead and Truhlar in 1979, in their study of the electronic structure of molecules under the Born-Oppenheimer approximation.[556, 557]

A very large Nernst effect may be of interest for the conversion of thermal power to electrical power. The Nernst effect, if it is large, may be of interest for the following reason. When the electric field is transverse to the temperature gradient (Nernst geometry), it is conceivable to make a very long strip of the material and obtain a large voltage (since it is the integral of the electric field over this long dimension). Materials with **perpendicular magneto-crystalline anisotropy** (PMA) would be particularly suitable, since the gradient and the electric field would be in the plane of the film. This is merely a geometrical voltage enhancement that may be of interest for applications. It should be noted that transverse transport effects because they can take advantage of the instrinsic anisotropic electrical and thermal conductances of some materials.[558]

Dyakonov and Perel realized in the early 1970s that driving electrons through a material with strong spin-orbit coupling could result in a build-up of spin polarization at the surface of the conductor, an effect now called the **spin Hall effect** (SHE).[559] The fundamental reason why the effect might arise from the intrinsic properties of the electronic structure of a conductor was established using the framework of gauge theory.[175] The formalism was later applied to analyze the intrinsic spin Hall effect expected in topological materials.[560] Alternatively, the interband transitions caused by an electric field were shown to imply a spin Hall effect.[561] As the first observations of the spin Hall effect were published, [562,563] a controversy arose as to whether the observed effect was intrinsic or extrinsic. A review of both theory and experiments may help clarify which of these contributions dominates in the systems that were studied.[564] In this chapter, the intrinsic spin Hall effect is illustrated for the case of GaAs and a heuristic argument is given for the extrinsic spin Hall effect based on skew scattering.

11.2 INTRODUCTION TO ANOMALOUS VELOCITIES

In Chapters 9 and 10, we saw how the Boltzmann equation can describe transport phenomena in a non-magnetic material. Starting in §9.3.2 and §9.3.3, we worked with a semi-classical description of electrons. We assumed that we were working with electrons in Bloch states of definite $\boldsymbol{k}$, for which we prescribed dynamical equations that were simply those of Newtonian mechanics, with $\hbar \dot{\boldsymbol{k}} = \dot{\boldsymbol{p}} = \boldsymbol{F}$ and $\dot{\boldsymbol{r}} = \boldsymbol{v} = \boldsymbol{p}/m$. In this section, we introduce a correction to this simple picture. We find that it requires modifying these constitutive equations. Their introduction in the Boltzmann's transport equation leads to important corrections.

11.2.1 BLOCH STATES AND VELOCITY

The AHE is critically dependent on interband mixing and non-commutativity effects. To create a semi-classical picture that accounts for these effects, we must find a way to include interband mixing in the description of the semi-classical dynamics and thus modify the simply dynamic picture according to which $\boldsymbol{v} = \dot{\boldsymbol{r}}$ and $\boldsymbol{F} = \dot{\boldsymbol{p}}$.

Let us consider a crystal with simple translational symmetry and assume that the single electron picture applies. Equivalent atoms in the crystals are located at positions given by $\boldsymbol{R}$ with,

$$\boldsymbol{R} = n_1 \boldsymbol{a}_1 + n_2 \boldsymbol{a}_2 + n_3 \boldsymbol{a}_3 \qquad (n_1, n_2, n_3 \text{ integers}) \tag{11.2}$$

where $(\boldsymbol{a}_1, \boldsymbol{a}_2, \boldsymbol{a}_3)$ are the shortest, linearly independent vectors connecting equivalent atoms in the primitive unit cell. The Hamiltonian of an electron (of mass m) in this crystal is given by,

$$\mathcal{H}_0 = \frac{-\hbar^2 \boldsymbol{\nabla}_r^2}{2m} + \mathcal{V}(\boldsymbol{r}) \tag{11.3}$$

where $\mathcal{V}(\boldsymbol{r})$ is the potential experienced by the electron. Owing to the crystal symmetry, we have,

$$\mathcal{V}(\boldsymbol{r} + \boldsymbol{R}) = \mathcal{V}(\boldsymbol{r}) \tag{11.4}$$

Bloch's theorem states that the eigenfunctions of $\mathcal{H}_0$ can be labelled by the wave vector $\boldsymbol{k}$. It has the following properties,

$$\Psi_{\boldsymbol{k}}(\boldsymbol{r} + \boldsymbol{R}) = e^{i\boldsymbol{k}\cdot\boldsymbol{R}} \Psi_{\boldsymbol{k}}(\boldsymbol{r})$$
$$\Psi_{\boldsymbol{k}}(\boldsymbol{r}) = \frac{1}{\sqrt{V}} e^{i\boldsymbol{k}\cdot\boldsymbol{r}} u_{\boldsymbol{k}}(\boldsymbol{r}) \qquad u_{\boldsymbol{k}}(\boldsymbol{r} + \boldsymbol{R}) = u_{\boldsymbol{k}}(\boldsymbol{r}) \tag{11.5}$$

The proof of Bloch's theorem can be found in solid state physics textbooks. The proof consists of introducing translation operators. Since these operators commute with the Hamiltonian, the eigenstates of the Hamiltonian can be written as linear combinations of the eigenstates of the translation operators. Let us look at the action of $\mathcal{H}_0$ on a Bloch function. The non-trivial terms come from the kinetic energy, as shown below,

$$\boldsymbol{\nabla}\left(e^{i\boldsymbol{k}\cdot\boldsymbol{r}} u_{\boldsymbol{k}}(\boldsymbol{r})\right) = i\boldsymbol{k} e^{i\boldsymbol{k}\cdot\boldsymbol{r}} u_{\boldsymbol{k}}(\boldsymbol{r}) + e^{i\boldsymbol{k}\cdot\boldsymbol{r}} \boldsymbol{\nabla} u_{\boldsymbol{k}}(\boldsymbol{r})$$
$$\boldsymbol{\nabla}^2 \left(e^{i\boldsymbol{k}\cdot\boldsymbol{r}} u_{\boldsymbol{k}}(\boldsymbol{r})\right) = (i\boldsymbol{k})^2 e^{i\boldsymbol{k}\cdot\boldsymbol{r}} u_{\boldsymbol{k}}(\boldsymbol{r}) + (i\boldsymbol{k}) e^{i\boldsymbol{k}\cdot\boldsymbol{r}} \boldsymbol{\nabla} u_{\boldsymbol{k}}(\boldsymbol{r})$$
$$\qquad + e^{i\boldsymbol{k}\cdot\boldsymbol{r}} \boldsymbol{\nabla}^2 u_{\boldsymbol{k}}(\boldsymbol{r}) + i\boldsymbol{k} e^{i\boldsymbol{k}\cdot\boldsymbol{r}} \boldsymbol{\nabla} u_{\boldsymbol{k}}(\boldsymbol{r}) \tag{11.6}$$
$$= (\boldsymbol{\nabla} + i\boldsymbol{k})^2 e^{i\boldsymbol{k}\cdot\boldsymbol{r}} u_{\boldsymbol{k}}(\boldsymbol{r}) = \frac{-1}{\hbar^2}(-i\hbar\boldsymbol{\nabla} + \hbar\boldsymbol{k})^2 e^{i\boldsymbol{k}\cdot\boldsymbol{r}} u_{\boldsymbol{k}}(\boldsymbol{r})$$

Thus, we find,

$$e^{-i\boldsymbol{k}\cdot\boldsymbol{r}} \mathcal{H}_0(\boldsymbol{p}, \boldsymbol{r}) e^{i\boldsymbol{k}\cdot\boldsymbol{r}} = \mathcal{H}_0(\boldsymbol{p} + \hbar\boldsymbol{k}, \boldsymbol{r}) \equiv \mathcal{H}_{\boldsymbol{k}} \tag{11.7}$$

which defines the $\boldsymbol{k}$-dependent Hamiltonian $\mathcal{H}_{\boldsymbol{k}}$,

$$\mathcal{H}_{\boldsymbol{k}} \equiv \left(\frac{1}{2m}(-i\hbar\boldsymbol{\nabla}_r + \hbar\boldsymbol{k})^2 + \mathcal{V}(\boldsymbol{r}) \right) \tag{11.8}$$

When the Bloch form (11.5) is applied to the eigenvalue equation for the Hamiltonian $\mathcal{H}_0$,

$$\mathcal{H}_0 \Psi_{\boldsymbol{k}}(\boldsymbol{r}) = \varepsilon_{\boldsymbol{k}} \Psi_{\boldsymbol{k}}(\boldsymbol{r}) \tag{11.9}$$

we find an eigenvalue equation for the Hamiltonian $\mathcal{H}_{\boldsymbol{k}}$,

$$\left(\frac{1}{2m} \left(-i\hbar \boldsymbol{\nabla}_{\boldsymbol{r}} + \hbar \boldsymbol{k}\right)^2 + \mathcal{V}(\boldsymbol{r}) \right) u_{\boldsymbol{k}}(\boldsymbol{r}) = \varepsilon_{\boldsymbol{k}} u_{\boldsymbol{k}}(\boldsymbol{r}) \tag{11.10}$$

As an aside, consider the following: If $\boldsymbol{k}$ where a function of time $\boldsymbol{k}(t)$, the equality (11.10) would represent an expression of what is called the **adiabatic approximation**. A general consequence of the adiabatic approximation is that the eigenstates, which would be $u_{\boldsymbol{k}(t)}(\boldsymbol{r})$, evolve in time. Their evolution comprises a dynamical phase term and a **geometrical phase** term, also known as the **Berry phase** term.[557–566]

In this chapter, we assume that the electric field is weak, so that the adiabatic approximation holds. In other words, the electron acceleration is small enough that, at any time, the electron is in a Bloch state. Its $\boldsymbol{k}(t)$ vector changes slowly in time, so that the eigenstates are Bloch states that can be indexed by $\boldsymbol{k}(t)$.

In order to introduce the notion of anomalous velocity, we prefer to work in the Heisenberg representation rather than in the Schroedinger representation. That is, we will seek the time evolution of the operator $\boldsymbol{r}$. Let us first note that, in general, there can be several solutions for a given $\boldsymbol{k}$, known as energy "bands." Hence, generally, we should label the eigenstates of $\mathcal{H}_0$ as $\Psi_{\boldsymbol{k}}^{(n)}(\boldsymbol{r})$. Thus, we now examine the expectation value of the velocity for a single band (of index n).[567] First, we start from the Heisenberg picture for the evolution of $\boldsymbol{r}(t)$, considered as an operator,

$$\frac{d\boldsymbol{r}}{dt} = \frac{i}{\hbar}[\mathcal{H}_0, \boldsymbol{r}] \tag{11.11}$$

We seek the expectation value of the velocity $\boldsymbol{v}_{\boldsymbol{k}}^{(n)}$ of an electron in the Bloch state $\Psi_{\boldsymbol{k}}^{(n)}(\boldsymbol{r})$,

$$\boldsymbol{v}_{\boldsymbol{k}}^{(n)} = \frac{i}{\hbar} \langle \Psi_{\boldsymbol{k}}^{(n)} | [\mathcal{H}_0, \boldsymbol{r}] | \Psi_{\boldsymbol{k}}^{(n)} \rangle \tag{11.12}$$

As shown below, a few manipulations lead to the result,

$$\boldsymbol{v}_{\boldsymbol{k}}^{(n)} = \frac{\partial \varepsilon_{\boldsymbol{k}}}{\partial (\hbar \boldsymbol{k})} \equiv \frac{1}{\hbar} \boldsymbol{\nabla}_{\boldsymbol{k}} \varepsilon_{\boldsymbol{k}} \tag{11.13}$$

The first equality reminds us of the thermodynamic point of view, which holds that velocity is the conjugate of the momentum.

To derive (11.13), we initially note that,

$$\boldsymbol{\nabla}_{\boldsymbol{k}} \left(e^{-i\boldsymbol{k}\cdot\boldsymbol{r}} \mathcal{H}_0 e^{i\boldsymbol{k}\cdot\boldsymbol{r}} \right) = i\, e^{-i\boldsymbol{k}\cdot\boldsymbol{r}} [\mathcal{H}_0, \boldsymbol{r}] e^{i\boldsymbol{k}\cdot\boldsymbol{r}} \tag{11.14}$$

as can be verified by applying the derivative to the exponentials on the left-hand side of this equation. Therefore, starting from (11.12), applying the translation (11.7), and the eigenvalue equation (11.10), we get successively,

$$\begin{aligned}
\boldsymbol{v}_{\boldsymbol{k}}^{(n)} &= \frac{i}{\hbar} \langle u_{\boldsymbol{k}}^{(n)} | e^{-i\boldsymbol{k}\cdot\boldsymbol{R}} [\mathcal{H}_0, \boldsymbol{r}] e^{i\boldsymbol{k}\cdot\boldsymbol{R}} | u_{\boldsymbol{k}}^{(n)} \rangle = \frac{1}{\hbar} \langle u_{\boldsymbol{k}}^{(n)} | \boldsymbol{\nabla}_{\boldsymbol{k}} \left(\mathcal{H}_0(\boldsymbol{p} + \hbar \boldsymbol{k}, \boldsymbol{r}) \right) | u_{\boldsymbol{k}}^{(n)} \rangle \\
&= \frac{1}{\hbar} \boldsymbol{\nabla}_{\boldsymbol{k}} \langle u_{\boldsymbol{k}}^{(n)} | \mathcal{H}_0(\boldsymbol{p} + \hbar \boldsymbol{k}, \boldsymbol{r}) | u_{\boldsymbol{k}}^{(n)} \rangle \\
&\quad - \langle \nabla_{\boldsymbol{k}} u_{\boldsymbol{k}}^{(n)} | \mathcal{H}_0(\boldsymbol{p} + \hbar \boldsymbol{k}, \boldsymbol{r}) | u_{\boldsymbol{k}}^{(n)} \rangle - \langle u_{\boldsymbol{k}}^{(n)} | \mathcal{H}_0(\boldsymbol{p} + \hbar \boldsymbol{k}, \boldsymbol{r}) | \nabla_{\boldsymbol{k}} u_{\boldsymbol{k}}^{(n)} \rangle \\
&= \frac{1}{\hbar} \boldsymbol{\nabla}_{\boldsymbol{k}} \langle u_{\boldsymbol{k}}^{(n)} | \mathcal{H}_0(\boldsymbol{p} + \hbar \boldsymbol{k}, \boldsymbol{r}) | u_{\boldsymbol{k}}^{(n)} \rangle - \varepsilon_{\boldsymbol{k}} \left(\langle u_{\boldsymbol{k}}^{(n)} | \nabla_{\boldsymbol{k}} u_{\boldsymbol{k}}^{(n)} \rangle + \langle \nabla_{\boldsymbol{k}} u_{\boldsymbol{k}}^{(n)} | u_{\boldsymbol{k}}^{(n)} \rangle \right)
\end{aligned} \tag{11.15}$$

The last term vanishes, since it is equal to $\nabla_{\boldsymbol{k}} \langle u_{\boldsymbol{k}}^{(n)} | u_{\boldsymbol{k}}^{(n)} \rangle$ and $|u_{\boldsymbol{k}}^{(n)}\rangle$ is normalized to 1. Thus,

$$\boldsymbol{v}_{\boldsymbol{k}}^{(n)} = \frac{1}{\hbar} \nabla_{\boldsymbol{k}} \langle u_{\boldsymbol{k}}^{(n)} | \mathcal{H}_0(\boldsymbol{p} + \hbar \boldsymbol{k}, \boldsymbol{r}) | u_{\boldsymbol{k}}^{(n)} \rangle \tag{11.16}$$

and we can conclude with (11.13) by applying in reverse the translation (11.7). In the next section, we will make use of the first line of (11.15), which we denote here for convenience as,

$$\boldsymbol{v}_{\boldsymbol{k}}^{(n)} = \frac{1}{\hbar} \langle u_{\boldsymbol{k}}^{(n)} | \nabla_{\boldsymbol{k}} \mathcal{H}_{\boldsymbol{k}} | u_{\boldsymbol{k}}^{(n)} \rangle \tag{11.17}$$

In the simplest approach, we could use an effective mass approximation whereby we write that $\varepsilon_{\boldsymbol{k}} = \hbar^2 k^2 / (2m)$ where m is the effective mass. Then, the time derivative of (11.13) gives us the acceleration, and we conclude for this case that $\hbar \dot{\boldsymbol{k}} = m d\boldsymbol{v}_{\boldsymbol{k}}^{(n)}/dt$, which is the force. So we expect $\boldsymbol{k}$ to evolve. Therefore, in the eigenvalue problem for $|u_{\boldsymbol{k}}^{(n)}\rangle$, we have a Hamiltonian with a parameter that varies in time.

11.2.2 SYSTEM SUBJECT TO A SLOWLY TIME-VARYING DRIVE

Let us now add to the stationary problem (11.10) a perturbation $\mathcal{H}_1$ that does not break the crystal symmetry. We will see later on how to treat the case of an electric field using a gauge that allows us to maintain the crystal symmetry. We make the assumption that the perturbation changes slowly, so that it does not induce transitions in the system. We say in this case that the evolution is **adiabatic**, or that the adiabatic approximation holds. Let us agree that $t = 0$ is some time far in the past when we slowly activated the perturbation.

Before the perturbation takes on any significance, the eigenstates of the unperturbed Hamiltonian are Bloch states of wave vector $\boldsymbol{k}$, which is a constant. Since the perturbation maintains the crystal symmetry, the perturbed eigenstates can be taken as Bloch states again but, evidently, due to the excitation $\mathcal{H}_1$, the wave vector $\boldsymbol{k}$ now varies in time. It would be too cumbersome to write the wave vector as $\boldsymbol{k}(t)$, with the initial condition $\boldsymbol{k}(0) = \boldsymbol{k}$. Thus, using the notation $\mathcal{H}_{\boldsymbol{k}} \equiv \mathcal{H}_0(\boldsymbol{p} + \hbar \boldsymbol{k}, \boldsymbol{r})$, we have the total Hamiltonian,

$$\mathcal{H} = \mathcal{H}_{\boldsymbol{k}} + \mathcal{H}_1 \tag{11.18}$$

The eigenvalue equations for $\mathcal{H}$ and $\mathcal{H}_{\boldsymbol{k}}$ are,

$$\begin{aligned} \mathcal{H} |\tilde{u}_{\boldsymbol{k}}^{(n)}\rangle &= \tilde{\varepsilon}_{\boldsymbol{k}}^{(n)} |\tilde{u}_{\boldsymbol{k}}^{(n)}\rangle \\ \mathcal{H}_{\boldsymbol{k}} |u_{\boldsymbol{k}}^{(n)}\rangle &= \varepsilon_{\boldsymbol{k}}^{(n)} |u_{\boldsymbol{k}}^{(n)}\rangle \end{aligned} \tag{11.19}$$

To the first order of perturbation, we write,

$$\begin{aligned} \tilde{\varepsilon}_{\boldsymbol{k}}^{(n)} &\approx \varepsilon_{\boldsymbol{k}}^{(n)} + \delta\varepsilon_{\boldsymbol{k}}^{(n)} \\ |\tilde{u}_{\boldsymbol{k}}^{(n)}\rangle &\approx |u_{\boldsymbol{k}}^{(n)}\rangle + |\delta u_{\boldsymbol{k}}^{(n)}\rangle \end{aligned} \tag{11.20}$$

where $|\delta u_{\boldsymbol{k}}^{(n)}\rangle$ is the first-order correction to the zeroth-order approximation $|u_{\boldsymbol{k}}^{(n)}\rangle$. Likewise, $\varepsilon_{\boldsymbol{k}}^{(n)}$ is the zeroth-order approximation for the energy, and $\delta\varepsilon_{\boldsymbol{k}}^{(n)}$ the first-order correction.

Now, let us assume non-degenerate unperturbed states. When we combine these equations, we get,

$$(\mathcal{H}_{\boldsymbol{k}} + \mathcal{H}_1) \left(|u_{\boldsymbol{k}}^{(n)}\rangle + |\delta u_{\boldsymbol{k}}^{(n)}\rangle \right) = \left(\varepsilon_{\boldsymbol{k}}^{(n)} + \delta\varepsilon_{\boldsymbol{k}}^{(n)} \right) \left(|u_{\boldsymbol{k}}^{(n)}\rangle + |\delta u_{\boldsymbol{k}}^{(n)}\rangle \right) \tag{11.21}$$

Multiplying this equation by $\langle u_{\boldsymbol{k}}^{(m)}|$ at the left, with $m \neq n$, and applying the orthogonality $\langle u_{\boldsymbol{k}}^{(m)} | u_{\boldsymbol{k}}^{(n)} \rangle = \delta_{mn}$, we get,

$$\langle u_{\boldsymbol{k}}^{(m)} | \delta u_{\boldsymbol{k}}^{(n)} \rangle = \frac{\langle u_{\boldsymbol{k}}^{(m)} | \mathcal{H}_1 | u_{\boldsymbol{k}}^{(n)} \rangle}{\varepsilon_{\boldsymbol{k}}^{(n)} - \varepsilon_{\boldsymbol{k}}^{(m)}} \langle u_{\boldsymbol{k}}^{(m)} | u_{\boldsymbol{k}}^{(m)} \rangle \qquad (m \neq n) \tag{11.22}$$

We have yet to determine $\langle u_{\boldsymbol{k}}^{(n)} | \delta u_{\boldsymbol{k}}^{(n)} \rangle$. A great simplification is obtained by making a special choice for the normalization of the wave function $|\tilde{u}_{\boldsymbol{k}}^{(n)}\rangle$.[181] Namely, we take,

$$\langle u_{\boldsymbol{k}}^{(n)} | \tilde{u}_{\boldsymbol{k}}^{(n)} \rangle = 1 \tag{11.23}$$

Under this condition, (11.20) yields,

$$\langle u_{\boldsymbol{k}}^{(n)} | \delta u_{\boldsymbol{k}}^{(n)} \rangle = 0 \tag{11.24}$$

Thus, from (11.22) and (11.20) we deduce,

$$|\delta u_{\boldsymbol{k}}^{(n)} \rangle = \sum_m \langle u_{\boldsymbol{k}}^{(m)} | \delta u_{\boldsymbol{k}}^{(m)} \rangle | u_{\boldsymbol{k}}^{(m)} \rangle = \sum_{m \neq n} \frac{\langle u_{\boldsymbol{k}}^{(m)} | \mathcal{H}_1 | u_{\boldsymbol{k}}^{(n)} \rangle}{\varepsilon_{\boldsymbol{k}}^{(n)} - \varepsilon_{\boldsymbol{k}}^{(m)}} | u_{\boldsymbol{k}}^{(m)} \rangle \tag{11.25}$$

To first order, we have the Schroedinger equation,

$$\mathcal{H}_1 | u_{\boldsymbol{k}}^{(n)} \rangle = -i\hbar \frac{\partial}{\partial t} | u_{\boldsymbol{k}}^{(n)} \rangle = -i\hbar | \partial_t u_{\boldsymbol{k}}^{(n)} \rangle \tag{11.26}$$

Thus, the perturbation of the state reads,[1]

$$|\delta u_{\boldsymbol{k}}^{(n)} \rangle = -i\hbar \sum_{m \neq n} \frac{| u_{\boldsymbol{k}}^{(m)} \rangle \langle u_{\boldsymbol{k}}^{(m)} |}{\varepsilon_{\boldsymbol{k}}^{(n)} - \varepsilon_{\boldsymbol{k}}^{(m)}} | \partial_t u_{\boldsymbol{k}}^{(n)} \rangle \tag{11.27}$$

We can now estimate the expectation value of the velocity for this time-dependent case. Using the first order correction $|\tilde{u}_{\boldsymbol{k}}^{(n)} \rangle = | u_{\boldsymbol{k}}^{(n)} \rangle + | \delta u_{\boldsymbol{k}}^{(n)} \rangle$ in the definition given in (11.17) of the velocity expectation value, we get,

$$
\begin{aligned}
\boldsymbol{v}_{\boldsymbol{k}}^{(n)} &= \frac{1}{\hbar} \langle u_{\boldsymbol{k}}^{(n)} + \delta u_{\boldsymbol{k}}^{(n)} | \boldsymbol{\nabla}_{\boldsymbol{k}} \mathcal{H}_{\boldsymbol{k}} | u_{\boldsymbol{k}}^{(n)} + \delta u_{\boldsymbol{k}}^{(n)} \rangle \\
&= \frac{1}{\hbar} \langle u_{\boldsymbol{k}}^{(n)} | \boldsymbol{\nabla}_{\boldsymbol{k}} \mathcal{H}_{\boldsymbol{k}} | u_{\boldsymbol{k}}^{(n)} \rangle + \frac{1}{\hbar} \langle u_{\boldsymbol{k}}^{(n)} | \boldsymbol{\nabla}_{\boldsymbol{k}} \mathcal{H}_{\boldsymbol{k}} | \delta u_{\boldsymbol{k}}^{(n)} \rangle + \frac{1}{\hbar} \langle \delta u_{\boldsymbol{k}}^{(n)} | \boldsymbol{\nabla}_{\boldsymbol{k}} \mathcal{H}_{\boldsymbol{k}} | u_{\boldsymbol{k}}^{(n)} \rangle
\end{aligned}
\tag{11.28}
$$

In the last expression, we now apply (11.13) for its first term, and (11.27) for its other two,

$$
\begin{aligned}
\boldsymbol{v}_{\boldsymbol{k}}^{(n)} &= \frac{1}{\hbar} \boldsymbol{\nabla}_{\boldsymbol{k}} \varepsilon_{\boldsymbol{k}} \\
&- i \sum_{m \neq n} \left(\frac{\langle u_{\boldsymbol{k}}^{(n)} | \boldsymbol{\nabla}_{\boldsymbol{k}} \mathcal{H}_{\boldsymbol{k}} | u_{\boldsymbol{k}}^{(m)} \rangle}{\varepsilon_{\boldsymbol{k}}^{(n)} - \varepsilon_{\boldsymbol{k}}^{(m)}} \langle u_{\boldsymbol{k}}^{(m)} | \partial_t u_{\boldsymbol{k}}^{(n)} \rangle - \frac{\langle u_{\boldsymbol{k}}^{(m)} | \boldsymbol{\nabla}_{\boldsymbol{k}} \mathcal{H}_{\boldsymbol{k}} | u_{\boldsymbol{k}}^{(n)} \rangle}{\varepsilon_{\boldsymbol{k}}^{(n)} - \varepsilon_{\boldsymbol{k}}^{(m)}} \langle \partial_t u_{\boldsymbol{k}}^{(n)} | u_{\boldsymbol{k}}^{(m)} \rangle \right)
\end{aligned}
\tag{11.29}
$$

This equation becomes much simpler after moving the $\boldsymbol{\nabla}_{\boldsymbol{k}}$ operator outside the expectation values. Notice that the last large parenthesis contains some expression minus its complex conjugate. So, we focus on the first term. We have,

$$\langle u_{\boldsymbol{k}}^{(n)} | \boldsymbol{\nabla}_{\boldsymbol{k}} \mathcal{H}_{\boldsymbol{k}} | u_{\boldsymbol{k}}^{(m)} \rangle = \boldsymbol{\nabla}_{\boldsymbol{k}} \left(\langle u_{\boldsymbol{k}}^{(n)} | \mathcal{H}_{\boldsymbol{k}} | u_{\boldsymbol{k}}^{(m)} \rangle \right) - \langle \boldsymbol{\nabla}_{\boldsymbol{k}} u_{\boldsymbol{k}}^{(n)} | \mathcal{H}_{\boldsymbol{k}} | u_{\boldsymbol{k}}^{(m)} \rangle - \langle u_{\boldsymbol{k}}^{(n)} | \mathcal{H}_{\boldsymbol{k}} | \boldsymbol{\nabla}_{\boldsymbol{k}} u_{\boldsymbol{k}}^{(m)} \rangle \tag{11.30}$$

---

[1][567], p. 447.

Using (11.10), the orthogonality of the eigenfunction $u_{\boldsymbol{k}}^{(n)}$ and the fact that $n \neq m$ in the sum, the first term vanishes and,

$$\langle u_{\boldsymbol{k}}^{(n)} | \, \boldsymbol{\nabla}_{\boldsymbol{k}} \mathcal{H}_{\boldsymbol{k}} \, | u_{\boldsymbol{k}}^{(m)} \rangle = -\varepsilon_{\boldsymbol{k}}^{(m)} \, \langle \boldsymbol{\nabla}_{\boldsymbol{k}} u_{\boldsymbol{k}}^{(n)} | u_{\boldsymbol{k}}^{(m)} \rangle - \varepsilon_{\boldsymbol{k}}^{(n)} \, \langle u_{\boldsymbol{k}}^{(n)} | \boldsymbol{\nabla}_{\boldsymbol{k}} u_{\boldsymbol{k}}^{(m)} \rangle \tag{11.31}$$

Because of the orthogonality, we can write,

$$\boldsymbol{\nabla}_{\boldsymbol{k}} \langle u_{\boldsymbol{k}}^{(n)} | u_{\boldsymbol{k}}^{(m)} \rangle = 0 = \langle \boldsymbol{\nabla}_{\boldsymbol{k}} u_{\boldsymbol{k}}^{(n)} | u_{\boldsymbol{k}}^{(m)} \rangle + \langle u_{\boldsymbol{k}}^{(n)} | \boldsymbol{\nabla}_{\boldsymbol{k}} u_{\boldsymbol{k}}^{(m)} \rangle \tag{11.32}$$

Thus,

$$\langle u_{\boldsymbol{k}}^{(n)} | \, \boldsymbol{\nabla}_{\boldsymbol{k}} \mathcal{H}_{\boldsymbol{k}} \, | u_{\boldsymbol{k}}^{(m)} \rangle = \left(-\varepsilon_{\boldsymbol{k}}^{(m)} + \varepsilon_{\boldsymbol{k}}^{(n)} \right) \langle \boldsymbol{\nabla}_{\boldsymbol{k}} u_{\boldsymbol{k}}^{(n)} | u_{\boldsymbol{k}}^{(m)} \rangle \tag{11.33}$$

If this result is put into (11.29), the singularity at $m = n$ disappears, we can add n to the sum, and we can apply the closure relation $\sum_m | u_{\boldsymbol{k}}^{(m)} \rangle \langle u_{\boldsymbol{k}}^{(m)} | = 1$. What remains of (11.29), owing to the normalization of the eigenfunctions $u_{\boldsymbol{k}}^{(n)}$, is,

$$\boldsymbol{v}_{\boldsymbol{k}}^{(n)} = \frac{1}{\hbar} \boldsymbol{\nabla}_{\boldsymbol{k}} \varepsilon_{\boldsymbol{k}} - i \left(\langle \nabla_{\boldsymbol{k}} u_{\boldsymbol{k}}^{(n)} | \partial_t u_{\boldsymbol{k}}^{(n)} \rangle - \langle \partial_t u_{\boldsymbol{k}}^{(n)} | \nabla_{\boldsymbol{k}} u_{\boldsymbol{k}}^{(n)} \rangle \right) \tag{11.34}$$

11.2.3 ELECTRON IN A CRYSTAL SUBJECT TO AN ELECTRIC FIELD

Let us apply (11.34) to the case of an electron in a uniform electric field. The approach taken in Chapter 9 was to write the electric field as $\boldsymbol{E} = -\boldsymbol{\nabla}_{\boldsymbol{r}} \varphi$ (see (9.24)). This would mean that the Hamiltonian contains a term that breaks the crystal symmetry. This problem can be circumvented by using the so-called **radiation gauge**.[170] In this gauge, the electrostatic potential vanishes. The vector potential $\boldsymbol{A}(t)$ is defined so that,

$$\boldsymbol{E} = -\frac{\partial \boldsymbol{A}(t)}{\partial t} \tag{11.35}$$

The Hamiltonian in (11.10) is modified in the presence of a vector potential as follows,

$$\mathcal{H}_{\boldsymbol{k}} \left(\boldsymbol{p}, \boldsymbol{r}, t \right) = \frac{1}{2m} \left(\boldsymbol{p} + \hbar \boldsymbol{k} + e \boldsymbol{A}(t) \right)^2 + \mathcal{V}(\boldsymbol{r}) \tag{11.36}$$

Since the crystal periodicity is maintained, $\boldsymbol{k}$ is still a constant of motion, i.e., $\partial \boldsymbol{k}/\partial t = 0$. To make use of the result (11.34) for the velocity expectation value, we define the wave vector $\boldsymbol{q}$ such that,

$$\hbar \boldsymbol{q} = \hbar \boldsymbol{k} + e \boldsymbol{A}(t) \tag{11.37}$$

In view of (11.35), the momentum $\hbar \boldsymbol{q}$ evolves according to,

$$\hbar \frac{\partial \boldsymbol{q}}{\partial t} = -e \boldsymbol{E} \tag{11.38}$$

This is the first semi-classical equation we will use in §11.3 to describe transport with the Boltzmann. In order to apply (11.34), we first note the following derivation chain rules,

$$\frac{\partial}{\partial \boldsymbol{k}} = \frac{\partial \boldsymbol{q}}{\partial \boldsymbol{k}} \frac{\partial}{\partial \boldsymbol{q}} = \frac{\partial}{\partial \boldsymbol{q}} \qquad \frac{\partial}{\partial t} = \frac{\partial \boldsymbol{q}}{\partial t} \frac{\partial}{\partial \boldsymbol{q}} = \frac{\partial \boldsymbol{q}}{\partial t} \frac{\partial}{\partial \boldsymbol{k}} = \frac{-e}{\hbar} \boldsymbol{E} \cdot \frac{\partial}{\partial \boldsymbol{k}} = \frac{-e}{\hbar} E_\beta \frac{\partial}{\partial k_\beta} \tag{11.39}$$

In the last expression, we have used the Einstein summation convention. In (11.39), we see that the time derivative is transformed into a derivative with respect to $\boldsymbol{k}$. Applying this transformation, we can rewrite the velocity in (11.34). To avoid confusion in the notation, we consider the α component of the velocity,

$$
\begin{aligned}
v_{k_\alpha}^{(n)} &= \frac{1}{\hbar}\frac{\partial \varepsilon_{\boldsymbol{k}}}{\partial k_\alpha} - i\left(\langle \frac{\partial u_{\boldsymbol{k}}^{(n)}}{\partial k_\alpha}|\frac{\partial}{\partial t}u_{\boldsymbol{k}}^{(n)}\rangle - \langle\frac{\partial}{\partial t}u_{\boldsymbol{k}}^{(n)}|\frac{\partial u_{\boldsymbol{k}}^{(n)}}{\partial k_\alpha}\rangle\right) \\
&= \frac{1}{\hbar}\frac{\partial \varepsilon_{\boldsymbol{k}}}{\partial k_\alpha} - i\frac{(-e)}{\hbar}\left(\langle\frac{\partial u_{\boldsymbol{k}}^{(n)}}{\partial k_\alpha}|\frac{\partial u_{\boldsymbol{k}}^{(n)}}{\partial k_\beta}\rangle - \langle\frac{\partial u_{\boldsymbol{k}}^{(n)}}{\partial k_\beta}|\frac{\partial u_{\boldsymbol{k}}^{(n)}}{\partial k_\alpha}\rangle\right)E_\beta
\end{aligned}
\tag{11.40}
$$

Thus, we find a correction to the thermodynamics definition $\partial \varepsilon_{\boldsymbol{k}}/\partial(\hbar k_\alpha)$ of the velocity as conjugate to the momentum. Let us write (11.40),

$$
v_{k_\alpha}^{(n)} = \frac{1}{\hbar}\frac{\partial \varepsilon_{\boldsymbol{k}}}{\partial k_\alpha} + \frac{e}{\hbar}\Omega_{\alpha\beta}E_\beta
\tag{11.41}
$$

where $\Omega_{\alpha\beta}(\boldsymbol{k})$ is an antisymmetric tensor called the **Berry curvature**,

$$
\Omega_{\alpha\beta} = i\left(\langle\frac{\partial u_{\boldsymbol{k}}^{(n)}}{\partial k_\alpha}|\frac{\partial u_{\boldsymbol{k}}^{(n)}}{\partial k_\beta}\rangle - \langle\frac{\partial u_{\boldsymbol{k}}^{(n)}}{\partial k_\beta}|\frac{\partial u_{\boldsymbol{k}}^{(n)}}{\partial k_\alpha}\rangle\right)
\tag{11.42}
$$

The product of an antisymmetric tensor and a three-dimensional vector can be written as a vectorial product. After some straightforward calculations, the vector built from the antisymmetric tensor $\Omega_{\alpha\beta}(\boldsymbol{k})$ can be found to have the form,

$$
\boldsymbol{\Omega}^{(n)}(\boldsymbol{k}) = i\,\langle\boldsymbol{\nabla}_{\boldsymbol{k}}u_{\boldsymbol{k}}^{(n)}|\times|\boldsymbol{\nabla}_{\boldsymbol{k}}u_{\boldsymbol{k}}^{(n)}\rangle
\tag{11.43}
$$

In vectorial form (11.41) reads,

$$
\boldsymbol{v}_{\boldsymbol{k}}^{(n)} = \frac{1}{\hbar}\boldsymbol{\nabla}_{\boldsymbol{k}}\varepsilon_{\boldsymbol{k}} - \frac{e}{\hbar}\boldsymbol{E}\times\boldsymbol{\Omega}^{(n)}
\tag{11.44}
$$

where $\boldsymbol{\Omega}^{(n)}$ is called the Berry curvature vector of band n.[2]

Owing to (11.39), we can write (11.38) as $\hbar\partial_t\boldsymbol{k}\equiv\hbar\dot{\boldsymbol{k}} = -e\boldsymbol{E}$. Then, our last result reads,

$$
\boldsymbol{v}_{\boldsymbol{k}}^{(n)} = \frac{1}{\hbar}\boldsymbol{\nabla}_{\boldsymbol{k}}\varepsilon_{\boldsymbol{k}} + \dot{\boldsymbol{k}}\times\boldsymbol{\Omega}^{(n)}
\tag{11.45}
$$

Since we found that $\hbar\dot{\boldsymbol{k}}$ equals the force applied to the electrons when an electric field is applied, let us assume that the same is true if we add the Lorentz force. Thus, we write,

$$
\hbar\dot{\boldsymbol{k}} = -e\boldsymbol{E} - e\boldsymbol{v}_{\boldsymbol{k}}^{(n)}\times\boldsymbol{B}
\tag{11.46}
$$

The results expressed in (11.46) and (11.45) are used in §11.3 to carry out a Boltzmann transport analysis that accounts for the anomalous velocity found in (11.45).

11.3 ANOMALOUS TRANSPORT IN A BOLTZMANN APPROACH

Research on the topological properties of materials, i.e., the topological properties of the possible orbits in $\boldsymbol{k}$-space,[535] has shown that the relevant quantity to account for the AHE and the ANE is the Berry curvature near the Fermi level. Thus, the crystal space group and details of the band structure are relevant. For example, AHE and ANE were observed in Cd_3As_2.[568,569] Strong AHE and ANE have also been found in non-collinear antiferromagnets like Mn_3Sn and Mn_3Ge.[570,571]

---

[2]The definition (11.5) of Bloch states implies that the functions $u_{\boldsymbol{k}}^{(n)}$ are unitless. Thus, the components of the vector $\boldsymbol{\Omega}^{(n)}$ have units of $1/k^2$, i.e., a length squared.

11.3.1 BOLTZMANN EQUATION WITH BERRY CURVATURE

We want to apply the Boltzmann theory of transport, applying the semi-classical equations (11.45) and (11.46) which contain a Berry curvature correction. Thus, we start with the semi-classical equations, written here in terms of the momentum $\boldsymbol{p}$,

$$\dot{\boldsymbol{r}} = \frac{\partial E(\boldsymbol{p})}{\partial \boldsymbol{p}} + \dot{\boldsymbol{p}} \times \boldsymbol{\Omega}/\hbar$$
$$\dot{\boldsymbol{p}} = e\boldsymbol{\nabla}\varphi - e\dot{\boldsymbol{r}} \times \boldsymbol{B} \tag{11.47}$$

We continue to designate $\boldsymbol{v}$ as the conjugate of the momentum, $\boldsymbol{v} = \partial E/\partial \boldsymbol{p}$, as in (9.27). The term $\dot{\boldsymbol{p}} \times \boldsymbol{\Omega}/\hbar$ in (11.47) is the anomalous velocity. Let us first solve (11.47) for $\dot{\boldsymbol{r}}$ and $\dot{\boldsymbol{p}}$. Combining both equations, we obtain,

$$\dot{\boldsymbol{r}} = \boldsymbol{v} + e\boldsymbol{\nabla}\varphi \times \boldsymbol{\Omega}/\hbar + e(\boldsymbol{\Omega}/\hbar \cdot \boldsymbol{B})\dot{\boldsymbol{r}} - e(\dot{\boldsymbol{r}} \cdot \boldsymbol{\Omega}/\hbar)\boldsymbol{B} \tag{11.48}$$
$$\dot{\boldsymbol{p}} = e\boldsymbol{\nabla}\varphi - e\boldsymbol{v} \times \boldsymbol{B} + e(\boldsymbol{B} \cdot \boldsymbol{\Omega}/\hbar)\dot{\boldsymbol{p}} - e(\boldsymbol{B} \cdot \dot{\boldsymbol{p}})\boldsymbol{\Omega}/\hbar \tag{11.49}$$

In the last term of (11.48), we now use $\boldsymbol{v}$ for $\dot{\boldsymbol{r}}$ so as to keep the anomalous correction to first order.[3] Likewise, we now use $-e\boldsymbol{E}$ in (11.49) for $\dot{\boldsymbol{p}}$, so as to retain only the first order terms in $\boldsymbol{B}$. Hence, we have semi-classical equations that are now expressed separately in terms of $\boldsymbol{v}$,[572]

$$\dot{\boldsymbol{r}} = (1 - (e/\hbar)\boldsymbol{\Omega} \cdot \boldsymbol{B})^{-1}\left[\boldsymbol{v} + (e/\hbar)\boldsymbol{\nabla}\varphi \times \boldsymbol{\Omega} - (e/\hbar)(\boldsymbol{\Omega} \cdot \boldsymbol{v})\boldsymbol{B}\right]$$
$$\dot{\boldsymbol{p}} = (1 - (e/\hbar)\boldsymbol{\Omega} \cdot \boldsymbol{B})^{-1}\left[e\boldsymbol{\nabla}\varphi - e\boldsymbol{v} \times \boldsymbol{B} - (e/\hbar)(\boldsymbol{B} \cdot e\boldsymbol{\nabla}\varphi)\boldsymbol{\Omega}\right] \tag{11.50}$$

The Boltzmann equation (9.18) should be considered with the newly found expression for $\dot{\boldsymbol{p}}$ instead of Newton's law (9.16). Thus, we write,

$$\frac{\partial f(\boldsymbol{p},\boldsymbol{r},t)}{\partial t} + \frac{\partial f(\boldsymbol{p},\boldsymbol{r},t)}{\partial \boldsymbol{r}}\dot{\boldsymbol{r}} + \frac{\partial f(\boldsymbol{p},\boldsymbol{r},t)}{\partial \boldsymbol{p}}\dot{\boldsymbol{p}} = -\frac{f(\boldsymbol{p},\boldsymbol{r},t) - f_0(\boldsymbol{p},\boldsymbol{r})}{\tau} \tag{11.51}$$

where, now, $\dot{\boldsymbol{r}}$ and $\dot{\boldsymbol{p}}$ are given by (11.50).

11.3.2 ANOMALOUS HALL EFFECT

In order to bring out the **_anomalous Hall effect_** corresponding to the Berry curvature term in (11.50), we analyze the consequences of the modified semi-classical equations (11.50) when $\boldsymbol{B} = 0$. Thus, we consider,

$$\dot{\boldsymbol{r}} = \boldsymbol{v} + (e/\hbar)\boldsymbol{\nabla}\varphi \times \boldsymbol{\Omega}$$
$$\dot{\boldsymbol{p}} = e\boldsymbol{\nabla}\varphi \tag{11.52}$$

We see that the Berry curvature implies an extra velocity,[551]

$$\boldsymbol{V} = (e/\hbar)\boldsymbol{\nabla}\varphi \times \boldsymbol{\Omega} \tag{11.53}$$

Let us now show that this anomalous velocity contributes to the Hall effect. To do this, we consider the linear response when the system is in a stationary state. The Boltzmann distribution $f(\boldsymbol{p},\boldsymbol{r},t)$ is adapted from (9.19) to account for the dynamics given by (11.50).

---

[3]From (2.34), we see that eB has SI units of kg s^{-1}. It is then easy to verify that $(e/\hbar)\Omega B$ is unitless.

The derivatives of f_0 were worked out in (10.77) and (10.78). We are considering a system free of temperature gradients and we assume $\nabla \mu = 0$. Then, what remains of (11.51) is,

$$\frac{\partial f_0\,(\boldsymbol{p}, \boldsymbol{r}, t)}{\partial \boldsymbol{p}} \dot{\boldsymbol{p}} = -\frac{f_1\,(\boldsymbol{p}, \boldsymbol{r}, t)}{\tau} \tag{11.54}$$

Therefore, the equation for f_1 is equivalent to (9.22), which we used to derive Ohm's law in §9.2.5. The electric current contribution of one electron is given by $-e\dot{\boldsymbol{r}}$. In view of the velocity given by (11.52), the integral (9.32) becomes,

$$\boldsymbol{j} = \frac{2}{(2\pi\hbar)^3}(-e) \int d^3p \left(\boldsymbol{v} + (e/\hbar)\nabla\varphi \times \boldsymbol{\Omega} \right) (f_0 + f_1) \tag{11.55}$$

The first term of the velocity, weighted by f_0, gives a vanishing contribution to the current, as usual. The first term of the velocity, weighted by f_1, gives the usual Ohm's law, with a conductivity σ. Since the second term is a perturbation, it is sufficient to take f_0 for the statistical weight. Thus, we have,[573]

$$\boldsymbol{j} = -\sigma\,\nabla\varphi - e^2 \frac{2}{(2\pi\hbar)^3} \int d^3p \left(\nabla\varphi \times \boldsymbol{\Omega}/\hbar \right) f_0 \tag{11.56}$$

Clearly, the Berry curvature introduces a conductivity $\sigma_{yx} \neq 0$. In a Hall measurement, we impose $j_y = 0$. To account for the transverse electric field $E_y = -\nabla_y\varphi$, we have to consider that there are in (11.56) two contributions to the transverse current, namely, we have, $j_y = \sigma(-\nabla_y\varphi) + \sigma_{yx}(-\nabla_x\varphi)$.

This method of accounting for the Hall effect is particularly relevant in topological materials.[574] The so-called Weyl and Dirac semimetals are materials in which the band structure has gapless electronic excitations. A gap cannot form because of symmetry and topology considerations. Thus, angle-resolved photoemession spectroscopy is one of the primary tools used to study these materials. These materials can also be characterized by transport measurements, typically by observing the anomalous Hall effect and anomalous longitudinal magnetoresistance, introduced below.[575]

11.3.3 ANOMALOUS LONGITUDINAL MAGNETORESISTANCE

Let us now mention one striking consequence of the Berry curvature that occurs in the presence of a magnetic induction field. According to (11.50), $\dot{\boldsymbol{p}}$ in (11.54) is modified by a term $(\boldsymbol{B} \cdot \nabla\varphi)\boldsymbol{\Omega}$. This implies that a magnetoresistance is expected when $\boldsymbol{B}$ and $\nabla\varphi$ are parallel, whereas the Lorentz force is inactive. This so-called *longitudinal magnetoresistance* is proportional to the magnitude of the Berry curvature.

Topological materials have very large values of the magnetoresistance that can be interpreted in terms of anomalous longitudinal magnetoresistance.[576–583]. This large magnetoresistance can happen with or without the chiral anomaly described below.[584]

Some insight about this effect and related anomalous transport phenomena can be gained within the framework of the transport equations (11.50) and (11.51).

For example, D. T. Son and B. Z. Spivak considered two Boltzmann distributions associated with charges near two Dirac cones of a Weyl semimetal, with each cone being characterized by a Berry curvature $\boldsymbol{\Omega}_i$.[585] Then, the evolution equation (11.51) for each distribution reads,

$$\frac{\partial f_i\left(\boldsymbol{p},\boldsymbol{r},t\right)}{\partial t} + \frac{\partial f_i\left(\boldsymbol{p},\boldsymbol{r},t\right)}{\partial \boldsymbol{r}}\left[\boldsymbol{v} + (e/\hbar)\boldsymbol{\nabla}\varphi \times \boldsymbol{\Omega}_i - (e/\hbar)(\boldsymbol{\Omega}_i \cdot \boldsymbol{v})\boldsymbol{B}\right]\left(1 - (e/\hbar)\boldsymbol{\Omega}_i \cdot \boldsymbol{B}\right)^{-1}$$

$$+ \frac{\partial f_i\left(\boldsymbol{p},\boldsymbol{r},t\right)}{\partial \boldsymbol{p}}\left[e\boldsymbol{\nabla}\varphi - e\boldsymbol{v} \times \boldsymbol{B} - (e/\hbar)(\boldsymbol{B} \cdot e\boldsymbol{\nabla}\varphi)\boldsymbol{\Omega}_i\right]\left(1 - (e/\hbar)\boldsymbol{\Omega}_i \cdot \boldsymbol{B}\right)^{-1}$$

$$= -\frac{f_i\left(\boldsymbol{p},\boldsymbol{r},t\right) - f_{0i}\left(\boldsymbol{p},\boldsymbol{r}\right)}{\tau} \qquad (i = 1,2)$$

$$(11.57)$$

Son and Spivak show that the evolution equations (11.57) for f_i ($i = 1,2$), imply the continuity equations,

$$\frac{\partial N_i}{\partial t} + \boldsymbol{\nabla} \cdot \boldsymbol{j}_i = k^i \frac{e^2}{4\pi^2\hbar^2}(\boldsymbol{B} \cdot \boldsymbol{\nabla}\varphi) - \frac{N_i - N_{i0}}{\tau} \qquad (11.58)$$

where the current density $\boldsymbol{j}_i$ conforms to our definition (9.32). In other words, we have,

$$\boldsymbol{j}_i = \frac{2(-e)}{(2\pi\hbar)^3}\int d^3p\frac{\left(\boldsymbol{v} + (e/\hbar)\boldsymbol{\nabla}\varphi \times \boldsymbol{\Omega}_i - (e/\hbar)(\boldsymbol{\Omega}_i \cdot \boldsymbol{v})\boldsymbol{B}\right)}{(1 - (e/\hbar)\boldsymbol{\Omega}_i \cdot \boldsymbol{B})}f_i \qquad (11.59)$$

and $k_i = \pm 1$. The continuity equations (11.58) present a source term proportional to $(\boldsymbol{B} \cdot \boldsymbol{\nabla}\varphi)$. The presence of this current source term is known as the ***chiral anomaly*** and was first derived in the framework of quantum field theory.[586] The contribution of chiral anomaly to the total current $\boldsymbol{j}_1 + \boldsymbol{j}_2$ is estimated by calculating integrals in $\boldsymbol{k}$-space centered at $\pm \boldsymbol{k}_i$, the positions of the two Dirac cones. This yields a current,

$$\boldsymbol{j}|_{chiral} = \frac{e^2}{4\pi^2\hbar^2}\boldsymbol{B}(k_1\mu_1 + k_2\mu_2) \qquad (11.60)$$

At equilibrium, the chemical potential μ_1 and μ_2 are equal and the chiral current vanishes. However, in the presence of an electric field in the $\boldsymbol{B}$ direction, there is a charge imbalance, and the chemical potentials μ_1 and μ_2 are no longer equal. For a formal, quantum mechanical formulation of the chiral anomaly, see the much earlier work presented in [587].

11.4   SPIN HALL EFFECT

The ***spin Hall effect*** (SHE) refers to the spin current density that can be induced by applying an electric field. This spin current density is perpendicular to the electric field and the spin polarization is perpendicular to both the electric field and the spin current density. Thus, we predict a spin current,

$$j_j^i = \sigma_{SHE}\epsilon^{ijk}E_k \qquad (11.61)$$

As mentioned in the introduction, there is abundant literature on the spin Hall effect. First, we will see how a spin Hall effect can arise from the band structure, i.e., as an intrinsic phenomenon. Second, we will see how collisions at impurities can give rise to a spin Hall effect, as well.

11.4.1   INTRINSIC SPIN HALL EFFECT

Here, we wish to develop an intuition for the intrinsic SHE effect using the specific example presented by Murakami et al. in 2003.[588] The III–V semiconductors such as GaAs and InP have the band structure shown in Fig. 11.1.

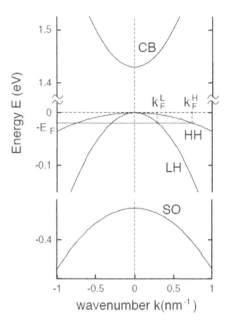

Figure 11.1 Band structure schematics for GaAs. The heavy hole (HH) and light hole (LH) bands are nearly degenerate whereas the split-off (SO) band is clearly separated.

This electronic structure is due to the spin-orbit interaction. The $\boldsymbol{k}\cdot\boldsymbol{p}$ method can be used to infer the existence of bands corresponding to light holes (LH) and heavy holes (HH).[72]

At the Γ-point, i.e., at $\boldsymbol{k}=0$, the state is four-fold degenerate. An effective Hamiltonian can account for this structure, written in terms of the spin $3/2$ operator $\boldsymbol{S}$,

$$\mathcal{H}_0 = \frac{\hbar^2}{2m}\left(\left(\gamma_1 + \frac{5}{s}\gamma_2\right)k^2 - 2\gamma_2\left(\boldsymbol{k}\cdot\boldsymbol{S}\right)^2\right) \tag{11.62}$$

The dimensionless helicity operator,

$$\lambda = \frac{1}{\hbar}\frac{\boldsymbol{k}}{k}\cdot\boldsymbol{S} \tag{11.63}$$

commutes with $\mathcal{H}_0$. It is possible to find eigenstates of $\mathcal{H}_0$ that are also eigenstates of λ. The eigenvalues of λ are $\pm 3/2$. The HH and LH dispersions are given by,

$$E_H = \frac{\hbar^2 k^2}{2m_H} \qquad E_L = \frac{\hbar^2 k^2}{2m_L} \tag{11.64}$$

where

$$m_H = \frac{m}{\gamma_1 - 2\gamma_2} \qquad m_L = \frac{m}{\gamma_1 + 2\gamma_2} \tag{11.65}$$

Both the LH and HH bands are two-fold degenerate, where the degeneracy is characterized by the helicity λ,

$$\text{HH}: \lambda = \pm\frac{3}{2} \qquad \text{LH}: \lambda = \pm\frac{1}{2} \tag{11.66}$$

In this discussion, we assume that the bands are spherical and we neglect the slight lifting of the degeneracy at the Γ point.

Now, we discuss the effect of applying an electric field $\boldsymbol{E}$. Thus, we consider the Hamiltonian,

$$\mathcal{H} = \mathcal{H}_0 + e\boldsymbol{E} \cdot \boldsymbol{x} \qquad (e > 0) \tag{11.67}$$

To clarify the upcoming calculations, we define the unitary transformation $U(\boldsymbol{k})$ that diagonalizes $\mathcal{H}_0$. From the expression (11.62) of $\mathcal{H}_0$, we see that we need the transformation that brings $\boldsymbol{k} \cdot \boldsymbol{S}$ into $k\hat{S}_z$, i.e.,

$$U(\boldsymbol{k}) (\boldsymbol{k} \cdot \boldsymbol{S}) U(\boldsymbol{k})^\dagger = k\hat{S}_z \tag{11.68}$$

Here, we anticipate the presentation of §19.4 on the representation of rotations with angular momenta. If $\boldsymbol{k}$ is given in cylindrical coordinates, i.e., $\boldsymbol{k} = k(\sin\theta\cos\phi, \sin\theta\sin\phi, \cos\theta)$, then the unitary matrix (which acts on the spin $\boldsymbol{S}$) is

$$U(\boldsymbol{k}) = \exp\left(i\theta\hat{S}_y\right) \exp\left(i\phi\hat{S}_z\right) \tag{11.69}$$

Under the unitary transformation shown in (11.68), the Hamiltonian $\mathcal{H}$ becomes,

$$\tilde{\mathcal{H}} = \frac{\hbar^2 k^2}{2m}\left(\gamma_1 + \frac{5}{2}\gamma_2 - 2\gamma_2\hat{S}_z^2\right) + U(\boldsymbol{k}) (e\boldsymbol{E} \cdot \boldsymbol{x}) U(\boldsymbol{k})^\dagger \tag{11.70}$$

The first term is diagonal and the eigenvalues of $\hat{S}_z$ are also the eigenvalues of the helicity operator λ. The unitary operator acts on functions of $\boldsymbol{k}$. In the Hilbert space of $\boldsymbol{k}$-functions, the action of the operator $\boldsymbol{x}$ is equivalent to $i\boldsymbol{\nabla}_{\boldsymbol{k}}$.

This notion is often learned in introductory quantum mechanics, where $\boldsymbol{p}$ and $\boldsymbol{x}$ have converse roles. Generally, it is stated that the operator $\boldsymbol{p}$ corresponding to linear momentum is the operator $-i\hbar\boldsymbol{\nabla}$ acting on the wave functions (understood as function of the spacial coordinates $\boldsymbol{x}$). Let us briefly review the steps of that reasoning. If Q_α is the α-component of the position operator vector, acting on the wave function $|\boldsymbol{x}\rangle$, then $\boldsymbol{x}$, $Q_\alpha|\boldsymbol{x}\rangle = x_\alpha|\boldsymbol{x}\rangle$. Just as in Lagrangian mechanics,[557] linear momentum is the generator of translation, meaning that $|\boldsymbol{x} + \boldsymbol{a}\rangle = \exp\left(-i\boldsymbol{a} \cdot \boldsymbol{p}/\hbar\right)|\boldsymbol{x}\rangle$. Therefore, when $\boldsymbol{a}$ has a small modulus, we have,[589]

$$\begin{aligned}\psi(\boldsymbol{x} + \boldsymbol{a}) &= \langle \boldsymbol{x} + \boldsymbol{a}|\psi\rangle = \langle\boldsymbol{x}|\exp\left(i\boldsymbol{a} \cdot \boldsymbol{p}/\hbar\right)|\psi\rangle \\ &\approx \langle\boldsymbol{x}|\left(1 + i\boldsymbol{a} \cdot \boldsymbol{p}/\hbar\right)|\psi\rangle \equiv \psi(\boldsymbol{x}) + i\boldsymbol{a} \cdot (\boldsymbol{p}/\hbar)\,\psi(\boldsymbol{x})\end{aligned} \tag{11.71}$$

Comparing this expression with a Taylor expansion leads to the statement that the operator $\boldsymbol{p}$ "is" $-i\hbar\boldsymbol{\nabla}$.

In the present discussion, we are working in $\boldsymbol{k}$-space and in (11.70) we have the operator $\boldsymbol{x}$ operating on functions of $\boldsymbol{k}$. As shown in [589], this implies that, in the momentum representation in which $P_\alpha|\boldsymbol{p}\rangle = \boldsymbol{p}|\boldsymbol{p}\rangle$, we have,

$$\boldsymbol{x} \equiv i\boldsymbol{\nabla}_{\boldsymbol{k}} \tag{11.72}$$

Thus, in (11.70), $\boldsymbol{x}$ should be considered as a differential operator. Therefore, when considering $\mathcal{H}|\psi\rangle$, the elementary rule of derivation implies that two terms appear:

$$U(\boldsymbol{k}) (e\boldsymbol{E} \cdot \boldsymbol{x}) U(\boldsymbol{k})^\dagger |\psi\rangle = e\boldsymbol{E} \cdot \left((i\boldsymbol{\nabla}_{\boldsymbol{k}}) + U(\boldsymbol{k})(i\boldsymbol{\nabla}_{\boldsymbol{k}})U(\boldsymbol{k})^\dagger\right)|\psi\rangle \tag{11.73}$$

These two terms have the form of a material derivative that we will denote $\boldsymbol{D}$. Thus, having done the rotation defined by $U(\boldsymbol{k})$, we have a Hamiltonian $\tilde{\mathcal{H}}$ that contains a kinetic term that is diagonal, and a perturbation $V(\boldsymbol{D})$, with

$$\tilde{\mathcal{H}} = \frac{\hbar^2 k^2}{2m}\left(\gamma_1 + \frac{5}{2}\gamma_2 - 2\gamma_2\hat{S}_z^2\right) + e\boldsymbol{E} \cdot \tilde{\boldsymbol{D}} \tag{11.74}$$

where

$$\tilde{D} = -i\boldsymbol{\nabla}_{\boldsymbol{k}} - \tilde{\boldsymbol{A}} \qquad (11.75)$$

$$\tilde{\boldsymbol{A}} = -iU(\boldsymbol{k})\boldsymbol{\nabla}_{\boldsymbol{k}}U(\boldsymbol{k})^\dagger \qquad (11.76)$$

Let us now work out an explicit expression for $\tilde{\boldsymbol{A}}$ in spherical coordinates. From the definition (11.69) of $U(\boldsymbol{k})$, we have,

$$U(\boldsymbol{k})^\dagger = \exp\left(-i\phi\hat{S}_z\right)\exp\left(-i\theta\hat{S}_y\right) \qquad (11.77)$$

Therefore, we can write $\tilde{\boldsymbol{A}}$ as,

$$\tilde{\boldsymbol{A}} = -\exp\left(i\theta\hat{S}_y\right)\exp\left(i\phi\hat{S}_z\right)\left[\boldsymbol{\nabla}_{\boldsymbol{k}}\phi\hat{S}_z + \boldsymbol{\nabla}_{\boldsymbol{k}}\theta\hat{S}_y\right]\exp\left(-i\phi\hat{S}_z\right)\exp\left(-i\theta\hat{S}_y\right) \qquad (11.78)$$

We have to be careful when rearranging the terms because $\hat{S}_x$ and $\hat{S}_y$ do not commute. Nonetheless, we can make the following simplification,

$$\tilde{\boldsymbol{A}} = -\exp\left(i\theta\hat{S}_y\right)\hat{S}_z\exp\left(-i\theta\hat{S}_y\right)\boldsymbol{\nabla}_{\boldsymbol{k}}\phi - \exp\left(i\phi\hat{S}_z\right)\hat{S}_y\exp\left(-i\phi\hat{S}_z\right)\boldsymbol{\nabla}_{\boldsymbol{k}}\theta \qquad (11.79)$$

We will learn in Chapter 19 that an expression like this one can be readily simplified using the notion that angular momenta are generators of rotations. (The general property of angular momentum operators what we use her is given by (19.28)). We thus obtain,

$$\tilde{\boldsymbol{A}} = -\left[\cos\theta\hat{S}_z - \sin\theta\hat{S}_x\right]\boldsymbol{\nabla}_{\boldsymbol{k}}\phi - \left[\sin\phi\hat{S}_x + \cos\phi\hat{S}_y\right]\boldsymbol{\nabla}_{\boldsymbol{k}}\theta \qquad (11.80)$$

Applying the formula for calculating a gradient in spherical coordinates, we find,

$$\boldsymbol{\nabla}_{\boldsymbol{k}}\phi = \hat{e}_\phi\frac{1}{k\sin\phi} \qquad \boldsymbol{\nabla}_{\boldsymbol{k}}\theta = \frac{1}{k}\hat{e}_\theta \qquad (11.81)$$

Using (11.81) in (11.80), and multiplying this by $d\boldsymbol{k} = k\sin\theta d\phi\hat{e}_\phi + kd\theta\hat{e}_\theta$, yields,

$$\tilde{\boldsymbol{A}}\cdot d\boldsymbol{k} = -[-\sin\theta d\phi + \sin\phi d\theta]\hat{S}_x - \cos\phi d\theta\hat{S}_y - \cos\theta d\phi\hat{S}_z \qquad (11.82)$$

We can find the matrices for $\hat{S}_x$, $\hat{S}_y$ and $\hat{S}_z$ in the basis where $\hat{S}_z$ is diagonal by using the well-known rules $\hat{S}_z\left|j,m\right\rangle = m\left|j,m\right\rangle$ and $\hat{S}_\pm = \sqrt{j(j+1)-m(m\pm 1)}\left|j,m\pm 1\right\rangle$. Hence, $\hat{S}_x$ and $\hat{S}_y$ have non-zero elements on the first diagonal above and below the main diagonal. As we will see below, retaining only the diagonal elements, i.e., those coming from $\hat{S}_z$, captures the physics we want to explain here. That is to say, we consider from now on only the contribution $-\cos\theta\, d\phi\hat{S}_z$. In other words, we have,

$$\tilde{\boldsymbol{A}} = -\hat{S}_z\cos\theta\boldsymbol{\nabla}_{\boldsymbol{k}}\phi \qquad (11.83)$$

Now, we want to revert back to $\mathcal{H}$, which we write in $\boldsymbol{k}$-representation as,

$$\mathcal{H} = \frac{\hbar^2 k^2}{2m_\lambda} + e\boldsymbol{E}\cdot\boldsymbol{D} \qquad (11.84)$$

When $\lambda = \pm 1/2$, $m_\lambda = m_L$, and when $\lambda = \pm 3/2$, $m_\lambda = m_H$.

In the present discussion, $\boldsymbol{D}$ is used as the $\boldsymbol{k}$-space representation of $\boldsymbol{x}$. This is similar to saying that $\boldsymbol{p}$ "is" $-i\hbar\boldsymbol{\nabla}$. We recall that $U(\boldsymbol{k})$ was defined to rotate $\boldsymbol{k}$ onto the z axis.

With the inverse operation, we will bring $\hat{S}_z$ in (11.80) onto S along k. Thus, A will contain trigonometric functions and all three operators, $\hat{S}_x$, $\hat{S}_y$ and $\hat{S}_z$. As a result, we find that $[D_i, D_j] \neq 0$ if $i \neq j$ because $\hat{S}_x$, $\hat{S}_y$ and $\hat{S}_z$ do not commute with one another. Let us agree to use the symbol x_i for D_i from now on. So, we have $[x_i, x_j] \neq 0$ if $i \neq j$. Thus, we have here an example of what is called **non-commutative** geometry.[4] A similar algebraic structure is found in the quantum Hall effect. These commutation relations are ubiquitous in planar systems subjected to a strong magnetic field and a weak electrostatic potential $V(x, y)$.[5]

The non-commutation $[x_i, x_j] \neq 0$ leads to the intrinsic spin Hall effect for the band structure illustrated in Fig. 11.1. In order to calculate a spin current, we need the spin-dependent equivalent to the kinetic relations (11.47) that we used for the spinless case. In particular, we need to work out $\dot{x}$ and will see that it contains an anomalous term. The Heisenberg evolution equation reads $\dot{x} = (i/\hbar)[\mathcal{H}, x]$. So, we see that we need to consider commutators such as $[x_i, x_j]$. To calculate this commutator, Murakami et al. [588] draw the analogy with the example Berry gave in his seminal paper.[565] Thus, they define the antisymmetric tensor $\mathbf{\Omega}^\lambda$,

$$\left(\mathbf{\Omega}^\lambda\right)_{ij} = \Omega_{ij}^\lambda = i[x_i, x_j] = \varepsilon_{ijk}\lambda \frac{(\boldsymbol{k})_k}{k^3} \tag{11.85}$$

Recall that, here, λ is the helicity operator (11.63), which depends on the spin S. The potential energy specified by the Hamiltonian (11.84) implies that the evolution of k and x is given by,

$$\hbar \dot{k}_i = -eE_i$$
$$\dot{x}_i = \frac{\hbar k_i}{m_\lambda} + \Omega_{ij}^\lambda k_j \tag{11.86}$$

Figure 11.2 shows the real-space trajectories predicted by these equations.[588] The deviations resemble the effect of a Lorentz force applied to charges of opposite sign, which is in fact the change in helicity λ in (11.86). This transverse drift of charges of opposite helicity implies that a spin current $\boldsymbol{j}_s$ flows transverse to the charge current. Thus, we have obtained for the band structure of Fig. 11.1 the spin Hall effect with the geometry corresponding to (11.61).

Sinova et al. pointed out that an intrinsic spin Hall effect is to be expected in any electronic system in which the Rashba coupling (6.38) is effective.[591, 592] Let us consider, as in [591], a two-dimensional electron gas for which the Hamiltonian reads,

$$\mathcal{H} = \frac{p^2}{2m} - \frac{\lambda_{SO}}{\hbar}\boldsymbol{\sigma} \cdot (\hat{\boldsymbol{z}} \times \boldsymbol{p}) \tag{11.87}$$

Our goal is simply to conduct a heuristic argument. So, we consider the Rashba term as a Zeeman term with a momentum-dependent magnetic induction field proportional to $\hat{\boldsymbol{z}} \times \boldsymbol{p}$. When we look at the spin-locked states of Fig. 6.7, we see that the spins are arranged parallel to this momentum-dependent field. Now, we consider that an electric field is applied in the x-direction, causing a change in momentum Δp_x. This causes an extra contribution to this momentum-dependent field, namely $\hat{\boldsymbol{z}} \times \Delta \boldsymbol{p}$. We only consider what happens in between collisions. During this short time, the spins precess. This causes a z-component of the spin to develop in the direction $z > 0$ for $p_y > 0$ and conversely in the direction $z < 0$ for

---

[4]We note that we have $[k_i, k_j] = 0$ and $[x_i, k_j] = i\delta_{ij}$, as usual.

[5]See, [181] §11-3 and [590].

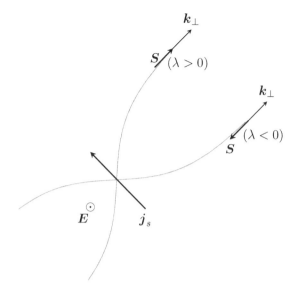

Figure 11.2 Trajectories in the plane normal to the applied electric field, obtained by integrating the kinetic equations (11.86) (figure adapted from [588]). A charge injection with momentum $k_\perp$ is assumed. The deviations from straight lines for charges in different helicity states imply a spin current j_s transverse to the charge current and to the electric field E.

$p_y < 0$ (Fig. 11.3). Therefore, a spin current develops in the y-direction, with spins in the x-directions.

11.4.2 EXTRINSIC SPIN HALL EFFECT

So far, we have examined the ***intrinsic*** spin Hall effect. We considered how it may arise in a crystal with a particular electronic band structure, which is a property of the perfect crystal. Impurity scattering can also lead to a spin Hall effect, which is then qualified as ***extrinsic***. There are two aspects of impurity scattering that lead to a spin Hall effect, the ***side jump*** and the ***skew scattering***. The side jump mechanism may seem like a strange concept if we rely too closely to a Drude-like picture, where charge carriers are localized objects. In a metal however, the eigenstates of the unperturbed crystal are Bloch states. To recover the physical intuition of point-like charges, we have to describe charge carriers as wave packets. We can work out the collision of each wave component with a point defect and then sum the resulting scattered waves to see what the wave packet becomes after a collision. We can thus find that the trajectory of the wave packet center has shifted with respect to the trajectory of the incoming wave packet. In other words, the center of the wave packet makes a side jump as a result of its collision with the impurity. Here, we are going to focus on the notion of skew scattering, which is due to a spin-orbit coupling.

From the spin-orbit coupling definition (2.43), we can infer that an electron with a spin $\boldsymbol{\sigma}$ passing by an impurity generating the electric field $-\boldsymbol{\nabla}V(r)$ has the spin-orbit energy,

$$V_{\rm SO} = \lambda_{\rm SO}\left(-\boldsymbol{\nabla}V(r) \times \boldsymbol{p}\right) \cdot \boldsymbol{\sigma} = \lambda_{\rm SO}\boldsymbol{\nabla}V(r) \cdot (\boldsymbol{\sigma} \times \boldsymbol{p}) \qquad (11.88)$$

Since the spin-orbit interaction is weak, we can assume that the momentum in the x direction is roughly unperturbed. From $V_{\rm SO}$, we can deduce a force on the electron, given by $-\boldsymbol{\nabla}V_{\rm SO}$.

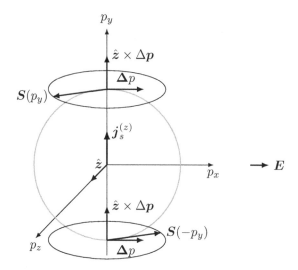

Figure 11.3 Without any applied electric field, the spins are tangent to the Fermi circle in the (p_x, p_y) plane and the action of the field $\hat{z} \times \boldsymbol{p}$ vanishes. Between two momentum collisions, the electrons are accelerated by the electric field applied in the x direction, causing a momentum change $\Delta \boldsymbol{p}$. This adds a momentum-dependent field $\hat{z} \times \Delta \boldsymbol{p}$. A spin precession is expected, causing the spin at $p_y < 0$ to precess below the plane of the figure, i.e., it develops a negative polarization in the z-direction. To the contrary, for the $p_y > 0$, the spins precess and acquire a polarization in the $+z$ direction. The net result is a spin current $\boldsymbol{j}_s^{(z)}$ in the y direction, of spin in the z direction.

Let us consider the configuration sketched in Fig. 11.4, where the spins are either in the $+z$ or $-z$ direction and the momentum is essentially in the $+x$ direction. Near a nucleus, the potential $V(r)$ is of the approximate form $-1/r$, and its second derivative, being proportional to $-1/r^3$, is everywhere negative. Therefore, regardless of whether the electron passes to the right or to the left of the nucleus, it is deviated to the $+y$ direction if its spin is up, and to the $-y$ direction if its spin is down (the force is proportional to $-\partial^2 V_{\text{SO}}/\partial y^2$). Hence, such collisions induce a spin current in the $-y$ direction.

The spin current we inferred from this qualitative argument can be expressed in a generalized form by the transport equation (11.61). Unlike in the previous section, the spin Hall effect here is extrinsic, being due to collisions with impurities. There is no applied magnetic field and no charge current since equal charges move equally in the y and the $-y$ directions.

If we were to apply a Boltzmann description of transport, we would calculate scattering cross-sections in order to estimate the collision time. If $\mathcal{H}_0$ is the crystal potential, V_1 the impurity potential, and V_2 the spin-orbit interaction, then under the Born approximation, we would calculate $|\langle \Phi_a| V_1 + V_2 |\Phi_b\rangle|^2$. This level of approximation would fail to capture any skew scattering because the cross-section would be time-invariant, whereas the process depicted in Fig. 11.4 is not. At a higher level of perturbation, we must take as wave functions those obtained from working out the dynamics under $\mathcal{H}_0 + V_1$. As a result, the scattering amplitude contains two terms: f_{ab}, which we would have in the absence of spin-orbit scattering, and an additional term g_{ab}, which is due to V_2. Thus, we have the scattering cross-section,[593]

$$\sigma_{ab} = |f_{ab} + g_{ab}|^2 \tag{11.89}$$

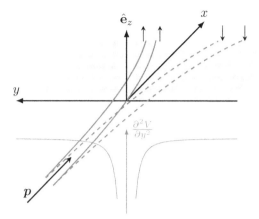

Figure 11.4 Schematics showing that electrons with spin up and spin down are deviated transversally with respect to the initial momentum $\boldsymbol{p}$, in opposite directions, no matter the sign of the impact parameter. Underneath in shaded gray: sketch of $\partial^2 V/\partial y^2$, showing that it is everywhere negative. The transverse force is proportional to this derivative and to the sign of the projection of $\boldsymbol{\sigma} \times \boldsymbol{p}$ on the y axis.

where for the spin-orbit coupling,[181]

$$g_{ab} \propto \langle \Phi'_a | \, \boldsymbol{\sigma} \, | \Phi'_b \rangle \cdot (\boldsymbol{k}_a \times \boldsymbol{k}_b) \tag{11.90}$$

Thus, g_{ab} changes sign when the process is reversed in time, i.e., $g_{ab} = -g_{ba}$. Consequently, $\sigma_{ab} \neq \sigma_{ba}$. This inequality is characteristic of skew scattering. In the present discussion, we have seen how g_{ab} leads to a transverse spin current, but the same mechanism also leads to a Hall effect when spin transport is not relevant.

As mentioned in the introduction §11.1, the spin Hall effect was conceptualized in the 1970s as a means to obtain spin polarization by driving a current through a conductor in which the spin-orbit coupling is strong. Typically, a spin accumulation is expected along the edge of the sample. Since there is no concomitant charge accumulation, it is quite challenging to detect this small effect. Awshalom's group detected spin accumulation in the 2D electron gas of a GaAs heterostructure, using Kerr microscopy (Fig. 11.5).[594] This group considered that their observation was probably due to impurities scattering rather than an intrinsic effect.[595] When a spin Hall effect is observed, it may be difficult to determine if it is due to spin-orbit scattering (i.e., extrinsic) or an intrinsic mechanism,[563] like the one discussed in 11.4.1. When Ir is doped in a Cu film, the spin Hall effect is made deliberately extrinsinc.[596]

In summary, we can distinguish three types of Hall effects (Fig. 11.6),[597]

- the ordinary Hall effect: a Hall voltage is detected that depends on the magnetic field, but there is no spin polarization,
- the anomalous Hall effect: a Hall voltage contribution is found that depends on magnetization, and there is some spin polarization,
- the spin Hall effect: a spin accumulation, but no Hall voltage, is found without application of a magnetic field.

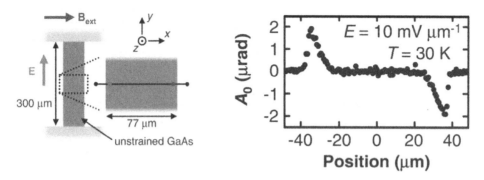

Figure 11.5 Left: GaAs Hall bar with an electric field E applied to drive a current in the longitudinal direction, inducing a spin current in the transverse direction that results in a spin accumulation at the edges of the Hall bar. Right: the maximum Kerr angle is detected for a line scan across the Hall bar. The Kerr rotation is at a maximum at the edges and changes sign, indicating opposite spin orientations normal to the Hall bar at the edges.[595]

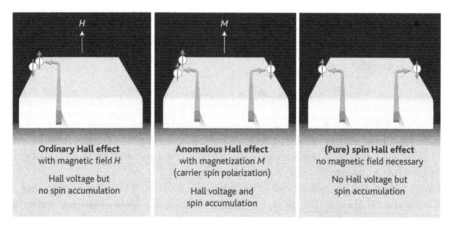

Figure 11.6 Three types of Hall effects: without or with spin polarization, in the presence or without an applied field.[597]

11.4.3 INVERSE SPIN HALL EFFECT

We have written (11.61), assuming that an electric field was imposed to the system and the spin current is a response to this electric field. Conversely, we can consider a case where the spin current is injected into the material and, in this case (11.61), predicts that the injected spin current will induce an electric field. Among other methods, the inverse spin Hall effect has been evidenced using ferromagnetic resonance (see Chapter 15). Under this approach, a material with strong spin-orbit effects is deposited on top of a ferromagnet. When the resonance of the magnetization is excited, the magnetization precesses about the effective field (7.20). In a stationary state, there is a balance between the microwave excitation that tends to open up the precession cone and the relaxation mechanisms that tend to close the precession cone (Fig. 11.7). In this stationary state, microwave energy is converted into a spin current that enters the metal electrode. This phenomenon is known as **spin pumping**. We defer a description of the spin current induced by spin pumping to section §17.6 on magnetic resonance. The existence of an inverse spin Hall effect means that

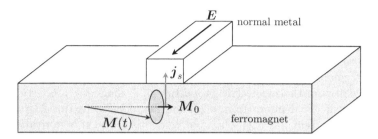

Figure 11.7 Schematic slice of a ferromagnetic thin film has its magnetization excited by a microwave field (not shown), resulting in the precession of the magnetization $M(t)$ about its equilibrium position M_0. In the stationary regime, this implies that a spin polarization parallel to M_0 is injected into the electrode, i.e., there is a spin current j_S (see §17.6 for this spin pumping effect). By way of the inverse spin Hall effect, this spin current implies an electric field E perpendicular to both M_0 and j_S.

an electric field develops in the electrode. This electric field is perpendicular to both the spin current and the spin polarization.[598] Spin pumping using antiferromagnetic resonance (§18.3) is also possible.[599]

11.5   FURTHER READINGS

Spin and orbital Hall effects

The SHE can be used to inject a ***spin current into an insulator***. It can be a ferrimagnet like YIG,[600] or an antiferromagnet like hematite.[457] The spin injection can be verified by non-local measurements (see §8.5). Typically, a current is driven in a Pt strip on top of the insulator. A few microns away, another Pt strip is used as a detector. The SHE in Pt drives a spin current into the insulator. If the material had a high damping, this spin polarization would relax right away, in between the electrodes. But in a low loss material, a spin accumulation develops, leading to a diffusive spin current flowing in the other Pt strip, giving rise to an inverse SHE.

When an electric field induces a transverse current of orbital moment L, the effect is called an orbital Hall effect.[601]

The possibility of having a ***spin Hall magnetoresistance*** has been explored for structures with a heavy non-magnetic metal on an oxide ferromagnet [602–607]. A non-equilibrium proximity effect has also been investigated.[608]

Spin-orbit torque

Spin injection via the spin Hall effect is an efficient way of acting on magnetization.[609,610] This method is referred to as ***spin-orbit torque*** (SOT).[342, 346, 611] SOT has also been observed in rare earth alloys,[612–614] in topological insulators,[615] or at the interface between a topological insulator and a ferromagnet.[616]

Berry phase formalism

This chapter presents an introduction to transport phenomena that can be explained using the Berry curvature formalism. Several review articles explore other phenomena and the details of anomalous transport. [573, 617]. The anomalous Nernst effect can be anticipated on the basis of expression (11.55) for the current. To find f_1, which is needed to calculate the current, we can start with (11.51), apply (10.77) for $\partial f_0 / \partial r$, and find that f_1 is proportional to the temperature gradient. Thermoelectric measurements, in addition to anomalous longitudinal magnetoresistance, provide evidence for chiral anomalies.[568] In the case of a magnetic material, the transport current must be calculated as the difference between the microscopic currents and the bound currents, $\nabla \times \boldsymbol{M}$.[618]

Berry phase formalism has been used to describe magnetic resonance,[619] AHE in magnetic materials,[620–626] thermoelectric properties,[627] magnetic properties (P. Bruno in [81]), in particular the DMI interaction, [628] and magnons in textured ferromagnets.[629] A *non-linear Hall effect* has been observed, which is linked to a derivative of the Berry curvature.[630]

In molecular dynamics, the adiabatic approximation is known as the ***Born-Oppenheimer approximation***. When the electron spin is included in the description of the molecular states, the spin-dependent geometrical phase that derives from the adiabatic approximation may lead to a non-trivial spin evolution.[557,631]

We have not touched on the ***quantum anomalous Hall effect***, the description of which requires an advanced quantum mechanical framework.[632] Experimental evidence for this effect has been found in topological insulators.[633–635] PMA materials are also of interest to attempts at attaining the quantum Hall effect without applying a very large magnetic field.[636,637] Evidence for a quantum spin Hall effect was obtained with graphene.[638]

Chirality

Some antiferromagnets like Mn_3Sn have magnetization sublattices with a chiral structure; hence, these materials are called ***chiral antiferromagnets***. Mn_3Sn presents a large ANE.[570] Skyrmion crystals can also be expected to have a large ANE.[639] Weyl semimetals present ANE.[532,553,640–642]

When a charge current is driven through a chiral crystal, the electron spins are polarized parallel to the chirality axis.[643] If an electrode with a strong spin-orbit coupling covers this material, it acts as a spin sink due to its strong spin-orbit relaxation. As a result, a spin current normal to the interface enters the electrode. In view of the inverse spin Hall effect, we can expect an electric field to be induced in this electrode (Fig. 11.8). This transverse spin current and its associated inverse spin Hall voltage was observed in a chiral crystal containing heavy nuclei.[644]

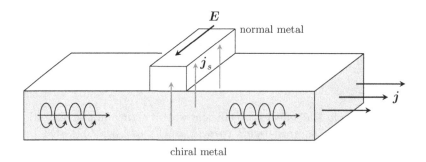

Figure 11.8   When a current density j is applied through a metallic (non-magnetic) chiral crystal, the charge carriers are spin-polarized.[644] Relaxation into a heavy metal electrode causes a diffusive spin current j_s. This spin current j_s induces an electric field $\boldsymbol{E}$ in the electrode, which is normal to the spin polarization and the direction of the spin current. The symbols comprising oriented arcs are meant to represent the non-symorphic crystalline structure. Both materials are non-magnetic.

Chirality-induced spin selectivity (CISS) has been observed in a variety of organic molecules usings several experimental methods.[281,282,548] CISS has been studied in carbon nanotubes also. One way of inducing chirality is to wrap a DNA molecule around a carbon nanotube.[645,646] Carbon nanotubes can be thought of as rolled-up graphene sheets. Depending on how the edges of the sheets are connected, the carbon nanotube itself may be chiral. Since curvature in graphene sheets induces a spin-orbit interaction,[358] one can expect a CISS effect in chiral carbon nanotubes.[647] This effect have been investigated mainly in helical molecules.[648–650] Group theory brings insight into this spin selectivity.[651,652]

Spin transport in the absence of strong spin-orbit coupling

In §11.4.1 and §11.4.2, we saw that strong spin-orbit effects result in spin Hall effects. However, spin transport can be induced in a non-magnetic material even if spin-orbit effects are small. In §12.7, we discuss the possibility that the fermionic character of electrons itself allows for spin-flip scattering in the metallic regime of conductance. Fransson considered a hopping model of conductance for helical organic molecules in which electron correlations were included .[653] A model molecule was attached to a ferromagnetic contact at one end and a non-magnetic one at the other. Fransson found that the presence of some spin-orbit coupling and the chirality of the molecules were essential to inducing a necessary symmetry breaking, but on-site Coulomb repulsion was necessary to obtain a spin current polarization $(j_\uparrow - j_\downarrow)/(j_\uparrow + j_\downarrow)$.[654]

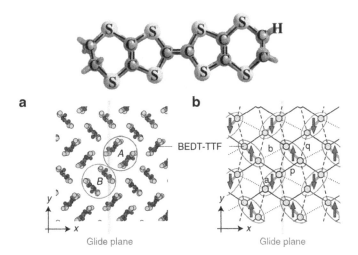

Figure 11.9 Top: BEDT-TTF in a ball-and-stick model. Bottom: BEDT-TTF dimers in the κ-phase crystal structure, with its glide plane symmetry in the high-temperature phase (a), which is broken in the low-temperature phase (b) because of the local moment. In b), three types of electron transfer bonds are indicated. [655]

In Naka et al., [655] the material is made of organic molecules called bis(ethylene dithio)tetrathiafulvalene (BEDT-TTF) (Fig. 11.9). When combined with, e.g., Cu[N(CN)$_2$]Cl, a charge transfer occurs. The salt can be crystallized in a so-called κ structure (Fig .11.9 a).[656] One way to produce these charge transfer salts is by an electrocrystallization.[657] Spin-orbit effects are very small in BEDT-TTF compounds, since the Z in (2.42) is small for all atoms of the molecule.[266] BEDT-TTF materials have been studied intensively because they become superconductors at sizable temperatures and with reasonable pressure applied.

At low temperatures, local moments build up on the dimers of the κ phase, that may give rise to a spin liquid phase.[658] In the antiferromagnetic phase considered in [655], the glide symmetry is broken (Fig. 11.9 b). The antiferromagnetic state is described in [655] using a Hubbard Hamiltonian that contains no spin-orbit coupling.[6] A key feature that leads to spin splitting is the anisotropy of the transfer integrals b_{ij} (Fig. 11.9 b).

---

[6]For a definition of a Hubbard Hamiltonian; see, e.g., [156]

Section IV

SPIN RELAXATION

12 Principles of Spin Relaxation

Alfred G. Redfield (1929–2019)

A. G. Redfield became interested in NMR early in his career, when he started collaborating with C.P. Slichter, N. Bloembergen, and J.H. Van Vleck. In 1955, he wrote a paper introducing the notion of spin temperature in the rotating frame, which remains a classic today. While working at the IBM Watson Laboratory at Columbia University, he started developing his theory for the relaxation rates of out-of-equilibrium spins due to fluctuations in their surroundings. His 1957 paper was published in the *IBM Journal of Research and Development*. It has so many applications that it is still cited decades later.

Spintronics: A primer

J.-Ph. Ansermet (https://orcid.org/0000-0002-1307-4864)

Chapter 12. Principles of Spin Relaxation

Spin relaxation is first defined phenomenologically for a spin 1/2 nucleus. Relaxation by fluctuating magnetic fields is described using a density matrix under the simplifying assumption that initially, only one energy eigenstate is occupied. This shows that relaxation rates are determined by the values of spectral density functions at transition frequencies. The fluctuation-dissipation theorem is first derived in a quantum mechanical framework, then in a classical statistical description. The Bloch-Wangess-Redfield theory is presented, thus providing a formal treatment of indirect couplings, master equations, and the principle of detailed balance.

DOI: 10.1201/9781003370017-12

12.1 HISTORICAL INTRODUCTION

Spin relaxation is a very important concept in magnetic resonance, in spintronics and in spin-based quantum computing. Therefore, it is useful to understand the concept of spin relaxation and to develop an intuition for possible spin relaxation mechanisms.

During the early days of nuclear magnetic resonance, nuclear spin relaxation mechanisms had yet to be identified.[659] In this chapter, we shall see that the speed at which spins evolve toward an equilibrium with a bath determines the likelihood that spin resonance can be detected in the continuum wave mode. This relaxation is a process during which the energy of the system varies. It is called **spin-lattice relaxation**, a term retained from the early studies of magnetic resonance.

When examining the precession of spins in an isolated ensemble, their energy remains constant. So long as the relative phases among spins remain constant, the system is said to maintain **coherence**. In magnetic resonance, the **transverse relaxation** time characterizes the decay of coherence.

The Bloch-Wangness-Redfield theory addresses both types of relaxation processes. As the formal derivation leads to expressions that can be complex, we will move progressively from some simple cases into more complex ones in order to bring out the physical meaning of this formalism. Namely, we will examine:

- the relaxation to thermal equilibrium of a two-level system coupled to a bath,
- a semi-classical description in which the bath is described classically and the initial state is a very peculiar form, allowing us to bring out the essential aspects of the general relaxation theory with relatively simple mathematical expressions,
- a full quantum mechanical description, but limited to a spin 1/2.

The relaxation of the spin of the electron is a central issue of spintronics because devices based on electron spins must be small enough that the electrons maintain their spin polarization as they go from one contact of the device to another. The **spin diffusion length** characterizes the length over which an electron maintains its spin polarization. Because it tends to be in the range of tens of nanometers, spintronics devices are of nanoscopic dimensions.

12.2 RELAXATION OF A TWO-LEVEL SYSTEM

Consider an ensemble of spin 1/2 magnetic moments described by the Zeeman Hamiltonian (7.5). Under the effect of a static homogeneous magnetic field $B_0 = \hat{k} B_0$ the energy levels $m = 1/2$ and $m = -1/2$ are split (Fig. 12.1). We apply an alternative field $B_1(t) = B_{xo} \cos(\omega t)\hat{x}$. As a consequence, the Hamiltonian describing the dynamics of the system contains a time-dependent term, $\mathcal{H}_1(t) = -\gamma \hbar B_{xo} \cos(\omega t) I_x$. If we were to apply time-dependent perturbation theory, we would get a probability of transition per unit of time given by Fermi's golden rule,

$$P_{a \to b} = \frac{2\pi}{\hbar} |\langle a| \mathcal{H}(t) |b\rangle|^2 \delta (E_a - E_b - \hbar\omega) \tag{12.1}$$

This result implies that the transition rates going from a to b and from b to a are equal,

$$P_{a \to b} = P_{b \to a} \equiv W \tag{12.2}$$

We want to analyze the time evolution of the populations of both levels, N_+ and N_-. Under the influence of the oscillating field, N_+ and N_- vary. Since the transition rates are equal,

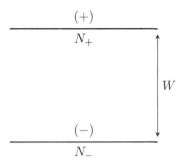

Figure 12.1 Two-level system corresponding to an ensemble of spin $1/2$ moments in a magnetic induction field of strength B_0. N_- and N_+ are the number of spins in state $-1/2$ and $+1/2$.

we call them both W. The ***principle of detailed balance***,[1] applied to this simple case, reads,

$$\frac{dN_+}{dt} = W(N_- - N_+)$$
$$\frac{dN_-}{dt} = W(N_+ - N_-)$$

(12.3)

The total population $N = N_+ + N_-$ is constant, whereas the ***spin polarization*** $n = N_+ - N_-$ evolves according to:

$$\frac{dn}{dt} = -2Wn$$

(12.4)

This differential equation implies the exponential decay,

$$n(t) = n_o e^{-2Wt}$$

(12.5)

Therefore, if the initial spin polarization is $n(0)$, then the effect of the oscillating field is to equalize the populations or, in other words, to decrease the polarization to zero. The absorption of energy by the ensemble of the system is given by:

$$\frac{dE}{dt} = N_+ W \hbar \omega - N_- W \hbar \omega = W n \hbar \omega$$

(12.6)

Therefore, we expect that the absorption of energy drops to zero in a time of the order of $1/W$.

If there is no $\boldsymbol{B}_1(t)$ field applied, there is no evolution of n. However, we know that when we apply a static field $\boldsymbol{B}_0$, the spins will eventually align in the field and $N_+ > N_-$. If we start with $N_+ = N_-$, we need a net excess of transitions $(-) \to (+)$. Where does the energy go in that process?

Our model so far is missing a ***coupling to a thermal bath***. We have to describe the process by which N_+/N_- decreases until it reaches the equilibrium value given by the Boltzmann ratio:

$$\frac{N_+^0}{N_-^0} = \exp\left(-\frac{\Delta E}{kT}\right)$$

(12.7)

---

[1]The term is used here for the first time, but a similar reasoning was already used to write (1.6), (9.93), and (9.106). §12.6 includes a formal derivation of the principle of detailed balance (12.171).

We must therefore assume a coupling of the spins to another system that constitutes a bath. We represent the effect of the coupling to the bath by the probability per unit of time W_U of transitions from $(-) \to (+)$ and W_D for the transition rate of the converse $(+) \to (-)$ process (Fig. 12.2).

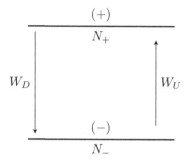

Figure 12.2 Two-level system corresponding to an ensemble of spin $1/2$ moments in a magnetic induction field of strength B_0. N_- and N_+ are the number of spins in state $-1/2$ and $+1/2$. The arrows indicate distinct transition rates from $+$ to $-$ and $-$ to $+$.

We again apply the principle of detailed balance and look at one rate equation only, since the other is immediately deduced from $N = N_+ + N_-$, where N is constant. Thus, we have,

$$\frac{dN_+}{dt} = W_D N_- - W_U N_+ \tag{12.8}$$

At equilibrium, $dN_+/dt = 0$ sets a condition on the ratio of the rates in order to reach the proper equilibrium populations:

$$\frac{N_+^0}{N_-^0} = \frac{W_D}{W_U} = e^{-\Delta E/kT} \tag{12.9}$$

If the transition rates W_U and W_D were determined by a matrix element as in (12.1), they would be equal. However, we will see below when developing the Bloch-Wangness-Redfield formalism, that these rates are determined not only by matrix elements of the Hamiltonian describing the interaction of the bathed system to the bath, but also by the statistical weights of the bath states.

From (12.8), we can deduce the evolution of n by using,

$$N_- = \frac{1}{2}(N - n) \qquad N_+ = \frac{1}{2}(N + n) \tag{12.10}$$

This yields:

$$\frac{dn}{dt} = \frac{n_0 - n}{T_1} \tag{12.11}$$

with

$$\frac{1}{T_1} = W_U + W_D \quad \text{and} \quad n_0 = \frac{W_D - W_U}{W_U + W_D} N \tag{12.12}$$

Here, n_0 is the equilibrium value of n. T_1 is known as the **_spin-lattice relaxation time_**. Integration of (12.11) is straightforward, yielding the function,

$$n(t) = n_o \left(1 - e^{-t/T_1}\right) \tag{12.13}$$

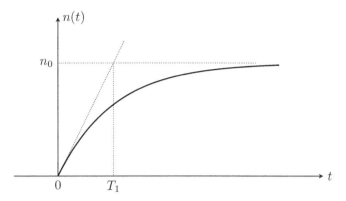

Figure 12.3   Relaxation according to (12.13). The asymptote at very short times, $n(t) \approx t/T_1$, crosses the long time asymptote $n(t) \approx n_0$ at the time T_1.

shown in Fig 12.3 with its asymptotes.

Now we combine the alternating field and the coupling to the bath. The rates add,

$$\frac{dn}{dt} = -2Wn + \frac{n_o - n}{T_1} \tag{12.14}$$

This model predicts a stationary state $dn/dt = 0$ with n given by,

$$n = \frac{n_o}{1 + 2WT_1} \tag{12.15}$$

Therefore, it is possible to maintain n near the equilibrium value, provided that $W \ll 1/(2T_1)$. That is to say, the excitation magnetic field must not be too intense. The rate of energy absorption is, in this case:

$$\frac{dE}{dt} = Wn\hbar\omega = \frac{n_o W\hbar\omega}{1 + 2WT_1} \approx n_o W\hbar\omega \tag{12.16}$$

When W is large, the energy absorption is independent of W. The system is said to have reached **saturation**. We have in this case:

$$\frac{dE}{dt} \approx \frac{n_o\hbar\omega}{2T_1} \tag{12.17}$$

This result expresses, in essence, a concern of the initial researchers looking for an NMR signal. They knew that if the relaxation time T_1 were too long, detection would be nearly impossible.

12.3   RELAXATION BY FLUCTUATING FIELDS

Here, we describe a system of nuclear spins using a density matrix. The thermal bath is treated classically, i.e., the effect of the bath is to produce a fluctuating field at the site of each nucleus. By assuming a density matrix of the initial state with only one non-vanishing matrix element, it is possible to carry out a calculation without too much trouble.[2] This simple example is sufficient to introduce the essential quantities that appear in the complete theoretical treatment of Bloch, Wangness and Redfield. To get started, we need a general result of second order perturbation theory.

---

[2]See, [236], chapter 5.

12.3.1   TIME EVOLUTION OF THE DENSITY MATRIX

Let us consider an ensemble of spins subjected to a large static homogeneous magnetic field. The interaction of the spins with the applied field dominates the dynamics of the spins, but the spins are also coupled to one another and to a thermal bath. Hence, this system is described by a Hamiltonian of the form:

$$\mathcal{H} = \mathcal{H}_0 + \mathcal{H}_1(t) \tag{12.18}$$

where $\mathcal{H}_0$ is time independent. $\mathcal{H}_1(t)$ is assumed to be a small perturbation. We use the so-called **interaction representation**. This approach is used to derive the response of a spin system to a perturbation. We use it here to describe the weak coupling of the system to a bath. Thus, we consider the evolution of the density matrix ρ^I defined by,

$$\rho = e^{-i\mathcal{H}_0 t/\hbar} \rho^I e^{i\mathcal{H}_0 t/\hbar} \quad \text{or} \quad \rho^I = e^{i\mathcal{H}_0 t/\hbar} \rho e^{-i\mathcal{H}_0 t/\hbar} \tag{12.19}$$

Note that ρ^I would be independent of time if the perturbation vanished. Calculating the time derivative of $d\rho/dt$ yields, after developing with care to respect non-commutativities:

$$\frac{d\rho^I}{dt} = \frac{i}{\hbar}\left[\rho^I, \mathcal{H}_1^I\right] \text{ with } \mathcal{H}_1^I(t) = e^{i\mathcal{H}_0 t/\hbar}\mathcal{H}_1(t)e^{-i\mathcal{H}_0 t/\hbar} \tag{12.20}$$

We assume that the perturbation is turned on at time $t = 0$. Integration of (12.20) gives:

$$\rho^I(t) = \rho^I(0) + \frac{i}{\hbar}\int_0^t dt' \left[\rho^I(t'), \mathcal{H}_1^I(t')\right] \tag{12.21}$$

We have not solved anything since we do not know $\rho^I(t')$ $(0 < t' < t)$. As a first approximation, we replace $\rho^I(t')$ in the integrand by $\rho^I(0)$,

$$\rho^I(t) = \rho^I(0) + \frac{i}{\hbar}\int_0^t dt' \left[\rho^I(0), \mathcal{H}_1^I(t')\right] \tag{12.22}$$

We now apply this approximation to calculated $\rho^I(t')$ inside the integrand of (12.21):

$$\rho^I(t) = \rho^I(0) + \frac{i}{\hbar}\int_0^t dt' \left[\rho^I(0) + \frac{i}{\hbar}\int_0^{t'} dt'' \left[\rho^I(0), \mathcal{H}_1^I(t'')\right], \mathcal{H}_1^I(t')\right]$$

$$= \rho^I(0) + \frac{i}{\hbar}\int_0^t dt' \left[\rho^I(0), \mathcal{H}_1^I(t')\right] + \left(\frac{i}{\hbar}\right)^2 \int_0^t dt' \left[\int_0^{t'} dt'' \left[\rho^I(0), \mathcal{H}_1^I(t'')\right], \mathcal{H}_1^I(t')\right] \tag{12.23}$$

Therefore, to the second order of perturbation, the equation of evolution for ρ^I is given by,

$$\frac{d\rho^I(t)}{dt} = \frac{i}{\hbar}\left[\rho^I(0), \mathcal{H}_1^I(t)\right] + \left(\frac{i}{\hbar}\right)^2 \int_0^t dt' \left[\left[\rho^I(0), \mathcal{H}_1^I(t')\right], \mathcal{H}_1^I(t)\right] \tag{12.24}$$

We must be careful that the last term has t as a parameter, not t'!

12.3.2 RELAXATION FROM FULL POLARIZATION IN ONE ENERGY EIGENSTATE

We now restrict the analysis to the case where the initial condition $\rho(0)$ has a very special form. We assume that only one state is occupied, which is an eigenstate of the unperturbed Hamiltonian. Let $|k>$ be this state, and let $|m> (m \neq k)$ designate the other eigenstates of $\mathcal{H}$. This means that the initial density matrix is diagonal, with,

$$\langle k| \rho(0) |k\rangle = 1 \qquad \langle m| \rho(0) |m\rangle = 0 \quad (m \neq k) \tag{12.25}$$

This very peculiar choice of the density matrix represents the statistical state of a system that is 100 % polarized in the energy eigenstate $|k>$. The advantage of this choice is that the transitions rates W_{km} can readily be computed as,

$$\frac{d \langle m| \rho |m\rangle}{dt} = W_{km} \tag{12.26}$$

From the definition of ρ^I, it follows that:

$$\begin{aligned}
\langle m| \rho |m\rangle &= \langle m| e^{-iH_o t/\hbar} \rho^I e^{iH_o t/\hbar} |m\rangle = e^{-iE_m t/\hbar} \langle m| \rho^I |m\rangle e^{iE_m t/\hbar} \\
&= \langle m| \rho^I |m\rangle
\end{aligned} \tag{12.27}$$

The equation of evolution (12.24) of ρ^I to second order gives:

$$\begin{aligned}
&\frac{d \langle m| \rho |m\rangle}{dt} = \frac{d \langle m| \rho^I |m\rangle}{dt} = \\
&\frac{i}{\hbar} \sum_n (\langle m| \rho^I(0) |n\rangle \langle n| \mathcal{H}_1^I(t) |m\rangle - \langle m| \mathcal{H}_1^I(t) |n\rangle \langle n| \rho^I(0) |m\rangle) + \\
&\left(\frac{i}{\hbar}\right)^2 \int_0^t dt' \sum_n (\langle m| \rho^I(0) |n\rangle \langle n| \mathcal{H}_1^I(t') \mathcal{H}_1^I(t) |m\rangle \\
&\qquad\qquad - \langle m| \mathcal{H}_1^I(t) \mathcal{H}_1^I(t') |n\rangle \langle n| \rho^I(0) |m\rangle) \\
&- \left(\frac{i}{\hbar}\right)^2 \int_0^t dt' \sum_n \sum_{m'} (\langle m| \mathcal{H}_1^I(t') |n\rangle \langle n| \rho^I(0) |m'\rangle \langle m'| \mathcal{H}_1^I(t) |m\rangle \\
&\qquad\qquad + \langle m| \mathcal{H}_1^I(t) |n\rangle \langle n| \rho^I(0) |m'\rangle \langle m'| \mathcal{H}_1^I(t') |m\rangle)
\end{aligned} \tag{12.28}$$

The terms of the first and second lines give zero because of the choice of the initial matrix. The terms of the last line are non-zero only if $m' = n = k$, in which case the matrix element is just 1. Taking this into account, we are left with only:

$$\begin{aligned}
&\frac{d \langle m| \rho |m\rangle}{dt} = \\
&\frac{1}{\hbar^2} \int_0^t dt' (\langle m| \mathcal{H}_1^I(t') |k\rangle \langle k| \mathcal{H}_1^I(t) |m\rangle + \langle m| \mathcal{H}_1^I(t) |k\rangle \langle k| \mathcal{H}_1^I(t') |m\rangle)
\end{aligned} \tag{12.29}$$

From the definition (12.20) of $\mathcal{H}_1^I(t)$, we get,

$$\langle m| \mathcal{H}_1^I(t) |n\rangle = e^{iE_m t/\hbar} \langle m| \mathcal{H}_1(t) |n\rangle e^{-iE_n t/\hbar} \tag{12.30}$$

We apply this inside (12.29) and get:

$$\frac{d \langle m| \rho |m\rangle}{dt} =$$

$$\frac{1}{\hbar^2} \int_0^t dt' \left(\langle m| \mathcal{H}_1(t') |k\rangle \langle k| \mathcal{H}_1(t) |m\rangle \, e^{i(E_m - E_k)(t'-t)} \right. \tag{12.31}$$

$$\left. + \langle m| \mathcal{H}_1(t) |k\rangle \langle k| \mathcal{H}_1(t') |m\rangle \, e^{-i(E_m - E_k)(t'-t)} \right)$$

So far, all we have said of $\mathcal{H}_1(t)$ is that it is a perturbation. Now we introduce the *random* character of $\mathcal{H}_1(t)$ and the *irreversibility of the evolution*. In order to do so, we consider an ensemble of spin systems that are equivalent in the sense that they have the same Hamiltonian $\mathcal{H}_0$ and the same initial state $\rho(0)$. $\mathcal{H}_1(t)$ is statistically distributed among the systems of this ensemble and we take a statistical average over this ensemble. Thus, we replace terms such as $\langle m| \mathcal{H}_1(t') |k\rangle \langle k| \mathcal{H}_1(t) |m\rangle$ by their ensemble average, which is noted,

$$\overline{\langle m| \mathcal{H}_1(t') |k\rangle \langle k| \mathcal{H}_1(t) |m\rangle} \tag{12.32}$$

The perturbation is assumed to be *stationary*, which means that it does not depend on when it was measured, but only on the time difference $\tau = t' - t$ (so $dt' = d\tau$). Therefore, we can define the function:

$$G_{mk}(\tau) = \overline{\langle m| \mathcal{H}_1(t + \tau) |k\rangle \langle k| \mathcal{H}_1(t) |m\rangle} \tag{12.33}$$

$G_{mk}(\tau)$ is called the *correlation function* of $\mathcal{H}_1(t)$. The ensemble-averaged equation of evolution (12.31) is given by,

$$\frac{d \langle m| \rho |m\rangle}{dt}$$

$$= \frac{1}{\hbar^2} \int_0^t dt' \left(G_{mk}(\tau) e^{-i(E_m - E_k)\tau} + G_{mk}(-\tau) e^{i(E_m - E_k)\tau} \right) \tag{12.34}$$

$$= \frac{1}{\hbar^2} \int_{-t}^t dt' \, G_{mk}(\tau) e^{-i(E_m - E_k)\tau}$$

We now show that $G_{mk}(\tau)$ is real. First, we note that the stationary condition implies,

$$G_{mk}(\tau) = \overline{\langle m| \mathcal{H}_1(t + \tau) |k\rangle \langle k| \mathcal{H}_1(t) |m\rangle}$$
$$= \overline{\langle k| \mathcal{H}_1(t - \tau) |m\rangle \langle m| \mathcal{H}_1(t) |k\rangle} = G_{km}(-\tau) \tag{12.35}$$

By definition of a hermitian conjugate, and because $\mathcal{H}_1(t)$ is hermitian, we have:

$$G^*_{mk}(\tau) = \overline{\langle m| \mathcal{H}_1(t) |k\rangle \langle k| \mathcal{H}_1(t + \tau) |m\rangle}$$
$$= \overline{\langle m| \mathcal{H}_1(t - \tau) |k\rangle \langle k| \mathcal{H}_1(t) |m\rangle} = G_{mk}(-\tau) \tag{12.36}$$

We must also have $G_{mk}(-\tau) = G_{mk}(\tau)$ because of the time-translation symmetry of the correlation function. Thus, we have $G^*_{mk}(\tau) = G_{mk}(-\tau) = G_{mk}(\tau)$, and this means that the correlation function $G_{mk}(\tau)$ is real.

A consideration that will be useful shortly thereafter is that the quantity

$$G_{mk}(0) = \overline{|\langle k| \mathcal{H}_1(t) |m\rangle|^2} \tag{12.37}$$

is independent of time.

We now introduce an additional assumption. It is crucial, because it accounts for the irreversibility of the time evolution of the system coupled to a bath. We assume that we examine the evolution at times t much greater than some **correlation time**. The correlation time τ_c characterizes the time scale over which the correlation (12.33) dies out, as shown schematically on Fig. 12.4. Under this assumption, the integration limits in the evolution

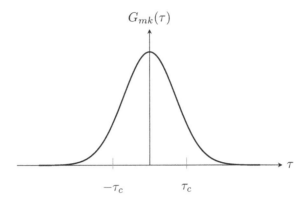

$G_{mk}(\tau)$

$-\tau_c$ τ_c

τ

Figure 12.4 Typical shape of a correlation function.

equation (12.34) can be extended to infinity without changing the result. Thus, we get the **transition rate**, i.e., the probability per unit of time for the system to make the transition from state $|k>$ to state $|m>$:

$$W_{km} = \frac{1}{\hbar^2} \int_{-\infty}^{\infty} dt' G_{mk}(\tau) e^{-i(E_m - E_k)\tau} \qquad (12.38)$$

This result should be contrasted with Fermi's golden rule, which states,

$$W_{km} = \frac{2\pi}{\hbar} |\langle k| V |m\rangle|^2 \rho(E_f) \qquad (12.39)$$

Fermi's golden rule refers to a situation involving a harmonic excitation and a spread in the energy levels of the final states, characterized by the distribution $\rho(E_f)$. Here, the situation leading to (12.38) is different. The energy levels are well-defined, but the fluctuations cover a spread in frequency, characterized by the **spectral density function**:

$$J_{km}(\omega) = \int_{-\infty}^{\infty} d\tau \, G_{mk}(\tau) e^{-i\omega\tau} \qquad (12.40)$$

The inverse Fourier transformation of the spectral density function gives the correlation function:

$$G_{km}(\tau) = \frac{1}{2\pi} \int_{-\infty}^{\infty} d\omega \, J_{mk}(\omega) e^{i\omega\tau} \qquad (12.41)$$

In view of the definition (12.40) of the spectral density function, we obtain the following key result:

$$W_{km} = \frac{1}{\hbar^2} J_{km}\left(\frac{E_m - E_k}{\hbar}\right) \qquad (12.42)$$

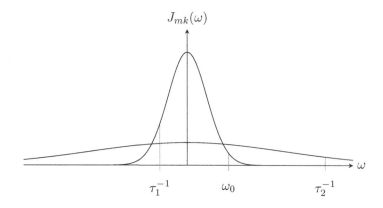

$J_{mk}(\omega)$

τ_1^{-1} ω_0 τ_2^{-1} ω

Figure 12.5 Spectral density function J_{mk} sketched for two values of the correlation time, τ_1 and τ_2. The value $J_{mk}(\omega_0)$ at the resonance frequency changes as the correlation time changes.

This can be considered as a very general result, although we have derived it by starting from a special initial state.

It is possible to understand qualitatively some aspects of relaxation processes with the following argument. The strength of the fluctuation is directly linked to the integral under the spectral density:

$$G_{km}(0) = \frac{1}{2\pi} \int\limits_{-\infty}^{\infty} d\omega \, J_{mk}(\omega) = \overline{|\langle k| \, \mathcal{H}_1(t) \, |m\rangle|^2} \qquad (12.43)$$

Let us suppose that the time constants of $\mathcal{H}_1(t)$ change, for example, because the temperature of the bath is changed. $G_{km}(0)$ does not change, i.e., the area under the spectral density function does not change when the correlation time changes (Fig 12.5). We see that, if the correlation time is very short, the spectral density is frequency-independent over broad frequency range and it is small. When the correlation time is long, at the frequency of interest $(E_m - E_k)/\hbar$, it is also small. In between, there is a value of the correlation time for which J_{mk} and consequently W_{mk} reach a maximum.

12.4 LINEAR RESPONSE

At this point of the chapter, we need to define linear response functions. We will then be able to describe the relaxation of a quantum mechanical system coupled to a bath in reference to these linear response functions.

For the sake of clarity, we introduce linear response functions with explicit reference to a magnetization subject to an external field. But this introduction is valid for the linear response pertaining to any other physical property. The starting hypothesis is that the system has a linear response, i.e., the magnetization produced by the sum of two weak fields is the sum of the magnetization contributions produced by each field.

Let us consider a field of the type shown in Fig. 12.6. The magnetization at time t due to this field is proportional to this field strength. If the field were twice as large, so too would be the response, i.e., $\Delta M \propto H(t')$. Likewise, the response is proportional to the duration, i.e., $\Delta M \propto \Delta t$, since the pulse can be decomposed into two juxtaposed pulses of half the

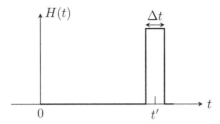

Figure 12.6 Schematics of a magnetic field impulse function.

width. The proportionality constant can be a function of $(t - t')$ only. Hence, we write,[3]

$$\Delta M = m(t - t')H(t')\Delta t \tag{12.44}$$

Because of linearity, the magnetization at any time is the sum of the response functions, hence the integral,

$$M(t) = \int_{-\infty}^{t} m(t - t')H(t')dt' \tag{12.45}$$

Suppose that $H(t') = \delta(t')$. The response to this excitation is given by,

$$M_\delta(t) = \int_{-\infty}^{t} m(t - t')\delta(t')dt' = m(t) \tag{12.46}$$

The last equality means that the function $m(t)$ introduced in (12.44) can be thought of as the magnetization response to a delta function at $t = 0$.

We now define the harmonic response function. We need to specify how the field is turned on at an infinitely remote time. This can be understood by considering the behavior of a harmonic oscillator with a high quality factor. Let us observe a pendulum made of a point mass at the end of a spring. The point mass oscillates in air or water. The pendulum is excited at a frequency near the resonance. In air, if the excitation is turned on rapidly, beats are observed that arise from the combination of the excitation and the eigenmode of the pendulum. If instead the excitation is turned on progressively, i.e., in a time of the order of the decay time of the free pendulum, then there are no beats.

We express the slow turning on of the excitation with $H_1(t) = He^{i\omega t}e^{st}$ (s>0) so that $t \to -\infty \Rightarrow H_1 \to 0$. For this excitation, the harmonic response is given by,

$$M_\omega(t) = H \int_{-\infty}^{t} m(t - t')e^{i\omega t'}e^{st'}dt'$$

$$= He^{i\omega t}e^{st} \int_{-\infty}^{t} m(t - t')e^{i\omega(t'-t)}e^{s(t'-t)}dt' \tag{12.47}$$

$$= H_1(t) \int_{0}^{\infty} m(\tau)e^{-i\omega \tau}e^{-s\tau}d\tau$$

where $\tau = t - t'$.

---

[3]See, [236], chapter 2.

The susceptibility appears as the Fourier transform of the impulse response m(t):

$$M_\omega(t) = \chi(\omega)He^{i\omega t} \tag{12.48}$$

where

$$\chi(\omega) = \lim_{s \to o} \int_0^\infty m(\tau)e^{-i\omega\tau}e^{-s\tau}d\tau \tag{12.49}$$

The magnetic response to the harmonic field is, e.g., given by the magnetization in the x direction when the field is applied in the x direction. The response function is then the susceptibility χ_{xx}. We could also consider the magnetization in the y direction as the field is applied in the x direction. The linear response would be given by χ_{yx}. Likewise for all other directions of the magnetization and the field.

Practically, it may be that $m(t)$ comprises a step function at zero time. Indeed, consider a step excitation as shown in Fig. 12.7. The response in this case can be written as,

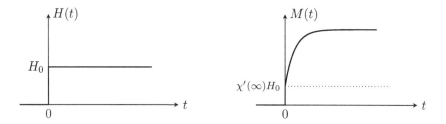

Figure 12.7   Field step at $t = 0$ (left) and magnetization response with a step response at $t = 0$ (right).

$$M_s(t) = \int_{-\infty}^t m(t-t')H(t')dt' = \int_0^\infty m(\tau)H(t-\tau)d\tau = \int_0^t m(\tau)d\tau \tag{12.50}$$

The response to such a step can have a discontinuity at $t = 0$ (Fig. 12.7), which corresponds practically to a process occurring on a much shorter time scale, or at a much higher frequency, than any considered here. Therefore, we should consider $m(t)$ of the form $m(t) = m_{\text{smooth}}(t) + c\delta(t)$. Hence, the response can be written as,

$$\chi(\omega) = \int_0^\infty m_{\text{smooth}}(t)e^{-i\omega\tau}d\tau + \int_0^\infty c\delta(t)e^{-i\omega\tau}d\tau \tag{12.51}$$

The second term is equal to c, which is usually noted $\chi'(\infty)$. The first term can then be written in a way that does not specify that $m(t)$ is the smooth response:

$$\chi(\omega) - \chi'(\infty) = \int_0^\infty m(t)e^{-i\omega\tau}d\tau \tag{12.52}$$

It is often useful to extend the definition of the susceptibility to negative frequencies. Since $m(t)$ is real, the following relations must hold,

$$\begin{aligned} \chi''(-\omega) &= -\chi''(\omega) \\ \chi'(-\omega) - \chi(\infty) &= \chi'(\omega) - \chi(\infty) \end{aligned} \tag{12.53}$$

This implies,

$$\chi^*(\omega) = \chi(-\omega) \tag{12.54}$$

We will often refer to the real part and the imaginary part of the susceptibility, defined by $\chi(\omega) = \chi'(\omega) - i\chi''(\omega)$. Then we have,

$$\chi'(\omega) - \chi(\infty) = \int_0^\infty m(t)\cos(\omega\tau)d\tau$$

$$\chi''(\omega) = \int_0^\infty m(t)\sin(\omega\tau)d\tau \tag{12.55}$$

Recall that, in these expressions, $m(t)$ is the smooth response to a delta function at $t = 0$.

The real and imaginary parts of the response function are linked by the so-called **Kramers-Kronig relations**. These relations are consequences of the analytical properties of functions of complex variables, of the linearity hypothesis and causality.[236] The Kramers-Kronig relations are:

$$\chi'(\omega) - \chi'(\infty) = \frac{1}{\pi}P\int_{-\infty}^\infty \frac{\chi''(\omega')d\omega'}{\omega' - \omega} = \frac{1}{\pi}\lim_{R\to 0}\left\{ \int_{-\infty}^{\omega-R} \frac{\chi''(\omega')d\omega'}{\omega' - \omega} + \int_{\omega+R}^\infty \frac{\chi''(\omega')d\omega'}{\omega' - \omega} \right\}$$

$$\chi''(\omega) = \frac{1}{\pi}P\int_{-\infty}^\infty \frac{\chi'(\omega') - \chi'(\infty)}{\omega' - \omega}d\omega' \tag{12.56}$$

The P in these equations means that we are to calculate the integrals as limits about the singularity at $\omega' = \omega$. Thus, for example,

$$P\int_{-\infty}^\infty \frac{\chi''(\omega')d\omega'}{\omega' - \omega} = \lim_{R\to 0}\left\{ \int_{-\infty}^{\omega-R} \frac{\chi''(\omega')d\omega'}{\omega' - \omega} + \int_{\omega+R}^\infty \frac{\chi''(\omega')d\omega'}{\omega' - \omega} \right\} \tag{12.57}$$

As an example, suppose that $\chi''(\omega)$ has some peak at positive frequencies. The extension to negative frequencies requires a symmetric, negative peak at the frequencies of opposite sign. The Kramers-Kronig relations then imply that $\chi'(\omega)$ must have the pattern shown in Fig. 12.8.

The imaginary part of the response function describes a dissipative process. This can be illustrated by looking at the resistive loss of an inductor containing a material of susceptibility $\chi(\omega)$. The impedance of the coil, the inductance of which is L_0 when empty, is given by,

$$Z = i\omega L = i\omega L_o(1 + \chi) = i\omega L_o(1 + \chi') + \omega L_o\chi'' \tag{12.58}$$

Thus, we see that the impedance has a resistive component due to χ''. The current in the coil and the field H_1 produced by it are linked by the relation:

$$\frac{1}{2}L_o i_0^2 = \frac{\mu_0}{2}H_1^2 V \tag{12.59}$$

Therefore, the power dissipated in the inductor due to the energy absorption in the material placed inside the coil is given by:

$$\frac{1}{2}R i_0^2 = \frac{1}{2}\omega L_0\chi''\frac{\mu_0}{2}H_1^2 V\frac{2}{L_0} = \frac{\mu_0}{2}\omega H_1^2 V\chi'' \tag{12.60}$$

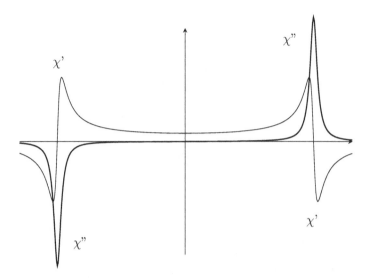

Figure 12.8 Real (thin) and imaginary (thin) parts of the harmonic response $\chi(\omega)$ using (15.19).

12.5 FLUCTUATION-DISSIPATION THEOREM

The application of the fluctuation-dissipation theorem can be very fruitful. R.M. White uses it in several instances in his book on magnetism. An in-depth study of this theorem can be found in the treatise of Kubo et al. on out-of-equilibrium statistical physics.[660] The seminal paper of Callen and Welton, [661] and the following papers,[662, 663] explain well the significance of the fluctuation-dissipation theorem.

12.5.1 QUANTUM STATISTICS

Let us consider a system of spins, the dynamics of which is determined by the Hamiltonian $\mathcal{H}_0$.[156] Suppose that, at a time infinitely remote in the past, the state of the system is given by the density matrix ρ_0. In order to calculate the susceptibility, we apply an oscillating field $\boldsymbol{B}_1(t) = \hat{\boldsymbol{x}} B_1 \exp{(i\omega t)}$ to the system. We suppose that this field produces only a perturbation, and we calculate the linear response of the spins. This calculation will provide us with several useful mathematical expressions of the susceptibility. We will also be able to introduce the so-called fluctuation-dissipation theorem.

At $t \rightarrow -\infty$, we assume that the field is turned on slowly. The Hamiltonian is $\mathcal{H} = \mathcal{H}_0 + \mathcal{H}_1$, where the Hamiltonian $\mathcal{H}_1 = -\gamma \hbar \hat{I}_x B_1 e^{i\omega t}$ represents the interaction of the magnetization $M_x = \left(\gamma \hbar \hat{I}_x / V \right)$ with the applied field $B_1 e^{i\omega t}$. We treat the oscillating field as a perturbation and look for the linear response $M(t) = \langle M_x(t) - M_x(-\infty) \rangle = Tr\{M_x(\rho(t) - \rho_o)\}$, from which we deduce $\chi_{xx}(\omega)$.

The density matrix ρ of this system of spins has a time evolution given by,

$$\frac{d\rho}{dt} = \frac{i}{\hbar} [\rho, \mathcal{H}_0 + \mathcal{H}_1] \tag{12.61}$$

We can make use of the result obtained in (12.22), with the difference that, here, the initial time is at $t \rightarrow -\infty$. Thus, we have in the interaction representation,

$$\rho^I(t) - \rho^I(-\infty) = \frac{i}{\hbar} \int\limits_{-\infty}^{t} dt' \left[\rho^I(t'), \mathcal{H}_1^I(t') \right] \tag{12.62}$$

Since we are looking for the linear response, we take $\rho^I(t')$ to be $\rho^I(-\infty)$ in the integral. Indeed, the difference between $\rho^I(t')$ and $\rho^I(-\infty)$ is due to the perturbation, so it is of the order of magnitude of the perturbation. But we want the right-hand side of the equation of evolution to be of the first order in the perturbation only, so we need to leave out the perturbation of $\rho^I(-\infty)$ in the integrand.

At $t \to -\infty$, we assume that the system is at thermal equilibrium. The result we are about to derive depends strongly on this hypothesis. Therefore, the initial value of the density matrix is given by,

$$\rho^I(-\infty) = \rho(-\infty) = \rho_o = \frac{1}{Z}e^{-\mathcal{H}_o/kT} \tag{12.63}$$

The change of variable $t' = t - \tau$ yields,

$$\rho^I(t) - \rho_o = \frac{i}{\hbar}\int_0^\infty d\tau \left[\rho_o, \mathcal{H}_1^I(t-\tau)\right] \tag{12.64}$$

Now, we convert this result back for the matrix ρ. Thus, we find,

$$\rho(t) - \rho_o = \frac{i}{\hbar}\int_0^\infty d\tau \left[\rho_o, e^{-i\mathcal{H}_0\tau/\hbar}\mathcal{H}_1(t-\tau)e^{i\mathcal{H}_0\tau/\hbar}\right] \tag{12.65}$$

Since we want to calculate χ_{xx}, we need $M_x(t)$. The linear response is in fact the change in the magnetization expectation value which is due to the perturbation. Hence, we are looking for:

$$
\begin{aligned}
M(t) &= \langle M_x(t) - M_x(-\infty)\rangle = Tr\left\{M_x(\rho(t)-\rho_o)\right\} \\
&= Tr\left\{\frac{\gamma\hbar\hat{I}_x}{V}\frac{i}{\hbar}\int_0^\infty \left[\rho_o, e^{-i\mathcal{H}_0\tau/\hbar}\left(-\gamma\hbar B_1\right)\hat{I}_x e^{i\mathcal{H}_0\tau/\hbar}\right]e^{i\omega(t-\tau)}d\tau\right\}
\end{aligned}
\tag{12.66}
$$

We can swap the trace and the integral, use permutations in the trace, and use the commutation of ρ_0 with $\exp(i\mathcal{H}_0\tau/\hbar)$ to obtain:

$$
\begin{aligned}
Tr&\left\{\hat{I}_x\left[\rho_o, e^{-i\mathcal{H}_0\tau/\hbar}\hat{I}_x e^{i\mathcal{H}_0\tau/\hbar}\right]\right\} \\
&= Tr\left\{\hat{I}_x\rho_0 e^{-i\mathcal{H}_0\tau/\hbar}\hat{I}_x e^{i\mathcal{H}_0\tau/\hbar} - \hat{I}_x e^{-i\mathcal{H}_0\tau/\hbar}\hat{I}_x e^{i\mathcal{H}_o\tau/\hbar}\rho_0\right\} \\
&= Tr\left\{e^{i\mathcal{H}_0\tau/\hbar}\hat{I}_x e^{-i\mathcal{H}_0\tau/\hbar}\rho_0\hat{I}_x - \rho_0 e^{i\mathcal{H}_0\tau/\hbar}\hat{I}_x e^{-i\mathcal{H}_0\tau/\hbar}\hat{I}_x\right\}
\end{aligned}
\tag{12.67}
$$

We define,

$$\hat{I}_x(t) = e^{i\mathcal{H}_o\tau/\hbar}\hat{I}_x e^{-i\mathcal{H}_o\tau/\hbar} \tag{12.68}$$

This would be the time evolution of $\hat{I}_x$ in the Heisenberg representation, under the influence of $\mathcal{H}_0$ only. We manipulate the trace above a little further by writing,

$$Tr\left\{\hat{I}_x(t)\rho_o\hat{I}_x - \rho_o\hat{I}_x(t)\hat{I}_x\right\} = Tr\left\{\hat{I}_x\hat{I}_x(t)\rho_o - \rho_o\hat{I}_x(t)\hat{I}_x\right\} = Tr\left\{\rho_o\left[\hat{I}_x, \hat{I}_x(t)\right]\right\} \tag{12.69}$$

This allows us to write, finally,

$$M(t) = \frac{-\gamma^2\hbar^2}{V}\frac{i}{\hbar}B_1 e^{i\omega t}\int_0^\infty Tr\left(\rho_o\left[\hat{I}_x, \hat{I}_x(\tau)\right]\right)e^{-i\omega\tau}d\tau \tag{12.70}$$

In view of the general results of linear response theory (12.50) and (12.52), we see that we have obtained,

$$
\chi_{xx}(\omega) = \frac{-\gamma^2 \hbar^2}{V} \frac{i}{\hbar} \int\limits_{0}^{\infty} Tr \left(\rho_o \left[\hat{I}_x, \hat{I}_x(\tau) \right] \right) e^{-i\omega\tau} d\tau
$$

$$
m(t) = \frac{-\gamma^2 \hbar^2}{V} \frac{i}{\hbar} Tr \left(\rho_o \left[\hat{I}_x, \hat{I}_x(t) \right] \right)
$$

(12.71)

It is often convenient to modify expression (12.71) in order to have the integral extending from $-\infty \to \infty$. Let us proceed as follows. In view of (12.54), we can calculate a new expression for $\chi''(\omega)$, namely,

$$
\chi''(\omega) = \frac{\chi(\omega) - \chi^*(\omega)}{-2i} = \frac{\chi(\omega) - \chi(-\omega)}{-2i} =
$$

$$
\frac{1}{-2i} \frac{-\gamma^2 \hbar^2}{V} \frac{i}{\hbar} \left(\int\limits_{0}^{\infty} Tr \left(\rho_o \left[\hat{I}_x, \hat{I}_x(\tau) \right] \right) e^{-i\omega\tau} d\tau - \int\limits_{0}^{\infty} Tr \left(\rho_o \left[\hat{I}_x, \hat{I}_x(\tau) \right] \right) e^{+i\omega\tau} d\tau \right)
$$

(12.72)

In the second integral, we change variable ($\tau = -t'$) and transform the integrand using permutations in the trace:

$$
\int\limits_{0}^{-\infty} Tr \left(\rho_0 \left[\hat{I}_x, \hat{I}_x(-t') \right] \right) e^{-i\omega t'} dt' = \int\limits_{0}^{-\infty} Tr \left(\rho_0 \hat{I}_x e^{-i\mathcal{H}_0 t'/\hbar} \hat{I}_x e^{i\mathcal{H}_0 t'/\hbar} - (...) \right) e^{-i\omega t'} dt'
$$

$$
= \int\limits_{0}^{-\infty} Tr \left(\rho_0 \left[\hat{I}_x(t'), \hat{I}_x \right] \right) e^{-i\omega t'} dt' = \int\limits_{-\infty}^{0} Tr \left(\rho_0 \left[\hat{I}_x, \hat{I}_x(t') \right] \right) e^{-i\omega t'} dt'
$$

(12.73)

We can then regroup both integrals into one, thus finding,

$$
\chi''(\omega) = \frac{1}{2\hbar} \frac{\gamma^2 \hbar^2}{V} \int\limits_{-\infty}^{\infty} Tr \left(\rho_o \left[\hat{I}_x, \hat{I}_x(t) \right] \right) e^{-i\omega t} dt
$$

(12.74)

The expression (12.74) for $\chi''(\omega)$ is nothing but a mathematical relationship between the time evolution of the operator $\hat{I}_x$ and its associated susceptibility. This result can be used to treat a variety of phenomena that can be described by a Hamiltonian evolution. However, we can also reformulate this result so as to identify in the integrand a function that we can identify (from physical arguments, not by mathematical derivation) as a correlation function. A correlation function would be defined with symmetrized operators:

$$
Tr \left(\rho_o \frac{\hat{I}_x \hat{I}_x(t) + \hat{I}_x(t) \hat{I}_x}{2} \right)
$$

(12.75)

We can write

$$
\chi''(\omega) = \frac{1}{\hbar} \frac{\gamma^2 \hbar^2}{2V} \cdot (A - B)
$$

(12.76)

where A and B correspond to the two terms of the commutator in the trace. In order to express the susceptibility in terms of a correlation function, we need $A + B$ expressed in terms of $\chi''(\omega)$. As we show below, there is a special relationship between A and B when

ρ_0 is given by the thermal equilibrium (12.63). This condition has significant implications from a physical standpoint. It imposes that the fluctuation-dissipation theorem that we are about to derive is valid when a system is near thermal equilibrium.

Operating cyclic permutations in the trace, we can write, successively,

$$A = \int_{-\infty}^{\infty} Tr\left(\rho_0 \hat{I}_x \hat{I}_x(t)\right) e^{-i\omega t} dt = \int_{-\infty}^{\infty} Tr\left(\frac{1}{Z} e^{-\mathcal{H}_0/kT} \hat{I}_x e^{i\mathcal{H}_0 t/\hbar} \hat{I}_x e^{-i\mathcal{H}_0 t/\hbar}\right) e^{-i\omega t} dt$$

$$= \int_{-\infty}^{\infty} Tr\left(e^{i\mathcal{H}_0 t/\hbar} \hat{I}_x e^{-i\mathcal{H}_0 t/\hbar} e^{-\mathcal{H}_0/kT} \hat{I}_x \frac{1}{Z}\right) e^{-i\omega t} dt \tag{12.77}$$

Now, we extend the integral in the complex time plane, writing,

Figure 12.9 Path in the complex plane to calculate (12.78).

$$A = \int_{-\infty}^{\infty} Tr\left(\frac{1}{Z} e^{-\mathcal{H}_0/kT} e^{i\mathcal{H}_0(t-i\hbar/kT)/\hbar} \hat{I}_x e^{-i\mathcal{H}_0(t-i\hbar/kT)/\hbar} \hat{I}_x\right) e^{-i\omega t} dt$$

$$= \int_{-\infty}^{\infty} Tr\left(\frac{1}{Z} e^{-\mathcal{H}_0/kT} e^{i\mathcal{H}_0(t-\frac{i\hbar}{kT}/\hbar)} \hat{I}_x e^{-i\mathcal{H}_0(t-i\hbar/kT)/\hbar} \hat{I}_x\right) e^{-i\omega(t-\frac{i\hbar}{kT})} e^{-i\omega\frac{i\hbar}{kT}} dt \tag{12.78}$$

$$= \int_{-\infty}^{\infty} Tr\left(\rho_0 \hat{I}_x(\tau) \hat{I}_x\right) e^{-i\omega\tau} d\tau e^{-i\omega\left(\frac{i\hbar}{kT}\right)} = e^{\frac{\hbar\omega}{kT}} B$$

In the last step, we consider the extension into the complex plane of the time integral that defines A. We use the fact that the integral over the path $P_1 P_2$ is equal to the integral over the path $P_3 P_4$ (Fig. 12.9) This is the case because the integrand is well-behaved in the loop (there are no poles), so the integral over the whole path $P_1 P_2 P_3 P_4$ vanishes. The integrand over the paths $P_2 P_3$ and $P_4 P_1$ is assumed to vanish in the limit of infinite times, since correlations over large times die out. Hence, we have that the integral over $P_1 P_2$ plus that over $P_3 P_4$ vanishes, so the integral from P_1 to P_2 is equal to that from P_3 to P_4.

In conclusion, the imaginary part of the susceptibility $\chi''(\omega)$ can be written as,

$$\chi''(\omega) = C(A - B) = CB\left(e^{\frac{\hbar\omega}{kT}} - 1\right) \tag{12.79}$$

where

$$A = \frac{\chi''(\omega) e^{\frac{\hbar\omega}{kT}}}{C\left(e^{\frac{\hbar\omega}{kT}} - 1\right)} \qquad B = \frac{\chi''(\omega)}{C\left(e^{\frac{\hbar\omega}{kT}} - 1\right)} \qquad C = \frac{1}{2\hbar}\frac{\gamma^2\hbar^2}{V} \tag{12.80}$$

Thus, we get:

$$C\left(A+B\right) = \chi^{''}\left(\omega\right)\left(\frac{e^{\frac{\hbar\omega}{kT}}}{\left(e^{\frac{\hbar\omega}{kT}}-1\right)} + \frac{1}{\left(e^{\frac{\hbar\omega}{kT}}-1\right)}\right)$$

$$= \frac{1}{\hbar}\frac{\gamma^2\hbar^2}{V}\int\limits_{-\infty}^{\infty} Tr\left(\rho_o\frac{\hat{I}_x\hat{I}_x(t)+\hat{I}_x(t)\hat{I}_x}{2}\right)e^{-i\omega t}dt \tag{12.81}$$

This can be rewritten in terms of the magnetization $M_x(t) = \gamma\hbar\hat{I}_x(t)/V$ as,

$$\chi^{''}\left(\omega\right)\coth\left(\frac{\hbar\omega}{2kT}\right) = \frac{V}{\hbar}\int\limits_{-\infty}^{\infty} Tr\left(\rho_o\frac{M_x(\tau)M_x+M_xM_x(\tau)}{2}\right)e^{-i\omega\tau}d\tau \tag{12.82}$$

The trace can be thought of as a correlation function, and this result is known as the **fluctuation-dissipation theorem**. We will encounter correlation functions again in §12.6, where the coupling to a thermal bath is analyzed in detail. As a first illustration, let us consider that the correlation function in (12.82) is a simple exponential, namely,

$$M_o^2\exp\left(\frac{-|t|}{\tau_c}\right) \tag{12.83}$$

Then, we have for the imaginary part of the susceptibility,

$$\chi^{''}\left(\omega\right)\coth\left(\frac{\hbar\omega}{2kT}\right) = \frac{1}{V\hbar}\int\limits_{-\infty}^{\infty} M_o^2\exp\left(\frac{|t|}{\tau_c}\right)e^{-i\omega\tau}d\tau \tag{12.84}$$

The integration can be decomposed in an integration over positive t and negative t. Over each domain of integration, the integration is straightforward, resulting in,

$$\chi^{''}\left(\omega\right) = \frac{\hbar\omega}{2kT}2M_o^2\frac{1/\tau_c}{\left|-1/\tau_c+i\omega\right|^2} \tag{12.85}$$

In magnetic resonance spectroscopy, this response function is known as a **lorentzian** line shape. Had we assumed a gaussian correlation function, we would have obtained a gaussian line shape. For a proper description of a resonance, we should first consider a dominant evolution determined by the applied field and a precession at ω_0, and then a small perturbation describing the fluctuations. In this case, ω in (12.85) would be replaced by $\omega - \omega_0$.

As another sample application of the fluctuation-dissipation theorem, let us consider the high-temperature approximation, which consists in replacing the definition (12.63) of ρ_0 by the approximation $\rho_0 = (1 - \mathcal{H}_0/(k_BT))/Z$. It may seem strange that we can do this, considering that we have a great number of spins. The fluctuation-dissipation theorem makes clear, however, that what is in fact relevant is the energy of one spin, i.e., $\hbar\omega$, compared to k_BT.[4]

12.5.2   CLASSICAL STATISTICS

In this section, the fluctuation-dissipation theorem is demonstrated in a classical framework. By showing both classical and quantum mechanical versions of the theorem, we illustrate

---

[4]See, [236], appendix E.

the notion that the key ingredient that leads to the physical property expressed by this theorem is the nature of the bath. Namely, the bath has a broad distribution of states, so it hosts small excitations.

Let us consider a dynamical system that is characterized by a Hamiltonian $\mathcal{H}_0(x)$, which is a function of some dynamical variable x. Owing to the coupling of the system to a bath, the variable x fluctuates in time. We represent this aspect of the system by assuming known transition rates,[664]

$$P(x', t | x, 0) \tag{12.86}$$

In view of the following discussion, it is important to note that these rates are defined for the dynamical system when it is not subjected to a force field.

Let us assume that, in the remote past, a force field was applied, modifying the Hamiltonian as follows,

$$\mathcal{H}(x) = \mathcal{H}_0(x) - f_0 \Theta(-t) x \tag{12.87}$$

The Heaviside function $\Theta(-t)$ accounts for the fact that this field is presumed to be turned off at $t = 0$. At $t \leq 0$, the system has had time to come to equilibrium in the presence of the force field. Hence, the statistical distribution $W(x, 0)$ at $t = 0$ is given by the Boltzmann distribution,

$$W(x, 0) = \frac{\exp(-\beta \mathcal{H}(x))}{\int dx' \exp(-\beta \mathcal{H}(x'))} \tag{12.88}$$

where $\beta = 1/(k_B T)$. Hence, at times $t > 0$, i.e., when the force field is turned off, the average value $< x(t) >$ is given by,

$$< x(t) > = \int dx' \int dx\, x'\, P(x', t | x, 0) W(x, 0) \qquad (t > 0) \tag{12.89}$$

This definition can be understood as follows. For each value of x, we work out the probability of the state at t to have the dynamical variable at the value x'. This probability multiplied by x', integrated over all possible values of x', gives the average $< x(t) >$. Notice that we keep track of the restriction $t > 0$ in (12.89), i.e., the statistical nature of $x(t)$ is calculated when the force field is turned off.

Let us now take into account the fact that the force field is a small perturbation. We proceed with a Taylor expansion of $W(x, 0)$ as follows,

$$W(x, 0) = \frac{\exp(-\beta \mathcal{H}_0(x) + \beta f_0 x(0))}{\int dx' \exp(-\beta \mathcal{H}_0(x') + \beta f_0 x')} = \frac{\exp(-\beta \mathcal{H}_0(x))(1 + \beta f_0 x(0))}{\int dx' \exp(-\beta \mathcal{H}_0(x'))(1 + \beta f_0 x')} \tag{12.90}$$

In the above expression, we distinguish the distribution in the absence of force field,

$$W_0(x) = \frac{\exp(-\beta \mathcal{H}_0(x))}{\int dx' \exp(-\beta \mathcal{H}_0(x'))} \tag{12.91}$$

Thus, we have,

$$W(x, 0) = W_0(x) \frac{(1 + \beta f_0 x(0))}{\left(1 + \frac{\beta f_0 \int dx' x' \exp(-\beta H_0(x'))}{\int dx' \exp(-\beta H_0(x'))}\right)} \tag{12.92}$$

In the denominator, we recognize the average value of x in the absence of a force field,

$$< x >_0 = \int dx' x' W_0(x') \tag{12.93}$$

Finally, the Taylor development under the assumption $\beta f_0 x \ll 1$ yields,

$$W(x, 0) = W_0(x) (1 + \beta f_0 x(0) - \beta f_0 < x >_0) \tag{12.94}$$

We now use this expression of $W(x,0)$ in the definition (12.89) of $< x(t) >$ to find,

$$< x(t) >= \int dx' \int dx x' P(x',t|x,0)W_0(x)$$
$$+ \beta f_0 \int dx' \int dx x' P(x',t|x,0)W_0(x)(x(0)- < x >_0) \tag{12.95}$$

Since $W_0(x)$ is the equilibrium distribution, it is stationary, i.e., $P(x',t|x,0)W_0(x) = W_0(x)$. Hence, the first term in (12.95) yields $< x >_0$. Since the average of $x(0)- < x >_0$ vanishes, in the second term of (12.95), we can replace x' by $x'- < x >_0$. Therefore, we find,

$$< x(t) >=< x >_0 + \beta f_0 \left\langle \left(x(t)- < x >_0 \right) \left(x(0)- < x >_0 \right) \right\rangle_0 \qquad (t > 0) \tag{12.96}$$

Thus, we find that that average $< x(t) >$ is given in terms of the correlation function,

$$A(t) = \left\langle \left(x(t)- < x >_0 \right) \left(x(0)- < x >_0 \right) \right\rangle_0 \tag{12.97}$$

This correlation function is assumed to be stationary, i.e., to depend only on the absolute value of the time difference between the two times at which the deviation $(x(t)- < x >_0)$ is evaluated. Hence, $A(t)$ is a real, symmetric function of t.

Let us now link this result with linear response theory. If a field $f(t)$ is turned on adiabatically at a time in the remote past, the response of the system can be characterized by,

$$< x(t) >=< x >_0 + \int_{-\infty}^{t} f(\tau)\chi(t - \tau)d\tau \tag{12.98}$$

Thus, we have from (12.96) and (12.98),

$$f_0 \int_{-\infty}^{t} \Theta(-\tau)\chi(t - \tau)d\tau = \beta f_0 A(t) \qquad (t > 0) \tag{12.99}$$

The left-hand side can be rewritten with the change of variable $t' = t - \tau$ to yield,

$$f_0 \int_{0}^{\infty} dt'\Theta(t' - t)\chi(t') = \beta f_0 A(t) \qquad (t > 0) \tag{12.100}$$

Taking the derivative with respect to the time t, given that the derivative of the Heaviside function is the Dirac delta function, we find,

$$\chi(t) = -\beta \frac{dA(t)}{dt}\Theta(t) \tag{12.101}$$

where $\Theta(t)$ is introduced to ensure, in the following development, that we do not omit the condition $t > 0$. We define the Fourier transform of $\chi(t)$ by,

$$\hat{\chi}(\omega) = \int_{-\infty}^{\infty} \chi(t) \exp(-i\omega t)dt \tag{12.102}$$

Thus, from (12.101) we find,

$$\hat{\chi}(\omega) = -\beta \int_{-\infty}^{\infty} \frac{dA(t)}{dt}\Theta(t) \exp(-i\omega t)dt = -\beta \int_{0}^{\infty} \frac{dA(t)}{dt} \exp(-i\omega t)dt \tag{12.103}$$

Integration by part yields,

$$\hat{\chi}(\omega) = \beta A(0) - i\omega\beta \int_0^\infty A(t)\exp(-i\omega t)dt \tag{12.104}$$

As $A(t)$ is real and symmetric, we have,

$$\hat{\chi}^*(\omega) = \beta A(0) + i\omega\beta \int_0^\infty A(t)\exp(i\omega t)dt = \beta A(0) + i\omega\beta \int_{-\infty}^0 A(t)\exp(-i\omega t)dt \tag{12.105}$$

We combine the last two results to obtain,

$$\mathrm{Im}\,(\hat{\chi}(\omega)) = \frac{-\omega}{2k_BT} \int_{-\infty}^\infty A(t)\exp(-i\omega t)dt \tag{12.106}$$

Thus, within a classical approach, we find that there is a link between the imaginary part of the susceptibility at any frequency ω and the Fourier transform of the correlation function (in the absence of a force field) estimated at the same frequency.

12.6   RELAXATION OF A QUANTUM MECHANICAL SYSTEM COUPLED TO A BATH

In this section, we address in general terms the statistical description of a quantum mechanical system coupled to a bath.[665] We treat the bath itself as a quantum mechanical system. We will also consider the situation in which the bathed system has not yet reached, but approaches, equilibrium. This formalism was initially developed as a description of magnetic resonance phenomena. It is in essence the formalism of Bloch [666] – Wangness[667] – Redfield [668,669] for the theory of relaxation. It was revisited recently in a more up-to-date presentation.[301] The developments presented in this section provide some basic tools such as the principle of detailed balance and the probability of transition in the form of a Fermi golden rule with statistical weights. This was already used to get (12.9). This formalism also gives a solid framework for deriving indirect interactions such as the RKKY coupling (§5.3).

Quantum mechanical description of a composite system

Let us consider an isolated system consisting of two subsystems A and B. The subsystem are described by the Hilbert spaces $\mathbb{H}_A$ and $\mathbb{H}_B$. The whole system is described by the tensor product $\mathbb{H}_A \otimes \mathbb{H}_B$. For example, we could take subsystem A to be an ensemble of local moments, and subsystem B, an ensemble of conduction electrons. The Hilbert spaces of subsystems A and B have the orthonormal basis $\{|a\rangle, a \in A\}$ and $\{|b\rangle, b \in B\}$. The dynamics of the system is determined by a time-independent Hamiltonian of the form:

$$\mathcal{H} = \mathcal{H}_0 + \mathcal{H}_{\mathrm{int}} \tag{12.107}$$

where $\mathcal{H}_0$ describes the evolution of the subsystems independently from one another and $\mathcal{H}_{\mathrm{int}}$ describes the interactions between subsystems A and B. $\mathcal{H}_0$ is composed of two terms that describe the evolution of each subsystem when they are not coupled to one another:

$$\mathcal{H}_o = \mathcal{H}_A \otimes \mathbb{1}_B + \mathbb{1}_A \otimes \mathcal{H}_B \tag{12.108}$$

where $\mathbb{1}_A$ is the identity operator in the Hilbert space of A, and likewise for B. Let $\rho(t)$ be the density matrix describing the time evolution of the system. A physical quantity of subsystem A is characterized by a hermitian operator of the form $O_A \otimes \mathbb{1}_B$. Its expectation value is given by:

$$\langle O_A \otimes 1_B \rangle = Tr\,(\rho\,(O_A \otimes 1_B)) \tag{12.109}$$

The trace operation Tr can be written as:

$$\langle O_A \otimes 1_B \rangle = \sum_{a,a'} (\rho_A)_{aa'} \langle a' | O_A | a \rangle \tag{12.110}$$

where,

$$(\rho_A)_{aa'} = \sum_b \langle b | \otimes \langle a | \rho | a' \rangle \otimes | b \rangle \tag{12.111}$$

The terms $(\rho_A)_{aa'}$ turn out to be the matrix elements of a density matrix that describes the subsystem A itself. This density matrix is obtained by the so-called **partial trace**, denoted:

$$\rho_A = Tr_B(\rho) \tag{12.112}$$

The same consideration, with operators of the subsystem B, leads to $\rho_B = Tr_A(\rho)$. It is useful to write the density matrix of the whole system as:

$$\rho(t) = \rho_A(t) \otimes \rho_B(t) + \eta_{AB}(t) \tag{12.113}$$

The operator $\rho_A(t) \otimes \rho_B(t)$ is a density matrix. When $\rho(t) = \rho_A(t) \otimes \rho_B(t)$, the expectation value of an operator of the form $O_A \otimes O_B$ is simply given by:

$$Tr(\rho(O_A \otimes O_B)) = \langle O_A \rangle \langle O_B \rangle \tag{12.114}$$

Hence, the operator $\eta_{AB}(t)$ expresses the statistical correlations between the subsystems.[5] The definition of $\eta_{AB}(t)$ implies:

$$Tr_B(\eta_{AB}(t)) = 0 \quad \text{and} \quad Tr_A(\eta_{AB}(t)) = 0 \tag{12.115}$$

The interaction representation

The time evolution of the whole system, in the Schroedinger representation, is given by:

$$\rho(t) = e^{-i\mathcal{H}t/\hbar} \rho(0) e^{i\mathcal{H}t/\hbar} \tag{12.116}$$

where $\rho(0)$ is the density matrix at time $t = 0$. It may seem surprising that the description of spins coupled to a bath of electrons is done with a time-independent Hamiltonian. While the Hamiltonian $\mathcal{H}_B$ is not explicitly time-dependent, it describes the entire bath dynamics, for example the dynamics of all the electrons, including the vibration of atoms, collisions in a fluid, etc. So,

If $\mathcal{H}_{\text{int}}$ were zero, we would have:

$$\rho(t) = e^{-i\mathcal{H}_o t/\hbar} \rho(0) e^{i\mathcal{H}_o t/\hbar} \tag{12.117}$$

We define the density matrix in the interaction representation by:

$$\rho(t) = e^{-i\mathcal{H}_o t/\hbar} \rho^I(t) e^{i\mathcal{H}_o t/\hbar} \Leftrightarrow \rho^I(t) = e^{i\mathcal{H}_o t/\hbar} \rho(t) e^{-i\mathcal{H}_o t/\hbar} \tag{12.118}$$

The expectation value of an operator $O(t)$ is given by,

$$\langle O(t) \rangle = Tr(\rho^I(t) O^I(t)) \tag{12.119}$$

where

$$O^I(t) = e^{i\mathcal{H}_o t/\hbar} O(t) e^{-i\mathcal{H}_o t/\hbar} \tag{12.120}$$

---

[5]See, e.g., [169] complement E_{III}, Eq. 70.

The equation of evolution of the density matrix in the interaction representation is:

$$\frac{d\rho^I}{dt} = \frac{i}{\hbar} \left[\rho^I(t), \mathcal{H}^i_{\text{int}}(t) \right] \tag{12.121}$$

where

$$H^I_{\text{int}}(t) = e^{i\mathcal{H}_o t/\hbar} \mathcal{H}_{\text{int}} e^{-i\mathcal{H}_o t/\hbar} \tag{12.122}$$

It is possible to assign a density matrix to the subsystem A. This density matrix in the interaction representation is given by:

$$\rho^I_A = Tr_B \rho^I(t) \tag{12.123}$$

This results from,

$$Tr_B \left(\rho^I(t) \right) = e^{i\mathcal{H}_A t/\hbar} \, Tr_B \left(\rho(t) \right) \, e^{-i\mathcal{H}_A t/\hbar} \tag{12.124}$$

Linear response of B to its coupling to A

The interactions between subsystems A and B can be written as,

$$\mathcal{H}_{\text{int}} = \sum_\alpha A_\alpha \otimes B_\alpha \tag{12.125}$$

In the interaction representation, the interaction is given by,

$$\mathcal{H}_{\text{int}}{}^I = \sum_\alpha A_\alpha{}^I(t) \otimes B_\alpha{}^I(t) \tag{12.126}$$

where

$$A_\alpha{}^I(t) = e^{i\mathcal{H}_A t/\hbar} \, A_\alpha{}^I \, e^{-i\mathcal{H}_A t/\hbar} \tag{12.127}$$

and likewise for subsystem B. Now we proceed further, considering situations where subsystems A and B are such that:

- the interaction is weak,
- the correlations are weak,
- subsystem B fluctuates around an equilibrium.

Because the interaction is weak, the time evolution of the density matrix $\rho^I_A(t)$ is estimated by a development to the second order, as it is done in time-dependent perturbation theory. To express that the correlations are weak, we assume that, at some time t_{00}, subsystems A and B were brought together. At that time, the correlations were zero. At later times, the correlations are assumed to not grow much. The fluctuations of subsystem B are expressed by taking the density matrix of subsystem B as:

$$\rho_B(t) = \rho^o_B + \delta\rho_B(t) \tag{12.128}$$

with the condition,

$$[\rho^o_B, \mathcal{H}_B] = 0 \tag{12.129}$$

In the decomposition (12.128), $\delta\rho_B(t)$ expresses the contributions due to the interaction with subsystem A.

We can calculate the evolution of $\delta\rho_B(t)$ to the first order as,

$$\delta\rho^I_B(t) = \frac{i}{\hbar} \int_{t_{oo}}^{t} dt' \sum_\beta [\rho^o_B, B^I_\beta(t')] \langle A_\beta \rangle (t') \tag{12.130}$$

Thus, the response of subsystem B due to the coupling to subsystem A is given by:

$$
\langle B_\alpha \rangle (t) - b_\alpha^o = Tr \left(\delta\rho_B^I(t) B_\alpha^I(t) \right)
$$

$$
= \int_{t_{oo}}^{t} dt' \sum_\beta Tr \left(\rho_B^o \frac{i}{\hbar} \left[B_\beta^I(t'), B_\alpha^I(t) \right] \right) \langle A_\beta \rangle (t') \tag{12.131}
$$

We define the correlation functions,

$$
c_{\alpha\beta}(t' - t'') = c_{\beta\alpha}(t'' - t')^* = Tr \left(\rho_B^o B_\alpha^I(t') B_\beta^I(t'') \right) \tag{12.132}
$$

The fact that these functions depend only on the difference $(t' - t'')$ comes from the commutation (12.129) of ρ_B^o with $\mathcal{H}_B$.

We note that:

$$
Tr \left(\rho_B^o \frac{i}{\hbar} \left[B_\beta^I(t'), B_\alpha^I(t) \right] \right) = \frac{2}{\hbar} \mathrm{Im} \left(c_{\alpha\beta}(t - t') \right) \tag{12.133}
$$

Assuming that t_{00} is so far in the past that it can be taken as $-\infty$, the response (12.131) can be written as,

$$
\langle B_\alpha \rangle (t) - b_\alpha^o = \int_{-\infty}^{\infty} dt' \sum_\beta \frac{2}{\hbar} \mathrm{Im} \left(c_{\alpha\beta}(t - t') \right) \Theta \left(t' - t \right) \langle A_\beta \rangle (t') \tag{12.134}
$$

We have thus introduced **retarded response functions**,

$$
c_{\alpha\beta}^{ret}(t - t') = \frac{2}{\hbar} \mathrm{Im} \left(c_{\alpha\beta}(t - t') \right) \Theta \left(t - t' \right) \tag{12.135}
$$

The real part of $c_{\alpha\beta}(t' - t'')$ is the quantum equivalent of a **correlation function**, defined as,

$$
\mathrm{Re}\, c_{\alpha\beta}(t' - t'') = \frac{1}{2} Tr \left(\rho_B^o \left\{ B_\beta^I(t'), B_\alpha^I(t) \right\} \right) \tag{12.136}
$$

Note that an anti-commutator $\{ \ \}$ appears in the definition. Thus, we are finding the linear response of the operator B_α relative to a time change of A_α due to the coupling between them described by $\mathcal{H}_{int}$. Proceeding further, we will get expressions of the **susceptibility** of the bath. The susceptibility here is defined as the response to the excitation defined by the coupling of subsystem A to the subsystem B, from now on considered as a bath. The results below can be thought of as an expansion of the introduction to linear response in §12.4, where the bath was implicit and the excitation was an external field. This case can be included here in the response of subsystem B to its coupling to subsystem A by assuming an interaction of the form $H(t) B_\alpha$, i.e., $\mathcal{H}_{int} = H(t) \mathbb{1} \otimes B_\alpha$.

In §12.5, we assumed that the bathed system was at equilibrium with the bath. Here, our working hypotheses allow us to go further. Subsystem A may be in the process of reaching equilibrium with subsystem B, but it is not at equilibrium with subsystem B yet. The full calculation, carried out in [301], shows that it is useful to introduce the Fourier transforms:

$$
\chi_{\alpha\beta}(\omega) = \frac{1}{\hbar} \int_{-\infty}^{\infty} c_{\alpha\beta}(\tau) e^{i\omega\tau} d\tau
$$

$$
\bar{\chi}_{\alpha\beta}(\omega) = \frac{i}{\hbar} \int_{-\infty}^{\infty} \varepsilon(\tau) c_{\alpha\beta}(\tau) e^{i\omega\tau} d\tau \tag{12.137}
$$

where

$$\varepsilon(\tau) = \begin{cases} -1, \tau < 0 \\ 0, \tau = 0 \\ 1, \tau > 1 \end{cases} \tag{12.138}$$

In view of the special role of the real and imaginary parts of the functions $c_{\alpha\beta}(\tau)$, the Fourier transforms (12.137) are decomposed as follows:

$$\chi_{\alpha\beta}(\omega) = \chi'_{\alpha\beta}(\omega) + i\chi''_{\alpha\beta}(\omega)$$

$$\chi'_{\alpha\beta}(\omega) = \frac{1}{\hbar} \int_{-\infty}^{\infty} \mathrm{Re}\left(c_{\alpha\beta}(\tau)\right) e^{i\omega\tau} d\tau \tag{12.139}$$

$$\chi''_{\alpha\beta}(\omega) = \frac{1}{\hbar} \int_{-\infty}^{\infty} \mathrm{Im}\left(c_{\alpha\beta}(\tau)\right) e^{i\omega\tau} d\tau$$

We do likewise for $\bar{\chi}_{\alpha\beta}$, so, for example,

$$\bar{\chi}''_{\alpha\beta}(\omega) = \frac{1}{\hbar} \int_{-\infty}^{\infty} \varepsilon(\tau) \mathrm{Im}(c_{\alpha\beta}(\tau)) e^{i\omega\tau} d\tau \tag{12.140}$$

These functions follow the Kramers-Kronig relations (12.56). Notably, we have,

$$\bar{\chi}''_{\alpha\beta}(\omega) = \frac{-1}{\pi} \mathbf{P} \int_{-\infty}^{\infty} \frac{\chi''_{\alpha\beta}(\omega')}{\omega - \omega'} d\omega' \tag{12.141}$$

The definitions (12.139) and the relation (12.141) imply the following symmetries:

$$\chi'_{\alpha\beta}(\omega)* = \chi'_{\beta\alpha}(\omega)$$
$$\chi''_{\alpha\beta}(\omega)* = -\chi''_{\beta\alpha}(\omega) \tag{12.142}$$
$$\bar{\chi}''_{\alpha\beta}(\omega)* = -\bar{\chi}''_{\beta\alpha}(\omega)$$

We now develop expressions for the susceptibility. The retarded response functions (12.135) can be expressed as:

$$c^{ret}_{\alpha\beta}(t) = \frac{1}{\hbar} \left(\mathrm{Im}\left(c_{\alpha\beta}(t)\right) + \varepsilon(t) \mathrm{Im}\left(c_{\alpha\beta}(t)\right) \right) \tag{12.143}$$

The harmonic response is given by the Fourier transform of $c^{ret}_{\alpha\beta}(t)$, which can be written in terms of the susceptibilities (*a priori* all are complex functions):

$$\kappa^{ret}_{\alpha\beta}(\omega) = \chi''_{\alpha\beta}(\omega) - i\bar{\chi}''_{\alpha\beta}(\omega) \tag{12.144}$$

We also decompose the harmonic response function into a real part and an imaginary part:

$$\kappa^{ret}_{\alpha\beta}(\omega) = \kappa'_{\alpha\beta}(\omega) - i\kappa''_{\alpha\beta}(\omega) \tag{12.145}$$

Note the choice of sign used here to conform to a common convention. It is straightforward to write the real and imaginary parts of $\kappa^{ret}_{\alpha\beta}(\omega)$ as:

$$\kappa'_{\alpha\beta}(\omega) = \frac{1}{2} \left(\kappa^{ret}_{\alpha\beta}(\omega) + \kappa^{ret}_{\alpha\beta}(\omega)* \right)$$
$$\kappa''_{\alpha\beta}(\omega) = \frac{-1}{2i} \left(\kappa^{ret}_{\alpha\beta}(\omega) - \kappa^{ret}_{\alpha\beta}(\omega)* \right) \tag{12.146}$$

Using the definitions (12.139) and the symmetry relations (12.142) yields:

$$\kappa''_{\alpha\beta}(\omega) = -i\chi''_{\alpha\beta}(\omega) \tag{12.147}$$

We can find explicit expressions for the real and imaginary parts of the susceptibility $\kappa^{ret}_{\alpha\beta}$ by expressing the trace in (12.136) in terms of a set of basis states of subsystem B. We choose for this set the eigenfunctions of the Hamiltonian $\mathcal{H}_B$, i.e.,

$$\mathcal{H}_B |b\rangle = E_B |b\rangle \tag{12.148}$$

Then, the Fourier transforms $\chi_{\alpha\beta}(\omega)$ and $\bar{\chi}_{\alpha\beta}(\omega)$ can be written as,

$$\chi_{\alpha\beta}(\omega) = \frac{2\pi}{\hbar} \sum_{b,b'} p_b \langle b| B_\alpha |b'\rangle \langle b'| B_\beta |b\rangle \delta(\omega - \omega_{bb'})$$

$$\bar{\chi}_{\alpha\beta}(\omega) = \frac{-2}{\hbar} \sum_{b,b'} p_b \langle b| B_\alpha |b'\rangle \langle b'| B_\beta |b\rangle \frac{1}{\omega - \omega_{b'b}} \tag{12.149}$$

When all the results above are combined, the imaginary part of the harmonic response can be written as,

$$\omega \kappa''_{\alpha\beta}(\omega) = \frac{\pi}{\hbar} \sum_{b,b'} (p_b - p_{b'}) \omega_{bb'} \langle b| B_\alpha |b'\rangle \langle b'| B_\beta |b\rangle \delta(\omega - \omega_{bb'}) \tag{12.150}$$

When $\omega \neq 0$, we can write,

$$\kappa''_{\alpha\beta}(\omega) = \frac{\pi}{\hbar} \sum_{b,b'} (p_b - p_{b'}) \langle b| B_\alpha |b'\rangle \langle b'| B_\beta |b\rangle \delta(\omega - \omega_{bb'}) \tag{12.151}$$

The diagonal terms are,

$$\kappa''_{\alpha\alpha}(\omega) = \frac{\pi}{\hbar} \sum_{b,b'} (p_b - p_{b'}) |\langle b| B_\alpha |b'\rangle|^2 \delta(\omega - \omega_{bb'}) \tag{12.152}$$

These diagonal terms have an intuitive interpretation: the dissipation is given by the overall result of transitions and the associated changes in the populations of the bath energy levels. Making use of the Kramers-Kronig relations, it is possible to draw from this result the real part of the response function:[6]

$$\kappa'_{\alpha\beta}(\omega) = \frac{1}{\hbar} \sum_{b,b'} \frac{(p_b - p_{b'})}{(\omega - \omega_{bb'})} \langle b| B_\alpha |b'\rangle \langle b'| B_\beta |b\rangle \tag{12.153}$$

When the calculation of the evolution of subsystem A (outlined below) is carried out in full and the evolution of the energy is assessed, it is found that the subsystem A tends toward an equilibrium with subsystem B, provided the probability of occupation of the energy levels of subsystem B decreases with increasing energy.[301] Furthermore, it is found under this condition for subsystem B that the fluctuations exactly compensate the dissipation. In other words, the formalism yields the fluctuation-dissipation theorem, as it should. In the present notation, the theorem reads,

$$\kappa''_{\alpha\beta}(\omega) = \tanh\left(\frac{\hbar\omega}{2kT}\right) \frac{1}{\hbar} \int_{-\infty}^{\infty} Tr\left(\rho_B^0 \frac{B_\alpha^I(\tau)B_\beta^I(0) + B_\beta^I(0)B_\alpha^I(\tau)}{2}\right) e^{i\omega\tau} d\tau \tag{12.154}$$

---

[6]See, e.g., [319].

Evolution of the bathed subsystem A

After a rather lengthy calculation,[301] the fundamental hypotheses of weak coupling, weak correlations, and subsystem B fluctuating around an equilibrium, the following time evolution of subsystem A can be found:

$$
\rho_A^I(t) = \rho_A^I(t_o) + \frac{i}{\hbar} \int_{t_o}^{t} dt' \left[\rho_A^I(t_o), \mathcal{H}_A^I(t) \right] + \left(\frac{i}{\hbar} \right)^2 \int_{t_o}^{t} dt' \int_{t_{oo}}^{t'} dt''
$$

$$
\sum_{\alpha\beta} \left[\rho_A^I(t_o) A_\beta^I(t''), A_\alpha^I(t') \right] c_{\alpha\beta}(t' - t'')^* - \left[A_\beta^I(t'') \rho_A^I(t_o), A_\alpha^I(t') \right] c_{\alpha\beta}(t' - t'')
$$

(12.155)

The first-order term expresses the evolution of subsystem A under the Hamiltonian,

$$
\mathcal{H}_A^I(t) = \sum_\alpha A_\alpha^I(t) b_\alpha^o
$$

(12.156)

with

$$
b_\alpha^o = Tr \left(\rho_B^o B_\alpha^I(t) \right) = Tr \left(\rho_B^o B_\alpha \right)
$$

(12.157)

The Hamiltonian $\mathcal{H}_A^I(t)$ expresses a first-order effect, corresponding to the evolution of subsystem A under the static effect of subsystem B. This effect can be, for example, the so-called Knight shifts of nuclear magnetic resonance.

The second-order term in (12.155) is determined entirely by the response functions $c_{\alpha\beta}(t' - t'')$. This second-order term can be simplified further when making explicit use of the assumption that subsystem B is a bath. By this we mean that:

- Subsystem B fluctuates around a statistical equilibrium described by a density matrix ρ_B^o, which commutes with the Hamiltonian $\mathcal{H}_B$, and consequently we limit the description of the effect to the first order only in the coupling to subsystem A,
- the functions $c_{\alpha\beta}(t' - t'')$ present a characteristic correlation time $\tau_{\alpha\beta}^{corr}$ that is very short compared to the characteristic time τ_A^{evol} of evolution of the subsystem A,
- the energy difference among different transitions between pairs of energy levels of subsystem A admits a minimum value $\hbar\omega_A$. This is always the case if subsystem A has a finite number of energy levels.

These assumptions make it possible to work out the evolution with a **coarse graining** approximation. That is, the time evolution is averaged out around any time t as follows:

$$
D_A^I(t) = \frac{1}{\Delta t} \int_{t - \Delta t/2}^{t + \Delta t/2} \rho_A^I(t') dt'
$$

(12.158)

where Δt is a time interval much smaller than the reciprocal of the largest frequency of transitions between eigenstates of subsystem A, although this Δt is still much larger than the correlation time. There are many physical situations implicating such a regime, so the model is relevant in many physical circumstances.

Furthermore, we usually consider cases where the evolution time τ_A^{evol} is long compared to the differences between any two transition frequencies, in the sense that:

$$
\tau_A^{evol} >> 2\pi/\Delta\omega_A \qquad \Delta\omega_A = \inf\left\{ \left| \omega_{a_1 a_1} - \omega_{a_1' a_1'} \right| \neq 0 \right\}
$$

(12.159)

Here, we use the short hand notation $\omega_{aa'} = (E_a - E_{a'})/\hbar$.

Often, the contributions of the oscillations can be omitted altogether. This is typically expressed by stating,

$$\omega_{a_1 a_1} - \omega_{a_1' a_1'} = 0 \tag{12.160}$$

This condition is not some energy conservation principle, but a consideration about the time scales over which the evolution is observed compared to the differences in transition frequencies. The coarse graining with this additional condition leads to an evolution for the density matrix of subsystem A that can be recast into a differential equation of the form:

$$\frac{d \langle a_1 | D_A^I(\bar{t}) | a_2 \rangle}{d\bar{t}} = \sum_{\left\{ a_1', a_2' \middle| \omega_{a_1', a_2'} = \omega_{a_1, a_2} \right\}} \Gamma_{a_1, a_2}^{a_1', a_2'} \langle a_1' | \bar{\rho}_A^I(t) | a_2' \rangle \tag{12.161}$$

This result is often referred to as a set of **master equations**. The explicit expression of the coefficients $\Gamma_{a_1 a_2}^{a_1' a_2'}$, in terms of the functions $c_{\alpha\beta}(t' - t'')$, is rather involved.[301] Let us only mention here that they have the symmetry property,

$$\Gamma_{a_1 a_2}^{a_1' a_2'} = \left(\Gamma_{a_2 a_1}^{a_2' a_1'} \right)^* \tag{12.162}$$

Some terms can be grouped together that contribute to a Hamiltonian evolution, and other terms contribute to the evolution with an anti-commutator. Namely, the coefficients of the master equations can be written as ([301], Eq. 23):

$$\Gamma_{a_1 a_2}^{a_1' a_2'} = \Gamma_{0 a_1 a_2}^{a_1' a_2'} - \frac{1}{\hbar} \left(\delta_{a_1}^{a_1'} G_{a_2' a_2} + \delta_{a_2}^{a_2'} G_{a_1 a_1'} \right) + \frac{i}{\hbar} \left(\delta_{a_1}^{a_1'} \Delta H_{a_2' a_2} + \delta_{a_2}^{a_2'} \Delta H_{a_1 a_1'} \right) \tag{12.163}$$

where,

$$\Gamma_{0 a_1 a_2}^{a_1' a_2'} = \frac{2\pi}{\hbar} \sum_{b, b''} p_b \langle b, a_2' | \mathcal{H}_{int} | a_2, b'' \rangle \langle b'', a_1 | \mathcal{H}_{int} | a_1', b \rangle \delta \left(E_{a_1'} + E_b - E_{a_1} - E_{b''} \right)$$

$$G_{aa'} = \frac{\hbar}{2} \sum_{a''} \Gamma_{0 a'' a''}^{a' a}$$

$$\Delta H_{aa'} = \frac{-1}{2} \sum_{a''} \sum_{\alpha, \beta} \langle a | A_\alpha | a'' \rangle \bar{\chi}_{\alpha\beta}(\omega_{a' a''}) \langle a'' | A_\beta | a' \rangle$$

$$\tag{12.164}$$

The presence of the Kronecker functions in (12.163) impose that the Hamiltonian terms $\Delta H_{aa'}$ as well as the $G_{aa'}$ terms affect degenerate levels ($\omega_{a_1 a_1'} = 0$). If we use the expression (12.149) of $\chi_{\alpha\beta}(\omega)$ in terms of the eigenstates of the bath B, we find,[301]

$$\Delta H_{aa'} = \sum_{a''} \sum_{b, b'} p_b \frac{\langle a, b | \mathcal{H}_{int} | a'' b'' \rangle \langle a'' b'' | \mathcal{H}_{int} | a' b \rangle}{E_a + E_b - E_{a''} - E_{b''}} \tag{12.165}$$

This is not quite the same as the standard second-order perturbation theory because of the presence of the statistical weights p_b in this expression. From this result, it is possible to infer the expectation of indirect coupling among elements of subsystem A via their coupling to the bath B (see §5.3).

With the decomposition (12.163), after going from the interaction representation back to the initial von Neuman evolution, the master equations (12.161) can be written synthetically as ([301], Eq. 27)

$$\frac{dD_A(\bar{t})}{d\bar{t}} = \frac{i}{\hbar} [D_A(\bar{t}), \mathcal{H}_A + \Delta H] - \frac{1}{\hbar} \{ D_A(\bar{t}), G \} + \Gamma_0 [D_A] \tag{12.166}$$

where the last term, expressed in components, is equivalent to the sum in (12.161). The G terms can be thought of as playing the role of friction coefficients. The Γ_0 terms contribute to the decay of coherences. NMR relaxation times can be accounted for with this formalism. For a given spin, say, a spin $5/2$, the definitions (12.164) imply relations among the different relaxation rates of coherences. [670] Further insight into the meaning of the Γ_0 terms can be drawn from the following considerations.

Principle of detailed balance

Consider a simple situation in which the energy splittings $\omega_{aa'}$ are all different. This means that the value of a splitting unequivocally specifies the energy levels E_a and $E_{a'}$. Furthermore, for the sake of keeping the notation simple, we consider that all of the energy levels are non-degenerate. Then the condition $\omega_{a_1' a_1} = \omega_{a_2' a_2}$, applied to the case where $a_1 = a_2$, implies $a_1' = a_2'$. Therefore, the diagonal elements of the density matrix follow the evolution equation:

$$\frac{d \langle a| D_A^I(t) |a \rangle}{dt} = \sum_{a'} \Gamma_{aa}^{a'a'} \langle a'| D_A^I(t) |a' \rangle \tag{12.167}$$

By inspection, we see that:

$$\Gamma_{a,a}^{a',a'} = \Gamma_{o\,a,a}^{a',a'} \tag{12.168}$$

Given that

$$G_{aa} = \frac{\hbar}{2} \sum_{a''} \Gamma_{o\,a'',a''}^{a,a} \tag{12.169}$$

we have,

$$\Gamma_{a,a}^{a,a} = \Gamma_{o\,a,a}^{a,a} - \sum_{a''} \Gamma_{o\,a'',a''}^{a,a} = - \sum_{a'' \neq a} \Gamma_{o\,a'',a''}^{a,a} \tag{12.170}$$

From these relations, which ensure that the trace of the density matrix remains equal to 1, follows the **principle of detailed balance**:

$$\frac{d \langle a| D_A^I(t) |a \rangle}{d\bar{t}} = \sum_{a' \neq a} \Gamma_{a' \to a} \langle a'| D_A^I(t) |a' \rangle - \Gamma_{a \to a'} \langle a| D_A^I(t) |a \rangle \tag{12.171}$$

where,

$$\Gamma_{a \to a'} = \Gamma_{o\,a',a'}^{a,a} \tag{12.172}$$

Fermi's golden rule

We can use the basis set of subsystem B to find explicit expressions for G and the matrix elements of Γ as we did for ΔH. Thus, we find,

$$G_{aa'} = \pi \sum_{a''} \sum_{b,b'} p_b \langle ab| \mathcal{H}_{\text{int}} |a''b'' \rangle \langle a''b''| \mathcal{H}_{\text{int}} |a'b \rangle \, \delta \left(E_a + E_b - E_{a''} - E_{b''} \right) \tag{12.173}$$

Under the matching condition $\omega_{a_1' a_1} = \omega_{a_2' a_2}$, we find,

$$\Gamma_{0\,a_1 a_2}^{a_1' a_2'} = \frac{2\pi}{\hbar} \sum_{b,b'} p_b \langle a_2'b| \mathcal{H}_{\text{int}} |a_2 b'' \rangle \langle a_1 b''| \mathcal{H}_{\text{int}} |a_1' b \rangle \, \delta \left(E_{a_1'} + E_b - E_{a_1} - E_{b''} \right) \tag{12.174}$$

In particular, for the transition rates of the detailed balance, we have,

$$\Gamma_{a \to a'} = \Gamma_{0\,a',a'}^{a\,a} = \frac{2\pi}{\hbar} \sum_{b,b'} p_b |\langle ab| \mathcal{H}_{\text{int}} |a'b' \rangle|^2 \delta \left(E_a + E_b - E_{a'} - E_{b'} \right) \tag{12.175}$$

That is, the transition rates $\Gamma_{a \to a'}$ are determined by Fermi's golden rule, with the statistical weights p_b included. We recall that the energy conserving effect of the δ function is a consequence of the coarse-graining procedure.

Suppose that the bath is a thermal bath:

$$p_b = \frac{1}{Z_B} e^{-E_B/kT} \tag{12.176}$$

Then the energy conservation allows us to write:

$$
\begin{aligned}
\Gamma_{a \to a'} &= \frac{2\pi}{\hbar} \sum_{b,b'} \frac{1}{Z_B} e^{-E_B/kT} |\langle ab| \mathcal{H}_{\text{int}} |a'b'\rangle|^2 \delta\left(E_a + E_b - E_{a'} - E_{b'}\right) \\
&= \frac{2\pi}{\hbar} \sum_{b,b'} \frac{1}{Z_B} e^{(E_a - E_{a'} - E_{b'})/kT} |\langle ab| \mathcal{H}_{\text{int}} |a'b'\rangle|^2 \delta\left(-E_a - E_b + E_{a'} + E_{b'}\right)
\end{aligned} \tag{12.177}
$$

This implies that,

$$\Gamma_{a \to a'} = e^{(E_a - E_{a'})/kT} \Gamma_{a' \to a} \quad \Leftrightarrow \quad \frac{\Gamma_{a \to a'}}{\Gamma_{a' \to a}} = \frac{\exp\left(\dfrac{E_a}{kT}\right)}{\exp\left(\dfrac{E_{a'}}{kT}\right)} \tag{12.178}$$

The equilibrium condition becomes:

$$\sum_{a' \neq a} \Gamma_{a' \to a} \left(\langle a'| \rho_A^I(\bar{t}) |a'\rangle - \exp\left(\frac{E_{a'} - E_a}{kT}\right) \langle a| \rho_A^I(\bar{t}) |a\rangle \right) = 0 \tag{12.179}$$

Therefore, we have:

$$\langle a| \rho_A^I(\bar{t}) |a\rangle = \frac{1}{Z_A} \exp\left(\frac{-E_a}{kT}\right) \tag{12.180}$$

with $Z_A = \sum_a \exp\left(-E_a/kT\right)$. This result expresses the notion that subsystem A tends toward a thermal equilibrium characterized by a temperature equal to that of the bath. Our rate equations correctly predict this effect because our correlation functions are defined in terms of ρ_B^0, i.e., they have the appropriate statistical weights built into them.

12.7   FURTHER READINGS

Fundamentals of spin relaxation

Mathematical physicists have developed rigorous master equations, as in Lindblad,[671], Gorini et al. [672] and Havel.[673] The conventional Bloch-Wangness-Redfield approach, which is commonly used in magnetic resonance, is referred to as inhomogeneous master equations. The Lindblad form of master equations were used in dynamic nuclear polarization (DNP) theory by Karbanov and Koekenberger,[674] and for the theory of quantum-rotor-induced polarization by Annabestani and Cory.[675] Experiments on hyperpolarization and NMR experiments on spin isomers brought forth some shortcomings of the inhomogeneous master equations. Homogeneous master equations were developed by Jeener,[676] and by Levitt and Dibari.[677,678] A novel approach by Bengs and Levitt provides an appropriate description of NMR in solids and liquids that avoid the troubles of the inhomogeneous master equations. It enables the description of a spin system far from equilibrium.[679]

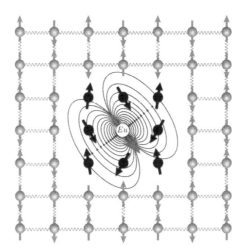

Figure 12.10 Europium-doped orthosilicate, Eu:Y_2SiO_5. A strong magnetic field induces a large dipole moment at the Eu ion that creates an inhomoegenous dipolar field acting on the nearby Y nuclei. As a consequence, these nearby Y nuclei constitute a "frozen core" of Y nuclei that are not on "speaking terms" with the bulk Y nuclei.[680]

Relaxation and quantum computing

In 1998, Loss and DiVincenzo suggested that an electron spin may not only be a conceptual model for qubits, but that the electron spins of semiconductor quantum dots could actually function as qubits.[681]

There are major efforts under way to use spins in quantum computing.[682, 683] DiVincenzo defined a set of five criteria that qubits must fulfill.[684] One criterion is the coherence time, which must be long enough so that several gate operations can be accomplished before decoherence sets in. For this reason, there is much research in finding ways to create qubits with a long coherence time. For example, a report indicates that a transition in the nuclear spins of europium-doped yttrium orthosilicate can be made to have a coherence time of six hours at 2 K (Fig. 12.10).[680]

A more practical spin system might be that of ^{31}P-doped silicon.[685] Controlled location of the phosphorus atoms by ion implantation in silicon oxide is possible, and single spin detection has been demonstrated, showing a T_2^* of 6 seconds in a field of 1.5 T.[686] Furthermore, it has been possible to locate precisely two phosphorus atoms in silicon using an STM, thus securing a well-defined coupling between the phosphorus nuclei.[687] These are crucial steps since isolated spins were previously argued to be the only way to implement spin-based quantum computing.[688] The possibility of using molecules as spin qubits is explored,[689] and the operation of 12-qubit molecules in liquid has been demonstrated.[690, 691]

From master equations to the semi-classical Boltzmann theory of transport

In Chapter 9, we introduced the classical Boltzmann theory of transport. Compared to the thermodynamical approach presented in Chapter 8, the Boltzmann approach allowed us to include microscopic mechanisms in the collision terms of the Boltzmann transport equation, such as the scattering of electrons with impurities (§9.3.2) or with phonons (§9.3.3). This was particularly useful when treating spin-dependent phenomena, as we could distinguish processes with or without spin flips (§9.4.2). In §11.5, we alluded to the possibility of having spin-dependent transport in the absence of strong spin-orbit interaction in situations where electron correlations are important. It is possible to expand the master equations of §12.6 and describe a system characterized by correlated electrons that are coupled to a bath.[692–694] In this series of papers, the authors start from master equations and define the conditions under which these equations lead to quantum Boltzmann equations. From those results, they use coherent states to derive semi-classical Boltzmann equations. The relaxation terms of these Boltzmann equations contain spin flips without spin-orbit coupling. Instead, these spin flips are associated with electron correlations.

13 Mechanisms of Spin Flip

Nicolaas Bloembergen (1920–2017)

Soon after N. Bloembergen arrived at Harvard University, E.M. Purcell and his students, H.C. Torrey and R.V. Pound, detected nuclear magnetic resonance for the first time. Together, they wrote a seminal paper in 1957 describing the basic principles of nuclear spin relaxation. Bloembergen also worked on microwave spectroscopy and the maser concept. In 1981, he received a Nobel prize with Arthur Schawlow and Kai Siegbahn for their development of laser spectroscopy.

Spintronics: A primer

J.-Ph. Ansermet (https://orcid.org/0000-0002-1307-4864)

Chapter 13. Mechanisms of Spin Flip

Spin-orbit scattering is the leading mechanism of electron spin relaxation in many metals and, consequently, often determines the spin-diffusion length. An argument is given to explain the scattering rate dependence on core electronic states. The *sd* model describes the exchange interaction between conduction electrons and local moments. This interaction can lead both to a magnetization relaxation process (in which case the electrons are a bath) and to spin flip without momentum scattering (spin mixing), a relaxation mechanism that can impact the magneto-thermopower of magnetic nanostructures. The Elliott-Yafet and Dyakonov-Perel relaxation mechanisms for conduction electron spins are defined qualitatively.

13.1   HISTORICAL INTRODUCTION

In the 1950s, some 10 years after the discovery of magnetic resonance, Bloembergen at Harvard [695] and Kittel at Berkley [696] investigated electron spin resonance in ferromagnets, in other words, ferromagnetic resonance (see Chapter 15). The picture soon arose that, in magnetic metals, magnetization does not efficiently relax directly to the lattice. Instead, the localized electrons that carry the magnetization couple strongly to conduction electron spins, and these spins relax to the lattice via spin-orbit coupling (Fig. 13.1).

If conduction electrons spin-lattice relaxation is fast compared to relaxation processes taking place between the local moments and conduction electron spins, then the conduction electron spins are polarized in the field resulting from the applied field and the exchange field. This is the typical situation considered in spintronics. In paramagnetic systems, it can be that cross relaxation between local moments and conduction electron spins is faster than the spin-lattice relaxation of conduction electrons. The analysis of the magnetization dynamics in this so-called bottleneck regime must consider the sum of the contributions of the local moments and conduction electron spins to the magnetization.[697]

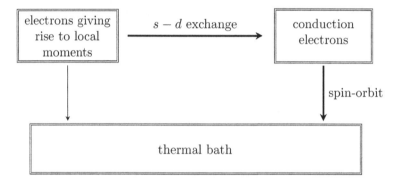

Figure 13.1   Relaxation pathway for local magnetic moments in a ferromagnetic metal. The direct relaxation of local moments to the lattice can be weak, but relaxation occurs via the coupling of local moments to conduction electrons, which relax efficiently to the lattice thanks to the spin-orbit coupling. A so-called bottleneck regime occurs when the cross-relaxation processes driven by the coupling between the local moments and the conduction electron spins are fast compared to the conduction electron spin relaxation to the lattice.

Winter explained this matter in his 1971 treatise on magnetic resonance in metals.[698] He combined ideas of Hasegawa and Dyson to account for the coupling between the dynamics of local magnetic moments and conduction electron spins.[699, 700] This was a long time before spintronics research explored the effect of a charge-driven spin current on magnetization dynamics. Winter wrote coupled equations for the magnetizations M_s and M_d. In a thermodynamic approach to the dynamics of s-d coupling, Hasegawa pointed out that there are 3 systems that need to be considered: M_s, M_d, and L, the translational motion. L is known to be a very fast subsystem. The issue before Hasegawa was whether we can assume that M_s is strongly coupled to L or whether it is M_s and M_d that are most strongly coupled. Hasegawa concluded that, in most cases, the situation is midway between these two limits. So, he proposed Bloch equations to allow for all cases. Winter took over these conclusions and added a diffusion term of the form $D\nabla^2 M_s$ to account for the diffusion of the spin-polarized conduction electrons.

In this chapter, we begin with a description of electron spin relaxation by collisions at point defects where spin-orbit effects dominate. We proceed with an account of the relaxation of magnons by conduction electron spins, and then the converse effect, the spin mixing effect, i.e., electron spin flips without moment relaxation (see, §8.2.2 and §9.4.2). The chapter ends with a definition of the Elliot-Yafet and Dyakonov-Perel relaxation mechanisms.

13.2 RELAXATION OF CONDUCTION ELECTRON SPINS

13.2.1 SCATTERING AT POINT DEFECTS

Let us now consider a non-magnetic metal, i.e., a conductor in which the spin-up and spin-down bands have the same structure. Consequently, the density of states at the energy level E is $\rho'(E) = \frac{1}{2}\rho(E)$ for both spin directions, where $\rho(E)$ is the total density of states.

To evaluate the relaxation rate, we analyze how fast electrons reach equilibrium if they have been initially prepared in an out-of-equilibrium state (Fig. 13.2). The population of up spins, N_+, differs from that of down spins, N_-. This initial state can be achieved, theoretically, by switching off an applied field in a short amount of time compared to the relaxation time of the electrons. In spintronics, such a fast turn-off of the field could model what happens when spins that have been polarized in a ferromagnetic metal are injected in a normal metal, e.g., electrons going from cobalt (Co) to copper (Cu) in a Co/Cu multi-layer.

The electrons are in states $|\boldsymbol{k}s\rangle$ of energy $E_{\boldsymbol{k}\,s}$. As the electrons follow the Fermi-Dirac statistics, we have the spin-dependent distributions,

$$f_{\pm}(E) = \frac{1}{1 + e^{(E-\mu_\pm)/kT}} \tag{13.1}$$

Here, $\mu_\pm$ correspond to the spin-dependent chemical potential and is defined by,

$$N_\pm = \sum_{\boldsymbol{k}} f_\pm(\mu_{\boldsymbol{k}}) = \int\limits_0^\infty dE\rho'(E) \int\limits_{4\pi} d\Omega \frac{1}{1 + e^{(E_{\boldsymbol{k}}-\mu_\pm)/kT}} \tag{13.2}$$

By having two distinct chemical potentials in the out-of-equilibrium case of Fig. 13.2, we imply, as we did when studying spin-dependent transport, that the electrons of each spin-band are brought to a thermal equilibrium with the bath at temperature T much faster than the spins of separate spin bands relax with one another. When equilibrium is reached, they have the same chemical potential, which is equal to the Fermi energy E_F. By applying the principle of detailed balance to the populations N_- and N_+ as we did in (12.3), we find a spin-lattice relaxation rate $1/T_1 = 2W$, where W is the probability per unit of time of flipping the spin of one electron, and T_1 is the *spin-lattice relaxation time*. Here, we neglect the difference between the rates $W_\downarrow$ and $W_\uparrow$ in (12.12) because we assume that the temperature T is sufficiently high. Also, we are not concerned here with the value of the spin polarization at equilibrium.

Let us now seek an expression for W in terms of the scattering of single electrons interacting with the potential V of an impurity. The probability per unit of time for an electron in a state $|\boldsymbol{k}s\rangle$ to scatter into a state $|\boldsymbol{k}'s'\rangle$ is calculated with the *Born formula*,

$$W(\boldsymbol{k}s, \boldsymbol{k}'s') = \frac{2\pi}{\hbar}|\langle \boldsymbol{k}s| V |\boldsymbol{k}'s'\rangle|^2 \delta(E_{\boldsymbol{k}s} - E_{\boldsymbol{k}'s'}) \tag{13.3}$$

The Born formula and Fermi's golden rule are equivalent, because the scattering can be understood as a time-dependent perturbation.[181, 484]

Because we are dealing with electrons, we must take care of the exclusion principle in doing our statistics. Therefore, a scattering event from state $|\boldsymbol{k}s\rangle$ to a state $|\boldsymbol{k}'s'\rangle$ occurs

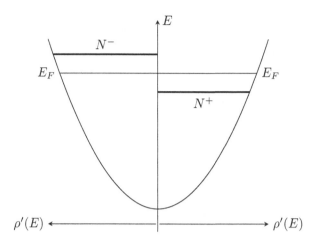

Figure 13.2 Density of states for down spins (left) and up spins (right). E_F is the Fermi level, N_+ and N_- are the occupation numbers of the spin bands when the spins are out of equilibrium.

with a probability proportional to the probability that the state $|ks\rangle$ is occupied, and the final state $|k's'\rangle$ is not, which is given by the product :

$$f_s(E_k)(1 - f_{s'}(E_{k'}))\qquad(13.4)$$

In the detailed balance (12.3), the spin-flip rate WN_- feeding N_+ can be expressed as:

$$WN_- = \sum_{k,k'} W(k-, k'+) f_-(E_k)(1 - f_+(E_{k'}))\qquad(13.5)$$

The rate of the converse process can be expressed as:

$$
\begin{aligned}
WN_+ &= \sum_{k,k'} W(k+, k'-) f_+(E_k)(1 - f_-(E_{k'})) \\
&= \sum_{k',k} W(k'+, k-) f_+(E_{k'})(1 - f_-(E_k))
\end{aligned}\qquad(13.6)
$$

The last equality is obtained by a change of variables in the double summation. We now use $W(k-, k'+) = W(k'+, k-)$, which is due to the symmetry of the scattering rate (13.3). Thus, we get,

$$\frac{dN_+}{dt} = WN_- - WN_+ = \sum_{k,k'} W(k-, k'+)(f_-(E_k) - f_+(E_{k'}))\qquad(13.7)$$

To find an explicit expression for T_1, we assume parabolic bands and, therefore, spherical Fermi surfaces. The sums are transformed into integrals over spherical surfaces of constant energy:

$$
\frac{dN_+}{dt} =
$$
$$
\int dE \int dE' \int_{4\pi} \frac{d\Omega}{4\pi} \int_{4\pi} \frac{d\Omega'}{4\pi} W(k-, k'+)\rho'(E)\rho'(E')(f_-(E) - f_+(E'))\qquad(13.8)
$$

The Born formula (13.3) imposes $E - E' = 0$. The term $(f_- (E) - f_+ (E))$ is non-zero over only a narrow range of energy (Fig. 9.6). Over this range, the rest of the integrand can be considered approximately constant. That is, this integrand can be estimated at the Fermi level. Hence, we have :

$$\frac{dN_+}{dt} = \frac{2\pi}{\hbar} \langle V^2 \rangle_F \rho'(E_F)^2 (E_- - E_+) \tag{13.9}$$

with,

$$\langle V^2 \rangle_F = \frac{2\pi}{\hbar} \int\limits_{4\pi} \frac{d\Omega}{4\pi} \int\limits_{4\pi} \frac{d\Omega'}{4\pi} |\langle \boldsymbol{k}_F - | V | \boldsymbol{k}_F' + \rangle|^2 \tag{13.10}$$

Now, since $N = N_+ + N_-$ is constant, we have $dN_-/dt = -dN_+/dt$ and $d(N_+ - N_-)/dt = 2dN_+/dt$.

The expression (12.10) of the detailed balance contains a term proportional to $N_+ - N_-$. We have,

$$N_+ - N_- = \sum_{\boldsymbol{k}} (f_+ (E_{\boldsymbol{k}}) - f_- (E_{\boldsymbol{k}})) = -(E_- - E_+) \rho'(E_F) \tag{13.11}$$

Therefore, we can write (13.9) as,

$$\frac{d(N_+ - N_-)}{dt} = 2\frac{dN_+}{dt} = \frac{-1}{T_1}(N_+ - N_-) = -\frac{2\pi}{\hbar} \langle V^2 \rangle_F \rho(E_F)(N_+ - N_-) \tag{13.12}$$

We conclude that,

$$\frac{1}{T_1} = \frac{2\pi}{\hbar} \langle V^2 \rangle_F \rho(E_F) \tag{13.13}$$

If there is an atomic concentration c of impurities, and N electrons ($\rho(E_F)$ is defined for one electron), then the relaxation rate becomes,

$$\frac{1}{T_1} = Nc\frac{2\pi}{\hbar} \langle V^2 \rangle_F \rho(E_F) \tag{13.14}$$

This last step amounts to assuming that the scattering events are independent (incoherent scattering), which is the case at sufficiently low impurity concentrations. The expression (13.14) can describe various mechanisms of relaxation. In the next section, we analyze $\langle V^2 \rangle_F$ due to spin-orbit scattering.

13.2.2 SPIN-ORBIT SCATTERING

As we have seen in part II, spin-orbit effects are ubiquitous in magnetism. This is also the case in spintronics. The link between spin-lattice relaxation and spin-dependent transport processes was shown in §9.4. Let us use the result (13.13) for the spin-lattice relaxation for a case where the interaction potential V is given by the spin-orbit coupling (2.44),

$$V = \lambda(r)\,\boldsymbol{S} \cdot \boldsymbol{l} \tag{13.15}$$

To estimate $\langle V^2 \rangle_F$, we follow the calculations presented by T. Asik in his thesis.[701] We can decompose the work into a spin part and an orbital part,[166]

$$
\begin{aligned}
|\langle \boldsymbol{k}_F - | V | \boldsymbol{k}_F' + \rangle|^2 &= \langle \boldsymbol{k}_F - | V | \boldsymbol{k}_F' + \rangle \langle \boldsymbol{k}_F' + | V | \boldsymbol{k}_F - \rangle = \\
&\langle + | \boldsymbol{S} | - \rangle \langle - | \boldsymbol{S} | + \rangle \langle \boldsymbol{k}_F | \lambda(r)\boldsymbol{l} | \boldsymbol{k}_F' \rangle \langle \boldsymbol{k}_F' | \lambda(r)\boldsymbol{l} | \boldsymbol{k}_F \rangle = \\
&\sum_{\substack{\alpha\beta \\ =x,y}} \langle + | S_\alpha | - \rangle \langle - | S_\beta | + \rangle \langle \boldsymbol{k}_F | \lambda(r)l_\alpha | \boldsymbol{k}_F' \rangle \langle \boldsymbol{k}_F' | \lambda(r)l_\beta | \boldsymbol{k}_F \rangle
\end{aligned} \tag{13.16}
$$

Since $\langle V^2 \rangle_F$ requires summing over all possible $\boldsymbol{k}_F$, we write this sum as:

$$\frac{1}{2} \sum_{\substack{\alpha\beta \\ =x,y}} \langle +| S_\alpha |-\rangle \langle -| S_\beta |+\rangle \tag{13.17}$$

$$\left(\langle \boldsymbol{k}_F| \lambda(r) l_\alpha |\boldsymbol{k}_F'\rangle \langle \boldsymbol{k}_F'| \lambda(r) l_\beta |\boldsymbol{k}_F\rangle + \langle \boldsymbol{k}_F| \lambda(r) l_\beta |\boldsymbol{k}_F'\rangle \langle \boldsymbol{k}_F'| \lambda(r) l_\alpha |\boldsymbol{k}_F\rangle \right)$$

Since $\langle +| S_\alpha |+\rangle = \langle -| S_\alpha |-\rangle = 0 \, (\alpha = x, y)$, we can add $\langle +| S_\alpha |+\rangle \langle -| S_\beta |-\rangle$ to this sum without changing it. Then, we make use of the closure relation to replace the spin part with $\langle +| S_\alpha S_\beta |+\rangle$. Again, since the sum is carried over both α and β, we can write it as:

$$\frac{1}{4} \sum_{\substack{\alpha\beta \\ =x,y}} \langle +| S_\alpha S_\beta + S_\beta S_\alpha |+\rangle \left(\begin{array}{c} \langle \boldsymbol{k}_F| \lambda(r) l_\alpha |\boldsymbol{k}_F'\rangle \langle \boldsymbol{k}_F'| \lambda(r) l_\beta |\boldsymbol{k}_F\rangle \\ + \langle \boldsymbol{k}_F| \lambda(r) l_\beta |\boldsymbol{k}_F'\rangle \langle \boldsymbol{k}_F'| \lambda(r) l_\alpha |\boldsymbol{k}_F\rangle \end{array} \right) \tag{13.18}$$

As the matrix elements of $S_\alpha S_\beta + S_\beta S_\alpha$ vanish if $\alpha \neq \beta$, the sum reduces to:

$$\frac{1}{4} \left\{ \langle +| 2S_x^2 |+\rangle \left(2 \left| \langle \boldsymbol{k}_F| \lambda(r) l_x |\boldsymbol{k}_F'\rangle \right|^2 \right) + \langle +| 2S_y^2 |+\rangle \left(2 \left| \langle \boldsymbol{k}_F| \lambda(r) l_y |\boldsymbol{k}_F'\rangle \right|^2 \right) \right\} \tag{13.19}$$

This expression is to be integrated over all possible $\boldsymbol{k}_F$. Since the l_x and the l_y terms give the same contributions, we have,

$$\langle V^2 \rangle_F = \int_{4\pi} \frac{d\Omega}{4\pi} \int_{4\pi} \frac{d\Omega'}{4\pi} \frac{1}{4} \left\{ \langle +| 2S_x^2 + 2S_y^2 |+\rangle \left(2 \left| \langle \boldsymbol{k}_F| \lambda(r) l_x |\boldsymbol{k}_F'\rangle \right|^2 \right) \right\} \tag{13.20}$$

Finally, using $\langle +| S_x^2 + S_y^2 |+\rangle = \langle +| S^2 - S_z^2 |+\rangle = 1/2$, we get,

$$\langle V^2 \rangle_F = \frac{1}{2} \int_{4\pi} \frac{d\Omega}{4\pi} \int_{4\pi} \frac{d\Omega'}{4\pi} \left| \langle \boldsymbol{k}_F| \lambda(r) l_x |\boldsymbol{k}_F'\rangle \right|^2 \tag{13.21}$$

13.2.3   SCATTERING CALCULATIONS FOR AN IMPURITY WITH NO EXCESS CHARGE

In the following, we make a simplifying assumption about the charge of an impurity and we calculate the orbital part of $\langle V^2 \rangle_F$ expressed in (13.21). Thus, at this stage of the calculation, the problem is one of solid state physics, namely, the electronic structure around a point defect. We need to know the $|\boldsymbol{k}_F\rangle$ states, which are conduction electron states, accurately enough near the nucleus of the impurity because this is where the scattering is the strongest.

Let us consider an impurity with no excess charge, so that no screening effects need to be taken into account. Our goal here is only to show that $\langle V^2 \rangle_F$ depends strongly on core states. Our choice is suggested by the following considerations.[1] Generally, the tight binding approach is acceptable if there are not too many overlaps. Hence, the scheme works well for narrow bands. When the bands broaden to the point that they overlap, the simplest scheme,

$$\phi_k(\boldsymbol{r}) = \sum_n e^{i\boldsymbol{k}\boldsymbol{R}_n} \phi_a(\boldsymbol{r} - \boldsymbol{R}_n) \tag{13.22}$$

---

[1]See, [702], p. 93ff.

must be modified. The method of orthogonalized plane wave (OPW) is a better approximation. We will use this method to explain why core states matter in describing scattering at impurities. The method consists in using the tight binding approach for core orbitals:

$$b_{tk}(\boldsymbol{r}) = \sum_l e^{ik\boldsymbol{R}_l} b_t(\boldsymbol{r} - \boldsymbol{R}_l) \tag{13.23}$$

where the $b_t(\boldsymbol{r} - \boldsymbol{R}_l)$ are strongly localized core orbitals. These states are solution of Schroedinger's equation and form a deep, narrow band that is fully occupied. We then consider states that are orthogonal to the $b_{tk}(\boldsymbol{r})$ states. As we know that these states must look like plane waves, we use trial functions of the form:

$$\chi_k(\boldsymbol{r}) = e^{ik\boldsymbol{r}} - \sum_t \beta_t b_{tk}(\boldsymbol{r}) \tag{13.24}$$

with $\beta_t = \langle b_{tk}(\boldsymbol{r})| e^{ik\boldsymbol{r}}\rangle$ so as to satisfy the orthogonality condition. The wave functions $\chi_k(\boldsymbol{r})$ look like plane waves outside the cores, and at the core of the atomic sites, they are orthogonal to the cores states. To complete the calculation and get numerical estimates, we could solve Schroedinger's equation with a linear combination of these OPWs, typically using a variational method, and find β_t for all t.

Here, we assume that the β_t coefficients are known and consider an impurity at $\boldsymbol{R}_1 = \boldsymbol{0}$. In this case, (13.23) implies

$$b_{tk}(\boldsymbol{r}) = b_t(\boldsymbol{r} - \boldsymbol{R}_1) = b_t(\boldsymbol{r}) \tag{13.25}$$

We further assume that the impurity has s and p core states. The s states will not contribute to the spin-orbit coupling, so we discard them. We consider only the core p-states $|b_t\rangle = |p_{x_t}\rangle = x_t f(\boldsymbol{r})$ $x_t = x, y, z$. The directions x, y, and z refer to crystalline directions. The conduction electron states are then taken to be:

$$|\boldsymbol{k}_F> = |e^{ik\boldsymbol{r}}> - \sum_i \beta_{ik}|x_i> \quad \text{with} \quad \beta_{ik} = <e^{ik\boldsymbol{r}}|x_i> \tag{13.26}$$

Consider coordinate axes X, Y, and Z such that the Z axis is along $\boldsymbol{k}_F$. We can write,

$$|x_i> = \sum_j c_{ij}|X_j> \tag{13.27}$$

Then

$$\beta_{ik} = <e^{ikZ}|x_i> = \sum_j c_{ij} <e^{ikZ}|X_j> \tag{13.28}$$

But the following are equal:

$$a = <e^{ikx}|xf> = <e^{iky}|yf> = <e^{ikz}|zf> \tag{13.29}$$

and we have,

$$<e^{ikz}|xf> = <e^{ikz}|yf> = 0 \tag{13.30}$$

Then $\beta_{ik} = a\, c_{ik}$. Let $\cos(i, \boldsymbol{k})$ be the cosine of the angle between the direction designated by i and the direction of $\boldsymbol{k}_F$. As $zf(r) = Z\cos(i, \boldsymbol{k})f(r)$, we have,

$$c_{ik} = cos(i, \boldsymbol{k}) \quad \text{and} \quad |\boldsymbol{k}_F> = |e^{ik\boldsymbol{r}}> - \sum_i a\cos(i, \boldsymbol{k})|x_i> \tag{13.31}$$

In calculating $\langle V^2 \rangle_F$, 4 terms appear:

- plane wave in, plane wave out,
- core state in, core state out,
- cross terms involving one plane wave and one core state.

Calculations show that the core-core terms are by far the largest.[703] We will neglect the others. Thus:

$$\langle V^2 \rangle_F = \frac{1}{2} \int_{4\pi} \frac{d\Omega}{4\pi} \int_{4\pi} \frac{d\Omega'}{4\pi}$$

$$\sum_{i i' i'' i'''} a^4 c_{ik} c_{i'k'} c_{i''k} c_{i'''k'} \langle x_{i'} | \lambda(r) l_x | x_i \rangle \langle x_{i''} | \lambda(r) l_x | x_{i'''} \rangle \qquad (13.32)$$

If i and i'' differ, we have $\int_{4\pi} \frac{d\Omega}{4\pi} c_{ik} c_{i''k} = 0$, whereas if $i = i''$, then $\int_{4\pi} \frac{d\Omega}{4\pi} c_{ik}{}^2 = \frac{1}{3}$. Therefore, we have,

$$\langle V^2 \rangle_F = \frac{1}{2} \frac{a^4}{9} \sum_{i i'} |\langle x_{i'} | \lambda(r) l_x | x_i \rangle|^2 \qquad (13.33)$$

Thus, we find that the scattering is expressed in terms of a matrix element with **core states**. The spin-orbit parameters such as $a^2 \lambda(r)$ are tabulated from measurements of X-ray spectra.[704]

13.3   MAGNETIC SCATTERING

Starting around 1999, research on spin transport was concerned with the effect of a spin current on the magnetization of a ferromagnetic metal, either the excitation of spin waves or flipping the magnetization. As for what concerns the time evolution of spin-waves, a good introduction to the topic can be found in a paper written by Callen.[396] In that paper, the evolution of the magnon number is worked out. This result was used by Berger when he suggested that a spin polarized current can flip the magnetization.[56] Switching of magnetization due to charge-driven spin torque was introduced in §1.1.5. We note that long before spintronics research, this interaction of conduction electron and magnetization at an interface between a ferromagnet and a metal was studied by electron spin resonance.[705] Also, we remark that it is possible to induce a spin current by heat, in metals as well as in insulators.[470, 706] In view of these various phenomena that are due to the interplay of magnetization dynamics and spin transport, we want to analyze here the collisions of electrons spins with spin-waves (see §7.5). We start out by describing electron-magnon collisions. Then, we use our results to estimate magnetic relaxation by magnons and the spin mixing rate for conduction electrons.

Let us start with the s-d model presented in §6.5. Since the Hamiltonian (6.31)

$$\mathcal{H}(\mathbf{k'}, \mathbf{k}) = -2J_{sd} \left(\frac{1}{N} \sum_j e^{i(\mathbf{k}-\mathbf{k'})\mathbf{R}_j} \mathbf{S}_j \right) \cdot \mathbf{s}$$

commutes neither with $\sum_{i,j} \mathbf{S}_i \cdot \mathbf{S}_j$ nor with $\sum_{i,j} S_{i,z}$, it provides a coupling mechanism that can change the magnitude and the z-component of the localized moments. Therefore, this interaction specifies how magnetization is transferred between s and d electrons. The total magnetization of s and d electrons is conserved in this Hamiltonian description.[328] This conservation rule is dictated by the invariance under rotation of the scalar coupling $\mathbf{S}_j \cdot \mathbf{s}$.

13.3.1   RELAXATION OF MAGNONS BY CONDUCTION ELECTRONS

We can use Hamiltonian $\mathcal{H}(\boldsymbol{k}', \boldsymbol{k})$ to estimate the contribution of conduction electron scattering to magnon relaxation. An electron of wave vector $\boldsymbol{k}$ scatters into a state with wave vector $\boldsymbol{k}'$, undergoing a spin flip in the process. In doing so, the ferromagnetic electron system changes from the state $S = -S_{\max}$ to the state $S = -S_{\max} + 1$, corresponding to the generation of a spin wave of momentum $\boldsymbol{\kappa} = \boldsymbol{k} - \boldsymbol{k}'$. The invariance by translation imposes this moment conservation rule.[2]

Thus, we are trying to describe the scattering of delocalized electron states by unbound states of magnetization excitations. This is analogous to the scattering of electrons by phonons. The spin-wave state is conveniently described by creation and annihilation operators constructed according to the Holstein-Primakoff method (§7.6.3). The key idea of this method is to represent spin waves as harmonic oscillators, hence the analogy with phonon scattering. We define the operators of creation and annihilation for spin waves as,

$$a_j = \frac{1}{\sqrt{N}} \sum_{\boldsymbol{\kappa}} e^{i\boldsymbol{\kappa} \cdot \boldsymbol{R}_j} a_{\boldsymbol{\kappa}} \qquad (13.34)$$

Substituting these definitions into $\mathcal{H}(\boldsymbol{k}', \boldsymbol{k})$ yields:

$$\mathcal{H}(\boldsymbol{k}', \boldsymbol{k}) = -\sqrt{\frac{2S}{N}} J_{sd} \sum_{\boldsymbol{\kappa}} \left[s^+ a_{\boldsymbol{\kappa}} \delta_{\boldsymbol{k}'-\boldsymbol{k},\boldsymbol{\kappa}} + s^- a_{\boldsymbol{\kappa}}^* \delta_{\boldsymbol{k}-\boldsymbol{k}',\boldsymbol{\kappa}} \right]$$
$$- \frac{2J_{sd}}{N} s^z \sum_{\boldsymbol{\kappa},\boldsymbol{\kappa}'} a_{\boldsymbol{\kappa}'}^* a_{\boldsymbol{\kappa}} \delta_{\boldsymbol{k}-\boldsymbol{k}',\boldsymbol{\kappa}'-\boldsymbol{\kappa}} \qquad (13.35)$$

The first term describes single spin-wave processes, and the second term corresponds to two-magnon processes of second-order magnitude. This second term is important when describing the relaxation of magnons of very long wavelengths, i.e., processes that are involved in the magnetization reversal experiments analyzed in Chapter 14.

The single-magnon term yields the following time evolution for the occupation number $n_{\boldsymbol{\kappa}}$ of magnons of wave vector $\boldsymbol{\kappa}$:

$$\frac{dn_{\boldsymbol{\kappa}}}{dt} = \frac{2\pi}{\hbar} \left(\frac{2S}{N} J_{sd}^2 \right) \sum_{\boldsymbol{k}',\boldsymbol{k}} \delta_{\boldsymbol{k}-\boldsymbol{k}',\boldsymbol{\kappa}} \delta \left(E_{\boldsymbol{k}'}^- + \varepsilon_{\boldsymbol{\kappa}} - E_{\boldsymbol{k}}^+ \right)$$
$$\left[\left(1 - f_{\boldsymbol{k}'}^- \right) f_{\boldsymbol{k}}^+ \left(n_{\boldsymbol{\kappa}} + 1 \right) - \left(1 - f_{\boldsymbol{k}}^+ \right) f_{\boldsymbol{k}'}^- n_{\boldsymbol{\kappa}} \right] \qquad (13.36)$$

This formula is obtained by applying Fermi's golden rule. In it, we recognize the energy conservation terms, and the probability for the "initial" states to be occupied and the "final" states to be empty. The energy of the conduction electrons is given by:

$$E_{\boldsymbol{k}}^\pm = \frac{\hbar^2 k^2}{2m^*} \mp \mu_B H \qquad (13.37)$$

The energy of the spin waves of momentum $\boldsymbol{\kappa}$ is given by:

$$\varepsilon_{\boldsymbol{\kappa}} = 2SJ\kappa^2 a^2 - g\mu_B H \qquad (13.38)$$

In this expression of the magnon dispersion, J is the exchange coupling of the localized moments described by the Hamiltonian:

$$-2J \sum_{i,j} \hat{\boldsymbol{S}}_i \cdot \hat{\boldsymbol{S}}_j - g\mu_B H \sum_i \hat{S}_{i,z} \qquad (13.39)$$

---

[2]See, [72], appendix M.

The distributions describing the statistics of conduction electrons are the Fermi-Dirac functions:

$$f_{\boldsymbol{k}}^{\pm} = \left(1 + \exp\left(\frac{E_{\boldsymbol{k}}^{\pm} - E_F}{k_B T}\right)\right)^{-1} \tag{13.40}$$

Under the approximation that $k_B T$ is small compared to the Fermi energy, Mitchell deduces an exponential decay of the magnon number,[325]

$$\frac{dn_{\kappa}}{dt} = -\frac{n_{\kappa} - n_{\kappa}^0}{T_1} \tag{13.41}$$

with,

$$\frac{1}{T_1} = \frac{S J_{sd}^2 m^{*2} \Omega \varepsilon_{\kappa}}{\pi \hbar^5 \kappa}\left[1 + \exp\left(\frac{1}{4kT}\frac{\left(\hbar^2 \kappa^2/2m^* + \varepsilon_{\kappa} + 2\mu_B H\right)^2}{(\hbar^2 \kappa^2/2m^*)} - \frac{E_F}{kT}\right)\right]^{-1} \tag{13.42}$$

where the last parenthesis is nearly 1, as the Fermi energy is much larger than all the magnon energies involved. The prediction of this single-magnon process does not provide a relaxation path for magnons of long wavelengths, as the spin-up and spin-down band splitting is large in a ferromagnet compared to the magnon energy when $\kappa \to 0$.

Mitchell considered magnons that have a wavelength of the order of the microwave penetration depth. Thus, using (13.42), he predicted that in nickel, magnons relax with a characteristic time of about 5 ns. This turns out to be a value close to the switching time observed in nanowires subject to a very sharp current pulse.[707] Mitchell had assumed the following numerical values:

- frequency: 24.4 GHz,
- field: 5.2 kOe,
- resistivity ρ: 7.7 10^{-6} Ohm cm,
- 1/skin depth $= \sqrt{4\pi\omega\mu/(\rho c^2)} = 2\ 10^5$ cm$^{-1} \approx (50$ nm$)^{-1}$,
- a = 3.5 10^{-8} cm,
- J= 230 k_B = 2 10^{-2} eV,
- J_{sd} = 0.01 eV (a very small value because the sd interaction is screened in solids).

13.3.2   CONDUCTION ELECTRON SPIN FLIP BY MAGNONS (SPIN MIXING)

Here, we look at the converse effect of electron-magnon collisions, namely, the rate of electron spin flips with momentum conservation, which is the transport phenomenon known as *spin-mixing* (see §8.2.2 and §9.4.2).

The difference in spin mixing rates from majority to minority versus minority to majority bands is crucial in some thermoelectric measurements.[432, 440] The difference between the spin mixing rates from up- to down-spins, and inversely, is thought to determine the magnetic field dependence of the Seebeck coefficient.[440, 481]

The scattering of electrons with magnons was initially discussed by T. Kasuya,[708] I. Mannari [709] and D. A. Goodings.[710] They focused their attention on the contributions of electron-magnon scattering to momentum relaxation. Later, A. Fert emphasized the role of electron-magnon collisions in spin mixing.[711] We can find useful insights into how to treat this problem by examining the treatment of a similar problem: collisions of electrons with phonons.[712]

There is a distinction between the mean time between collisions and the transport relaxation time.[472] Here, we seek only the mean time between spin flips τ_{+-} and τ_{-+}, which are due to collisions with magnons, and ignore the small effect of these collisions on momentum relaxation. The time τ_{-+} refers to the process in which the conduction-electrons spins gain one Bohr magneton. In view of the form (13.35) of the electron-magnon interaction, this gain must correspond to a loss of one Bohr magneton on the part of the magnetization. This is equivalent to saying that one magnon is created (Fig. 13.3).

Since we need to keep track of the difference in the two directions of spin flips, we calculate the times τ_{+-} and τ_{-+} using (12.175), the Fermi golden rule that includes the statistical weights of the initial states. For electrons, we ensure that the final electron state is empty and the initial electron state is occupied with a given probability. Including the sum over all possible magnon wavevectors, this yields:

$$\frac{1}{\tau_{-+}} = \frac{2\pi}{\hbar} \int \frac{d^3q}{(2\pi)^3} |M_q|^2 N_q \, f(E_-) \, (1 - f(E_+)) \, \delta\left((E_+ + \hbar\omega_q) - E_-\right) \qquad (13.43)$$

In creating a magnon, the conduction electron loses some energy, so this integral contains a product of the form $f_0(E) \, (1 - f_0(E - \Delta E))$ $(\Delta E > 0)$ (Fig. 13.3). The term N_q in (13.43) stems from the same reasoning as that used in (9.98) for phonons. In the converse

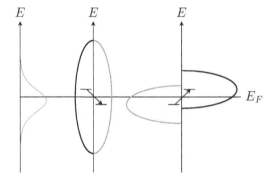

Figure 13.3   At the far left, a sketch of the product $f_0(E) \, (1 - f_0(E - \Delta E))$ showing the energy range where electron-magnon transitions may happen. To its right, successively, density of states of up and down electrons for conduction electrons, then d states.

process, a magnon is annihilated and the rate of such process is given by:

$$\frac{1}{\tau_{+-}} = \frac{2\pi}{\hbar} \int \frac{d^3q}{(2\pi)^3} |M_q|^2 (N_q + 1) \, f(E_+) \, (1 - f(E_-)) \, \delta\left((E_- - \hbar\omega_q) - E_+\right) \qquad (13.44)$$

The magnetization gives some energy to the electrons in this process, so the integrand contains a term of the form $f_0(E) \, (1 - f_0(E + \Delta E))$ $(\Delta E > 0)$. The term $(N_q + 1)$ in (13.44) comes from the action of the creation operator, as in (9.99). In Fig. 13.4, we see that the two products, namely $f_0(E) \, (1 - f_0(E - \Delta E))$ and $f_0(E) \, (1 - f_0(E + \Delta E))$ where $\Delta E > 0$, are noticeably different. At the Fermi level, the magnon creation process has a product of about 0.2, whereas the magnon annihilation process has a product of about 0.3. This simple consideration is enough to infer that the spin mixing rates $(\tau_{+-})^{-1}$ and $(\tau_{-+})^{-1}$ differ because they depend on whether transitions are from majority to minority spins or vice versa.

Let us now estimate the temperature dependence of the difference between $(\tau_{+-})^{-1}$ and $(\tau_{-+})^{-1}$. We group all coefficients in one constant K, including the matrix element, and

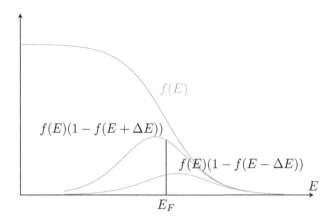

Figure 13.4 Fermi-Dirac distribution and the products $f(E)\left(1 - f(E - \Delta E)\right)$ and $f(E)\left(1 - f(E + \Delta E)\right)$ ($\Delta E > 0$) having different values at E_F.

assume the same for all q, which leaves us to estimate:

$$
\begin{aligned}
\frac{1}{\tau_{-+}} &= K \int \frac{d^3q}{(2\pi)^3} N_q \, f(E_-) \left(1 - f(E_- - \hbar\omega_q)\right) \\
&= K \frac{4\pi}{(2\pi)^3} \frac{1}{\hbar D} \int \hbar D q^2 dq \, \frac{\left(1 - \frac{1}{\left[\exp\left(\frac{E_- - \hbar\omega_q - E_F}{kT}\right) + 1\right]}\right)}{\left[\exp\left(\frac{\hbar\omega_q}{kT}\right) - 1\right]\left[\exp\left(\frac{E_- - E_F}{kT}\right) + 1\right]}
\end{aligned}
\tag{13.45}
$$

We assumed a magnon dispersion $\omega = Dq^2$. We want to evaluate the spin mixing rate at the Fermi level, i.e., at $E_- = E_F$. This gives,

$$
\frac{1}{\tau_{-+}} = K \frac{4\pi}{(2\pi)^3} \frac{1}{2\hbar D} \int \frac{\varepsilon_q dq}{\left(\exp\left(\frac{\varepsilon_q}{kT}\right) - 1\right)\left(1 + \exp\left(\frac{-\varepsilon_q}{kT}\right)\right)}
\tag{13.46}
$$

From the magnon dispersion, we deduce:

$$
\varepsilon_q = \hbar D q^2 \;\rightarrow\; q = \sqrt{\frac{\varepsilon_q}{\hbar D}} \;\rightarrow\; dq = \sqrt{\frac{1}{4\hbar D}} \frac{d\varepsilon_q}{\sqrt{\varepsilon_q}}
\tag{13.47}
$$

Thus, the spin mixing rate becomes,

$$
\frac{1}{\tau_{-+}} = K' \int \frac{\sqrt{\varepsilon_q} d\varepsilon_q}{\left(\exp\left(\frac{\varepsilon_q}{kT}\right) - 1\right)\left(1 + \exp\left(\frac{-\varepsilon_q}{kT}\right)\right)}
\tag{13.48}
$$

where K' contains all the constants.

Now, we have yet to establish the limits of integration. We assume that there is a splitting of the majority and minority bands, the energies of which are E_+, E_-. Then, the magnon energies must have at least the value of this gap in order for them to induce transitions between the two bands. We associate a temperature T_F with this splitting, writing,

$$
k_B T_F = E_+ - E_-
\tag{13.49}
$$

This is our lower bound. The largest q should be k_F, but this corresponds to magnon energies that have zero occupancy probability, so we can set the upper bound to infinity. Hence:

$$\frac{1}{\tau_{-+}} = K'' \int\limits_{kT_F}^{\infty} \frac{\sqrt{\varepsilon_q}d\varepsilon_q}{\left(\exp\left(\frac{\varepsilon_q}{kT}\right) - 1\right)\left(1 + \exp\left(\frac{-\varepsilon_q}{kT}\right)\right)}$$

$$= K'' \, T^{3/2} \int\limits_{T_F/T}^{\infty} \frac{\sqrt{x} \, dx}{\left(\exp\left(x\right) - 1\right)\left(1 + \exp\left(-x\right)\right)} \tag{13.50}$$

The dependence of this integral as a function of T/T_F is illustrated in Fig. 13.5, in which the mixing rate is divided by T^2 to show that the spin mixing rate has roughly a T^2 dependence at high temperatures. As T goes below T_F, the mixing rate drops sharply to zero.[711] Carrying out the same calculation for the reverse process, we deduce from (13.44),

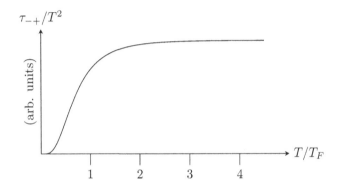

Figure 13.5 The spin mixing rate (13.50) divided by T^2 grows steeply at about T_F, a measure of the spin-dependent band splitting, and reaches an asymptote at about 3 T_F.

$$\frac{1}{\tau_{+-}} = K'' \int\limits_{kT_F}^{\infty} \left[\exp\left(\frac{\varepsilon_q}{kT}\right) - 1\right]^{-1} \left[1 + \exp\left(\frac{+\varepsilon_q}{kT}\right)\right]^{-1} \sqrt{\varepsilon_q} \, d\varepsilon_q \tag{13.51}$$

In the thermodynamic description of spin-dependent transport of §8.3, we saw that the difference between spin currents mixing $\Delta L = L_{+-} - L_{-+}$ contributes to the spin-dependent properties when the spins are out of equilibrium, . This difference is related to the spin mixing rates by,

$$\Delta L = L_{+-} - L_{-+} \propto \frac{\tau_{+-} - \tau_{-+}}{\tau_{-+}^2 \, \tau_{+-}^2} \tag{13.52}$$

The temperature dependence of the difference in spin mixing rates is illustrated in Fig. 13.6. Therefore, a transport measurement that is sensitive to spin mixing can be used to estimate T_F.

Here, we have examined the contribution of the electron-magnon collision mechanism to spin mixing. We did not consider its contribution to momentum relaxation. As Rauet et al. point out, *its contribution to the magnetic resistivity via electronic spin-flip transitions remains poorly known. One experiment has inferred a direct estimate of the pure spin-flip electronic scattering contribution to the resistivity in a temperature range where spin excitations, electron-phonon scattering and inter-electronic collisions coexist.*[713] Rauet et

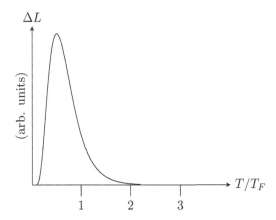

Figure 13.6 Spin mixing rate difference on a wide temperature scale that is normalized to the splitting T_F of the majority and minority bands.

al. deduced from their very high field magnetoresistance measurements that the freezing temperature (T_F) of the electron-magnon process is about 15 K for Nickel, and 250 K for cobalt. Thus, details of the band structures of these two metals give rise to very large differences in the temperature dependence of electron-magnon collision effects, even though their spin-wave dispersions are not very different.

13.4   ELLIOTT-YAFET MECHANISM

The Elliott-Yafet mechanism is based on the fact that, in a metallic crystal, the Bloch states are not spin eigenstates. The spin-orbit interaction in the ions mixes spin-up and spin-down states. Thus, if we were to prepare the system in the state of a given spin and could turn on the spin-orbit interaction adiabatically, the system would evolve toward a state that contains a component of the other spin. The small admixture of the other spin component would be of the order of $\lambda/\Delta E$, where λ is the spin-orbit coupling constant and ΔE is the conduction-electron band width.[714]

Elliott found that collisions due to spin-independent interactions could give rise to transitions from almost spin-up states to almost spin-down states, and vice versa, when the bands are defined with mixed-spin states and the collisons cause a large momentum change.[715] Large momentum changes can occur in collisions with surfaces, impurities and phonons. This idea is consistent with a remark we made about the sd model. We had noticed that, when k and k' differ, the Hamiltonian $\mathcal{H}(k, k')$ defined in (6.31) does not commute with s^2; thus, it contributes to the relaxation of the s spins. Therefore, strong scatterers contribute to the spin relaxation, as depicted in Fig. 13.7.

P. Monod and his group studied by electron spin resonance (ESR) the spin-lattice relaxation of conduction electrons in metals such as copper, silver and aluminium.[716] They observed that by increasing the number of microstructural grains in sheets of these metals, they reduced the relaxation time, owing to the strong scattering effect of grain boundaries. However, the dominant mechanism in a sample of reasonable purity is the spin-flip scattering due to impurities.[717, 718]

In semiconductors, *hole spins* relax much faster than electron spins because hole states have a strong admixture of spin-up and spin-down states.

The contribution of transverse and longitudinal *phonons* to the electron spin relaxation was also analyzed by A. Overhauser.[719] Longitudinal phonons contribute to the electron spin relaxation via spin-orbit coupling. Y. Yafet showed that the temperature dependence of the spin-lattice relaxation rate followed that of the resistance.[163] Overhauser also estimated the contribution of **electron-electron collisions** to the spin-lattice relaxation, including the spin-spin dipolar coupling and the coupling of spins to the electron charge current.[719] These mechanisms were also analyzed by P. Boguslawski.[720]

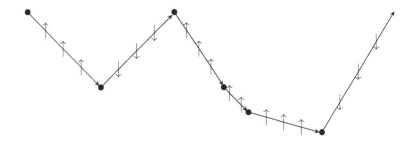

Figure 13.7 Schematics representing the Elliott-Yafet relaxation mechanism : at collisions with large changes in k vector, the spin has a high probability of flipping.

13.5 DYAKONOV-PEREL MECHANISM

In a crystal that lacks inversion symmetry, the spin-orbit interaction lifts the spin degeneracy (see §6.6.1). One way to think of this splitting is to attribute a k-dependent magnetic induction field $B(k)$. Thus, at every sudden change of k, the spin experiences a new field and precesses about it. This precession implies a non-vanishing probability of transitions from spin-up to spin-down states. This is the mechanism that was first envisaged by Dyakonov and Perel.[721]

Typically, the time τ between collisions that change k is short compared to the time for a spin to complete one revolution around the effective field (Fig. 13.8). Thus, this mechanism is in the motional narrowing regime, for which $1/T_1 = \omega^2 \tau$. In other words, the relaxation rate $1/T_1$ is proportional to the momentum relaxation time τ.

The motional narrowing formula can be understood as follows. If the dephasing is $\delta = \omega \tau$ between two collisions, after n collisions, the average dephasing $\Delta\varphi$ is given by $\Delta\varphi^2 = n\delta^2$, as in a random-walk problem. The number of collisions is $n = T_1/\tau$, and we take $\Delta\varphi = 1$ in a time T_1.

This relaxation mechanism is analogous to a relaxation mechanism proposed for electrons in composite materials comprising ferromagnetic clusters in a metallic but non-magnetic matrix.[722] The fluctuations occur as the electrons travel through space. This so-called jitterbug relaxation mechanism is in the strong collision regime, i.e., $T_1 \approx \tau$.

13.6 FURTHER READINGS

Long relaxation times are desirable in spintronics, notably in spin valves, when the spacer thickness can be made large enough for the layering to be reliable in rugged sensor applications. Long relaxation times are also important in dynamic nuclear polarization (§16) and spin-based quantum information processing. In this chapter, we have seen that spin-orbit coupling leads to electron spin relaxation. The strength of the coupling depends strongly on the atomic number Z (see Eq. (2.42)). Therefore, **graphene** is of particular interest when seeking to achieve long relaxation times. Both the Dyakonov-Perel and Elliott-Yafet mechanisms seem to play a role in graphene.[723] Based on the weak spin-orbit coupling,

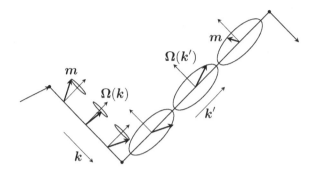

Figure 13.8 Schematics representing the Dyakonov-Perel relaxation mechanism : every time the k vector reorients, the spin m experiences a different effective field $\Omega(k)$. The collision with an impurity is assumed to be so sudden that the spin has no time to reorient during the collision. The precession angle increase is enlarged for the sake of clarity. In this relaxation mechanism, the angular change between collisions is very small (motional narrowing regime).

the relaxation time is estimated in the microsecond range. In effect, it is observed in the nanosecond range![724] This discrepancy could be due to defects, strain, impurities, lattice defects, adsorbed atoms, or corrugation.[725] Indeed, the curvature of a graphene sheet can induce an enhanced spin-orbit effect, as mentioned in §11.5.

We also looked at the fate of an electron spin inside of a metal. Further insight into the various types of spin-flip scattering mechanisms can be gained from spin-polarized electron energy loss spectroscopy (SPEELS). Deviation from Bragg reflection and energy loss may be due to spin waves or Stoner excitations.[726] SPEELS can also probe *spin flip exchange scattering*, a notion we have already encountered in §12.7.[727]

Spin relaxation times were predicted to be as long as 0.5 μs at room temperature, 2 ms at 4K, in quantum wells based on wurtzite AlN semiconductors.[728] GaN and ZnO also have the wurtzite structure. Experimentally, a relaxation time of 0.3 ns was reported in InGaN/GaN quantum wells.[729] Generally, *quantum wells* have much longer spin relaxation times owing to the suppression of the Dyakonov-Perel mechanism.[730] The wurtzite structure is interesting because its spin-orbit interaction is less than in a zinc blend structure (e.g., GaAs).[731] In bulk ZnO, a relaxation time of 20 ns was observed at 30K, but the relaxation time turned out to be much shorter in thin films.[732]

Electric-field control of the spin relaxation time can be achieved in a semiconductor quantum well, because the applied electric field may change the effective spin-orbit interaction. This is the case in GaAs quantum wells, where the application of an electric field increased the spin relaxation time considerably.[733, 734] In nanowires cut out in modulation-doped GaAs/AlGaAs structures, the spin relaxation time is found to depend strongly on the crystalline orientation of the wire.[735]

Quantum beats occur when a coherent superposition of two energy eigenstates is excited. The resulting non-stationary state has a time dependence at the frequency corresponding to the difference in energy of the two states.[181, 736] Quantum beats of the exciton fluorescence in quantum wells reflect spin splitting in a transverse magnetic field and depend critically on hole spin flips.[737]

14 Magnetic Relaxation

Louis E. F. Néel (1904–2000)

L.E.F. Néel was co-awarded the 1970 Nobel Prize in Physics for his discovery of antiferromagnetism and ferrimagnetism. He found that magnetic grains below a certain size are single-domain and studied the effect of thermal fluctuations on magnetization evolution. His work helped to interpret geological findings regarding the evolution of the earth's magnetic field.

Spintronics: A primer
J.-Ph. Ansermet (https://orcid.org/0000-0002-1307-4864)
Chapter 14. Magnetic Relaxation
The micromagnetics approximation, according to which the local magnetization has a constant modulus, implies that magnetic relaxation undergoes a rotational Brownian motion when the magnetization relaxes. Starting with a Fokker-Planck equation, magnetic relaxation is expressed as a normal mode problem from which the Néel relaxation time is deduced.

DOI: 10.1201/9781003370017-14

14.1 HISTORICAL INTRODUCTION

Magnetic relaxation was studied in bulk materials a long time ago. Motivated by magnetic recording technology, there has always been an interest in the magnetic relaxation of nanostructures, especially when a single particle can be observed at a time. As magnetic memory bits become smaller, the relaxation of a magnetic nanostructure becomes a critical issue. In bulk magnetic materials, domain wall creep is a dominant relaxation mechanism.[738]

In order to characterize intrinsic magnetic relaxations, older studies were conducted using nano-powders that were produced chemically.[739] The study of GMR (giant magneto-resistance) naturally led to the characterization of magnetic excitation and relaxation in multilayers.[740] Starting in the early 1990s, it became possible to observe the time evolution of single, isolated and well-defined nanostructures. In doing so, it became possible to distinguish whether a distribution of barrier heights was due to the presence of many particles with slightly different magneto-crystalline anisotropies, or whether stretched exponentials were an intrinsic property of individual particles.

When a ferromagnetic material is exposed to a sudden change of the applied field, its magnetization may be observed to relax to a new equilibrium value. By "sudden," we mean that the process takes place over a time that is short compared to the relaxation time. The fact that the magnetization evolves slowly after a sudden change is called the *magnetic after-effect*.

When a single barrier dominates the magnetic relaxation process, the first approach to describing magnetic relaxation is the *Néel-Brown model*. A more refined approach is based on the Langevin equation,[741–744] leading to the concept of integral relaxation time for single-domain particle.[745, 746]

Magnetic nanostructures have been made that present a single exponential relaxation behavior.[747] The other extreme in magnetic relaxation behavior is the *spin glass* regime.[748]

In this chapter, we introduce the Landau-Lifshitz-Gilbert (LLG) equation for magnetization dynamics. From the LLG equation, we work out a description of the rotational Brownian motion. On this basis, we show why magnetic relaxation can be thought of as an eigenvalue problem. This in turn allows us to introduce the Néel relaxation time.

14.2 ROTATIONAL BROWNIAN MOTION

Whether a magnetic system is in its ground state or a metastable state depends on the history of magnetic fields that the system has experienced. Let us now address the question of whether random fluctuations can lead to a jump from a metastable state to the ground state.

In §7.3, we introduced the magnetic energy $V(\boldsymbol{M})$ or $V(\theta, \varphi)$, from which we derived the effective magnetic field (7.20) appearing in the LLG equation (7.37) describing magnetization dynamics. When $V(\theta, \varphi)$ is very large compared to the thermal energy $k_B T$, we can describe quasi-static measurements reasonably well without considering thermal effects. The Gilbert equation (7.23) is sufficient. This is also the case when carrying out a magnetic resonance experiment. A change in temperature may simply cause a change in the value of the damping parameter but not the occurrence of switching.

When $V(\theta, \varphi)$ is small compared to $k_B T$, then the single domain particle is in the so-called *superparamagnetic state*. This is frequently the case with ferromagnetic nanostructures. For this reason, it is not possible to create magnetic memories with conventional ferromagnets of nanometer sizes. The term "superparamagnetic" expresses the notion that such a single domain particle has a magnetization that can be treated as a macroscopic paramagnetic moment. Much information about superparamagnetism can be found in the treatise of A. Aharoni.[183]

The case we want to analyze here is that of intermediate values of $V(\theta, \varphi)$ compared to $k_B T$. Experiments that characterize this dynamic regime are carried out as follows. First, a saturating magnetic field is applied for a time long enough for an ensemble of particles to reach equilibrium. Then, the field is brought to some intermediate value. The magnetization of this ensemble of particles is found to relax to a new equilibrium value. This is the after-effect. A logarithmic time dependence is often observed. An abundance of literature has been produced on this subject since the 1960s. Since the early 1990s, it has been possible to observe the magnetization of single, isolated particles. In such cases, what is observed is a jump of the magnetization from its metastable state to the ground state. The switching probability follows roughly an exponential behavior.[747]

This intermediate regime corresponds to a form of Brownian motion and is best described by a Fokker-Plank equation as described below. The prototypical example of Brownian motion is the jittery motion of tiny pollen particles in water, first observed by Brown, a 19th century botanist. The first theoretical account for Brownian motion was given by Einstein in 1905,[749] in what is considered as a seminal paper of statistical physics.[750] A collection of ensuing papers that laid the groundwork for statistical physics was prepared by Wax.[751] Another scientist named Brown, William Fuller Brown, Jr., extended the theory of Brownian motion to magnetic single domain particles.[752]

We consider the evolution of the magnetization as the motion of a point on a unit sphere. We want to write an equation of evolution for the statistical *distribution function* W of points on this unit sphere. First, we can write a continuity equation for W, of the form:

$$\frac{\partial W}{\partial t} = -\boldsymbol{\nabla} \cdot \boldsymbol{J} \tag{14.1}$$

By analogy with any kind of transport description, the current is defined by the product of a density and a velocity:

$$\boldsymbol{J} = W\dot{\mathbf{u}} \qquad \dot{\mathbf{u}} = \frac{1}{M_s}\frac{d\boldsymbol{M}}{dt} \tag{14.2}$$

The vector $\dot{\boldsymbol{u}}$ is tangent to the unit sphere. Now comes the question of introducing thermal fluctuations. This is done by adding a random field to $\boldsymbol{B}_{\text{eff}}$ in the Gilbert equation (7.34). This is equivalent to writing a Langevin equation for the magnetization.[741] Under reasonable stochastic properties of the fluctuation field, this modified Gilbert equation implies an equation of evolution for W that is known as a *Fokker-Planck equation*.[750]

In 1965, W. F. Brown pointed out that a lengthy calculation can be avoided by *a priori* adding a diffusion term to the current.[752] Let us recall from introductory physics that a gradient of concentration generates a flow proportional to this gradient and that this gradient tends to equalize the concentration. Thus, Brown writes:

$$\frac{\partial W}{\partial t} = -\boldsymbol{\nabla} \cdot (\boldsymbol{J} - k'\boldsymbol{\nabla}W) \tag{14.3}$$

Under this assumption, the equation of evolution (7.37) for $\dot{\boldsymbol{u}}$ in spherical coordinates implies,

$$\begin{aligned} J_\theta &= \left(-h'\frac{\partial V}{\partial \theta} + g'\frac{1}{\sin\theta}\frac{\partial V}{\partial \varphi}\right)W - k'\frac{\partial W}{\partial \theta} \\ J_\varphi &= \left(-g'\frac{\partial V}{\partial \theta} - h'\frac{1}{\sin\theta}\frac{\partial V}{\partial \varphi}\right)W - k'\frac{1}{\sin\theta}\frac{\partial W}{\partial \varphi} \end{aligned} \tag{14.4}$$

Finally, the equation of evolution for the distribution function W is obtained by taking the divergence of the probability current $\boldsymbol{J}$. In spherical coordinates, this reads,

$$\frac{\partial W}{\partial t} = -\frac{1}{\sin\theta}\frac{\partial}{\partial\theta}\left(\sin\theta\, J_\theta\right) - \frac{1}{\sin\theta}\frac{\partial}{\partial\varphi}\left(J_\varphi\right) \tag{14.5}$$

The Fokker-Plank equation in vector form reads:

$$\begin{aligned}\frac{\partial W}{\partial t} &= -\boldsymbol{\nabla}\cdot\left(W\left[-g'\left(\boldsymbol{u}\times\boldsymbol{\nabla}V\right)+h'\boldsymbol{u}\times\left(\boldsymbol{u}\times\boldsymbol{\nabla}V\right)\right]-k'\boldsymbol{\nabla}W\right)\\ &= +g'\boldsymbol{u}\cdot\left(\boldsymbol{\nabla}V\times\boldsymbol{\nabla}W\right)+h'\boldsymbol{\nabla}\left(W\boldsymbol{\nabla}V\right)+k'\nabla^2 W\\ &\quad + g'W\boldsymbol{\nabla}\left(\boldsymbol{u}\times\boldsymbol{\nabla}V\right)-h'\boldsymbol{\nabla}\cdot\left(W\left(\boldsymbol{u}\cdot\boldsymbol{\nabla}V\right)\boldsymbol{u}\right)\end{aligned} \tag{14.6}$$

The fourth term vanishes, as can be verified by writing it explicitly in spherical coordinates. The last term is zero because $\boldsymbol{\nabla}V$ has zero component along the radial unit vector. Therefore, we obtain ([741], Eq. 1.17.12),

$$\frac{\partial W}{\partial t} = +g'\mathbf{u}\cdot\left(\boldsymbol{\nabla}V\times\boldsymbol{\nabla}W\right)+h'\boldsymbol{\nabla}\left(W\boldsymbol{\nabla}V\right)+k'\nabla^2 W \tag{14.7}$$

In order to get an explicit expression for the coefficient k' that we introduced in (14.3), we impose that equation (14.7) implies at equilibrium the Boltzmann distribution,

$$W = W_0\exp\left(\frac{-V(\theta,\varphi)v}{k_B T}\right) \tag{14.8}$$

where v is the volume of the particle. Substituting this solution in (14.7) yields,

$$\begin{aligned}0 &= 0 + h'\boldsymbol{\nabla}W\boldsymbol{\nabla}V + h'\left(W\boldsymbol{\nabla}^2 V\right)+k'\nabla^2 W\\ 0 &= h'\frac{-v}{kT}(\boldsymbol{\nabla}V)^2 W + h'\left(W\boldsymbol{\nabla}^2 V\right)+k'\left(\frac{v}{kT}\right)^2(\boldsymbol{\nabla}V)^2 W + k'\left(\frac{-v}{kT}\right)\nabla^2 VW\end{aligned} \tag{14.9}$$

We need to be careful in evaluating $\nabla^2 W$: the operator $\boldsymbol{\nabla}$ must be operated twice separately to get the correct result. The second and fourth terms vanish because the magnetostatic potential V generally obeys the equation for magnetostatics given by, ([350], eq. 5.95, 5.96)

$$\nabla^2 V = -4\pi\rho_M \quad\text{where}\quad \rho_M = -\boldsymbol{\nabla}\cdot\boldsymbol{M} \tag{14.10}$$

The magnetostatic charge ρ_M vanishes for a single domain particle since by definition, the magnetization in the particle is uniform. Thus, we find for k',

$$k' = \frac{kT}{v}h' \tag{14.11}$$

In view of the definition (7.35) of h', we see that (14.11) is a kind of **Einstein relation** between the diffusion process and the dissipation in a rotational process. In other words, the relation (14.11) is a consequence of the fluctuation-dissipation theorem (see §12.5).

14.3   MAGNETIC RELAXATION AS AN EIGENVALUE PROBLEM

We can seek a solution of the Fokker-Plank equation for W of the generic form $W = T(t)F(\theta,\varphi)$ by posing,

$$W = W_0 + \sum_{n=1}^{\infty} A_n F_n\left(\theta,\varphi\right)\exp\left(-p_n t\right) \tag{14.12}$$

Now, we show that the relaxation rates p_n appear as eigenvalues of the differential equation for the spatial coordinates. For the sake of concreteness, we consider with Brown [752] an axially symmetric V and W. In this case, the probability currents are,

$$J_\theta = -h' \frac{\partial V}{\partial \theta} W - k' \frac{\partial W}{\partial \theta} \qquad J_\varphi = -g' \frac{\partial V}{\partial \theta} W \qquad (14.13)$$

and the Fokker-Plank equation reduces to,

$$\frac{\partial W}{\partial t} = \frac{1}{\sin\theta} \frac{\partial}{\partial \theta} \left(\sin\theta \left[h' \frac{\partial V}{\partial \theta} W + k' \frac{\partial W}{\partial \theta} \right] \right) \qquad (14.14)$$

We proceed to a change of variable, posing $x = \cos\theta$. For any function f,

$$\frac{\partial f}{\partial \theta} = \frac{\partial f}{\partial x} \frac{\partial x}{\partial \theta} = -\sin\theta \frac{\partial f}{\partial x} \qquad (14.15)$$

Therefore, the Fokker-Plank equation can be written as,

$$\frac{\partial W}{\partial t} = \frac{\partial}{\partial x} \left((1 - x^2) \left[h' \frac{\partial V}{\partial x} W + k' \frac{\partial W}{\partial x} \right] \right) \qquad (14.16)$$

We note that the equilibrium solution is easy to find in this case. We have,

$$(1 - x^2) \left[h' \frac{\partial V}{\partial x} W + k' \frac{\partial W}{\partial x} \right] = cste \qquad (14.17)$$

The second parenthesis is finite for all values of x. Therefore, (14.17) evaluated at $x = 1$ implies that the constant must be zero. Thus, we have,

$$h' \frac{\partial V}{\partial x} W + k' \frac{\partial W}{\partial x} = 0 \qquad (14.18)$$

This is equivalent to stating that the flow of probability on the unit sphere is zero. The solution to this equation is $W \propto \exp(-h'V/k')$. This yields the Boltzmann distribution, (14.8) provided the Einstein relation (14.11) applies.

By inserting the trial solution (14.12) into the Fokker-Plank equation (14.16), we obtain,

$$\sum_{n=1}^{\infty} A_n F_n (-p_n) \exp(-p_n t)$$
$$= \sum_{n=1}^{\infty} A_n \exp(-p_n t) \frac{d}{dx} \left[(1 - x^2) \left(h' \frac{dV}{dx} F_n + k' \frac{dF_n}{dx} \right) \right] \qquad (14.19)$$

W_0 does not appear in this equation since it is the equilibrium solution. We can also write,

$$h' \frac{dV}{dx} F_n + k' \frac{dF_n}{dx} = k' \left(\frac{h'}{k'} \frac{dV}{dx} F_n + \frac{dF_n}{dx} \right) = k' \exp\left(-\frac{h'V}{k'}\right) \frac{d}{dx} \left(\exp\left(\frac{h'V}{k'}\right) F_n \right) \qquad (14.20)$$

For each p_n and F_n, this equation yields,

$$\left(\frac{-p_n}{k'} \right) F_n = \frac{d}{dx} \left[(1 - x^2) \exp\left(-\frac{h'V}{k'}\right) \frac{d}{dx} \left(\exp\left(\frac{h'V}{k'}\right) F_n \right) \right] \qquad (14.21)$$

Applying the expression (14.11) for k', this gives,

$$\left(\frac{-p_n v}{kT h'} \right) F_n = \frac{d}{dx} \left[(1 - x^2) \exp\left(-\frac{vV}{kT}\right) \frac{d}{dx} \left(\exp\left(\frac{vV}{kT}\right) F_n \right) \right] \qquad (14.22)$$

Thus, we have expressed the magnetic relaxation dynamics as an eigenvalue problem for a ***Sturm-Liouville equation.***[753]

14.4   THE NÉEL RELAXATION TIME

Here, we follow W. F. Brown [752] and use the so-called Kramers method to further analyze the magnetic relaxation of a single domain particle. We want to consider the case in which the wells of the potential (Fig.4.2) are deep compared to $k_B T$. This implies that the occupied points of the unit sphere are concentrated very near the "North" or "South" poles. There are almost no points between the poles. However, a small current flows on the surface of the unit sphere, from one pole to the other. This flow is manifested as an after-effect: the system is practically stable in one configuration but if one waits long enough, the system progressively leaks, so to speak, to the more stable state.

Let us call I the probability current going from North to South on the unit sphere. If we call the populations at the poles n_1 and n_2, then $I = -\dot{n}_1 = \dot{n}_2$. Clearly, the quantity I tells us how the populations at the North and South pole vary. It is analogous to a relaxation rate of a two-level system (§12.2). By definition of the current density J_θ, this flow is given by $I_2 \pi \sin\theta J_\theta$, since $2\pi \sin\theta$ is the length of the parallel on the unit sphere. Therefore, we have,

$$J_\theta = -h'\frac{\partial V}{\partial\theta}W - k'\frac{\partial W}{\partial\theta} = \frac{I}{2\pi\sin\theta} \tag{14.23}$$

This expression can be rewritten as,

$$\frac{-I}{k'2\pi\sin\theta} = \frac{h'}{k'}\frac{\partial V}{\partial\theta}W + \frac{\partial W}{\partial\theta} = \exp\left(\frac{-h'V}{k'}\right)\frac{\partial}{\partial\theta}\left(\exp\left(\frac{h'V}{k'}\right)W\right) \tag{14.24}$$

Using the expression (14.11) of k', we have,

$$\frac{-Iv}{kTh'2\pi}\frac{\exp\left(\frac{vV}{kT}\right)}{\sin\theta} = \frac{\partial}{\partial\theta}\left(\exp\left(\frac{vV}{kT}\right)W\right) \tag{14.25}$$

We now integrate this expression between two values of the angle θ that define the neighborhoods of the occupied regions 1 and 2,

$$\frac{-Iv}{kTh'2\pi}\int_{\theta_1}^{\theta_2}\frac{\exp\left(\frac{vV}{kT}\right)}{\sin\theta}d\theta = \exp\left(\frac{vV}{kT}\right)W\bigg|_{\theta_1}^{\theta_2} \tag{14.26}$$

We expand V about the maximum,

$$V = V_m - \frac{1}{2}k_m(\theta - \theta_m)^2 \tag{14.27}$$

The approximation of a high-potential barrier allows us to replace $\sin\theta$ by $\sin\theta_m$, and let the bounds go to infinity. This gives us,

$$\frac{-Iv}{kTh'2\pi}\frac{\exp\left(\frac{vV_m}{kT}\right)}{\sin\theta_m}\sqrt{\frac{2\pi kT}{vk_m}} = \exp\left(\frac{vV}{kT}\right)W\bigg|_{\theta_1}^{\theta_2}$$
$$= W_2\exp\left(\frac{vV_2}{kT}\right) - W_1\exp\left(\frac{vV_1}{kT}\right) \tag{14.28}$$

Since the occupied phase space points are either at the North or South poles, W_1 and W_2 are proportional to n_1 and n_2. Hence, we can write a master equation of the form,

$$\dot{n}_1 = -I = -\dot{n}_2$$
$$= c_{21}\exp\left(\frac{-v(V_m - V_2)}{kT}\right)n_2 - c_{12}\exp\left(\frac{-v(V_m - V_1)}{kT}\right)n_1 \tag{14.29}$$

In this master equation, the coefficients of n_1 and n_2 are the jump rates for going from 1 to 2, and vice versa. W. F. Brown [752] gives the analytical expressions for the pre-exponential factors c_{12} and c_{21}. In practice, these prefactors are estimated as typical magnetization frequencies.

Garanin has shown how this simple approach breaks down when the potential wells are not very deep. [745]

14.5 FURTHER READINGS

Magnetic nanostructures

The magnetic relaxation is often observed to be proportional to the logarithm of the time elapsed since the last change in applied field. This is likely to happen when monitoring the relaxation of an ensemble of magnetic particles with a broad size distribution. It has been shown that a broad distribution of relaxation times leads to a log-of-time behavior. Intermediate distribution spreads are common and may lead to misinterpretation in terms of magnetic tunneling.[754]

The interpretation of magnetic relaxation becomes more challenging when the magnetization of a single nanoparticle is measured as a function of time at a set temperature. One possible measurement method is to rely on the effect called ***anisotropic magnetoresistance*** (AMR). AMR refers to the dependence of the electrical resistance of a magnetic conductor on the current flow direction with respect to the magnetization orientation. This effect can be used to monitor the evolution of magnetization in a nanowire.[755] Alternatively, a magnetic force scanning microscope can be applied when processes are slow. In the case of thin films, the anomalous Hall effect (11.1) can be used to characterize the magnetization orientation as a function of time.

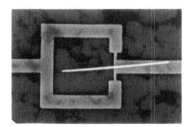

Figure 14.1   Nickel nanowire on a micro-SQUID, about 2 micrometers on a side.[747]

Also, a superconducting quantum interference device with micron size dimensions (micro-SQUID) can detect the magnetization of single isolated nanostructures. A single exponential behavior was observed in the case of a magnetic nanowire.[747, 756] Data collected by repeatedly observing the behavior of a single nanostructure, such as the one shown in Fig. 14.1, can challenge simple models of magnetic relaxation.[754] In spin valves, there are typically two layers of nanometer-sized dimensions that can be coupled. Typically, one layer has a magnetization that can jump depending on what field or current is applied. Back-and-forth jumps between the two equilibrium states have also been observed when the magnetic layers are maintained out of equilibrium by a spin-polarized current driven through the spin valve.[63]

The search for ***macroscopic quantum tunneling*** in magnetic objects has led to ever smaller magnetic particles at ever lower temperatures.[757–759] The ultimate size was reached when considering organometallic assemblies called ***molecular magnets***.[760–763] These magnets are so small that a macrospin approximation can account for the observed dependence of tunneling rate on magnetic field.[764] This research received the Agilent Prize of the European Physical Society. The smallest molecular antiferromagnet contains just two antiparallel localized moments. This configuration was achieved using two high-spin state iron ions located in an organic complex.[765] The antiferromagnetic coupling, which could be superexchange coupling (§2.4), dominates the magnetic susceptibility below 50 K.

Disordered materials

Ferrofluids are suspensions of sub-micron magnetic particles in oil. The particles can be so small that they are superparamagnetic. Thus, the magnetic relaxation behavior in ferrofluids may include a reorientation of the particles in the fluid as well as a reorientation of the magnetization in each particle.[766] A transformation from superparamagnetic to ferromagnetic behavior was observed at an oil-water interface due to the jamming of a monolayer of particles located at this interface.[767] Relaxation of ferrofluids made of antiferromagnetic particles was also investigated.[768]

Materials characterized by a *random anisotropy* on a nanometer scale have a peculiar relaxation behavior.[769] Such materials can be constituted by condensing nanoparticles from a metal vapor, collecting them, and then compacting them into a solid.[770–772] Quasi-static magnetic properties were investigated experimentally in these materials.[773, 774]

Domain walls

The physics of *domain wall motion* is very rich, calling for a modern theoretical approach to tackle the complexity of effects like the creep phenomenon.[775] Given the natural randomness of magnetic materials, domain wall motion can give rise to jumps that are known to cause the *Barkhausen effect*.[776]

Domain walls where studied experimentally in rare earth magnets where the wall width was very small.[777] Random inhomogeneities can be imparted so as to slightly trap walls in regions where the composition departs from stoichiometry.[778] The idea of a quantum mechanical tunneling regime for domain wall motion,[779, 780] has motivated a plethora of experimental observations.[781]

Domain wall positions can be controlled with precision given the right material and structure. It is then possible to create *logic gates* based on domain wall motion.[782, 783] For information processing applications, high-speed domain wall motion is necessary. In normal transition metal ferromagnets, the so-called *Walker breakdown* occurs above a critical speed. Instead, in *chiral magnetic material*, the DMI interaction (§2.7) stabilizes the domain walls and high speed can be reached. Walls can be pushed by a current, owing to spin transfer torque effects.[784] In chiral magnetic materials, the domain wall speed has been found asymmetric regarding the direction of the current.[785, 786]

Domain walls are also found in antiferromagnets. The spin orbit torque is expected to be an efficient means of pushing walls in antiferromagnets.[787] A ferrimagnet has part of its magnetization in one direction, part in the opposite one. At some temperature, the resulting magnetization vanishes. In this state, walls can be moved fast using a current.[788] Likewise, ultra-fast magnetization switching can be achieved at the compensation point of magnetic heterostructures.[789] In multiferroic materials, switching can be induced by an electric field.[790]. We note in passing that laser pulses can trigger *ultra-fast spin dynamics*. Several mechanisms may be at play on a very short time scale.[791, 792]

Section V

SPIN RESONANCE

15 Magnetic Resonance

Felix Bloch (1905–1983)

F. Bloch attended courses given by P. Debye, P. Scherrer, H. Weyl, and E. Schroedinger, among others, while studying in Zurich. As a professor at Stanford University, in 1945, F. Bloch detected a nuclear free-induction decay and reported his finding in a remarkably short letter to the *Physical Review*.

Spintronics: A primer

J.-Ph. Ansermet (https://orcid.org/0000-0002-1307-4864)

Chapter 15. Magnetic Resonance

The rotating frame picture of spin precession clarifies the notion that a resonant radio frequency magnetic pulse imposes a rotation to precessing spins. Adding to the spin precession a longitudinal and a transverse relaxation yields the Bloch equations. Adiabatic fast passage helps to understand spin evolution when an electron crosses a domain wall. Transmission electron spin resonance relies on spin diffusion across a thin film. NMR spectroscopy is illustrated with applications to magnetism.

DOI: 10.1201/9781003370017-15

15.1 HISTORICAL INTRODUCTION

Thanks to the work of J. Larmor, since 1896, physicists have known that a magnetic dipole density undergoes a precession when it is subject to a magnetic field. Later, I.I. Rabi detected the precession of nuclei using a molecular beam method.[793]

At Kazan University in 1941, Y. Zavoisky succeeded in detecting the first nuclear magnetic resonance in condensed matter, namely, the resonance of protons in water.[794] In 1942, C. J. Gorter and J. F. Broer reported on their failure to detect nuclear magnetic resonance in solids.[795] They were faced with two possibilities : the relaxation time was too long, hence the energy absorption (see §12.2) was too weak to detect; or, alternatively, the resonance spectrum was broadened excessively due to the effect of the nuclear environment, giving rise to what is now known as an inhomogeneous broadening. In 1946, two letters to the Physical Review announced the detection of nuclear magnetic resonance. At Harvard, E. M. Purcell et al. used a continuous wave method, thus measuring the resonant absorption of protons in 850 cm^3 of paraffin.[796] At Stanford, F. Bloch et al. used crossed coils, one to tilt the spins away from the direction of the static magnetic field, the other to detect nuclear free induction in a variety of materials.[797]

Electron spin resonance was first detected by Y. Zavoisky in 1945 in paramagnetic salts using excitation in the MHz range.[798] The first commercial spectrometers became available in the 1950s, operating first in the GHz range and later in the tens of GHz range. In both cases, they were using continuous wave (CW) spectrometers. Time-resolved electron paramagnetic resonance (TREPR) proved to be an invaluable tool in the study of photochemistry.[799]

With the advent of superconducting magnets, electron spin resonance spectroscopy can now reach the range above 100 GHz,[800] while large scale NMR magnets allow high resolution proton resonance spectroscopy at 1 GHz and above.[801] In ferromagnetic materials, nuclear spins resonate at very high frequencies in the absence of an applied magnetic field, owing to the hyperfine interactions between the nuclei and correlated electrons of the ferromagnet. This technique is known as *ferromagnetic nuclear resonance* (FNR). An extensive compilation of FNR data was established by E.A. Turov and M.P. Petrov as early as 1972.[802] FNR has been used to characterize magnetic nanostructures, owing to the large signal enhancement provided by the coupling of nuclear spins to electrons.[803]

In this chapter, we examine the Bloch equation describing spin precession with longitudinal and transverse relaxation processes. We then address the resonance method known a fast adiabatic passage, which describes a spin dynamics regime that is relevant when electrons in a ferromagnet pass through, e.g., a domain wall. Transmission electron spin resonance is an electron spin resonance technique by which electrons travel through a metal film under test. Hence, this spectroscopic technique involves physical phenomena that are at the core of spintronics devices. Next, we show that, when the electron spin resonance is excited in a ferromagnetic metal capped with a non-magnetic metal with a strong spin-orbit effect, a current is induced from the ferromagnet to the normal metal (spin pumping). We end this chapter on magnetic resonance with a survey of some contributions of NMR to the investigation of magnetic materials.

15.2 ROTATING FRAME PICTURE

Let us consider one spin undergoing precession in a magnetic field, as expressed in (7.6). Here, we assume that the magnetic field is composed of a large static field $B_o\hat{z}$ and a field $\boldsymbol{B}_1(t)$ rotating near the precession frequency. This situation is typical of magnetic resonance experiments. Hence, we write,

$$\boldsymbol{B} = B_o \hat{\boldsymbol{z}} + \boldsymbol{B}_1(t)$$
$$\boldsymbol{B}_1(t) = B_1 \hat{\boldsymbol{x}} \cos(\omega t) + B_1 \hat{\boldsymbol{y}} \sin(\omega t) \tag{15.1}$$

We can write $\boldsymbol{B}_1(t)$ as (see Fig. 15.1),

$$\boldsymbol{B}_1(t) = B_1 \hat{\boldsymbol{u}} \qquad \hat{\boldsymbol{u}} = \hat{\boldsymbol{x}} \cos(\omega t) + \hat{\boldsymbol{y}} \sin(\omega t) \tag{15.2}$$

Here, $\hat{\boldsymbol{u}}$ is a unit vector rotating in the x, y plane at the angular velocity $\omega \hat{\boldsymbol{z}} \equiv \boldsymbol{\omega}$. The field $\boldsymbol{B}_1(t)$ is said to be circularly polarized. It is advantageous to work with projections of the vectors on the unit vectors $(\hat{\boldsymbol{u}}, \hat{\boldsymbol{v}}, \hat{\boldsymbol{z}})$, whereas the unit vectors $(\hat{\boldsymbol{x}}, \hat{\boldsymbol{y}}, \hat{\boldsymbol{z}})$ are fixed in the laboratory frame (Fig. 15.1). When the spin precession is represented by its projections on $(\hat{\boldsymbol{u}}, \hat{\boldsymbol{v}}, \hat{\boldsymbol{z}})$, we commonly speak of the **rotating frame** picture, though $(O, \hat{\boldsymbol{u}}, \hat{\boldsymbol{v}}, \hat{\boldsymbol{z}})$ is not treated as a reference frame, in the sense of classical mechanics.[557]

In considering time derivatives of the projections of any vector $\boldsymbol{R}$ on the rotating unit vector basis, we have the following derivatives:

$$\frac{d}{dt} (R_x \hat{\boldsymbol{u}} + R_y \hat{\boldsymbol{v}} + R_z \hat{\boldsymbol{z}}) = \frac{dR_x}{dt} \hat{\boldsymbol{u}} + \frac{dR_y}{dt} \hat{\boldsymbol{v}} + \frac{dR_z}{dt} \hat{\boldsymbol{z}} + R_x \frac{d\hat{\boldsymbol{u}}}{dt} + R_y \frac{d\hat{\boldsymbol{v}}}{dt} + R_z \frac{d\hat{\boldsymbol{z}}}{dt} \tag{15.3}$$

Since the unit vectors $(\hat{\boldsymbol{u}}, \hat{\boldsymbol{v}}, \hat{\boldsymbol{z}})$ are rotating with respect to the reference frame at the angular velocity $\boldsymbol{\omega} = \omega \hat{\boldsymbol{z}}$, we have,

$$\frac{d\hat{\boldsymbol{u}}}{dt} = \boldsymbol{\omega} \times \boldsymbol{u} \qquad \frac{d\hat{\boldsymbol{v}}}{dt} = \boldsymbol{\omega} \times \boldsymbol{v} \qquad \frac{d\hat{\boldsymbol{z}}}{dt} = 0 \tag{15.4}$$

So, the derivatives in (15.3) can be summarized with the notation,

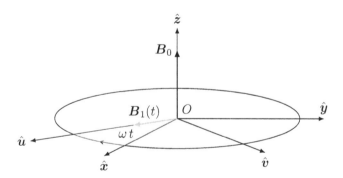

Figure 15.1 Reference frame $(O, \hat{\boldsymbol{x}}, \hat{\boldsymbol{y}}, \hat{\boldsymbol{z}})$ and rotating frame $(O, \hat{\boldsymbol{u}}, \hat{\boldsymbol{v}}, \hat{\boldsymbol{z}})$. Here, γ is assumed positive

$$\frac{d\boldsymbol{R}}{dt} = \frac{d'\boldsymbol{R}}{dt} + \boldsymbol{\omega} \times \boldsymbol{R} \tag{15.5}$$

Therefore, the evolution of the magnetic moment (7.8) in the rotating frame picture is given by:

$$\frac{d'\boldsymbol{m}}{dt} + \omega \hat{\boldsymbol{z}} \times \boldsymbol{m} = -\gamma (B_o \hat{\boldsymbol{z}} + B_1 \hat{\boldsymbol{u}}) \times \boldsymbol{m} \tag{15.6}$$

or

$$\frac{d'\boldsymbol{m}}{dt} = -((\gamma B_o + \omega) \hat{\boldsymbol{z}} + \gamma B_1 \hat{\boldsymbol{u}}) \times \boldsymbol{m} \tag{15.7}$$

In this rotating frame picture, m precesses around B_{eff} (Fig. 15.2), where the ***effective field*** is defined by,

$$B_{\text{eff}} = \left(B_o + \frac{\omega}{\gamma} \right) \hat{z} + B_1 \hat{u} \qquad (15.8)$$

If we take $\omega = -\gamma B_o \hat{z}$, then $B_{\text{eff}} = B_1 \hat{u}$. When a very large magnetic field is applied, the spin dynamics is dominated by the Zeeman coupling of the spins to this field. Thus, an advantage of the rotating frame picture is that it removes this very fast precession from the representation of the spin dynamics.

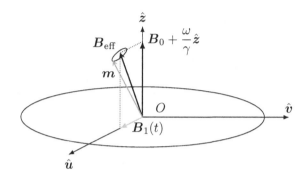

Figure 15.2   Precession cone corresponding to the precession of m about the effective field B_{eff}.

The gyromagnetic factor of protons (^{1}H) is about $\gamma \approx 2\pi \cdot 4.2$ MHz/kG. In a field of 7 T, $\omega \approx 2\pi \cdot 300$ MHz. In typical magnets, the resonance frequency is in the radio frequency range (5–600 MHz) for nuclei. Some groups have ventured into conducting proton NMR at a frequency of 1 GHz. The gyromagnetic factor of electrons is about $\gamma \approx 2\pi \cdot 2.8$ GHz/kG (g=2). Very often, electron resonance experiments are carried out at a frequency in the X-band (8–12 GHz) or the K-band (28–27 GHz). Experiments above 100 GHz may require such devices as gyrotrons for the source. Table 15.1 summarizes the proton and electron resonance frequencies at various applied magnetic fields.

Table 15.1

Magnetic induction field (in Tesla), proton nuclear magnetic resonance frequency (in MHz) and electron spin resonance (in GHz)

Field	Proton	Electron
Tesla	MHz	GHz
0.3	14	9
3.4	143	95
9.5	400	267
19.1	800	533

15.3   THE BLOCH EQUATIONS

When an ensemble of polarized spins is exposed to a magnetic field, the magnetization resulting from these spins precesses around the applied magnetic field. Simple experiments such as a free induction decay – first observed by Bloch – show that, with time, the magnetization does not maintain a constant angle with respect to the applied magnetic field. Instead, the transverse component of the magnetization decays to zero. Furthermore, the magnetization relaxes toward the equilibrium value of magnetization in the applied magnetic field. Alternatively, we know that initially unpolarized spins will over time build up a non zero spin polarization, or magnetization, along the applied field.

Let us consider a model for the evolution of a spin ensemble that takes into account the spin precession, the transverse decay and the relaxation to the equilibrium the spin polarization. This phenomenological model was first established by F. Bloch.

In a static field, the magnetization grows along the static applied field. This process is expressed by a single exponential relaxation, i.e., the evolution equation for this process is given by,

$$\frac{dM_z}{dt} = \frac{M_o - M_z}{T_1} \tag{15.9}$$

The relaxation time T_1 is the **spin-lattice relaxation time.**

Let us now turn to the transverse relaxation process. To account for this process, we write,

$$\frac{dM_x}{dt} = -\frac{M_x}{T_2} \qquad \frac{dM_y}{dt} = -\frac{M_y}{T_2} \tag{15.10}$$

This second relaxation time is referred to as the **transverse relaxation time**, when the decay is due to intrinsic mechanisms, not a distribution in the value of the applied magnetic field. It is also called the spin-spin relaxation time, as it depends on the interactions among the spins. The Bloch-Wangness-Redfield theory, which was presented in §12.6, offers a formal framework for describing these relaxation processes.

By taking into account the precession around the magnetic field and both relaxation processes, we obtain the **Bloch equations,**

$$\frac{d\boldsymbol{M}}{dt} = \gamma \boldsymbol{M} \times \boldsymbol{B} - \frac{M_x \hat{\boldsymbol{x}} + M_y \hat{\boldsymbol{y}}}{T_2} - \frac{M_z - M_o}{T_1}\hat{\boldsymbol{z}} \tag{15.11}$$

Let us suppose that a static field B_0 is applied along the z axis, and that there is also an applied rotating field $\boldsymbol{B}_1(t) = B_1\left(\cos(\omega t)\hat{\boldsymbol{x}} + \sin(\omega t)\hat{\boldsymbol{y}}\right)$. We use the notation $B_0 = -\omega_0/\gamma$ and $\omega_1 = -\gamma B_1$. In the frame rotating with the $\boldsymbol{B}_1$ field, the Bloch equations become:

$$\begin{aligned}
\frac{d' M_u}{dt} &= -\frac{M_u}{T_2} + \Delta\omega M_v \\
\frac{d' M_v}{dt} &= -\frac{M_v}{T_2} - \Delta\omega M_u - \omega_1 M_z \\
\frac{d' M_z}{dt} &= \omega_1 M_y - \frac{M_z - M_o}{T_1}
\end{aligned} \tag{15.12}$$

where $\Delta\omega = \omega - \omega_0$. Let us calculate the stationary solution, i.e., the response when $d'M_{(u,v,z)}/dt = 0$. Simple algebraic manipulations yield,

$$\begin{aligned}
M_u &= M_0 \frac{\Delta\omega\,\gamma T_2^2}{1 + \Delta\omega^2 T_2^2 + \omega_1{}^2 T_2 T_1} B_1 \\
M_v &= M_0 \frac{\gamma T_2}{1 + \Delta\omega^2 T_2^2 + \omega_1{}^2 T_2 T_1} B_1 \\
M_z &= M_0 \frac{1 + \Delta\omega\,{}^2 T_2^2}{1 + \Delta\omega^2 T_2^2 + \omega_1{}^2 T_2 T_1}
\end{aligned} \tag{15.13}$$

In view of these equations, we can understand a concern that arose when scientists initially sought to detect nuclear magnetic resonance. We see in the (15.13) for M_y' that the denominator can become very large if ω_1 is large enough. In electron spin resonance, T_2 is roughly equal to T_1, so the condition is $\omega_1 \gg T_1^{-1}$. To have a large B_1 value requires the application of a lot of microwave power. This condition is called a **saturation** of the resonance.

If B_1 is small or, alternatively, if T_1 is short, the terms arising from spin-lattice relaxation can be neglected. In that case, we obtain from (15.13),

$$M_u = M_0 \frac{\Delta\omega\,\gamma T_2^2}{1 + \Delta\omega^2 T_2^2} B_1$$
$$M_v = M_0 \frac{\gamma T_2}{1 + \Delta\omega^2 T_2^2} B_1 \tag{15.14}$$
$$M_z = M_0$$

When a resonance is characterized by the response given by (15.14), the corresponding line shape is called a **lorentzian**.

Let us go back to the laboratory frame with the transformation. For the magnetization, we have,

$$M_x = M_u \cos(\omega t) - M_v \sin(\omega t)$$
$$M_y = M_u \sin(\omega t) + M_v \cos(\omega t) \tag{15.15}$$

For the field $B_1(t)$, we simply have,

$$B_{1x} = B_1 \cos(\omega t) \qquad B_{2y} = B_1 \sin(\omega t) \tag{15.16}$$

We can define a susceptibility for the rotating field according to:

$$M_x = M_u \frac{e^{i\omega t} + e^{-i\omega t}}{2} - M_v \frac{e^{i\omega t} - e^{-i\omega t}}{2i}$$
$$= \frac{1}{2}(M_z + iM_v)e^{i\omega t} + \frac{1}{2}(M_u - iM_v)e^{-i\omega t} \tag{15.17}$$
$$= \chi_R(-\omega)B_1 e^{i\omega t} + \chi_R(\omega)B_1 e^{-i\omega t}$$

The susceptibility in the rotating frame is given by,

$$\chi_R(\omega) = \chi_R'(\omega) - i\chi_R''(\omega) = \frac{\omega_o \Delta\omega\, T_2^2}{1 + \Delta\omega^2 T_2^2 + \omega_1^2 T_2 T_1} \frac{\chi_o}{2} - i\frac{\omega_o T_2}{1 + \Delta\omega^2 T_2^2 + \omega_1^2 T_2 T_1} \frac{\chi_o}{2} \tag{15.18}$$

where $\chi_0 = M_0/B_0$. The susceptibility for a linearly polarized field is related to this susceptibility by:

$$\chi' = \chi_R'(\omega) + \chi_R'(-\omega)$$
$$\chi'' = \chi_R''(\omega) - \chi_R''(-\omega) \tag{15.19}$$

A sketch of the real and imaginary part of the susceptibility, χ' and χ'', is shown in Fig. 12.8.

15.4   ADIABATIC FAST PASSAGE

Adiabatic fast passage is a notion from the early days of magnetic resonance that also appears in the context of magnetoresistance, notably, when a spin-polarized current traverses a magnetic domain wall.[804] The same notion also explains one mechanism of current-driven spin torque.[520, 805]

The idea is that, when a spin precesses rapidly around a magnetic field that rotates slowly, the precessing spin follows (so to speak) the magnetic field. It is an adiabatic evolution in the sense explained in §11.2.1 in the context of an electronic band structure. Consider a magnetic induction $\boldsymbol{B}(t)$ undergoing a rotation in the xyz (Fig. 7.1). To calculate the time evolution of the $\boldsymbol{M}$, we apply the evolution equation (15.11) without the relaxation terms (hence the "fast" in the name of this process). Thus, we have a precession at an angular velocity $\boldsymbol{\Omega}$ given by (7.7). Let us project (15.11) in the frame $x'y'z'$ where z' tracks $\boldsymbol{B}(t)$ and $v' \equiv v$. The unit vectors $\hat{\boldsymbol{x}}'$ and $\hat{\boldsymbol{z}}'$ evolve according to,

$$\frac{d\hat{\boldsymbol{x}}'}{dt} = \dot{\boldsymbol{\phi}} \times \hat{\boldsymbol{x}}' \qquad \text{and} \qquad \frac{d\hat{\boldsymbol{z}}'}{dt} = \dot{\boldsymbol{\phi}} \times \hat{\boldsymbol{z}}' \tag{15.20}$$

where $\dot{\boldsymbol{\phi}} = \dot{\phi}\hat{\boldsymbol{v}}$. Thus, we have,

$$\frac{d\boldsymbol{M}}{dt} = \dot{M}_{x'}\hat{\boldsymbol{x}}' + \dot{M}_y\hat{\boldsymbol{y}} + \dot{M}_{z'}\hat{\boldsymbol{z}}' + \dot{\boldsymbol{\phi}} \times \boldsymbol{M} = \boldsymbol{\Omega} \times \boldsymbol{M} \tag{15.21}$$

which we can write as,

$$\begin{pmatrix} \dot{M}_{x'} \\ \dot{M}_v \\ \dot{M}_{z'} \end{pmatrix} = \left(\boldsymbol{\Omega} - \dot{\boldsymbol{\phi}}\right) \times \boldsymbol{M} = \begin{pmatrix} 0 & -\Omega & -\dot{\phi} \\ \Omega & 0 & 0 \\ \dot{\phi} & 0 & 0 \end{pmatrix} \begin{pmatrix} M_{x'} \\ M_v \\ M_{z'} \end{pmatrix} \tag{15.22}$$

Thus, we find that in the frame $x'y'z'$, the magnetization $\boldsymbol{M}$ precesses around the vector $\boldsymbol{\Omega}$, approximately, provided $\Omega \gg \dot{\phi}$. If $\boldsymbol{M}$ was nearly parallel to $\boldsymbol{\Omega}$ at the start of the process, the angle between $\boldsymbol{M}$ and $\boldsymbol{\Omega}$ remains small.

In a magnetic resonance experiment, it is not possible to rotate $\boldsymbol{B}(t)$. However, it is possible to sweep the $\boldsymbol{B}_1$ field (Fig. 15.2). Then, it is the effective field $\boldsymbol{B}_{\text{eff}}(t)$ in the rotating frame that changes direction with time. Hardy et al. have considered two ways to sweep the frequency of the $\boldsymbol{B}_1$ field.[806] In the conventional fast adiabatic experiment, the frequency is swept at a constant rate, and the angle θ_B between the effective field and the z axis has the time evolution showing in Fig. 15.3 (left).[1] But the frequency sweep can be done so that θ_B varies linearly in time. This second method allows for $\boldsymbol{M}$ to track $\boldsymbol{B}_{\text{eff}}$ much better (Fig. 15.3, right).

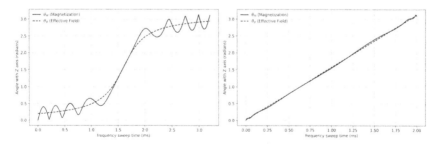

Figure 15.3 Polar angles of the magnetization (solid line), θ_M, and of the effective field (dotted line), θ_B, during adiabatic fast passage experiments. Left: the excitation frequency is swept at a constant rate (left). Right: the frequency sweep is designed so that θ_B changes linearly in time (right).

---

[1]Numerical simulation based on [806], by Artemiy Burov, Fachhochschule Nordwestschweiz

15.5 TRANSMISSION ELECTRON SPIN RESONANCE

There is a way of measuring electron spin resonance (ESR) that relies on spin diffusion in normal metals. In this method, electron precession is excited on one side of a thin film and the precessing spins diffuse to the other face of the film, where their radiation is detected. Hence, the term "transmission ESR" (TESR) is used to designate this detection method. Clearly, TESR has a close conceptual connection to spin injection and spin diffusion in normal metals.

As TESR relies on spin diffusion, it is an ancestor to all other spintronics experiments.[807] The film mounted across the cavity must be thin enough that when the electrons diffuse across the film, their precessional behavior does not decay completely. This means that the thickness must be of the same order of magnitude as the spin diffusion length. If the film is too thin, the direct electromagnetic wave transmission from the emitter (E) to the receiver (R) is too large, and the spin signal under investigation is buried in the large electromagnetic signal transmitted without the help of spin diffusion (Fig. 15.4).

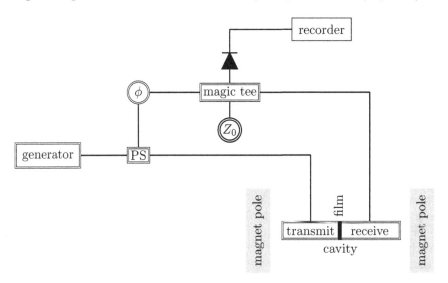

Figure 15.4   Principle of TESR. A microwave generator produces a signal that is split by a power splitter (PS) into two signals. A reference signal, of which the phase can be set (ϕ), is fed to the "magic tee." The other signal is fed into the "transmit" side of the cavity. The signal at the receiver side of the cavity is fed to the "magic tee," of which the output is rectified and recorded.[808]

TESR of a paramagnetic metal implanted with magnetic ions on one side resulted in an enhancement of the signal at a low temperature (1 K) by as much as 1000, indicating some form of dynamic polarization.[809] In 1980, Janossy considered a normal metal coated on both sides with a ferromagnet like permalloy, thus making a nanostructure akin to a spin valve.[808] This mechanism of spin injection is now called *spin pumping* (described in §17.6).

15.6 SPYING ON MAGNETISM WITH MAGNETIC RESONANCE

The following examples illustrate concepts of magnetism presented in part II, using observations made by NMR. The intra-atomic exchange gives rise to Hund's first rule, for example. This can be confirmed by measuring the strength of the hyperfine coupling determined by

NMR. Intra-atomic exchange also gives rise to a spin-dependent deformation of the electron cloud of inner shells when outer shells are spin-polarized. This gives rise to what is known in NMR as core polarization. The superexchange is the mechanism that predicts the correct hyperfine field at the location of the F nuclei in the antiferromagnet MnF_2. Indirect coupling were first observed by NMR in molecules. The oscillatory response of electron spins to a local magnetic dipole was also detected by NMR.

15.6.1   HUND'S RULE OBSERVED BY NMR

We want to use NMR to measure the magnetic moment at the europium (Eu) sites of ferromagnetic europium oxide (EuO).[810] In a first approximation, the oxygen strips Europium of two electrons. The valence electrons are then all on a 4f level. We saw in Chapter 3 that it is reasonable to assume that the 7 electrons on the 4f levels are "shielded" from the rest of the solid. Thus, it is reasonable to attempt to treat the Eu atomic sites as if they were isolated ions. There are 7 electrons on the 4f level. Hund's first rule implies all the f states of this level are occupied. Using Hund's second rule, we find that the total angular momentum vanishes.[72] Hence, according to this model, the magnetic moment is expected to be 7 Bohr magnetons.

The electrons create a hyperfine field H_{hf} at the nucleus position. We need not concern ourselves with the mechanism of this effect. It is sufficient to know that, at very low temperatures, the nuclei resonate (with zero applied field) at about 140 MHz. This tells us that, in the fully magnetized state, the hyperfine field is given by the relation 140 MHz $= \bar{\gamma} H_{hf}$, where $\bar{\gamma} = \gamma/(2\pi)$ is the gyromagnetic ratio of the Eu nucleus.

We want to look for this resonance in the paramagnetic state in the presence of an applied field. We consider that, in the paramagnetic state, the nuclear moment dynamics is so slow compared to the magnetization that the nuclei "see" only the average Eu moment at the temperature of the measurement. We use the Curie-Weiss law for the susceptibility to estimate this magnetic moment. If the applied field is H_a and the susceptibility is χ, then at a temperature T in the paramagnetic state, the hyperfine field is reduced by the ratio:

$$\frac{\chi H_a}{M_s} \tag{15.23}$$

where M_s is the saturation magnetization. Hence, we expect that the Eu nuclei resonate at the frequency given by:

$$\nu = \frac{\gamma \mu_0}{2\pi} \left(H_a + H_{hf} \frac{\chi H_a}{M_s} \right) \tag{15.24}$$

Now we show that the frequency depends on the the magnetic moment per Eu site. Indeed, the Curie-Weiss law states,[2]

$$\chi = \frac{N}{V} \frac{(g\mu_B)^2 S(S+1)}{3k_B (T - T_c)} \tag{15.25}$$

where N/V is the volume concentration of magnetic moments with $S = 7/2$. The saturation magnetization is given by:

$$M_s = \frac{N}{V} 7g\mu_B \tag{15.26}$$

Calculating N/V from the volume of the unit cell of EuO (15.26), yields a fairly accurate value for the NMR frequency in the paramagnetic state.[811]

---

[2]See, (3.17) and use $(T - T_C)$ instead of T.

15.6.2   INTRA-ATOMIC SPIN POLARIZATION: CORE POLARIZATION

The following example illustrates the exchange mechanism in atoms. We consider a 3d ion and assume that there is one valence electron in a 3d state (Fig. 15.5). We consider the effect of the exchange interaction between this 3d state and the core s states. NMR shows that, at the location of the nucleus, the probability of an electron being present becomes spin-dependent. This means that locally (on the scale of nuclear sizes), there is no longer perfect pairing of the spins. This phenomenon is referred to as **core polarization**: the core states are "polarized" (in spin) by the outer states, which themselves are assumed to be spin-polarized or magnetic because of mechanisms we need not consider here.

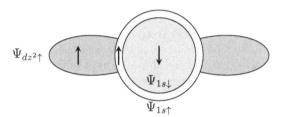

Figure 15.5   Exchange interaction between one d-state and two inner shell s-states may cause a slight change in the extent of these s-state wave functions. As a consequence, a hyperfine coupling of these core states is induced by the outer shell electrons.

Getting more than an estimate is a challenging computation because core polarization results from the nearly perfect cancellation of two large quantities, hence each quantity must be calculated very accurately. Nonetheless, NMR is a very sensitive probe of this effect for the following reason. The hyperfine interaction $\mathcal{H}_{\mathrm{hf}} = -\gamma_I \hbar \hat{\boldsymbol{I}} \cdot \boldsymbol{B}_{\mathrm{hf}}$ between a nucleus of moment $\gamma_I \hbar \hat{\boldsymbol{I}}$ and its electronic environment is characterized by the hyperfine field:

$$\boldsymbol{B}_{\mathrm{hf}} = \frac{\mu_0}{4\pi}(-g\mu_B)\left[\frac{8\pi}{3}\hat{\boldsymbol{S}}\delta(\boldsymbol{r}) - \left(\frac{\hat{\boldsymbol{S}}}{r^3} - \frac{3\boldsymbol{r}(\hat{\boldsymbol{S}}\cdot\boldsymbol{r})}{r^5}\right) + \frac{\hat{\boldsymbol{\ell}}}{r^3}\right] \tag{15.27}$$

The first term is called the contact term. It has a contribution from only s-states but, when it does, it is generally the dominant contribution to the hyperfine field. If the core s-states were fully spin-polarized, the field given by the contact term of the hyperfine field would be:

$$\boldsymbol{B}_s = -\frac{\mu_0}{4\pi}\frac{8\pi}{3}g\mu_B|\Phi(0)|^2\hat{\boldsymbol{S}} \tag{15.28}$$

Numerical estimate:

For an s-state wave function, a typical value for its radial extent would be 0.3 angstrom. This would imply a field of 240 T! In fact, the s-states are not fully spin-polarized but instead, both spins cancel each other out to a great extent. As a consequence, a slight deviation from a perfect spin pairing of the core states is enough to produce a large effect on the NMR frequency.[3]

In the 1950s, it became clear that a main contribution to the shift of the NMR frequency in paramagnetic as well as ferromagnetic compounds was the distortion of the electron cloud

---

[3]For convenience, we recall that $\mu_B = 9.2 \; 10^{-24}$ J/T, $\mu_0 = 4\pi \; 10^{-7}$ Tm/A $= 4\pi 10^{-7}$ T^2 m^3/J, noting that J/T = Am^2.

of the core s-state with a spin parallel to the d-state spin, compared to that of the s-state with the spin anti-parallel to the d-state spin.[4] These authors realized that it was necessary to assume different radial wave functions for each spin orientation of the core s-states in order to account for the core polarization effect.

As a more concrete example, we consider the hyperfine field of a hypothetical atom with the configuration $(2s^1, 1s^2)$. We will assume that the $2s$ state is fully spin polarized. We depart from the d-state of Fig. 15.5 to keep the calculations as simple as possible. Hence, we take:

$$\psi_{2s}(\boldsymbol{r}) = \frac{1}{\sqrt{8\pi a_0^3}} \left(1 - \frac{r}{2a_0} \right) e^{-r/2a_0} |u\rangle \tag{15.29}$$

As we saw in the calculation of the matrix $\mathcal{H}$, there will not be an exchange term between this $2s|u>$ state and the $1s|d>$ state, where $|u>$ and $|d>$ stand for up-spin and down-spin wave functions. So the antiparallel $1s$ state will simply be that of the hydrogen atom:

$$\psi_{1sa} = \frac{1}{\sqrt{\pi a_0^3}} e^{-r/a_o} |d\rangle \tag{15.30}$$

The $1s$ state with spin parallel to that of the $2s$ state is assumed to have a distortion. Thus, we write:

$$\psi_{1sp}(\boldsymbol{r}) = \frac{\alpha^{3/2}}{\sqrt{\pi a_0^3}} e^{-r\alpha/a_0} |u\rangle \tag{15.31}$$

Of course, we ensure the normalization of the state. The spatial extension of this wave function is given by a_0/α. We will see that exchange tends to bring the core state with parallel spin closer to the outer spin-polarized state. So we expect α to be somewhat smaller than 1. In order to estimate this distortion, we apply a variational approach. We build a 3-electron wave function Ψ, estimate the ground state energy of the system as $E_0 = <\Psi|\mathcal{H}|\Psi>$ and minimize with respect to α. If $\mathcal{H}_0$ is the Hamiltonian of the hydrogen atom for one electron, we have:

$$E_0 = \langle\Psi|\, \mathcal{H}_0(1) + \mathcal{H}_0(2) + \mathcal{H}_0(3) + \frac{e^2}{4\pi\varepsilon_0} \left[\frac{1}{|\boldsymbol{r}_1 - \boldsymbol{r}_2|} + \frac{1}{|\boldsymbol{r}_1 - \boldsymbol{r}_3|} + \frac{1}{|\boldsymbol{r}_3 - \boldsymbol{r}_2|} \right] |\Psi\rangle \tag{15.32}$$

Now, we build our wave function as a 3-electron-state Slater determinant:

$$|\Psi\rangle = \frac{1}{\sqrt{3!}} \begin{vmatrix} \psi_{2s}(1)u(1) & \psi_{2s}(2)u(2) & \psi_{2s}(3)u(3) \\ \psi_{1sa}(1)d(1) & \psi_{1sa}(2)d(2) & \psi_{1sa}(3)d(3) \\ \psi_{1sp}(1)u(1) & \psi_{1sp}(2)u(2) & \psi_{1sp}(3)u(3) \end{vmatrix} \tag{15.33}$$

Notice that this is not like the usual procedure for deriving Hartree-Fock equations,[72] because here, the spatial wave functions of a given configuration are assumed different. However, we see that the trial wave function Ψ has the following structure:

$$\begin{aligned} \Psi &= \varphi(1,2,3) - \varphi(2,1,3) + \varphi(3,1,2) \\ &= \varphi(1,2,3) + \varphi(2,3,1) + \varphi(3,1,2) \end{aligned} \tag{15.34}$$

If $\psi_{1sa} = \psi_{1sp}$, then the trial state has the singlet structure for the core states,

$$\varphi(1,2,3) = \psi_{2s}(1)u(1)\psi_{1s}(2)\psi_{1s}(3)\left[d(2)u(3) - d(3)u(2)\right] \tag{15.35}$$

---

[4]See, V. Heine et al.,[812] and A. J. Freeman et al.[813]

We expect α to be close to 1 and, hence, our trial state is close to a reasonable configuration. In calculating the estimate of the ground state energy $E_0 = <\Psi|\mathcal{H}|\Psi>$, there are two terms of importance: the Coulomb and exchange integrals between the $2s$ and $1sp$ states. So, we need to minimize $K_{2s1sp} - J_{2s1sp}$ (see (2.8) in §2.2) with respect to α where,

$$K_{2s1sp} = \int \int d^3 r_1 d^3 r_2\, \psi_{2s}^*(1)\psi_{1sp}^*(2)\frac{e^2}{4\pi\varepsilon_0\,|r_1 - r_2|}\psi_{2s}(1)\psi_{1sp}(2)$$

$$J_{2s1sp} = \int \int d^3 r_1 d^3 r_2\, \psi_{2s}^*(1)\psi_{1sp}^*(2)\frac{e^2}{4\pi\varepsilon_0\,|r_1 - r_2|}\psi_{2s}(2)\psi_{1sp}(1)$$

(15.36)

The problem was defined for s-states to keep the calculation simple. Both of these integrals can be thought of as integrals of the form:

$$\int d^3 r_1 f(r_1) \int d^3 r_2 \frac{g(r_2)}{|r_1 - r_2|}$$

(15.37)

in which the second integral is the electrostatic potential energy at r_1 of a spherical distribution of charges given by $g(r_2)$. Since the distribution is spherical, this potential depends on r_1 only. The electric field associated with this potential would be radial and of constant magnitude on any sphere centered at the origin. Hence, we have,

$$4\pi R^2 \left(\frac{-\partial\Phi}{\partial R}\right) = \int_0^R r^2 g(r) 4\pi dr$$

(15.38)

This gives us,

$$\int d^3 r_2 \frac{g(r_2)}{|r_1 - r_2|} = \int_{\|r_1\|}^{\infty} \frac{dR}{R^2} \int_0^R r^2 g(r) dr$$

(15.39)

The integrals are greatly simplified if we neglect the polynomial part of the $2s$-state. Given that the purpose of the present development is to demonstrate the exchange effect on the orbital expansion, we can reasonably assume that this simplification keeps the essential features of the calculation, for the following reason. Either the term in r in the polynomial is small, or when r is not small, the contribution of that term to the integral is small thanks to the exponential factor. This approximation yields:

$$\int d^3 r_1 \int d^3 r_2 \frac{e^{-ar_1}e^{-br_2}}{|r_1 - r_2|} = 8\pi \frac{(a + b)^2 + ab}{(ab)^2(a + b)^3}$$

(15.40)

Integrals involving polynomials of the radial coordinate can be found from this formula by deriving (15.40) with respect to the parameters a and b. Thus, we find,

$$K_{2s1sp} - J_{2s1sp} = \frac{e^2}{4\pi\varepsilon_0 a_0} p(\alpha)$$

(15.41)

where,

$$p(\alpha) = \frac{\alpha^3}{\pi}\left(\frac{(2\alpha + 1)^2 + 2\alpha}{4\alpha^2(2\alpha + 1)^3} - \frac{20}{(1 + 2\alpha)^5}\right)$$

(15.42)

presents a minimum near $\alpha = 0.5$ (Fig. 15.6). Thus, we find a tendency for the core $1s$-state with spin parallel to that of the $2s$-state to expand out toward the $2s$-state. This imbalance between the two 1s states causes a shift of the NMR frequency; this mechanism is known as core polarization. Via the hyperfine effect, the nuclei experience the difference in the core state spin densities.

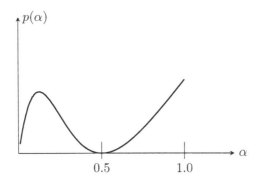

Figure 15.6   Plot of the function (15.42), presenting a minimum at $\alpha = 0.5$.

15.6.3   SUPEREXCHANGE PROBED BY NMR

The Heisenberg Hamiltonian $\mathcal{H} = -J\sum_i \boldsymbol{S}_i \cdot \boldsymbol{S}_j$ was soon considered quite successful in accounting for many properties of magnetic systems. Consequently, it was a surprise to find magnetic interactions among magnetic ions that are clearly separated from one another by diamagnetic ions. Furthermore, the solids were insulators.[157] The problem had not yet been elucidated in the 1950s, when P. W. Anderson made his major contributions in this field. He coined the coupling "superexchange." In the 1950s, soon after the discovery of NMR, people started investigating by NMR the paramagnetic state of compounds such as MnF_2, which are antiferromagnetic insulators. R. G. Shulman undertook the study of the ^{19}F-NMR.[814] For a long time, he could not find the resonance. The reason is quite revealing of the mechanism at play in superexchange. Since he worked in the paramagnetic state, he could use the Curie-Weiss law to estimate the average moment per manganese (Mn) site and sum the dipole fields seen at one fluorine (F) site by the neighboring Mn sites. Hence, he expected the resonance to be shifted by an amount given by,

$$\Delta H \propto H_0 \frac{g^2 \mu_B^2 S(S+1)}{k(T - T_c)} \sum_i \frac{(1 - 3\cos^2\theta_j)}{r_j^3} \tag{15.43}$$

Figure 15.7 shows that the actual resonance was far away compared to the prediction (15.43).[815] Then, R. G. Shulman successfully analyzed his data and similar data obtained by V. Jaccarino [816,817] Shulman's model is based on the idea that the $3d$-Mn unpaired electron forms an anti-bonding orbital with the s and p orbitals of the fluorine atom. An analysis of the data in these terms shows that the unpaired electron has a probability of 0.5% to be on the F site. Such ideas on superexchange were used, much later, by Nobel Prize winner R. Hoffmann and his group.[161] This model was also applied in a search for ferromagnetism in organometallic compounds.[818] From the standpoint of the nucleus, an effective hyperfine field arises from this admixture of orbitals, which is called the **transfer hyperfine field**.

15.6.4   INDIRECT SPIN-SPIN COUPLING

The NMR spectrum of protons in ethanol (CH_3CH_2OH) presents three lines when the spectrum is acquired with a moderate resolution. These different line positions reflect differences in the extent to which the electrons can shield the applied magnetic field. This effect was briefly encountered in §3.3 when discussing the notion of orbital moment quenching. The

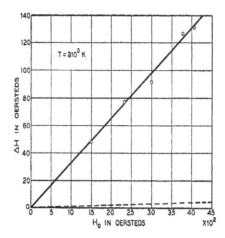

Figure 15.7   Prediction according to equation (15.43) (dotted line), actual data point and fit (continuous line).

values of these so-called chemical shifts depend on the immediate environment of the protons, i.e., the chemical shifts take on different values when the protons are in a CH_3, a CH_2 or an OH group. If the ethanol NMR spectrum is a high-resolution spectrometer, the spectrum appears to have a fine structure.[5] The line of the CH_2 group is split according to the different potential configurations of the protons in the neighboring CH_3 group. As there are 4 different spin configurations, there are 4 lines. Furthermore, these fine structure lines appear to be doubled. This splitting is due to the two states of the proton in the neighboring OH group. Likewise, the protons of the CH_3 group are split by the protons of the neighboring CH_2 group, with 3 possible configurations. This coupling between nuclei in a molecule, which is mediated by electrons, is called the *J-coupling* or the *scalar spin-spin coupling*.

In §5.3, we interpreted the RKKY coupling in terms of the oscillatory response of the bath to the presence of a local magnetic dipole moment. Solid state NMR was used to verify the spatial oscillation of electron spin polarization around a magnetic impurity. For example, if atoms of chromium (Cr) or iron (Fe) are dispersed in a non-magnetic metal such as copper (Cu), these impurities carry magnetic moments. The electron spins of the non-magnetic metal are coupled to the neighboring impurity moments. This response of the electronic spins oscillates as a function of the distance to the impurity. In turn, the electronic spins produce a hyperfine field at the location of the nuclei of the non-magnetic metal as shown in Fig. 15.8.[81] In this experiment, the nucleus can be thought of as a probe of the local electron spin polarization around the magnetic impurity.

15.7   FURTHER READINGS

Special resonance effects

The *Bloch-Siegert shift* of the resonance frequency observed by applying a rotating field of magnitude B_1 in a field B_0 is $B_1^2/(4B_0)^2$.[822] When this shift can be detected, it can be used in, for example, magnetic resonance imaging.[823] Evidently, this shift is significant if

---

[5]See, [819], figures 7–15 and 7–20.

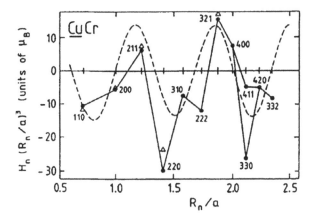

Figure 15.8 Local field value at copper sites near chromium impurities determined by NMR (data from Slichter et al. [820, 821] found in [81])

the static field B_0 is small. It has been studied in strongly driven resonances, i.e., when B_1 is large.[824] When an inductor and a capacitor forming a resonant circuit are put in the proximity of a superconducting qubit, the two resonances are strongly coupled (see §18.5) and the coupling strength can be thought of as a Bloch-Siegert shift.[825, 826]

The **spin-Dicke** effect is the spin equivalent of the Dicke effect. The Dicke effect is the narrowing of light emission by molecules that undergo several collisions in a distance shorter than the optical wavelength. This collision regime causes a narrowing of the spectral line. A narrowing can also take place when the spontaneous emission of nearby molecules couples them to one another. This can happen if the emitting molecules are spatially confined within a wavelength.[827] A spin-Dicke effect has been predicted when the triplet states of an organic light-emitting diode (OLED) is strongly driven with an AC field.[828] This prediction was confirmed.[829]

Single spin detection

Conventional electron spin resonance carried out at about 10 GHz (X-band) requires at least 10^{13} spins. Using inductive detection and Josephson parametric amplifiers, the detection of a few hundred spins was made possible.[830, 831] **Electrically detected magnetic resonance** (EDMR) is another highly sensitive detection method. EDMR is possible when the electrical conduction mechanism involves spins. This is the case, for example, when electrical conduction involves electron-hole recombination.[832] Another example is the nuclear resonance that affects conductance in the quantum Hall regime.[833] EDMR was also used to detect a single spin located at a Si/SiO_2 interface.[834]

Optically detected magnetic resonance allows the detection of a single spin located in the sample volume, albeit with no spatial resolution.[835, 836] Some localization can be achieved with a confocal microscope.[837] Mechanical detection of single spin resonance was demonstrated, requiring an outstanding detection strategy.[838]

Scanning tunneling microscopy can detect single spin resonance, e.g., Fe atoms on MgO.[839] Read and write operations were demonstrated on holmium (Ho) atoms, using one nearby Fe atom as a magnetic sensor.[840] The mechanism by which the resonance is excited is not straightforward, and the dominant mechanism may depend on the paramagnetic center.[841]

Kerr effect spectroscopy has been used to probe electron spin dynamics. Among the most appealing applications of this technique, let us mention the detection of nuclear magnetic resonance,[842] terahertz Kerr measurements in topological insulators [843, 844] and the detection of spin accumulation to the spin Hall effect.[562, 594]

Conventional electron spin resonance spectroscopy is normally conducted using a cavity, and nuclear magnetic resonance is done using a resonance LC circuit. In both cases, an AC magnetic field excites the resonance. In the early days of magnetic resonance, it was understood that an *electric field excitation* is possible with quadrupolar nuclei located at a non-centrosymmetric site. Indeed, in this case, a linear Stark effect can be expected.[845] This idea was applied in the detection of a single Sb spin in silicon.[846]

High frequency resonance spectroscopy

With increased magnetic field strength, the sensitivity of NMR spectroscopy improved steadily since its discovery. In the 1990s, fields of 9.4 T were commercially available, in which protons resonate at 400 MHz. If dynamic nuclear polarization (DNP, Chapter 16) is to be conducted in such a field, then the electron spin resonance must be excited at about 260 GHz. Thus, over time, there was a pressing need to be able to detect electron spin resonance at such frequencies. A particularly impressive early result was the electron-nuclear pulsed double resonance spectrometer (ENDOR), which was built to run at 140 GHz.[847]

Electrical detection of magnetic resonance at these high frequencies is advantageous because it allows the experimentalist to focus on ensuring the excitation of the resonance. For example, spins of a two-dimensional electron gas were excited at 94 GHz (W band) and EDMR was used for detection.[848] The force-detected magnetic resonance which was used at first for a single spin,[838] was later applied to a finite size sample in the case of a resonance in the W band (75–110 GHz).[849]

An electron spin resonance spectrometer can be designed to operate as a scanning probe. One notable advantage is that the user need not be concerned with the intricacies of high-Q resonators. Furthermore, sensitivity is improved over the standard technique, which relies on high-Q resonators,[850] in particular when using interferometric detection schemes.[851] At low frequencies, permanent magnets can be used, thus making it possible to configure the spectrometer into a "mouse" that is scanned over the surface to be tested.[852] The higher the frequency, the higher the spatial resolution one ought to be able to attain, because the sensitivity is greater. The scanning mode avoids many issues regarding wave propagation and cavity operation.[853] In a normal, large scale configuration, high-precision mechanical parts containing overmoded waveguides are required.[854]

16 Dynamic Nuclear Polarization

Albert Overhauser (1925–2011)

A. Overhauser is known for his ideas on dynamic nuclear polarization. He earned a PhD in physics at the University of California, Berkeley. He was a post-doctoral researcher at the University of Illinois in Urbana-Champaign, then was on the faculty at Cornell University until he joined Ford Motor Company. Later, he returned to academia as professor at Purdue University.

Spintronics: A primer
J.-Ph. Ansermet (https://orcid.org/0000-0002-1307-4864)
Chapter 16. Dynamic Nuclear Polarization

Dynamic nuclear polarization (DNP) enhances spin polarization, which can be necessary when performing multidimensional spectroscopy, or when using spin qubits in a quantum computer. Overhauser predicted that an out-of-equilibrium electron spin population can induce a nuclear spin hyperpolarization, as explained here with detailed balance arguments. When coupled spins do not move, other continuous wave methods such as the solid effect can induce DNP. Using the notion of spin temperature, adiabatic demagnetization is described both in a quasi-static field and in a resonant field.

DOI: 10.1201/9781003370017-16

16.1 HISTORICAL INTRODUCTION

Dynamic nuclear polarization

In a seminal paper, Overhauser predicted that nuclei in a metal could be spin-polarized well beyond the thermal equilibrium polarization if a strong enough excitation of the conduction electron spin resonance were applied to the sample.[855] This is one of the first instances of a set of methods referred to as ***dynamic nuclear polarization*** (DNP). Interestingly for the spintronics community, in an accompanying paper discussing paramagnetic relaxation in metals, Overhauser used the notion of ***spin-dependent chemical potential***.[719] He drew a Fermi-Dirac distribution for each spin (Fig. 16.1) and argued that, upon driving the electron spin resonance, the spin populations equalize, which means they are brought far from equilibrium. By an argument shown in section §16.2.3, we show that this out-of-equilibrium electron spin population induces a very strong nuclear polarization.

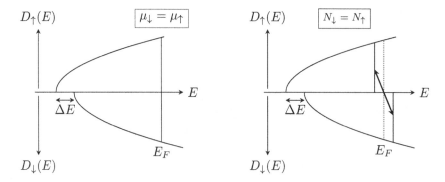

Figure 16.1 Spin-down and spin-up state occupations for conduction electron spins in a magnetic field B_0. At equilibrium (left), the chemical potentials of both spin species have equalized. Under microwave irradiation (right, double arrow), the spin populations equalize when the resonance is driven to saturation. This implies that their chemical potentials differ. This reflects the fact that the spin populations are now out of equilibrium.

Overhauser's prediction of nuclear polarization enhancement was verified by Carver and Slichter the same year that Overhauser published his idea.[856] Carver and Slichter's experiment was conducted at a remarkably low frequency. Nowadays, such experiments are conducted in the high fields produced by superconducting magnets.[857, 858]

In this chapter, the Overhauser effect is explained in detail, as it illustrates one of the many ways that spin polarization can be enhanced. The chapter ends with the the notion of spin temperature and describes one resonance method that can be used to cool spins by adiabatic demagnetization.

Spin temperature

The concept that an ensemble of spins can be characterized by a spin temperature, distinct from the lattice temperature, is itself an interesting application of the general concepts of statistical physics. Since it was first proposed, the concept has appeared in various physical situations.[859] We mention just a few here:

- Spins can be prepared in a state of negative temperature.[860]
- In the framework of spintronics, it has been proposed that the effect of spin-polarized currents on magnetization can be thought of in terms of spin temperature. Here, "spins" refers to local magnetic moments, and the incoming conduction electron

spins relax into this system, causing its temperature to rise.[62, 63] This spin temperature approach has been discarded in favor of the generally accepted spin transfer torque model.

- The spin Seebeck effect has been interpreted in terms of electron-magnon collisions, with a difference between the temperature of the lattice and that of the magnons.[482]
- A DNP mechanism known as "thermal mixing" also relies on the concept of spin temperature. It was extensively studied by T. Wenckebach.[861, 862]
- When a ferromagnet is excited on a very short time scale, it is possible to consider that the phonons (the lattice), the electrons, and the spins (the magnetic moments) each have a defined temperature (Fig. 16.2). Since these three systems are coupled, over time, their temperatures become equal, in other words, the three sub-systems thermalize.[863]

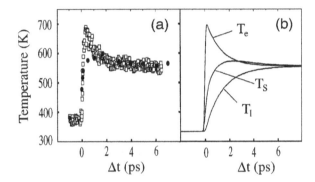

Figure 16.2 Estimated electron and spin temperatures in nickel following an ultra-short laser pulse (left); and modeling of the evolution of spin, electron and lattice temperatures (right) as a function of time.[863]

16.2   THE OVERHAUSER EFFECT

16.2.1   SIMPLIFIED DETAILED BALANCE

The relaxation of a 2-level system was described in §12.2. We pointed out that the rates W_U and W_D must be different, if the detailed balance equations are to predict the correct thermal equilibrium value (12.9) of the spin polarization.

However, when we try to work out the relaxation rate for $\hat{S}_z$, the difference between W_U and W_D is not significant. So, in the magnetic resonance literature, it is customary to neglect this difference. However, in order to predict the correct equilibrium value of $\hat{S}_z$, the equation of evolution deduced from the detailed balance is modified. A term is added so that the evolution equation predicts relaxation toward the correct, and intuitively known, equilibrium value.[1] To drive the point further, we notice that, generally, the detailed balance has the form:

$$\dot{N}_i = \sum_j A_{ij} N_j \tag{16.1}$$

---

[1] See, e.g., [864].

The equilibrium populations are given by,

$$0 = \sum_j A_{ij} N_j^0 \tag{16.2}$$

In order for this equation to give a meaningful result, the matrix elements A_{ij} must account for the coupling to a bath. However, if the equilibrium populations N_j^0 are already known, we can write the detailed balance as,

$$\dot{N}_i = \sum_j A_{ij}(N_j - N_j^0) \tag{16.3}$$

and use simpler, i.e., approximate, expressions for the matrix elements A_{ij}. Namely, we can now assume that $A_{ij} = A_{ji}$. We will use this approach below to explain the Overhauser effect.

16.2.2   PAIRS OF DISTINCT, COUPLED SPINS

We consider an ensemble of pairs of coupled spins, noted S and I spins, both being $1/2$ spins. The S spin corresponds to electronic spins, and the I spin to nuclear spins. The S spins are coupled to a bath. On the time scale of the nuclear spins, the electron spins fluctuate at a very fast rate and can be considered from the standpoint of nuclear dynamics as being at equilibrium. In other words, these electron spin fluctuations have a vanishingly small spectral density at the frequency of the nuclei.

However, we assume that the nuclei experience fluctuating fields at their nuclear spin frequency because the strength of the $I - S$ coupling fluctuates. These fluctuating fields experienced by the nuclei can cause nuclear spin relaxation. As we will see, the existence of this nuclear relaxation mechanism is an essential aspect of the DNP mechanism described in this section. In other words, this DNP mechanism works only if the $S - I$ coupling fluctuates at the Larmor frequency of the nuclei, i.e., the spectral density of its fluctuations has a significant contribution at the resonance frequency of the nuclei, as shown in §12.3.2. In view of the slow time scale of nuclei compared to electrons, we expect this picture to be relevant in the case of liquids. It is indeed the case. However, Overhauser first considered the case of conduction electrons (see §16.1), in which electrons "fly" by the nuclei, so to speak, causing a fluctuating coupling.

We now consider two types of coupling, scalar coupling and dipolar coupling. A simple **hyperfine coupling** is of the scalar form,

$$\mathcal{H}_{\mathrm{hf}} = A\boldsymbol{I} \cdot \boldsymbol{S} \tag{16.4}$$

In this particular case, the preliminary discussion above amounts to saying that A fluctuates over time, and the spectral density function of $A(t)$ has a significant component at the nuclear Larmor frequency.

As usual, we define $\hat{I}_{\pm} = \hat{I}_x \pm i\hat{I}_y$ and likewise for the S spins. We have $\hat{I}_x = \frac{1}{2}(\hat{I}_+ + \hat{I}_1)$, $\hat{I}_y = \frac{1}{2i}(\hat{I}_+ + \hat{I}_1)$ and likewise for S. Then, let us note that,

$$2(\hat{I}_x\hat{S}_x + \hat{I}_y\hat{S}_y) = (\hat{I}_+ + \hat{I}_-)(\hat{S}_+ + \hat{S}_-) - (\hat{I}_+ - \hat{I}_-)(\hat{S}_+ - \hat{S}_-) = 2(\hat{S}_+\hat{I}_- + \hat{S}_-\hat{I}_+) \tag{16.5}$$

The coupling term (16.5) can induce **mutual spin flips**, i.e., both spins flip simultaneously.

The **dipolar coupling** between I and S spins has the form,

$$\mathcal{H}_{dip} = \frac{\mu_0}{4\pi} \frac{\gamma_I \gamma_S}{r^3} \left[\boldsymbol{I} \cdot \boldsymbol{S} - 3\left(\boldsymbol{S} \cdot \hat{\boldsymbol{r}}\right)\left(\boldsymbol{I} \cdot \hat{\boldsymbol{r}}\right) \right] \tag{16.6}$$

We write the unit vector pointing from one spin to the other in spherical coordinates as,

$$\hat{r} = \begin{pmatrix} \sin\theta\cos\phi \\ \sin\theta\sin\phi \\ \cos\theta \end{pmatrix} \qquad (16.7)$$

The dipolar coupling then develops as,

$$\hat{I} \cdot \hat{S} \ - \ 3\,(\boldsymbol{S} \cdot \hat{r})\,(\boldsymbol{I} \cdot \hat{r}) =$$
$$\hat{S}_z \hat{I}_z + \hat{S}_+ \hat{I}_- + \hat{S}_- \hat{I}_+$$
$$-\frac{3}{4}\left[(\hat{S}_+ + \hat{S}_-)\sin\theta\cos\phi - i(\hat{S}_+ - \hat{S}_-)\sin\theta\sin\phi + 2\hat{S}_z\cos\theta\right]$$
$$\left[(\hat{I}_+ + \hat{I}_-)\sin\theta\cos\phi - i(\hat{I}_+ - \hat{I}_-)\sin\theta\sin\phi + 2\hat{I}_z\cos\theta\right] \qquad (16.8)$$

These various terms can be grouped according to the type of transitions they drive (zero-quantum, one-quantum and double-quantum transitions). This gives the terms,

$$A = (1 - 3\cos^2\theta)\hat{I}_z\hat{S}_z$$
$$B = \frac{-1}{4}(1 - 3\cos^2\theta)(\hat{S}_+\hat{I}_- + \hat{S}_-\hat{I}_+)$$
$$C = \frac{-3}{2}\sin\theta\cos\theta e{-i\phi}(\hat{I}_z\hat{S}_+ + \hat{I}_+Sz)$$
$$D = \frac{-3}{2}\sin\theta\cos\theta ei\phi(\hat{I}_z\hat{S}_- + \hat{I}_-Sz) \qquad (16.9)$$
$$E = \frac{-3}{4}\sin^2 e^{-2i}\phi\hat{I}_+\hat{S}_+$$
$$F = \frac{-3}{4}\sin^2 e^{2i}\phi\hat{I}_-\hat{S}_-$$

The A and B terms are sometimes called the *secular* terms of the dipolar interactions. They are the terms that commute with the Zeeman Hamiltonian of the $\hat{I}$ and $\hat{S}$ spins. In a calculation where the Zeeman interaction is large compared to the dipolar coupling, only the secular terms contribute to the first order perturbation.

16.2.3   SIMPLE PICTURE

In his seminal paper, Overhauser considered nuclei coupled to electron spins [855]. Here, we consider non-metals so as to avoid having to take into account Fermi statistics. We assume that the applied field is large enough that the energy levels are defined by the Zeeman terms. The energy levels for one couple of spins $1/2$ are therefore as shown in Fig. 16.3. We further assume that fluctuation of the hyperfine coupling induces zero-quantum transitions, which occur at the rate W_0.

Following the remark in the previous section concerning the equilibrium condition, we write the detailed balance with equal transition rates in both directions, up-to-down and down-to-up transitions. $N_1, \ldots, N_4$ are the populations of the levels. W_1 is the rate that determines the relaxation of the S spins with the lattice. We leave out nuclear relaxation processes that could occur independently of the coupling of the nuclear spins to the electron spins.

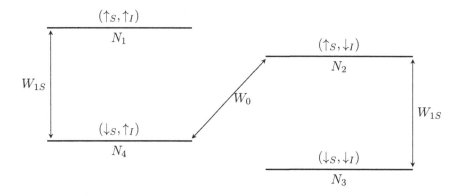

Figure 16.3 Energy levels for two spins 1/2. Rates are described in the text.

The detailed balance in this case reads:

$$
\begin{aligned}
\dot{N}_1 &= -W_1 N_1 + W_1 N_4 \\
\dot{N}_2 &= -(W_0 + W_1) N_2 + W_1 N_3 + W_0 N_4 \\
\dot{N}_3 &= W_1 N_2 - W_1 N_3 \\
\dot{N}_4 &= W_1 N_1 + W_0 N_2 - (W_0 + W_1) N_4
\end{aligned}
\tag{16.10}
$$

We define (see Fig. 16.3):

$$
\begin{aligned}
\hat{S}_z &= N_1 + N_2 - N_3 - N_4 \\
\hat{I}_z &= N_1 - N_2 - N_3 + N_4
\end{aligned}
\tag{16.11}
$$

We deduce from the detailed balance the evolution of $\hat{S}_z$ and $\hat{I}_z$:

$$
\dot{\hat{S}}_z = -2W_1(N_1 - N_3) - 2(W_0 + W_1)(N_2 - N_4)
\tag{16.12}
$$

$$
\dot{\hat{I}}_z = +2W_0(N_2 - N_4)
\tag{16.13}
$$

We notice that the definitions (16.11) imply:

$$
\begin{aligned}
\hat{S}_z + \hat{I}_z &= 2(N_1 - N_3) \\
\hat{S}_z - \hat{I}_z &= 2(N_2 - N_4)
\end{aligned}
\tag{16.14}
$$

Now, we introduce the equilibrium values in the detailed balance and take into account (16.14). This yields,

$$
\dot{\hat{S}}_z = -(W_0 + 2W_1)(\hat{S}_z - \hat{S}_0) + W_0(\hat{I}_z - \hat{I}_0)
\tag{16.15}
$$

$$
\dot{\hat{I}}_z = +W_0(\hat{S}_z - \hat{S}_0) - W_0(\hat{I}_z - \hat{I}_0)
\tag{16.16}
$$

This result leads to the following expression of the Overhauser effect. Consider the stationary regime obtained when the S resonance is saturated. We have described this concept using (12.16). Here, we cannot use (16.15) to infer a saturation effect, because we have not introduced the excitation term. However, under saturation condition we know that $\hat{S}_z = 0$. Then, the evolution of $\hat{I}_z$ deduced from (16.16) is given by,

$$
\hat{I}_z - \hat{I}_0 = -\hat{S}_0
\tag{16.17}
$$

If S corresponds to electron spins and I to nuclear spins, $\hat{S}_0 \gg \hat{I}_0$. Thus, we have found that the nuclear spins I have a polarization that is minus the polarization of the electron spins, a huge enhancement compared to the polarization $\hat{I}_0$.

16.2.4 FLUCTUATING DIPOLAR COUPLING

As we have seen, the dipolar coupling implies zero-quantum, one-quantum and double-quantum transitions. We now consider this case, illustrated in figure 16.4. Once again, we

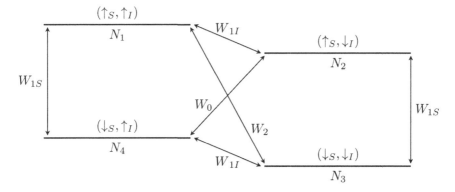

Figure 16.4 Energy levels for two spins 1/2, and transitions: zero-quantum (W_0), one-quantum (W_{1I}, W_{1S}) and double-quantum transitions (W_2).

do not distinguish up-to-down and down-to-up transitions, so the equilibrium must be added after we have worked out the evolution equations for $\hat{I}_z$ and $\hat{S}_z$. We should keep in mind that the transitions rates W_{1S} include both the transitions induced by the fluctuations of the dipolar coupling and the transitions due to the coupling of the S spins to the lattice.

In the case illustrated by figure 16.4, the detailed balance reads:

$$
\begin{aligned}
\dot{N}_1 &= -(W_{1S} + W_{1I} + W_2)N_1 + W_{1I}N_2 + W_2N_3 + W_{1S}N_4 \\
\dot{N}_2 &= W_{1I}N_1 - (W_{1S} + W_{1I} + W_0)N_2 + W_{1S}N_3 + W_0N_4 \\
\dot{N}_3 &= W_2N_1 + W_{1S}N_2 - (W_{1I} + W_{1S} + W_2)N_3 + W_{1I}N_4 \\
\dot{N}_4 &= W_{1S}N_1 + W_0N_2 + W_{1I}N_3 - (W_{1I} + W_{1S} + W_0)N_4
\end{aligned}
\tag{16.18}
$$

The matrix that defines the evolution of the population is symmetric and therefore diagonalizable. In principle, we could integrate using a few linear algebraic manipulations. Here, however, we simply seek to work out the evolution equation for $\hat{S}_z$ and $\hat{I}_z$. We have:

$$
\begin{aligned}
\dot{\hat{S}}_z &= -2(W_{1S} + W_2)(N_1 - N_3) - 2(W_{1S} + W_0)(N_2 - N_4) \\
\dot{\hat{I}}_z &= -2(W_{1I} + W_2)(N_1 - N_3) + 2(W_{1I} + W_0)(N_2 - N_4)
\end{aligned}
\tag{16.19}
$$

Using (16.14), and adding the equilibrium values, we thus find:

$$
\dot{\hat{S}}_z = -(2W_{1S} + W_2 + W_0)(\hat{S}_z - \hat{S}_0) + (W_2 + W_0)(\hat{I}_z - \hat{I}_0)
\tag{16.20}
$$

$$
\dot{\hat{I}}_z = -(W_2 - W_0)(\hat{S}_z - \hat{S}_0) - (2W_{1I} + W_2 + W_0)(\hat{I}_z - \hat{I}_0)
\tag{16.21}
$$

We now have the following expression of the Overhauser effect. If we saturate the S resonance in the steady state according to (16.21), we have:

$$
\hat{I}_z - \hat{I}_0 = \frac{(W_2 - W_0)}{(W_0 + 2W_{1I} + W_2)}\hat{S}_0
\tag{16.22}
$$

Thus, the Overhauser enhancement (16.22) is positive for the dipolar coupling.

16.3 THE SOLID EFFECT

Let us now consider an ensemble of spins I and S that are coupled to each other, where the coupling is independent of time. Thus, we consider pairs of fixed spins with the time-independent Hamiltonian,

$$\mathcal{H} = -\gamma_S \hbar B_0 \hat{S}_z - \gamma_I \hbar B_0 \hat{I}_z + \mathcal{H}_{IS} \tag{16.23}$$

The transition from $|++>$ to $|-->$ is forbidden. But when the coupling $\mathcal{H}_{IS}$ is a perturbation of the Zeeman interaction, then the eigenstates of the Zeeman Hamiltonian are perturbed by the coupling, and the eigenstates of $\mathcal{H}$ consist of admixtures of the unperturbed states and states of adjacent energy (Fig. 16.5).[865]

From an experimental standpoint, since it is only thanks to a perturbation that this transition is allowed, it is to be expected that driving this transition requires a lot of power. This point is especially critical because DNP by the solid effect requires that the transition rate W_{ext} dominate all other relaxation pathways.

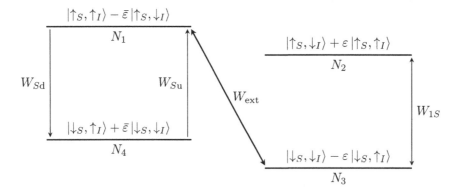

Figure 16.5 Energy levels for two spins 1/2 including a perturbation of the states due to the coupling of the spins. The uncoupled states are indicated by the magnetic numbers $|m_S, m_I|$. The eigenstates of the coupled spins are admixtures that make the transition W_{ext} possible, whereas this transition would be forbidden if the spins were uncoupled. Only the dominant transition rates are indicated, keeping track of the difference between up-to-down and down-to-up rates.

We now proceed to a detailed balance similar to that in (16.18). Here, however, the cause of the transitions is different, so the analogy in the 4-level diagram Figs. 16.5 and 16.3 does not mean that the mechanisms are the same. Expressing the detailed balance in this case, it is convenient to distinguish the relaxation rates (W_{Sd}) and (W_{Su}) for up-to-down and down-to-up transitions, respectively. In the stationary regime, we have,

$$\begin{aligned}
\dot{N}_1 &= 0 = -(W_{Sd} + W_{ext})N_1 + W_{ext}N_3 + W_{Su}N_4 \\
\dot{N}_2 &= 0 = -(W_{Sd})N_2 + W_{Su}N_3 \\
\dot{N}_3 &= 0 = W_{ext}N_1 + W_{Sd}N_2 - (W_{Su} + W_{ext})N_3 \\
\dot{N}_4 &= 0 = W_{Sd}N_1 - (W_{Su})N_4
\end{aligned} \tag{16.24}$$

We assume that the transition (1–3) with the rate W_{ext} is driven so strongly that terms containing W_{ext} dominate all other terms in the first and the third of these equations.

Then the two equations containing W_{ext} are both satisfied if $N_1 = N_3$. The remaining two equations can be written as,

$$\frac{N_4}{N_1} = \frac{W_{Sd}}{W_{Su}} = B_S \qquad \frac{N_3}{N_2} = \frac{W_{Sd}}{W_{Su}} = B_S \qquad (16.25)$$

where B_S is the Boltzmann factor for the S spin transitions. We can simplify the writing of the solution if we take $N_1 = 1$. Then $N_3 = 1$, $N_4 = B_S$, and $N_2 = 1/B_S$. The I spin polarization P_I can be calculated as,

$$P_I = \frac{N_1 + N_4 - N_2 - N_3}{N_1 + N_4 + N_2 + N_3} = \frac{B_S - 1/B_S}{2 + B_S + 1/B_S} \qquad (16.26)$$

In the high temperature approximation, we can write,

$$B_S = \exp\left(\frac{\Delta E_S}{k_B T}\right) \approx 1 + \frac{\Delta E_S}{k_B T} \qquad (16.27)$$

Hence, the I spin polarization can be approximated as,

$$P_I = \frac{\Delta E_S}{2 k_B T} \qquad (16.28)$$

In the absence of the driving of the (1–3) transition, the I spin polarization is given by,

$$P_{I0} = \frac{N_+ - N_-}{N_+ + N_-} = \frac{N_+/N_- - 1}{N_+ + N_-} = \frac{\exp\left(\frac{\Delta E_I}{k_B T}\right) - 1}{2} \approx \frac{\Delta E_I}{2 k_B T} \qquad (16.29)$$

Therefore, the I spin polarization is enhanced with respect to the equilibrium value by a coefficient,

$$\frac{P_I}{P_{I0}} = \frac{\Delta E_S}{\Delta E_I} = \frac{\gamma_S}{\gamma_I} \qquad (16.30)$$

Thus, the enhancement is given by the ratio of the gyromagnetic factors of the S and I spins. This ratio is very large when the S spins are electron spins and the I spins are nuclear spins.

16.4   SPIN DIFFUSION WITHOUT MASS TRANSPORT

Spin diffusion can take place without a spin carrier moving through space. Indeed, spin diffusion can take place when spins are coupled sufficiently strongly mutual spin flips to take place before spin-lattice relaxation processes occur. Consider the case of a one-dimensional chain with a nearest-neighbor Heisenberg coupling proportional to $\hat{S}_i \cdot \hat{S}_{i+1}$ (see (2.56)). This coupling can be written as,

$$\hat{S}_i \cdot \hat{S}_{i+1} = \hat{S}_{i,z} \hat{S}_{i+1,z} + \frac{1}{2}\left(S_+^i S_-^{i+1} + S_-^i S_+^{i+1}\right) \qquad (16.31)$$

The terms in parentheses induced mutual spin flips. In Fig. 16.6, we start from a state prepared with half of the spins up, and the other half down. We proceed with mutual spin flips between a few randomly chosen pairs and quickly find that this step in the spin-polarization profile is randomized, just like a step in a temperature profile along a rod would smooth out over time.

Figure 16.6 suggests a spin-diffusion process without mass transport. To be more explicit, let us work out the continuum limit for this process. We designate d as the inter-atomic

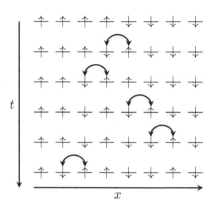

Figure 16.6 Eight atomic sites are represented as a function of position (x). Initially, the system is assumed with the left half of the sites having spins up, and the right half having spins down. Mutual spin flips (double arrows) are assumed to be possible, owing to the coupling among nearest neighbors. After five mutual spin flips, the spin configuration is nearly a random configuration. The average polarization does not change since mutual spin flips don't change the overall spin polarization.

distance along a chain of 1/2 spins, and work out a detailed balance for $n_+(x)$ at position x along the chain. There are mutual spin flips with neighbors than can either increase or decrease $n_+(x)$, according to,[866]

$$\frac{\partial n_+(x)}{\partial t} = W p_-(x) \left(p_+(x+d) + p_+(x-d)\right) - W p_+(x) \left(p_-(x+d) + p_-(x-d)\right) \quad (16.32)$$

where W is the mutual spin-flip rate. Defining for any time t the polarization at x as $p(x,t) = n_+(x) - n_(x) = 2n_+(x) - 1$, and neglecting terms that are quadratic in $p(x)$, we find:

$$\frac{\partial p(x,t)}{\partial t} = W \left(p(x+d,t) + p(x-d,t) - 2p(x,t)\right) \quad (16.33)$$

In the continuous limit, this gives the diffusion equation,

$$\frac{\partial p(x,t)}{\partial t} = W d^2 \frac{\partial^2 p(x,t)}{\partial x^2} \quad (16.34)$$

16.5   SPIN TEMPERATURE

In this section, we first analyze the concept of spin temperature. Very often, the spins are at equilibrium with a thermal bath and the spins are at the temperature of the bath. However, in spintronics, spins may be brought out of equilibrium because of a charge current or an optical excitation. There are also DNP techniques that can be understood in terms of a spin temperature. In spin-based quantum information processing, there is a need to lower the entropy of ancillia qubits, which can be achieved by creating local temperature gradients.

16.5.1   TEMPERATURE OF AN ISOLATED SYSTEM

Let us introduce the concept of a spin temperature by considering a thermodynamic system that is isolated. More precisely, we mean that the system can be considered isolated on the time scale over which it is observed. Over this time scale, we assume that there are strong enough couplings among the spins that the whole system can be assumed to have reached equilibrium. According to thermodynamics, the system has a definite energy E and its entropy has reached a maximum.[2] In statistical physics, following Gibbs, we would consider an ensemble of identical systems such as this one. This is called the ***microcanonical ensemble***.

We now argue that a temperature can be defined for the thermodynamic system. Since the system is large, it can be divided into two parts that are also large, so that thermodynamics can be applied to the subsystems, as well. The subsystems have the energies E_1 and $E_2 = E - E_1$, since energy is an extensive quantity. Likewise, entropy is extensive, so we can write,

$$S(E) = \hat{S}_1(E_1) + \hat{S}_2(E_2) \tag{16.35}$$

Since entropy reaches a maximum at equilibrium, we have $dS/dE_1 = 0$. This implies:

$$\frac{d\hat{S}_1(E_1)}{dE_1} - \frac{d\hat{S}_2(E - E_1)}{dE_2} = 0 \tag{16.36}$$

The thermodynamic temperature can be defined by,

$$\frac{dS}{dE} = \frac{1}{T} \tag{16.37}$$

Thus, (16.36) means that each part of the system has the same temperature.

Now we establish the probability for a constituent of the system to have an energy E_i. We start with the statistical definition of entropy for a system with a set of states that can be numbered:[386]

$$S = -k_B \sum_i f_i \ln(f_i) \tag{16.38}$$

We again consider a system with a definite energy. In conformity with the description of the microcanonical ensemble,[3] we consider that all microscopic configurations are equally probable. Let Ω be the number of configurations at equilibrium. Then $f_i = 1/\Omega$ and we have:

$$S = k \ln(\Omega) \tag{16.39}$$

We must remember that we are considering very large ensembles of spins. Let N be the number of spins. We now ask what is the probability $p(E_i)$ that one of these spins has a particular energy E_i. To avoid writing a normalization constant, we seek the ratio $p(E_i)/p(E_j)$. The probability $p(E_i)$ is proportional to the number of configurations of $N - 1$ spins with energy $E - E_i$, that is:

$$\frac{p(E_i)}{p(E_j)} = \frac{\Omega(N-1, E-E_i)}{\Omega(N-1, E-E_j)} \tag{16.40}$$

As $E \gg E_i$, $(\forall i)$, we can make a Taylor expansion using (16.39) as follows:

$$\frac{p(E_i)}{p(E_j)} = \frac{\exp\left[\frac{1}{k}S(N-1, E-E_i)\right]}{\exp\left[\frac{1}{k}S(N-1, E-E_j)\right]} = \frac{\exp\left[\frac{1}{k}\frac{dS(N-1,E)}{dE}(-E_i)\right]}{\exp\left[\frac{1}{k}\frac{dS(N-1,E)}{dE}(-E_j)\right]} \tag{16.41}$$

Applying the definition (16.37) of the temperature T, we obtain,

$$\frac{p(E_i)}{p(E_j)} = \exp\left[-\frac{E_i - E_j}{kT}\right] \tag{16.42}$$

---

[2]See, e.g.,[376, 867].

[3]See, e.g.,[868].

In other words, the probability of finding the system in the state of energy E_i is given by the **Boltzmann factor**,

$$p(E_i) = \frac{1}{Z} \exp\left(\frac{-E_i}{k_B T}\right) \qquad (16.43)$$

where the **partition function** Z ensures that the probabilities are normalized to 1.

16.5.2 SPIN TEMPERATURE IN THE LABORATORY FRAME

We consider N spins of angular momentum I ($I = 1/2, 1, 3/2, ...$).[4] The spins are subjected to a uniform field B_0. They are coupled among each other by dipolar interactions. The Hamiltonian of the Zeeman interaction (7.5) is written as,

$$\mathcal{H}_Z = -\gamma \hbar B_0 \sum_{i=1}^{N} \hat{I}_{z,i} \qquad (16.44)$$

The Hamiltonian accounting for the dipolar interactions is noted $\mathcal{H}_d$. The total Hamiltonian is $\mathcal{H} = \mathcal{H}_Z + \mathcal{H}_d$. The eigenstates of $\mathcal{H}$ are $|\psi_n\rangle$, of energy E_n, i.e.,

$$(\mathcal{H}_Z + \mathcal{H}_d)|\psi_n\rangle = E_n|\psi_n\rangle \qquad (16.45)$$

The states $|\psi_n\rangle$ are generally unknown. However, when we assume that these spins are characterized by a temperature θ, we can say that the eigenstates are occupied with probabilities:

$$p_n = \frac{1}{Z} \exp\left[\frac{-E_n}{k\theta}\right] \qquad \text{where} \qquad Z = \sum_n \exp\left[\frac{-E_n}{k\theta}\right] \qquad (16.46)$$

We now show that, when the notion of spin temperature can be used, average quantities such as the energy $\bar{E}$, the magnetization $\bar{M}_z$ and the entropy S can be calculated. By definition, we have,

$$\bar{E} = \sum_n p_n E_n$$

$$\bar{M}_z = \sum_n \gamma \hbar \langle n|\hat{I}_z|n\rangle \, p_n \qquad (16.47)$$

$$S = -k_B \sum_n p_n \ln(p_n)$$

The partition function can be written as a trace:

$$Z = \sum_n \langle n| \exp\left[-\frac{\mathcal{H}}{k_B \theta}\right] |n\rangle = Tr\left(\exp\left[-\frac{\mathcal{H}}{k_B \theta}\right]\right) \qquad (16.48)$$

The ability to write these quantities as traces is important, because a trace is independent of the state basis that we choose to span the Hilbert space, i.e., we do not need to determine the eigenstates $|\psi_n\rangle$. For example, we can calculate the trace with the eigenstates of $\mathcal{H}_Z$, which are readily known. Using the statistical physics relation $F = -k_B \theta \ln Z$, and referring to the Legendre transform $F = \bar{E} - \theta S$, we can also write the entropy in terms of a trace via the expression,

$$S = \frac{\bar{E} + k_B \theta \ln Z}{\theta} \qquad (16.49)$$

---

[4]See, C. P. Slichter,[236], chapter 6.

Thus, the entropy S can be written in terms of traces. We can also express $\bar{E}$ in terms of traces. Deriving (16.46) with respect to θ, we get,

$$\frac{1}{Z}\frac{\partial Z}{\partial \theta} = \sum_n \frac{1}{Z}\exp\left(-\frac{E_n}{k_B\theta}\right)\left(\frac{E_n}{k_B\theta^2}\right) = \sum_n p_n E_n \frac{1}{k_B\theta^2} \tag{16.50}$$

Thus, we have,

$$\bar{E} = k_B\theta^2 \frac{\partial}{\partial \theta}(\ln Z) \tag{16.51}$$

Finally, let us express $\bar{M}_z$ in terms of traces. First, we cast the expression of Z as,

$$Z = \sum_n \sum_{m_I^n} \langle\{m_I^n\}|\exp\left(\frac{\gamma\hbar B_0 \sum_r \hat{I}_{z,r} - \mathcal{H}_d}{k_B\theta}\right)|\{m_I^n\}\rangle \tag{16.52}$$

Deriving this expression of Z with respect to B_0, we get,

$$\frac{1}{Z}\frac{\partial Z}{\partial B_0} = \sum_n \sum_{m_I^n} \langle\{m_I^n\}|\frac{\gamma\hbar \sum_r \hat{I}_{z,r}}{k_B\theta}\frac{1}{Z}\exp\left(\frac{-\mathcal{H}}{k_B\theta}\right)|\{m_I^n\}\rangle = \frac{\bar{M}_z}{k_B\theta} \tag{16.53}$$

Thus, we have,

$$\bar{M}_z = k_B\theta \frac{\partial}{\partial B_0}(\ln Z) \tag{16.54}$$

In summary, we can calculate S, $\bar{E}$, and $\bar{M}_z$ from Z.

In many cases, especially when dealing with nuclei, the high temperature approximation applies. By Taylor expansion, we find,

$$Z = Tr\left(1 - \frac{\mathcal{H}}{k_B\theta} + \frac{1}{2}\left(\frac{\mathcal{H}}{k_B\theta}\right)^2\right) = (2I+1)^N + \frac{1}{2k_B^2\theta^2}Tr\left(\mathcal{H}^2\right) \tag{16.55}$$

since $Tr(\mathcal{H}) = 0$. Since $\mathcal{H}_Z$ contains the z-axis projection of the total angular momentum, it is proportional to the generator of rotations around the z-axis. The Hamiltonian $\mathcal{H}_d$ is invariant under these rotations, so $\mathcal{H}_Z$ and $\mathcal{H}_d$ commute. We can write,

$$Z = (2I+1)^N + \frac{1}{2k_B\theta^2}Tr\left(\mathcal{H}_Z^2 + \mathcal{H}_d^2 + 2\mathcal{H}_Z\mathcal{H}_d\right) \tag{16.56}$$

Let us write the last term as,

$$Tr(\mathcal{H}_Z\mathcal{H}_d) = \sum_{n,m} <n|\mathcal{H}_Z|n><m|\mathcal{H}_d|m> \tag{16.57}$$

We can take for $|m>$ eigenstates of $\mathcal{H}_Z$. We then have,

$$Tr(\mathcal{H}_Z\mathcal{H}_d) = \sum_n E_{Z,n} <n|\mathcal{H}_d|n> = \frac{1}{2}\sum_n E_{Z,n}\left(<n|\mathcal{H}_d|n> - <-n|\mathcal{H}_d|-n>\right) \tag{16.58}$$

where the notation $|-n>$ designates the state where all m_I^n numbers have changed to the opposite sign. These states can be obtained by mirror symmetry with respect to the plane normal to the z-axis. This is a plane of symmetry for $\mathcal{H}_d$, so the term in the last sum vanishes,

$$Tr(\mathcal{H}_Z\mathcal{H}_d) = 0 \tag{16.59}$$

Let us now address $Tr\left(\mathcal{H}_Z^2\right)$ by writing,

$$Tr\left(\mathcal{H}_Z^2\right) = \sum_n \sum_{m_I^n} < \{m_I^n\}| \left(-\gamma\hbar B_0 \sum_r \hat{I}_{z,r}\right)^2 |\{m_I^n\} > \qquad (16.60)$$

Thus, we find,

$$Tr\left(\mathcal{H}_Z^2\right) = \gamma^2\hbar^2 B_0^2 \frac{1}{3}NI(I+1)(2I+1)^N \qquad (16.61)$$

We now wish to express $Tr\left(\mathcal{H}_d^2\right)$ in terms of an effective field B_L that represents the strength of the dipolar coupling. To do so, we use the above result and define this average local dipolar field strength B_L using,

$$Tr\left(\mathcal{H}_d^2\right) = \gamma^2\hbar^2 B_L^2 \frac{1}{3}NI(I+1)(2I+1)^N \qquad (16.62)$$

Let us now use (16.54) to calculate the magnetization. In this calculation, we approximate $1/Z$ by $1/(2I+1)^N$. Thus, we find,

$$\bar{M}_z = \frac{\gamma^2\hbar^2}{k_B\theta}\frac{1}{3}NI(I+1)B_0 \qquad (16.63)$$

Using the approximation (7.41) and noting $B_0 = \mu_0 H_0$, we find a Curie law,

$$M = \frac{C}{\theta}H_0 \quad \text{with} \quad C = \frac{\gamma^2\hbar^2 NI(I+1)}{3k_B} \qquad (16.64)$$

Using (16.51), we find,

$$\bar{E} = -\frac{C\left(B_0^2 + B_L^2\right)}{\theta} \qquad (16.65)$$

And finally, the relation (16.49) for the entropy yields,

$$S = N\ln(2I+1) - \frac{C\left(B_0^2 + B_L^2\right)}{2}\frac{1}{\theta^2} \qquad (16.66)$$

Examples

Spin temperature, given a set energy
First, we consider a system that has a time independent Hamiltonian and a set energy E_0 before it reaches thermal equilibrium. The energy $\bar{E}$ will not change and, therefore, when the spin temperature is established (via the internal couplings), the temperature can be calculated from (16.65) to be:

$$\theta_f = -\frac{C\left(B_0^2 + B_L^2\right)}{E_0} \qquad (16.67)$$

From this result, we can predict the spin polarization to be:

$$M_f = \frac{C}{\theta_f}H_0 \qquad (16.68)$$

We must keep in mind that, if the Hamiltonian is made up of two parts that commute, the energy of each part is conserved and there will be two spin temperatures. This becomes relevant in experiments carried out with a continuous irradiation with radio-frequency or microwave fields (see below).

Negative spin temperatures

In this second example, we show how a negative temperature can be achieved in a spin system. Let us consider a system at thermal equilibrium with a defined spin temperature θ_i. At $t = 0$, the field is switched from a value $\boldsymbol{B}(0^-)$ to a value $\boldsymbol{B}(0^+) = -\boldsymbol{B}(0^-)$. Instantly, the state of the system cannot change; therefore, its magnetization cannot change. Applying the Curie law to the magnetization immediately prior to the switch yields,

$$\bar{E}(0^+) = -C\frac{\boldsymbol{B}(0^-) \cdot \boldsymbol{B}(0^+)}{\theta_i} - \frac{CB_L^2}{\theta_i} \tag{16.69}$$

With reasonable magnetic field strength, we can ensure that $\bar{E}(0^+)$ is positive. Now, if we wait long enough, the system at this energy will reach thermal equilibrium with the spin temperature found by applying (16.67),

$$\bar{E}_0(0^+) = -\frac{C\left(B_0^2(0^+) + B_L^2\right)}{\theta_f} \tag{16.70}$$

Therefore, θ_f must be negative. The spin temperature θ_f is given by,

$$\theta_f = \theta_i \frac{B_0^2(0^+) + B_L^2}{\boldsymbol{B}(0^-) \cdot \boldsymbol{B}(0^+) + B_L^2} \tag{16.71}$$

Purcell and Pound used these ideas to generate a system with a negative temperature.[860] They reversed the field at $t = 0$ so fast that, while the spins were in zero field, they could not undergo any significant precession around the local field.

Let us remark that a system with a negative temperature is very hot as it gives away energy to any system with a positive temperature. Also, a system at a negative temperature relaxes to an equilibrium with a positive temperature by going through a state of infinite temperature. It is more comfortable to think in terms of an axis of inverse temperatures and consider $1/\theta_f$ going through the origin of this axis.

16.6 ADIABATIC DEMAGNETIZATION

16.6.1 BASIC PRINCIPLE

In Figure 16.7, the statistical entropy $S = \sum_i f_i \ln f_i$ is calculated for a magnetic moment that can be in either of two states. The magnetic moment is assume to have a magnitdue of 8 Bohr magnetons. This last assumption sets the temperature scale for B values within experimental reach. The process starts with the system in state A, i.e., at thermal equilibrium with a bath. At first, the magnetic field is vanishingly small, then it is increased while the temperature is maintained until state B is reached. At this point, the system is thermally disconnected from the thermal bath and an adiabatic process takes place, which consists in lowering the magnetic field. Using thermodynamical arguments, it is possible to show that the temperature decreases.[376]

Let us now use the concept of spin temperature to describe adiabatic demagnetization in preparation for the next section, where we will see that a resonance method can induce adiabatic demagnetization. So, we consider an ensemble of spins subjected to a magnetic field that is changed in such a manner that the entropy is constant, i.e., the system is adiabatically closed and there are no irreversible processes taking place. We will call this an adiabatic process, for short. Then the expression (16.66) for the entropy implies:

$$\frac{\left(B_{0i}^2 + B_L^2\right)}{\theta_i{}^2} = \frac{\left(B_{0f}^2 + B_L^2\right)}{\theta_f{}^2} \tag{16.72}$$

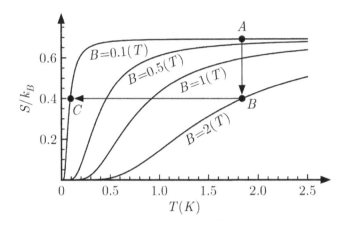

Figure 16.7 The entropy per magnetic moment is expressed as a function of temperature T and magnetic induction field $\boldsymbol{B}$. The numerical values are obtained for a fictitious magnetic moment of 8 Bohr magnetons μ_B that can only be either parallel or antiparallel to the magnetic induction field $\boldsymbol{B}$. The temperature is assumed to be low enough for the contribution of crystal vibrations to be negligible.

Consider the case where the initial field B_{0i} is much larger than the local field B_L, and the final field B_{0f} is much smaller than B_L. Then, the spin temperature decreases, with,

$$\frac{\theta_f}{\theta_i} = \frac{B_L}{B_{0i}} \tag{16.73}$$

Thus, the spin temperature is much lower than the initial temperature. Typically, the initial temperature is the bath temperature.

Let us note one curious phenomenon associated with adiabatic cooling. The magnetization predicted by the Curie law (16.64) is given by:

$$M_f = M_i \frac{B_f}{\sqrt{B_{0f}^2 + B_L^2}} \tag{16.74}$$

This implies that the adiabatic demagnetization process leads to a very small magnetization when the applied field is brought down to zero. As the process is adiabatic, the inverse process can bring the magnetization back to the initial value. There is nothing magical about this; the magnetization in zero applied field is along the local field B_L and, since the process is adiabatic and reversible, the magnetization can be rebuilt by again applying a field adiabatically.[869]

16.6.2 ADIABATIC DEMAGNETIZATION IN THE ROTATING FRAME

Let us once again consider a system of spins that can be characterized by a spin temperature. Here, we assume that the system is subjected to an oscillating field at a frequency close to the resonance frequency of the spins. A. Redfield showed that the concept of spin temperature also applies in this case.[870] Because of the oscillating field, the Zeeman Hamiltonian becomes time-dependent, so we write,

$$\mathcal{H} = \mathcal{H}_Z(t) + \mathcal{H}_d \tag{16.75}$$

The oscillating field is assumed to be a rotating field of amplitude B_1, and of angular frequency ω. When we consider this system in the rotating frame picture, the Hamiltonian becomes,

$$\mathcal{H}' = -\gamma\hbar\left[\left(B_0 - \frac{\omega}{\gamma}\right)\hat{I}_z + B_1\hat{I}_x\right] + \mathcal{H}_d^0 + f(\omega, 2\omega) \tag{16.76}$$

where $f(\omega, 2\omega)$ stands for all the terms oscillating at frequency ω or 2ω. The term $\mathcal{H}_d^0$ is the part of the dipolar coupling that commutes with $\hat{I}_z$. The first term can be seen as a Zeeman coupling to an effective field,

$$\boldsymbol{B}_{\text{eff}} = \left(B_0 - \frac{\omega}{\gamma}\right)\hat{\boldsymbol{z}} + B_1\hat{\boldsymbol{x}} \tag{16.77}$$

The typical field value in this rotating picture is of the order of the effective field or the dipolar fields that the spins exert on one another. So, it is typically much less than the applied field B_0. Consequently, the terms at ω and 2ω are very far from resonance, as perceived in the rotating frame, and cannot induce transitions among the energy levels in the rotating frame picture. We neglect these terms and are left with,

$$\mathcal{H} = -\gamma\hbar\boldsymbol{B}_{\text{eff}} \cdot \boldsymbol{I} + \mathcal{H}_d^0 = \mathcal{H}_Z + \mathcal{H}_d^0 \tag{16.78}$$

Here, the presence of the B_1 term implies that $\mathcal{H}_Z$ and $\mathcal{H}_d^0$ do not commute. This implies that these two parts exchange energy. Since the Hamiltonian thus defined is time-independent, the energy is conserved and we can reasonably expect the system to come to thermal equilibrium with a certain spin temperature θ.

How long should we wait to reach this state? Since the coupling between the two parts depends on the presence of the B_1 term, this time depends on the strength of the oscillating field. By analogy with the results (16.64), (16.65), and (16.66), we find for the energy,

$$\bar{E} = -\frac{C\left(B_{\text{eff}}^2 + B_L'^2\right)}{\theta} \tag{16.79}$$

where,

$$C = \frac{N\gamma^2\hbar^2 I(I+1)}{3k} \tag{16.80}$$

Using (7.41) and writing $\boldsymbol{B}_{\text{eff}} = \mu_0\boldsymbol{H}_{\text{eff}}$, we have the Curie law,

$$\boldsymbol{M} = \frac{C}{\theta}\boldsymbol{H}_{\text{eff}} \tag{16.81}$$

The local fields have a strength given by,

$$B_L'^2 = \frac{1}{Ck_B(2I+1)^N}Tr\left((\mathcal{H}_d^0)^2\right) \tag{16.82}$$

The entropy is,

$$S = N\ln(2I+1) - \frac{C}{2}\frac{\left(B_{eff}^2 + B_L'^2\right)}{\theta^2} \tag{16.83}$$

The local fields, denoted as B_L', are distinct from those in (16.62) because here, only the secular part of the dipolar coupling needs to be considered. Further calculations would show that $B_L'^2$ is equal to one third of the second moment of the resonance line.[236] The Curie law in (16.81) implies that, when we sweep the frequency of the oscillating field, the magnetization follows the field and at resonance, i.e., when $\boldsymbol{B}_{\text{eff}} = B_1\hat{\boldsymbol{x}}$, the magnetization is perpendicular to the applied field.

With these results, we can now envisage *adiabatic demagnetization in the rotating frame*. Let us assume that we initially let the system come to equilibrium with the lattice in the static field, so that the spin polarization takes the value $M = M_0\hat{z}$. We assume that an oscillating field is first applied at a frequency well above the resonance. Then the magnetization remains along B_0. We then sweep the frequency to approach resonance. The magnetization follows and, as in (16.74):

$$M_f = M_i \frac{B_f}{\sqrt{B_{0f}^2 + B_L^2}} \tag{16.84}$$

The spin temperature, according to (16.74) and (16.81), is given by,

$$\theta = \frac{C}{M_0}\sqrt{B_{eff}^2 + B_L^2} \tag{16.85}$$

To the extent that M_0 is large and the local fields are small, it is possible to obtain a low spin temperature in the rotating frame. For example, we can imagine a situation in which we cool electron spins. We can then allow an efficient coupling between electron and nuclear spins. This can be done by applying irradiation to both nuclei under the Hartmann-Hahn condition.[5]

16.6.3   A SPIN-BASED CARNOT COOLING CYCLE

W. T. Wenckebach pointed out a possible Carnot machine involving nuclear spins and electron spins.[861] This is more of a thought experiment than a realistic implementation of DNP, but it contains some key ingredients of DNP. The electronic resonance is usually very broad, so we consider a packet of spins that have approximately the same frequency. These electron spins are coupled to nearby nuclei very strongly, and mutual spin-flip among them are not possible.[862] There are many other nuclei in the samples that are further away from electron spins. Nuclear polarization diffuses among these nuclei by *spin diffusion*, of the type illustrated in Fig. 16.6.

Up to now, we have considered the spin system as totally decoupled from the outside world. Now, we take into consideration spin-lattice relaxation. In a Carnot cycle, entropy is carried by some medium (typically a gas) from the cold source to the hot source if the machine is a heat pump. In the experiment we want to imagine here, the electrons transport this entropy; the cold reservoir is the bath of bulk nuclei and the hot reservoir is the lattice. The four segments of the Carnot cycle are shown in Fig. 16.8. We have,

- an isothermal process: the electron spins are polarized in the applied field by spin lattice relaxation,
- an adiabatic demagnetization in the rotating frame using a microwave field, so that the electron spins establish a spin temperature significantly lower than the lattice temperature,
- an isothermal process, when, in the rotating frame picture, the energy differences between eigenstates approximate the values of the transition energies for the nuclei, so electron and nuclear spins are "on speaking terms," and thermal equilibrium takes place by transferring entropy from the nuclei to the colder electron spins,
- an adiabatic remagnetization performed by sweeping the frequency away from resonance.

---

[5][236], §7.12.

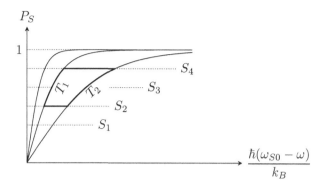

Figure 16.8 Carnot cycle drawn on a diagram showing spin polarization P_S as a function of the frequency offset for various temperatures (isotherms). A process at fixed P_S is adiabatic (dotted lines).[861]

In the most common form of DNP, the microwave field is applied continuously. Then, the adiabatic and isothermal processes occur together. There are experiments, however, where these two steps are clearly separated. One of these is the DNP method called *nuclear orientation via electron spin locking* (NOVEL).[871–873] The other is the *integrated solid effect*.[874–876]

16.7 FURTHER READINGS

Spin temperature

In the early days of nuclear magnetic resonance, the validity of the concept of spin temperature was verified experimentally.[878] Optical magnetometry measurements on nuclear spins showed that the spin temperature concept also applies to GaAs microcavities.[879] The extension of the spin temperature concept to translational motion of freedom was envisaged for ultracold quantum gases.[880] The concept of spin temperature applies to any given system only when there are strong enough interactions among the constituents of the system.[881, 882]

DNP and quantum computing

One of the essential ingredients of quantum information processing is the use of ancillary qubits, which provide the means for quantum error correction (QEC). When spins are considered for qubits, these ancillia can indeed provide error correction only if they are highly polarized. Hence, there is a need, in spin-based quantum computing, for some form of dynamic polarization of the ancillary qubits.[683]

Cycling through a cooling process is one possibility to achieve a high polarization of qubits. In the so-called *heat-bath algorithmic cooling*, entropy is taken away from target qubits and accumulated on an ensemble of "reset" qubits. These "reset" qubits can be connected to a bath, into which the accumulated entropy is dumped.[883, 884]

Current-induced spin polarization

As we have seen in §16.2, a large change in electron spin polarization may lead to an enhanced nuclear polarization. Electron spin polarization does change rapidly when spins are driven across the interface between two different materials.[885, 886] This long sought-after current-induced nuclear spin hyperpolarization was eventually observed in semiconductors.[887–889]

Optically-induced spin polarization

In this chapter, we focused on spectroscopic techniques by which thermal-equilibrium electron spin polarization can be transferred to nuclear spins. Enhancing the *electron spin polarization* may also be of interest. Notably, from the perspective of DNP, there is a strong interest for techniques that allow high electron spin polarization without the use of superconducting electromagnets and cryogenic equipment to reach ultra-low temperatures. One method is an optical excitation in III–V semiconductors. The selection rules for optical transitions between the conduction band (CB) and the heavy-hole and light-hole bands (Fig. 11.1) are such that, by using circularly polarized light, a spin polarization can be "pumped" into the conduction band. This phenomenon is referred to as *optical orientation*.[890] Applying ferromagnetic electrodes and tunnel junctions to optically-excited GaAs, it is possible to detect this effect using a typical spintronics strategy.[891,892] Optical orientation has also been used to test spin-dependent electron transfer at GaAs electrodes which were functionalized with chiral molecules.[893]

Point defects in diamond called *NV-centers*,[894] allow for *optically induced DNP*.[895, 896] A sophisticated combination of optical excitation and microwave irradiation over a broad frequency range allows for hyperpolarization of ^{13}C spins at room temperature in nanodiamonds of all orientations.[897] The transfer of polarization outside the hyperpolarized diamond has been demonstrated.[898] NV centers could become of use as spin qubits that can be hyperpolarized at room temperature. For example, a 50% polarization was obtained.[899] Using the paramagnetic color centers in *silicon carbide* nuclear hyperpolarization has been achieved.[900]

It is also possible to hyperpolarize nuclei via optical excitation, e.g., by exciting *triplets*,[901, 902] or using the optical orientation of III-V semiconductors.[890, 903–905] Nuclei in a *quantum dot* can be polarized in a few milliseconds.[906, 907] If the electron spin on the dot is removed immediately after polarization, the nuclear spin lifetime is in the seconds range. It can be made one or two orders of magnitude longer by applying a field of about 1 mT. This field strength is sufficient to quench the non-secular nuclear dipolar couplings.

DNP strategies

Spin polarization can be obtained in a non-magnetic material by applying a magnetic field. When a spin polarization is achieved than is higher than this equilibrium polarization, it is generally called a *hyperpolarization*. Many methods have been developed that generate nuclear or electron spin hyperpolarization.[908]

When Overhauser developed his ideas, he had in mind the coupling of electron spins and nuclear spins in a metal. Nowadays, the Overhauser effect is generally used to hyperpolarize nuclear spins in liquids, where each molecule undergoes collisions with the polarizing radical in solution. This method can become quite challenging at the high fields commonly used to carry you NMR spectroscopy.[857, 909] The Overhauser effect can also be implemented to transfer spin polarization from a nuclear spin with a large gyromagnetic factor (nuclear "gamma") to one with a much smaller gamma. This is the *nuclear Overhauser effect* (NOE). [910, 911]

In solids, an alternative to the solid effect is the *cross-effect*.[912] The cross-effect involves a biradical as polarizing agent. The DNP mechanism in this case involves therefore a three-spin system.[913, 914] The dynamics of the cross-effect changes considerably when the sample is rotating, which is something that is done in high-resolution NMR. An Overhauser effect was observed in dielectric solids.[915] The effect is thought to occur thanks to the hyperfine fluctuations due to molecular vibrations.[916]

In a non-magnetic solid, spin polarization diffuses in space from nuclei that are located near an electron spin. The mechanism that allows for this spin diffusion was first examined in the framework of *spectral spin diffusion*. This diffusion process is manifest in *spectral hole burning*. If a broad resonance is driven at saturation at a specific frequency in the spectrum, and the spectrum is measured immediately afterwards, there will be a "hole" at that frequency. Two mechanisms contribute to the relaxation of this hole. One is the spin-lattice relaxation. The other is spectral spin diffusion.[917, 918]

Using photoexcitation as a form of cooling engine, W. T. Wenckebach's group achieved a 5,000-fold signal increase at room temperature using a system of pentacene molecules embedded in a naphthalene crystal.[875, 877]

In view of the matching energy condition (12.160), nuclei with different gyromagnetic ratios cannot undergo mutual spin flips. As Slichter put it in his textbook, "you would have to apply different magnetic fields to different nuclei."[236] The solution is the *Hartmann-Hahn method*.[919] When a rotating field is applied at a frequency far from resonance, it is ineffective. So the radio-frequency field that excites one resonance does not affect the other. In their respective rotating frames (see §19.3), each nuclear spin species precesses around the effective field. The field strengths can be adjusted so that the precession frequencies are the same. This is the Hartmann-Hahn condition. The underlying phenomenon is the basis for another family of dynamic polarization methods known as *cross polarization*.[236, 920] NOVEL also applies the Hartman-Hahn method to connect electron and nuclear spins.[862] The Hartmann-Hahn condition imposes requirements that are often difficult to meet practially. An alternative was demonstrated, which is based on a nuclear spin-locking method.[921] This method was demonstrated at high fields, showing that a source of low power is sufficient. This is a great advantage when working at high fields, i.e., in the sub-THz frequency range, where high-power is difficult to obtain.[922]

DNP instrumentation

NMR spectrometers have gained in sensitivity over the years by increasing the applied field. As DNP experiments require the excitation of the electron spin resonance, the frequency reaches the sub-THz range in high-field NMR spectrometers. As a high power is generally needed for the excitation of the electron spin resonance, the use of cyclotron maser as a microwave source was envisaged as early as 1993.[923] These masers, called *gyrotrons*, became commercially available nearly 20 years later.

Since this is a very demanding technology, alternatives are considered. For example, the sample can be quickly brought from one region of low magnetic field, where it is hyperpolarized, to a region of high magnetic field for spectroscopy.[924] The method is referred to as *field cycling*. It is also possible to turn the sample into a liquid in the high field magnet.[925] This is a variation on *dissolution DNP*, a technique that consists of hyperpolarizing in a magnet at very low temperatures, then rapidly melting the sample by adding a hot solvent, and finally, transferring the resulting solution to another magnet where NMR spectroscopy is conducted.[926, 927]

NMR imaging may improve significantly in some areas thanks to the use of hyperpolarized nanoparticles.[931] Also, gases can be hyperpolarized by a variety of methods.[928] Xenon, for example, has been polarized by laser pumping.[929] Xenon is especially interesting because of its many applications in probing surfaces or imaging lungs.[930]

17 Ferromagnetic Resonance Spectroscopy

Prof. Charles Kittel (1916–2019)

C. Kittel started his research career at Bell Telephone Laboratories before becoming professor at the University of California, Berkeley. His theoretical interests included magnetism, ultrasounds and thermal properties of solids. With the aim of determining material properties, his research included electron spin resonance, nuclear spin resonance and resonance in ferromagnets. Together with M. Ruderman, he developed a model for the magnetic interaction now known as the RKKY coupling.

Spintronics: A primer

J.-Ph. Ansermet (https://orcid.org/0000-0002-1307-4864)

Chapter 17. Ferromagnetic Resonance Spectroscopy

Ferromagnetic resonance spectroscopy characterizes: the uniform mode excited at a frequency given by the Kittel formula; the magnetostatic modes occurring at long wavelengths; and exchange spin waves, such as the perpendicular stationary spin wave (PSSW) of thin films. The thermodynamics of irreversible processes is used to describe spin transfer torques in conducting ferromagnets, spin-pumping spin current, and spin-wave spin current.

DOI: 10.1201/9781003370017-17

17.1 HISTORICAL INTRODUCTION

In 1946, J. H. E. Griffiths reported in a letter to Nature an "Anomalous high-frequency resistance of ferromagnetic metals." [932] His intention was to measure the magnetic susceptibility μ of ferromagnetic metals at microwave frequencies. He electrodeposited a thin film of nickel at the bottom of a cylindrical cavity and applied an in-plane steady magnetic field B that was perpendicular to the microwave magnetic field. He monitored the quality factor of the resonator, which is proportional to the product $\mu\rho$ where ρ is the resistivity of the film, and found a non-monotonic change that he attributed to a resonance. The frequency was not at the Larmor frequency corresponding to the applied magnetic field. In 1947, Kittel explained the discrepancy by taking into account the demagnetizing field.[933] His derivation was made clear in his textbook and the formula is commonly called the *Kittel formula*.[79]

In 1957, L.R. Walker realized that modes other than Kittel's mode could be excited by a uniform microwave magnetic field in samples with dimensions similar to the electromagnetic excitation wavelength.[934] Hence, these *magnetostatic modes* are often called Walker modes.

In 1960, H. Juretschke realized that in any thin ferromagnetic film with magnetoresistance, the relationship $V = R(M)I$ between voltage V and current I is non-linear. For Juretschke, the non-linearity stems from the fact that the microwave irradiation induces eddy currents that cause a time evolution of the magnetization, which in turn causes a time evolution of the resistance $R(M)$. Thus, the resonant excitation of the magnetization M at a frequency f generates signals at $f = 0$ and at the second harmonic $2f$.[935] He verified his prediction using thin films of ferromagnetic nickel and paramagnetic gadolinium.[936] These experiments were the first *electrically detected magnetic resonance*, i.e., they were detected in the form of a voltage difference detected on two points of the sample itself.

To understand the spin transport across a ferromagnetic-paramagnetic bilayer observed in experiments using transmission electron spin resonance (see TESR in §15.5), a thermodynamics argument accounting for the spin current at the interface was proposed by M. Sparks and R. H. Silsbee.[937] At that time, R. H. Silsbee and M. Johnson had just written their foundational articles on spintronics.[429, 938] A description based on the thermodynamics of irreversible processes was presented later, when spin pumping became a frequently used method for generating spin currents.[939, 940]

Magnetization dynamics in the presence of unpaired conduction electron spins was first explored in the 1950s. This problem was explained by J. Winter in his 1971 treatise on magnetic resonance in metals.[698] He combined the ideas of H. Hasegawa and F. J. Dyson to account for the coupling of local magnetic moments and conduction electron spins.[698–700] This was long before spintronics research explored the effect of a charge-driven spin current on magnetization dynamics. Winter wrote coupled equations for the magnetizations M_s and M_d. In a thermodynamic approach to the dynamics of s-d coupling, Hasegawa pointed out that there are 3 systems that need to be considered: M_s, M_d, and L, the translational motion. L is known to be a very fast subsystem. The issue was whether it could be assumed that M_s is strongly coupled to L or whether M_s and M_d are more strongly coupled. Hasegawa concluded that, in most cases, the situation is midway between these two limits. So, he proposed Bloch equations to allow for all cases. In his 1971 treatise, J. Winter took over these conclusions and added a diffusion term of the form $D\nabla^2 M_s$ to account for the diffusion of the spin-polarized conduction electrons ([698], chapter 10). There was no such term for M_d because it describes localized spins, which do not diffuse. Hence, Winter wrote:

$$\frac{dM_s}{dt} = \gamma_s M_s \times B_0 + \lambda\gamma_s M_s \times M_d - \frac{1}{T_{sd}}M_s + \frac{1}{T_{ds}}M_d - \frac{1}{T_{sl}}\left(M_s - M_{s0}\right) + D\nabla^2 M_s$$

$$\frac{dM_d}{dt} = \gamma_d M_d \times B_0 + \lambda\gamma_d M_d \times M_s + \frac{1}{T_{sd}}M_s - \frac{1}{T_{ds}}M_d - \frac{1}{T_{dl}}\left(M_d - M_{d0}\right)$$

$$(17.1)$$

We recognize in these equation a detailed balance and spin-lattice relaxation. The phenomenological parameter λ is proportional to the exchange coupling J_{sd}. Following W. Heitler and E. Teller (1936),[941] and A. W. Overhauser (1953),[855] H. Hasegawa provided expressions for the spin-spin relaxation rates T_{sd}^{-1} and T_{ds}^{-1}, by analogy with the relaxation of nuclear spins through hyperfine coupling. He then linearized equations (17.1) in the absence of the diffusion term and solved for the normal mode problem. The terms containing T_{sl}^{-1} and T_{dl}^{-1} in (17.1) account for spin-lattice relaxation processes. These ideas were further developed by S.E. Barnes.[697]

The exchange coupling between ferromagnetic layers separated by non-magnetic layers was studied extensively in the 1980s. These novel nanostructures were investigated not only for their all-important giant magnetoresistance, but also for their magnetic excitations. The magnetostatic waves of a double layer were calculated by P. Gruenberg, who won the Nobel Prize with A. Fert for their discovery of giant magnetoresistance in magnetic multilayers.[942] The presence of magnetostatic modes in multilayers and their dependence on the spacing between adjacent magnetic layers was reviewed by J. Barnas.[943]

In this chapter, we derive Kittel's formula for the uniform mode and proceed with an introduction to magnetostatic modes, which are long wavelength excitations where exchange energy does not play a role. We then discuss the excitation of exchange spin waves, first introduced in Chapter 7. The chapter ends with a description of spin current and spin torques, based on a thermodynamic approach. First, the notion of spin transfer torque is explained. Then, we desribe the spin-pumping effect in terms of the generation of a spin current and finally, we point out that a spin current is associated with spin waves.

17.2   FERROMAGNETIC RESONANCE

The uniform precession of the magnetization of a ferromagnet around the magnetization equilibrium orientation is called **ferromagnetic resonance** (FMR). We start with a very well known formula for the resonance frequency. We then discuss why the bare electron gyromagnetic ratio appears in this formula.

The Kittel formula

Let us consider a ferromagnetic single domain magnetic particle that has an ellipsoidal shape. A uniform, static magnetic induction field $B_0 = B_0\hat{z}$ that is sufficiently large is applied. It can be shown that the magnetization is uniform in this case.[944] The value of the field when this happens is called the **saturation field**. Depending on the material, its strength is, typically, of about 0.1–2 T. The magnetic induction field that drives the magnetization dynamics is composed of the applied field B_0 and the magnetic induction associated with the **demagnetization field** H_{demag}. In a linear medium, the demagnetization field is a linear function of the magnetization M and can be written in tensorial form as follows,

$$\boldsymbol{H}_{demag} = -\mathsf{N} \cdot \boldsymbol{M} = - \begin{pmatrix} N_x & 0 & 0 \\ 0 & N_y & 0 \\ 0 & 0 & N_z \end{pmatrix} \begin{pmatrix} M_x \\ M_y \\ M_z \end{pmatrix} \tag{17.2}$$

where $Tr(\mathsf{N}) = 1$ and the coordinate axes are along the principal axes of the tensor N. This field is opposed to the magnetization, hence its name. The concept of a demagnetizing field is best understood by calculating the magnetization inside a sphere in a magnetic induction field that is homogeneous at a large distance from the sphere.[170, 350]

In the **Landau-Lifshitz equation** (7.21), the magnetic induction is simply the sum of the applied field and the magnetic induction field associated with the demagnetizing magnetic field,

$$\frac{d\boldsymbol{M}}{dt} = \gamma \, \boldsymbol{M} \times (\boldsymbol{B}_0 + \mu_0 \boldsymbol{H}_{demag}) \tag{17.3}$$

We consider the evolution of a small perturbation $\boldsymbol{m}$ from the equilibrium magnetization $\boldsymbol{M}_s$, i.e., $\boldsymbol{M} = \boldsymbol{M}_s + \boldsymbol{m}$. The equilibrium $\boldsymbol{M}_s$ is the solution of,

$$\boldsymbol{0} = \boldsymbol{M}_s \times (\boldsymbol{B}_0 - \mu_0 \mathsf{N} \boldsymbol{M}_s) \tag{17.4}$$

Straightforward calculations yield $\boldsymbol{M}_s = M_s \hat{\boldsymbol{z}}$. To the first order, the evolution of $\boldsymbol{m}$ is given by,

$$\frac{d\boldsymbol{m}}{dt} = -\gamma \boldsymbol{M}_s \times \mu_0 \mathsf{N} \cdot \boldsymbol{m} + \gamma \boldsymbol{m} \times (\boldsymbol{B}_0 - \mu_0 \mathsf{N} \cdot \boldsymbol{M}_s) \tag{17.5}$$

Let us assume that the perturbation $\boldsymbol{m}$ is of the form,

$$\begin{pmatrix} m_x \\ m_y \\ 0 \end{pmatrix} e^{i\omega t} \tag{17.6}$$

Substitution yields two linear equations for m_x and m_y that have non-trivial solutions if and only if,

$$\begin{vmatrix} i\omega & \gamma B_0 + \gamma \mu_0 M_s(N_y - N_z) \\ -\gamma B_0 - \gamma \mu_0 M_s(N_x - N_z) & i\omega \end{vmatrix} = 0 \tag{17.7}$$

If we were to apply a microwave field at the pulsation ω, we would be on resonance with the system, since ω is its natural frequency of precession, just as the on-resonance excitation frequency of a harmonic oscillator (without damping) is the frequency of its eigenmode. Thus, the resonance frequency is given by the following formula, known as the **Kittel formula**,[79]

$$\hbar\omega = g\mu_B \sqrt{(B_0 + (N_x - N_z)\mu_0 M_s)(B_0 + (N_y - N_z)\mu_0 M_s)} \tag{17.8}$$

where we use the definition (7.2) of γ and $M_s = |\boldsymbol{M}|$ is often referred to as the **saturation magnetization**.

The giant angular momentum approach

The evolution equation for $\boldsymbol{M}$ given by the Landau-Lifshitz formula (7.21) may be surprising, since it involves the gyromagnetic ratio of the free electron, defined in γ (7.2), whereas $\boldsymbol{M}$ refers to a collective state of the electrons of the ferromagnet. However, in 1948, J. M. Luttinger and C. Kittel showed that this semi-classical approach is consistent with a rigorous quantum mechanical approach.[945] Luttinger and Kittel started out with a classical expression for the energy associated with the demagnetizing fields, which gives the correct demagnetizing field according to (7.20),

$$V(\boldsymbol{M}) = \frac{\mu_0}{2}\left(N_x M_x^2 + N_y M_y^2 + N_z M_z^2\right) \tag{17.9}$$

from which they formed a Hamiltonian. They also included a Zeeman term by replacing the classical magnetic moment $\boldsymbol{M}V$ of the ellipsoid of volume V by an angular moment $\hbar\boldsymbol{J}$ such that,

$$-g\mu_B\hat{\boldsymbol{J}} = \boldsymbol{M}V \tag{17.10}$$

Thus, Luttinger and Kittel wrote,

$$\mathcal{H} = g\mu_B\hat{J}_z B + \frac{1}{2}\frac{g2\mu_B^2}{V}\left(N_x\hat{J}x^2 + N_y\hat{J}y^2 + N_z\hat{J}z^2\right) \tag{17.11}$$

In other words, they represented the angular momentum of the correlated state of the ferromagnet by a giant angular moment of some 10^{15} spins. At normal temperatures, the eigenvalue m of $\hat{J}z$ was estimated to be approximately 10^5 smaller than that of J. They worked out an evolution equation in this limit of large J and large m, with m much smaller than J. The resulting equation is analogous to that of a harmonic oscillator. The spacing between the energy levels corresponds exactly to the classical energy given in (17.8). A. H. Morrish relies on this result to justify his classical treatment of FMR.[946]

17.3 MAGNETOSTATIC MODES

Let us now consider non-uniform magnetization excitations in the limit of long wavelength. By "long wavelength," we mean that exchange energy plays a negligible role. We addressed cases in which exchange energy plays a role in §7.5 and will discuss excitation of exchange spin waves in §17.4. We want to consider modes that are not uniform, but have a spatial variation on the scale of the sample size. The typical frequency of such modes is near 10 GHz for a thin film sample, and the wavelength is of the order of the sample's thickness, say 1 µm. In free space, an electromagnetic wave of this wavelength would have a frequency given by $c/\lambda \approx 3 \cdot 10^{14}$ Hz. The frequency of the modes we want to describe in this section is much smaller, roughly 10^{10} Hz. These modes are called *magnetostatic modes*.

Because the frequency of these magnetostatic modes is so low, we need not consider radiative effects. Consequently, we can neglect the displacement current term of Maxwell's equations. Furthermore, we will only address cases where electromagnetic induction is negligible. This is certainly the case with insulators such as garnets and ferrites, where magnetostatic modes were first observed.

Since we are considering the magnetostatic limit of Maxwell's equations in the absence of currents, we have,

$$\begin{aligned}\boldsymbol{\nabla} \times \boldsymbol{H} &= \boldsymbol{0} \\ \boldsymbol{\nabla} \cdot \boldsymbol{B} &= \boldsymbol{0}\end{aligned} \tag{17.12}$$

The equation of evolution (7.21) for $\boldsymbol{M}$ can be used to derive the constitutive equation relating $\boldsymbol{M}$ and $\boldsymbol{H}$. By writing $\boldsymbol{B} = \mu_0(\boldsymbol{H} + \boldsymbol{M})$, the Landau-Lifshitz equation (7.21) becomes,

$$\frac{d\boldsymbol{M}}{dt} = \gamma\mu_0\,\boldsymbol{M} \times \boldsymbol{H} \tag{17.13}$$

Let us consider the linear response of the system when it is near an equilibrium assumed to be a uniform magnetization. Hence, we set:

$$\begin{aligned}\boldsymbol{M} &= M_s\hat{\boldsymbol{z}} + \boldsymbol{m}(t) & \boldsymbol{H} &= H_i\hat{\boldsymbol{z}} + \boldsymbol{h}(t) \\ \boldsymbol{m}(t) &= \boldsymbol{m}\,e^{i\omega t} & \boldsymbol{h}(t) &= \boldsymbol{h}\,e^{i\omega t}\end{aligned} \tag{17.14}$$

The notation H_i designates the large field inside the sample, not the outside applied field. Keeping terms of the first order in m and h, the evolution equation (17.13) for M gives,

$$\frac{i\omega}{\gamma\mu_0}\begin{pmatrix} m_x \\ m_y \\ m_z \end{pmatrix} = \begin{vmatrix} \hat{x} & m_x & h_x \\ \hat{y} & m_y & h_y \\ \hat{z} & M_s + m_z & H_i + h_z \end{vmatrix} \approx \begin{pmatrix} m_y H_i - M_s h_y \\ M_s h_x - m_x H_i \\ 0 \end{pmatrix} \qquad (17.15)$$

Here, we introduce the notation first proposed by R. W. Damon and J. R. Eshbach, converted to the International System of Units (SI units).[947] Thus, we define,

$$\frac{\omega}{\mu_0\gamma M_s} = \Omega \qquad\qquad \frac{H_i}{M_s} = \Omega_H$$

$$\kappa = \frac{\Omega_H}{\Omega_H{}^2 - \Omega^2} \qquad \nu = \frac{\Omega}{\Omega_H{}^2 - \Omega^2} \qquad (17.16)$$

The susceptibility tensor deduced from the linearized evolution equation above is:

$$\begin{pmatrix} m_x \\ m_y \end{pmatrix} = \begin{pmatrix} \kappa & -i\nu \\ i\nu & \kappa \end{pmatrix}\begin{pmatrix} h_x \\ h_y \end{pmatrix} \qquad (17.17)$$

This susceptibility tensor is known as the **Polder tensor**.

Since $H_i\hat{z}$ is uniform, we have $0 = \nabla \times H = \nabla \times h$. Therefore, there is a scalar potential function ϕ such that,

$$\nabla\phi = h \qquad (17.18)$$

Maxwell's equation $\nabla \cdot B = 0$ yields $\nabla \cdot (H + M) = 0$, which in components reads,

$$\nabla \cdot \left\{ \begin{pmatrix} h_x \\ h_y \\ h_z \end{pmatrix} + \begin{pmatrix} \kappa & -i\nu & 0 \\ i\nu & \kappa & 0 \\ 0 & 0 & 1 \end{pmatrix}\begin{pmatrix} h_x \\ h_y \\ h_z \end{pmatrix} \right\} = 0 \qquad (17.19)$$

In terms of the magnetic potential ϕ, this gives:

$$(1 + \kappa)\left(\frac{\partial^2\phi}{\partial x^2} + \frac{\partial^2\phi}{\partial y^2}\right) + \frac{\partial^2\phi}{\partial z^2} = 0 \qquad (17.20)$$

Outside the ferromagnet, $\kappa = 0$ since no magnetization m is induced by an applied magnetic field. In other words, the differential equation for the potential ϕ outside the sample is given by,

$$\frac{\partial^2\phi}{\partial x^2} + \frac{\partial^2\phi}{\partial y^2} + \frac{\partial^2\phi}{\partial z^2} = 0 \qquad (17.21)$$

The Maxwell equations in the magnetostatic regime (17.12) impose the following boundary conditions:

1. The parallel component of H is continuous across the interface between the material and its environment.
2. The component of B that is normal to the interface is continuous across the interface.

In terms of the magnetic potential, these boundary conditions translate into the following conditions:

1. The potential is continuous at the surface: $\phi^{in} = \phi^{out}$, and,
2. The normal derivative is continuous at the surface: $\frac{\partial\phi^{in}}{\partial n} = \frac{\partial\phi^{out}}{\partial n}$.

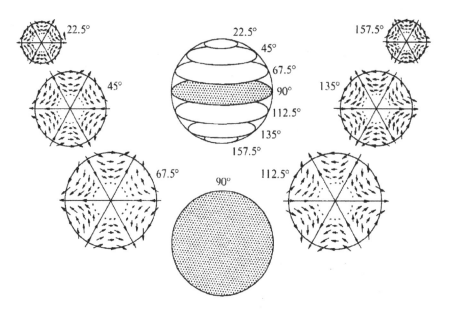

Figure 17.1 Geometrical representation of one Walker mode in a sphere.[948] Planes are defined at set latitudes (center top) and the transverse magnetization in these planes (left and right sides) is represented in a rotating frame picture (§15.2).

3. The potential is zero at infinity: $\phi^{out} \to 0 \quad \|\mathbf{r}\| \to \infty$.

While the derivation above provides the eigenmodes of a non-uniform precession, it does not say whether, or how, these modes can be excited. Inhomogeneous microwave fields should be used to excite these modes. This point is of practical relevance. It will be examined quantitatively in §17.4 for the case of exchange spin waves.

In the 1960s, L. R. Walker wrote a review on magnetostatic modes.[949] Hence, these modes are often referred to as **Walker modes**. L. R. Walker worked out the modes for a sphere (Figure 17.1),[934,948] and R. Plumier for a cylinder.[950] The magnetostatic modes of an infinite slab can be found, e.g., in Landau and Lifshitz's book on electromagnetism of continuous media.[951]

Surface magnetostatic modes are to be expected and were studied by Damon and Eshbach.[947] These modes are characterized by a wavevector which is perpendicular to the equilibrium mangnetization. These modes, now known as **Damon-Eshbach** (DE) modes.[1] DE modes are still a subject of research, for example, in the framework of studies on magnetization reversal.[953]

17.4   EXCHANGE SPIN WAVES EXCITATION

Exchange spin waves were introduced in §7.5. In the early days of ferromagnetic resonance, it was expected that strong field gradients would be necessary to excite such short wavelength modes. In 1958, Kittel pointed out that the spins at the surface ought to be pinned by the

---

[1]See, [952], chapter 5.

surface magnetic anisotropy.[954] In order to examine the extent to which these modes can be excited with a uniform microwave field, we consider the magnetic induction field to be:

$$
\begin{pmatrix} b_x \\ b_y \\ B_i \end{pmatrix}
\tag{17.22}
$$

Recall that in §7.5, we worked out the eigenmodes of the Landau-Lifshitz equation (7.51) with an exchange term. With the field as specified here, the equation of motion (7.51) reads,

$$
\begin{pmatrix} \partial_t m_x \\ \partial_t m_y \\ \partial_t m_z \end{pmatrix} = \begin{vmatrix} \hat{x} & \gamma m_x & b_x \\ \hat{y} & \gamma m_y & b_y \\ \hat{z} & \gamma M_s & B_i \end{vmatrix} + D \begin{vmatrix} \hat{x} & m_x & \nabla^2 m_x \\ \hat{y} & m_y & \nabla^2 m_y \\ \hat{z} & m_z & 0 \end{vmatrix}
\tag{17.23}
$$

We introduce the notation $b_\pm = b_x \pm i b_y$ and $m_\pm = m_x \pm i m_y$. Then (17.23) can be written as,

$$
\frac{\partial m_\pm}{\partial t} = i\gamma M_s b_\pm - i\gamma B_i m_\pm + iDM_s \nabla^2 m_\pm
\tag{17.24}
$$

We seek a solution of the form,

$$
m_+ = e^{i\omega t} \sum_p a_p \sin(k_p z)
\tag{17.25}
$$

If one side of the film is at $z = 0$ and the other is at $z = L$, then the pinning of the spins at the surface imposes,

$$
k_p = \frac{p\pi}{L}
\tag{17.26}
$$

The microwave field is assumed to be homogeneous, of the form $b_+ e^{i\omega t}$. So the amplitudes a_p must satisfy,

$$
\sum_p a_p (\omega + \omega_p) \sin(k_p z) = \gamma M_s b_+
\tag{17.27}
$$

The pulsation ω_p is given in terms of k_p by the dispersion (7.54). The modes $\sin(k_p z)$ form an orthogonal basis. To determine any one of the coefficients a_p, e.g., the coefficient a_m, we can calculate the projection of (17.27) on the basis given by $\sin(k_m z)$. To do so, we multiply by (17.27) by $\sin(k_m z)$ and integrate from 0 to L. This yields,

$$
a_m (\omega + \omega_m) \frac{L}{2} = \frac{\gamma M_s b_+}{\pi m/L}(1 - \cos(m\pi))
\tag{17.28}
$$

The modes with an even value for m are not excited. The oscillator strength of the odd modes is given by,

$$
a_m = \frac{4\gamma M_s}{(\omega + \omega_m)\pi m} b_+
\tag{17.29}
$$

The larger m is, the weaker is the resonance amplitude. This is intuitive: the higher m is, the smaller is the wavelength of the mode. As a consequence, the larger m is, the harder it is to excite the mode of index m with a uniform field. In a typical microwave absorption experiment, the frequency ω is kept constant for practical reasons. The applied field B_0 is swept instead.

Soon after Kittel suggested that spin waves could be excited in a thin film with a homogeneous field, the spin wave resonances of a permalloy thin film were detected (Figure 17.2).[955] With such a highly structured spectrum, a direct estimate of the

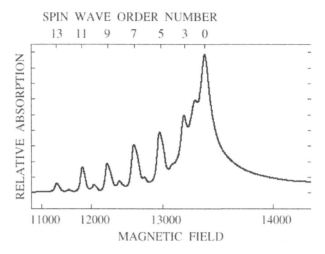

Figure 17.2 First modes of a thin permalloy film, about 100 nm in thickness. From [955]

exchange constant D of the dispersion (7.54) can be made using (17.29), provided M_s is known.

It may be useful to keep some orders of magnitudes in sight. For the experiment above, the film was about half a micrometer thick, and the spacing between two adjacent lines of the spectrum was of about 500 Gauss. This implies that the mode $m = 5$ has a wavelength of about 100 nm.

We see that these modes depend on both the exchange constant and the size of the sample. Instead, magnetostatic modes depend on the shape of the sample, but not on the exchange constant. The magnetostatic modes can be shown to be the long wavelength limit of the spin wave modes, as it should be.

17.5   SPIN TRANSFER TORQUE

We now turn to a description of the magnetization dynamics of two clearly distinct subsystems in a ferromagnetic metal: the local moments of the ferromagnet and the spin polarization of its delocalized electrons. In Chapter 8, we reviewed the thermodynamics of irreversible processes and saw that this thermodynamic appraoch can infer couplings among effects of different natures. So, we expect that the same descriptive framework provides invaluable relations between generalized currents for magnetization and generalized forces for two "substances": the localized magnetic moments, labelled with the index A; and the delocalized electrons spins, labelled B. The description below leads to a dynamic coupling between them, i.e., a spin transfer torque.[377, 956].

Continuity equations

Magnetization is a reduced extensive quantity (since magnetization is a magnetic moment per unit of volume). Therefore, we can write a continuity equation for magnetization which is similar to those we wrote for spin-up and spin-down electrons in (8.2). Since the continuity equation of any reduced extensive quantity comprises a current term and a source term, we can identify a magnetization current in the LLG equation (see §17.6 and §17.7).

Let us start with a quick review of continuity equations in continuous media. For a reduced scalar extensive quantity f of a fluid characterized by the velocity field $\boldsymbol{v}$, the continuity equation for f in the local rest frame can be written as,[376]

$$\dot{f} + (\boldsymbol{\nabla} \cdot \boldsymbol{v})f + \boldsymbol{\nabla} \cdot \boldsymbol{j}_f = \rho_f \tag{17.30}$$

where j_f is the current density of f (a diffusive current), and ρ_f the source of f. The term $\dot{f}$ is to be understood as the material derivative, defined as,

$$\dot{f} = \partial_t f + (\boldsymbol{v} \cdot \boldsymbol{\nabla})f \tag{17.31}$$

Hence, we can rewrite (17.30) as,

$$\partial_t f + \boldsymbol{\nabla} \cdot (f\boldsymbol{v} + \boldsymbol{j}_f) = \rho_f \tag{17.32}$$

We see two types of current densities in this continuity equation. The first is called a **convective current** density, whereas the second is a **diffusive current density**. These notions are clear when we think of, for example, a drop of ink in a glass of water. The motion of ink entrained by the motion of water is a convective term. The process that follows depositing a drop of ink at the bottom of a glass of water, when ink slowly spreads within the still water, is a diffusive process. Here, as in Chapter 8 on spin-dependent transport, there is no convective current and $\dot{f} = \partial_t f$.

For a reduced vectorial extensive quantity $\boldsymbol{f}$, the continuity equation for $\boldsymbol{f}$ in the local rest frame has the same form as 17.30, namely,

$$\dot{\boldsymbol{f}} + (\boldsymbol{v} \cdot \boldsymbol{\nabla})\boldsymbol{f} + \boldsymbol{\nabla} \cdot \mathsf{j}_{\boldsymbol{f}} = \boldsymbol{\rho}_{\boldsymbol{f}} \tag{17.33}$$

We have used a special font to designate the diffusive current $\mathsf{j}_{\boldsymbol{f}}$ to signal that it is a second-rank tensor represented by a 3×3 matrix. Let us now apply the continuity equation (17.33), first for the case of magnetic moments $\boldsymbol{m}_d$ fixed at atomic sites, and second, for the magnetic moments $\boldsymbol{m}_s$ associated with conduction electrons spins. Here, as in Chapter 8, there is no convection effect because the system is a solid, i.e., $\boldsymbol{v} = \boldsymbol{0}$ and $\dot{\boldsymbol{f}} = \partial_t \boldsymbol{f}$. In both cases, we will have to specify the vectorial source term $\boldsymbol{\rho}_{\boldsymbol{f}}$ from physical considerations.

Localized moments

Let us start with the magnetic moments $\boldsymbol{m}_d$ fixed at atomic sites. In the following, we do not consider a spin wave spin current, a notion that will be introduced in §17.7. Because the moments $\boldsymbol{m}_d$ are associated with atomic sites, there is no velocity field associated with these moments and no convective spin current is expected. Hence, the continuity equation for the localized moments must be of the form,

$$n_d \dot{\boldsymbol{m}}_d = \boldsymbol{\rho}_{M_d} = \gamma n_d (\boldsymbol{m}_d \times \boldsymbol{B}_d) - n_d \boldsymbol{m}_d \times \boldsymbol{\Omega}_d \tag{17.34}$$

where the magnetic induction field $\boldsymbol{B}_d$ has yet to be given explicitly. We have used $\dot{n}_d = 0$, which is the case because the localized moments are at atomic positions and their numbers do not change over time. As will become clear below, the term $-n_d \boldsymbol{m}_d \times \boldsymbol{\Omega}_d$ in (17.34) will account for relaxation processes. It is written in a general form that takes into account the micromagnetics approximation, according to which the modulus of the magnetization does not change. We learned in §7.3 how to define an effective field $\boldsymbol{B}_{\text{eff}}$ as a conjugate variable to the magnetic enthalpy. In the present problem, we have to add to the magnetic enthalpy (7.19) a term of the form,

$$V_{sd} = -A_{sd} n_s n_d \boldsymbol{m}_s \cdot \boldsymbol{m}_d \tag{17.35}$$

to account for the sd coupling (§6.5). Then, the magnetic induction $\boldsymbol{B}_d$ in (17.34) contains a sum of two terms,

$$\boldsymbol{B}_d = \boldsymbol{B}_{\text{eff}} + A_{sd} n_s \boldsymbol{m}_s \tag{17.36}$$

Thus, the continuity equation (17.34) becomes,

$$n_d \dot{\boldsymbol{m}}_d = \gamma n_d (\boldsymbol{m}_d \times \boldsymbol{B}_{\text{eff}}) - n_d \boldsymbol{m}_d \times \boldsymbol{\Omega}_d + \gamma_{sd} n_d (\boldsymbol{m}_d \times n_s \boldsymbol{m}_s) \tag{17.37}$$

with $\gamma_{sd} = \gamma A_{sd}$. The reader may be surprised not to see any reference to a charge-driven spin current in the evolution equation we obtained for the localized moments. It is the evolution of the electron spin under the effect of the charge current that will give rise to a non-vanishing m_s that acts on the localized moments.

Delocalized electron spins

The continuity equation for the delocalized electron spins is more complex. Applying (17.33) to the magnetization $M_s = n_s m_s$, we have,

$$(\dot{M}_s) + \boldsymbol{\nabla} \cdot \mathrm{j}_{M_s} = \rho_{M_s} \tag{17.38}$$

The source term ρ_{M_s} is analogous to that of $M_d = n_d m_d$. The source term contains a relaxation term characterized by $\boldsymbol{\Omega}_s$ and a precessional term that depends on a field B_s. The field B_s has two contributions. The first contribution is simply the magnetic induction B in the material. The other contribution arises from the coupling V_{sd} between conduction electrons and local moments. Hence, we have,

$$B_s = B + A_{sd} n_d m_d \tag{17.39}$$

and the continuity equation reads,

$$\rho_{M_s} = \gamma n_s (m_s \times B) - n_s m_s \times \boldsymbol{\Omega}_s + \gamma_{sd} n_s (m_s \times n_d m_d) \tag{17.40}$$

Let us now work on the left-hand side of the continuity equation (17.38). For the sake of the argument to come, relative to Galilean invariance, we revert back to the use of material derivatives, writing,

$$\dot{n}_s m_s + n_s \dot{m}_s + \boldsymbol{\nabla} \cdot \mathrm{j}_{M_s} = \rho_{M_s} \tag{17.41}$$

Applying the continuity equation (17.30), we have,

$$\dot{n}_s + \boldsymbol{\nabla} \cdot \boldsymbol{j}_{n_s} = 0 \tag{17.42}$$

Substituting for $\dot{n}_s$ in (17.41) yields,

$$n_s \dot{m}_s - (\boldsymbol{\nabla} \cdot \boldsymbol{j}_{n_s}) m_s + \boldsymbol{\nabla} \cdot \mathrm{j}_{M_s} = \rho_{M_s} \tag{17.43}$$

We now make use of the vectorial identity,

$$(\boldsymbol{\nabla} \cdot \boldsymbol{j}_s) m_s = \boldsymbol{\nabla}_j (m_s \odot \boldsymbol{j}_s) - (\boldsymbol{j}_s \cdot \boldsymbol{\nabla}) m_s \tag{17.44}$$

where the symbol $\boldsymbol{\nabla}_j$ means that the derivatives are carried out on current terms only and $\odot$ designates the symmetrized form of the tensorial product $\otimes$. Thus, we find,

$$n_s \dot{m}_s + (\boldsymbol{j}_s \cdot \boldsymbol{\nabla}) m_s + \boldsymbol{\nabla}_j \cdot \left(\mathrm{j}_{M_s} - m_s \odot \boldsymbol{j}_s \right) = \rho_{M_s} \tag{17.45}$$

The result (17.45) must hold for any current flow $\boldsymbol{j}_s$. This condition yields an explicit equation for the spin current,

$$\mathrm{j}_{M_s} = m_s \odot \boldsymbol{j}_s \tag{17.46}$$

This expression of j_{M_s} is quite natural. The current density $\boldsymbol{j}_s$ characterizes the diffusive motion of the conduction electrons and the magnetization current density j_{M_s} is a physical property that specifies the magnetic dipole orientation m_s of that current density. This structure has been dubbed the "anatomy of a spin current." [458]

Using the source term (17.40) in (17.45), the continuity equation for the electron spins is found to be,

$$\begin{aligned} n_s \dot{m}_s + (\boldsymbol{j}_s \cdot \boldsymbol{\nabla}) m_s = &\gamma n_s (m_s \times B) - n_s m_s \times \boldsymbol{\Omega}_s \\ &+ \gamma_{sd} n_s (m_s \times n_d m_d) \end{aligned} \tag{17.47}$$

Onsager approach

The thermodynamics of irreversible processes is an approach that takes into account the first and second laws of thermodynamics applied to a small volume, assumed to be locally at equilibrium. It also takes into account the continuity equations for energy, entropy and for the various substances in the system. For the particular case of conduction electrons and localized moments, we have continuity equations for the magnetization fields corresponding to the localized moments and the delocalized moments, m_d and m_s, respectively. After a lengthy development shown in [377], we can identify the forces associated with the current densities and the torques associated with Ω_d and Ω_s. Thus, the forces are found to be,

$$\begin{aligned} \boldsymbol{F}_d &= -\boldsymbol{\nabla}\mu_d - q_d\boldsymbol{\nabla}V + m_d\boldsymbol{\nabla}B_d \\ \boldsymbol{F}_s &= -\boldsymbol{\nabla}\mu_s - q_s\boldsymbol{\nabla}V + m_s\boldsymbol{\nabla}B_s \end{aligned} \tag{17.48}$$

where μ_d and μ_s are the chemical potentials, q_d and q_s the charges per unit of volume, and V the electrostatic potential. The terms $m_d\boldsymbol{\nabla}B_d$ and $m_s\boldsymbol{\nabla}B_s$ are the magnetic forces that the magnetic moments m_d and m_s experience in an inhomogeneous field B.[2] By the same development, the torques associated with Ω_d and Ω_s are found to be,

$$\boldsymbol{T}_d = \boldsymbol{m}_d \times \boldsymbol{B}_d \qquad\qquad \boldsymbol{T}_s = \boldsymbol{m}_s \times \boldsymbol{B}_s \tag{17.49}$$

By combining together the continuity equations, it is possible to find an expression for the local source of entropy. This expression is a sum of products in which two terms can be distinguished. In each product, there is a generalized current and a generalized force acting on that current. Onsager's approach consists in ensuring that the local source of entropy (expression shown in [377]) is a definite positive function of the generalized currents and forces. Thus, we write that the current densities of entropy, $\boldsymbol{j}_S$, local moments $\boldsymbol{j}_d = \boldsymbol{0}$, and delocalized spins $\boldsymbol{j}_s$, which are vectorial quantities, are linked to the generalized forces $\boldsymbol{F}_S = -\boldsymbol{\nabla}T$, $\boldsymbol{F}_d$, and $\boldsymbol{F}_s$ by the relationships,

$$\begin{aligned} \boldsymbol{j}_S &= \mathsf{L}_{SS}\cdot(-\boldsymbol{\nabla}T) + \mathsf{L}_{Sd}\cdot\boldsymbol{F}_d + \mathsf{L}_{SB}\cdot\boldsymbol{F}_s \\ \boldsymbol{j}_d &= \mathsf{L}_{dS}\cdot(-\boldsymbol{\nabla}T) + \mathsf{L}_{dd}\cdot\boldsymbol{F}_d + \mathsf{L}_{ds}\cdot\boldsymbol{F}_s \\ \boldsymbol{j}_s &= \mathsf{L}_{dS}\cdot(-\boldsymbol{\nabla}T) + \mathsf{L}_{sd}\cdot\boldsymbol{F}_d + \mathsf{L}_{ss}\cdot\boldsymbol{F}_s \end{aligned} \tag{17.50}$$

The relations (17.50) stem from the same approach as the two-current model equations (8.17). Of course, here, $\boldsymbol{j}_d = \boldsymbol{0}$ since d designates moments fixed at atomic sites. The Onsager approach applied to the pseudo-vectorial quantities yields,

$$\Omega_d = \mathsf{L}_{dd}^{(M)}(\boldsymbol{m}_d \times \boldsymbol{B}_d) + \mathsf{L}_{ds}^{(M)}(\boldsymbol{m}_s \times \boldsymbol{B}_s) \tag{17.51}$$

$$\Omega_s = \mathsf{L}_{sd}^{(M)}(\boldsymbol{m}_d \times \boldsymbol{B}_d) + \mathsf{L}_{ss}^{(M)}(\boldsymbol{m}_s \times \boldsymbol{B}_s) \tag{17.52}$$

The various Onsager tensors introduced here satisfy the Onsager reciprocity relations (see §10.7). Thus, defining $\boldsymbol{M} = n_d\boldsymbol{m}_d + n_s\boldsymbol{m}_s$, we have,

$$\mathsf{L}_{Sd}(\boldsymbol{M}) = \mathsf{L}_{dS}(-\boldsymbol{M}); \quad \mathsf{L}_{ds}(\boldsymbol{M}) = \mathsf{L}_{sd}(-\boldsymbol{M}); \quad \mathsf{L}_{ds}^{(M)}(\boldsymbol{M}) = \mathsf{L}_{sd}^{(M)}(-\boldsymbol{M}) \tag{17.53}$$

In the following, we make two simplifications. First, we neglect the cross terms characterized by $\mathsf{L}_{ds}^{(M)}$, i.e., we set $\mathsf{L}_{ds}^{(M)} = \mathsf{L}_{sd}^{(M)} = 0$. Second, we treat the Onsager tensors as isotropic, i.e., they can be written as, for example, $\mathsf{L}_{dd}^{(M)} = L_{dd}^{(M)}\mathbb{1}$, where $L_{dd}^{(M)}$ is a scalar, and likewise $\mathsf{L}_{ss}^{(M)} = L_{ss}^{(M)}\mathbb{1}$.

---

[2]See, e.g., [376], Eq. (9.34), the ith component of $\boldsymbol{a}\boldsymbol{\nabla}\boldsymbol{b}$ is equal to $\sum_j a_j\nabla_i b_j$.

Evolution of the localized moments

Let us now focus on the consequences of the Onsager approach for the time evolution of m_d. Using in (17.37) the Onsager relationship (17.51) and the definition (17.36) of B_d yields,

$$
\begin{aligned}
\dot{m}_d =\ & \gamma(m_d \times B_{\text{eff}}) - L_{dd}^{(M)} m_d \times (m_d \times B_{\text{eff}}) \\
& - \left(A_{sd} n_s L_{dd}^{(M)} \right) m_d \times (m_d \times m_s) \\
& + \gamma_{sd}(m_d \times n_B m_s)
\end{aligned}
\tag{17.54}
$$

Some remarks are called for about the three lines of (17.54). The terminology introduced below is illustrated in Fig. 17.3.

- The first line corresponds to the Landau-Lifshitz equation (7.34), which includes a damping term corresponding to the Gilbert damping term.
- The term containing $m_d \times (m_d \times m_s)$ has the structure of the **spin transfer torque**, first derived by J. Slonczewski when analyzing the magnetization dynamics in a spin valve.[57] It is sometimes called the antidamping torque or in-plane torque.[957, 958]
- The last term is called a **field-like** spin torque, because its form is analogous to the magnetic induction term expressing the Larmor precession in the magnetic induction field B. It is the direct consequence of the sd coupling and non-vanishing m_s as found when there is spin accumulation (see §8.4). The field-like spin torque is sometimes called the out-of-plane torque.
- The last two terms vanish when m_s vanishes. We need to look at the evolution of the delocalized electron spins for a given m_d to obtain m_s. The electron spin dynamics is much faster than the magnetization $n_s m_s$ dynamics. So, m_s can be calculated for a given magnetic configuration, the consequent evolution of m_d calculated for an infinitesimal time step, then m_s recalculated for this configuration, etc.

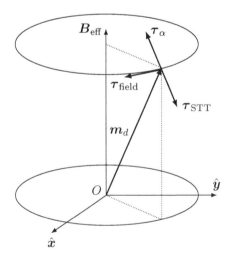

Figure 17.3 Illustration of the various torques that determine the evolution of m_d according to (17.54): τ_α is the damping torque, τ_{STT} is the spin transfer torque, and τ_f is the field-like torque.

Evolution of the delocalized electron spins

We now turn our attention to the time evolution of $\boldsymbol{m}_s$ and the spin current J_M. The continuity equation (17.47) for electron spin moments $\boldsymbol{m}_s$ can be written as,

$$\dot{\boldsymbol{m}}_s = \gamma(\boldsymbol{m}_s \times \boldsymbol{B}) - \boldsymbol{m}_s \times \boldsymbol{\Omega}_s + \gamma(\boldsymbol{m}_s \times A_{sd} n_d \boldsymbol{m}_d) - n_s^{-1}(\boldsymbol{j}_s \cdot \boldsymbol{\nabla})\boldsymbol{m}_s \qquad (17.55)$$

The magnetic induction $\boldsymbol{B}$ is typically many orders of magnitude less than the field derived from an exchange energy, because magnetic energy terms are usually much smaller than electrostatic energies. Hence, from now on, we neglect $\boldsymbol{B}$ with respect to $A_{sd} n_d \boldsymbol{m}_d$. Let us now use the the Onsager relation (17.52) (assuming the isotropy of the Onsager tensors) and the definition (17.39) of $\boldsymbol{B}_s$, of which we again keep only the exchange term. This yields,

$$\dot{\boldsymbol{m}}_s = \gamma(\boldsymbol{m}_s \times A_{sd}\boldsymbol{M}_d) - L_{ss}^{(M)} \boldsymbol{m}_s \times (\boldsymbol{m}_s \times A_{sd}\boldsymbol{M}_d)$$
$$- n_s^{-1}(\boldsymbol{j}_s \cdot \boldsymbol{\nabla})\boldsymbol{m}_s \qquad (17.56)$$

Let us now make use of the Onsager relation (17.50) for the current density $\boldsymbol{j}_s$. Keeping only the dominant term, we write,

$$\boldsymbol{j}_s = -\frac{1}{q}\boldsymbol{\sigma} \cdot \boldsymbol{\nabla}V \qquad (17.57)$$

which is the local form of Ohm's law (q being the electron charge). Recall that $\boldsymbol{j}_s$ is the current density of electrons (i.e., the number of electron moles crossing a unit of surface per unit of time). It follows that $q\,\boldsymbol{j}_s$ is an electric current density. The magnetoresistance effect tells us that the electrical conductivity is a function of the magnetic induction field. Here, the magnetic induction is essentially $A_{sd}\boldsymbol{M}_d$. We denote $\hat{\boldsymbol{M}}_d$ as the unit vector along $A_{sd}\boldsymbol{M}_d$. The conductivity tensor may have a contribution that is independent of magnetic induction field. Then, this contribution to the conductivity tensor is symmetric. The contribution that depends on magnetic field must be antisymmetric (see §10.7). Therefore, for any vector $\boldsymbol{a}$, we have the decomposition,

$$\boldsymbol{\sigma} \cdot \boldsymbol{a} = \sigma_\| \boldsymbol{a} + \sigma_\perp \left(\hat{\boldsymbol{M}}_d \times \boldsymbol{a} \right) \qquad (17.58)$$

Using this result, (17.56) becomes,

$$\dot{\boldsymbol{m}}_s = \gamma(\boldsymbol{m}_s \times A_{sd}\boldsymbol{M}_d) - L_{ss}^{(M)} \boldsymbol{m}_s \times (\boldsymbol{m}_s \times A_{sd}\boldsymbol{M}_d)$$
$$+ \frac{1}{n_s q}\sigma_\|(\boldsymbol{\nabla}V \cdot \boldsymbol{\nabla})\boldsymbol{m}_s + \frac{1}{n_s q}\sigma_\perp \left((\hat{\boldsymbol{M}}_d \times \boldsymbol{\nabla}V) \cdot \boldsymbol{\nabla} \right) \boldsymbol{m}_s \qquad (17.59)$$

The physical interpretation of this result is the following. The first two terms on the right-hand side of (17.59) are those of a Landau-Lifshitz equation with damping. They bring the magnetization (here, spin polarization) into alignment with the effective field $A_{sd}\boldsymbol{M}_d$ on a time scale much shorter than the time scale of the $\boldsymbol{M}_d$ evolution. In other words, these two terms vanish on a time scale that is significant for the time evolution of $\boldsymbol{M}_d$. We also assume that $\boldsymbol{\nabla}V$ is applied over a long time scale. This is the adiabatic regime discussed in §15.4; on this time scale, $\boldsymbol{m}_s$ is aligned with $\boldsymbol{M}_d$ and we have,

$$\dot{\boldsymbol{m}}_s = +\frac{m_s}{n_s q}\sigma_\|(\boldsymbol{\nabla}V \cdot \boldsymbol{\nabla})\hat{\boldsymbol{M}}_d + \frac{m_s}{n_s q}\sigma_\perp \left((\hat{\boldsymbol{M}}_d \times \boldsymbol{\nabla}V) \cdot \boldsymbol{\nabla} \right) \hat{\boldsymbol{M}}_d \qquad (17.60)$$

This equation tells us that the spin polarization $\boldsymbol{m}_s$ is produced (positively or negatively) where the magnetization orientation $\hat{\boldsymbol{M}}_d$ varies. This is the continuum version of the spin accumulation analyzed in §8.4, where the change in magnetization orientation was abrupt. The first term gives a contribution when the magnetization $\hat{M}_d$ varies in the direction of the electric field $-\nabla V$. The second term, which is associated with the Hall effect (17.58), contributes to $\dot{m}_s$ when there is a change in $\hat{M}_d$ that is perpendicular to both the main electron flow and the magnetization M_d.

If the adiabatic approximation of §15.4 is not fully satisfied, a correction called the non-adiabatic term must be added.[959–961] The effect of the non-adiabatic term was observed when analyzing spin torque effects in very narrow walls.[3]

Equation (17.60) also tells us that the magnetic moment m_s evolves. Therefore, a torque must be exerted on this subsystem. This torque comes from the m_d magnetic subsystem. Conversely, we can say that the current imposed by ∇V induces a torque on the magnetization M_d, leading to a domain wall excitation. Because we have a system composed of two subsystems, it is the self-consistent treatment of both (17.54) and (17.60) (or more precisely (17.59)) that provides the full picture.

17.6 SPIN-PUMPING SPIN CURRENTS

Let us consider that a ferromagnetic thin film is deposited on a substrate and that a thin strip of a paramagnetic metal is deposited on top of the thin magnetic film, in a direction perpendicular to the equilibrium magnetization (Fig. 11.7). When the ferromagnetic resonance is excited and the strip is a high-Z material such as platinum, a spin current is detected via the inverse spin Hall effect (ISHE, §11.4.3), giving rise to a voltage across the strip. A complete theoretical derivation and discussion on recent results can be found in the review of Y. Tserkovniak et al.[966] In this section, we lay out a simple argument for the notion that a spin current must take place at the interface between the ferromagnet and the non-magnetic metal when the FMR is excited.

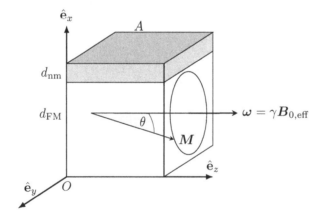

Figure 17.4 Schematics of a cubic piece of ferromagnet of area A and thickness d_{FM}, covered by a non-magnetic metal of thickness d_{nm}. Owing to a resonant microwave field, the magnetization M precesses around the applied static field $B_{0,\mathrm{eff}}$ at the angular velocity ω.

Let us consider the geometrical configuration of Fig. 17.4, and assume that the ferromagnet has a thickness d_{FM} sufficiently small such that, at any distance away from the interface

---

[3]See, e.g.,[962–965].

to the non-magnetic metal, the relaxation mechanism is dominated by the presence of this capping layer rather than by an intrinsic Gilbert-type damping mechanism. The assumption that d_{FM} is small allows us to avoid working out which mechanism dominates as a function of the distance x (Fig. 17.4). We assume that this damping mechanism is expressed in the usual Gilbert form, i.e., that the Landau-Lifshitz-Gilbert equation (7.23) holds. The effective field $\boldsymbol{B}_{\text{eff}}$ comprises a static effective field $\boldsymbol{B}_{0,\text{eff}}$ and a rotating $\boldsymbol{B}_1$ field. Thus, we have,

$$\frac{d\boldsymbol{M}}{dt} = \gamma \boldsymbol{M} \times \boldsymbol{B}_{0,\text{eff}} + \gamma \boldsymbol{M} \times \boldsymbol{B}_1 - \frac{\alpha_{\text{nm}}}{M_s} \boldsymbol{M} \times \left(\frac{d\boldsymbol{M}}{dt} \right) \tag{17.61}$$

Here, we write α_{nm} to indicate that the damping mechanism is not a local Gilbert damping mechanism but the result of damping due to the presence of a non-magnetic metal layer at the surface of the ferromagnet.[967]

The angular velocity $\boldsymbol{\omega}$ of the microwave field $\boldsymbol{B}_1$ satisfies the resonance condition,

$$\boldsymbol{\omega} = -\gamma \boldsymbol{B}_{0,\text{eff}} \tag{17.62}$$

In continuous-wave spectroscopy, the FMR is excited in a stationary regime. Hence, the magnetization $\boldsymbol{M}$ precesses around the static field $\boldsymbol{B}_{0,\text{eff}}$, i.e.,

$$\frac{d\boldsymbol{M}}{dt} = \boldsymbol{\omega} \times \boldsymbol{M} \tag{17.63}$$

In a first-order approximation, we can use this expression for $d\boldsymbol{M}/dt$ in the damping term of (17.61) to obtain,[4]

$$\gamma \boldsymbol{M} \times \boldsymbol{B}_1 = \alpha_{\text{nm}} M_s \boldsymbol{\omega} - \frac{\alpha_{\text{nm}}}{M_S} (\boldsymbol{M} \cdot \boldsymbol{\omega}) \boldsymbol{M} \tag{17.64}$$

This equation stems from (17.61), which can be considered a continuity equation for $\boldsymbol{M}$ since magnetization is generally a reduced extensive quantity. Therefore, (17.64) represents a balance in the stationary regime between the torque of the applied microwave field $\boldsymbol{B}_1$ and the damping mechanism. So, we can associate with this damping torque a current of angular momentum per unit of time per unit of area that is absorbed by the capping layer.

For the sake of clarity, we assumed above that the damping is due to only the non-magnetic capping metal on the ferromagnet. In reality, there is an intrinsic damping mechanism characterized by α_i, which represents the sink term in the continuity equation for $\boldsymbol{M}$, i.e., it does not intervene in our estimate of the magnetization current. What we designated as α_{nm} is the difference between damping when there is no capping layer, α_i and the overall damping α in the presence of the capping layer. Hence, $\alpha_{\text{nm}} = \alpha - \alpha_{\text{i}} \equiv \Delta \alpha$.

The precession angle θ is constant, and $(\boldsymbol{M} \cdot \boldsymbol{\omega})$ is equal to $M_s \omega \cos \theta$. Let us consider a small volume of the ferromagnet located adjacent to the metal film. This volume has a surface area A facing the contact and a thickness d_{FM}. The quantity $\gamma \boldsymbol{M} \times \boldsymbol{B}_1 A \, d_{\text{FM}} \, dt$ is, according to (17.64), equal to the magnetic moment that crosses the interface of area A into the non-magnetic metal in a time dt, where it is absorbed. Hence, we find that the magnetic moment projected onto the $\hat{\boldsymbol{z}}$ direction,

$$\left(\alpha_{\text{nm}} M_s \boldsymbol{\omega} - \frac{\alpha_{\text{nm}}}{M_S} (\boldsymbol{M} \cdot \boldsymbol{\omega}) \boldsymbol{M} \right) \cdot \hat{\boldsymbol{z}} \, A \, d_{\text{FM}} dt \tag{17.65}$$

crosses the surface area A, and is absorbed in the capping layer, in every time intervalle dt. We can define a spin current, which is the magnetic moment density in the z-direction that

---

[4] Apply the identity $\boldsymbol{a} \times (\boldsymbol{b} \times \boldsymbol{c}) = (\boldsymbol{a} \cdot \boldsymbol{c}) \, \boldsymbol{b} - (\boldsymbol{a} \cdot \boldsymbol{b}) \, \boldsymbol{c}$.

crosses a unit of area of the interface per unit of time, $\boldsymbol{j}^{(z)}$. It is a vector that points in the $\hat{\boldsymbol{x}}$ direction (Fig. 17.4), given by,

$$\boldsymbol{j}^{(z)} = \alpha_{\mathrm{nm}}\,\omega\,M_s d_{\mathrm{FM}}\,\sin^2\theta\,\hat{\boldsymbol{x}} \tag{17.66}$$

where the scalar $\omega = \boldsymbol{\omega}\cdot\hat{\boldsymbol{z}}$, being a projection, carries the sign corresponding to this projection.

To sum up, we have seen three ways of producing a spin current without a charge current: spin pumping, heat-driven spin current and diffusive spin currents obtained in non-local measurements. These pure spin currents can be detected by using them to excite an electron spin resonance.[968] Alternatively, they can be detected by a conversion from spin to charge using the ISHE (§11.4.3).

17.7   SPIN-WAVE SPIN CURRENTS

The magnetization $\boldsymbol{M}$ is, from a thermodynamic point of view, a reduced extensive quantity, i.e., an extensive quantity per unit of volume.[376] Therefore, the Landau-Lifshitz equation (7.51), which includes an effective field derived from exchange interactions, must conform to a continuity equation. The term $\gamma\boldsymbol{M}\times\boldsymbol{B}$ expresses local dynamics, namely, the torque that is exerted on the magnetization. It is the source term of this continuity equation.

The second term in the LLG equation (7.51) can be written as the divergence of a current $\mathsf{j}_{\boldsymbol{M}}$, thanks to the identity,

$$\boldsymbol{\nabla}\cdot(\boldsymbol{M}\times\boldsymbol{\nabla}\boldsymbol{M}) = \boldsymbol{M}\times\boldsymbol{\nabla}^2\boldsymbol{M} \tag{17.67}$$

In other words, the magnetization current $\mathsf{j}_{\boldsymbol{M}}$ is given by,

$$\mathsf{j}_{\boldsymbol{M}} = -D\,\boldsymbol{M}\times\boldsymbol{\nabla}\boldsymbol{M} \tag{17.68}$$

The quantity $\mathsf{j}_{\boldsymbol{M}}$ is called a ***spin-wave spin current***.[521, 969] It is a second-rank tensor represented by a 3×3 matrix of components,

$$(\mathsf{j}_{\boldsymbol{M}})_{ij} = -D\sum_{kl}\varepsilon_{ikl}M_k\nabla_j M_l \tag{17.69}$$

where ε_{ikl} is the Levi-Civita symbol.[5]

Consider a small volume in the ferromagnet, delimited by a surface S, and a surface element of area dS. Let $\hat{\boldsymbol{n}}$ be the unit vector normal to the surface element, pointing outward from the volume. The spin current flowing across this surface element is a torque $d\boldsymbol{\tau}$ exerted on the magnetic moment of the volume that is considered. The torque $d\boldsymbol{\tau}$ is given by,

$$d\boldsymbol{\tau} = -\mathsf{j}_{\boldsymbol{M}}\cdot d\boldsymbol{S} \tag{17.70}$$

where $d\boldsymbol{S} = dS\hat{\boldsymbol{n}}$.

To illustrate the definition (17.68) of a spin current, we consider a spin wave that is a plane wave propagating in the y direction, whereas the equilibrium magnetization is in the z direction. The local moment precession (Fig 7.3) takes place in the (x,y) plane. We have

$$\mathsf{j}_{\boldsymbol{M}} = -D\begin{pmatrix} 0 & M_y\nabla_y M_z - M_z\nabla_y M_y & 0 \\ 0 & M_z\nabla_y M_x - M_x\nabla_y M_z & 0 \\ 0 & M_x\nabla_y M_y - M_y\nabla_y M_x & 0 \end{pmatrix} \tag{17.71}$$

---

[5] ε_{ikl} is equal to 1 if $(i, k, l))$ is an even permutation of $(1, 2, 3)$, -1 if it is an odd permutation, and 0 if any of the values i, k, l is repeated.

Keeping only first-order terms in j_M, for any direction $\hat{n} = (n_x, n_y, n_z)$, we have,

$$d\boldsymbol{\tau} = D \begin{pmatrix} 0 & -M_z \nabla_y M_y & 0 \\ 0 & M_z \nabla_y M_x & 0 \\ 0 & 0 & 0 \end{pmatrix} \begin{pmatrix} n_x dS \\ n_y dS \\ n_z dS \end{pmatrix} = D\, dS\, n_y \begin{pmatrix} -M_z \nabla_y M_y \\ M_z \nabla_y M_x \\ 0 \end{pmatrix} \qquad (17.72)$$

Thus, we find that $d\boldsymbol{\tau}$ is maximum when $n_y = 1$, i.e., when $\hat{n}$ is in the direction of the spin wave propagation. Thus $d\boldsymbol{\tau}$ is parallel to the wave vector and perpendicular to the equilibrium magnetization.

17.8   FURTHER READINGS

FMR detection methods

With the use of vector network analyzers (VNA), the flip-chip method has become the most convenient way to detect FMR. In essence, a VNA measures the electromagnetic wave reflection and transmission through any two-port device. In this FMR application of a VNA, transmission through a microstrip line is monitored. A thin film deposited on a "chip" is placed upside down on the device and the applied magnetic field is swept stepwise, allowing for spectra to be measured for each field value.[970–973]

Among other FMR detection methods, let us highlight one other method, the use of time-resolved Kerr effect measurements.[974]

A very sensitive detection method relies on the non-linear resistive behavior of magnetic nanostructures. This method, first demonstrated by H. Juretschke (see §17.1), has appeared in the context of spintronics with the confusing, but frequently used, name of *spin diode effect*. Let us clarify the basic principle of this method. We assume that a material is characterized by a magnetoresistance $R(\boldsymbol{M})$. There are several circumstances under which the magnetization will depend on the applied current, $\boldsymbol{M}(I)$. It can be that a ferromagnetic metal is irradiated with microwaves that excite the FMR and induce a current in the ferromagnetic metal. It can also be that the current induces a spin transfer torque in a magnetic nanostructure. Then, according to Ohm's law, we have the following development,

$$V = R(\boldsymbol{M}(I))\, I = R(I)\, I = R_0 I + \frac{\partial R(\boldsymbol{M}(I))}{\partial I} I \qquad (17.73)$$

where R_0 is the part of the resistance that is independent of $\boldsymbol{M}$. The term $R_0 I$ oscillates at the frequency f of the current I. The second term in (17.73) has a Fourier component oscillating at $2f$ and a DC component at $f = 0$. This DC term has been coined the *spin diode effect*.[68]

This rectifying effect affords a very sensitive electrical detection of FMR.[975] The FMR signal of a nanoscopic spin valve (about 10^6 spins) has been detected in the measurement of its magnetoresistance.[976] Several modes have been detected in spin valves 100 nm in diameter.[977] In bulk materials, this effect may be compounded with the bolometric effect.[978, 979]

This FMR-induced voltaic effect can be used to probe the extent to which spin waves and conduction electrons are coupled, i.e., test the dependence $R(\boldsymbol{M})$ when $\boldsymbol{M}$ varies because of spin waves being excited.[980]

Spin-wave spectroscopy

In Fig. 17.2, we saw that the spectrum of standing waves excited normal to permalloy films. This experiment has been repeated recently with films of the *organometallic ferromagnet* V[TCNE]$_{x \approx 2}$.[981, 982] The quality factor of these resonances has been found similar to

312 Spintronics: A primer

that obtained with yttrium iron garnet (YIG), which is quite remarkable, especially given that this material is not crystalline.[981]

Spin waves excitation in chiral magnets have been investigated.[983–985] Spin wave can also propagate within environments that exhibit distinct magnetic textures, such as domain walls,[986, 987] stripes,[988] *skyrmions* or bulk helimagnetic structures.[190, 420, 983, 989]

Coupling of spin waves with acoustic waves

Spin waves can couple to surface acoustic waves for the following reason. Deformations in an elastic solid are described by a tensor.[990] This tensor in full generality has a symmetric part and an antisymmetric part. The latter represents a local rotation. Consequently, a surface acoustic wave can host locally non-zero angular momentum and hence, couple to ferromagnetic resonance or spin waves. For example, a surface acoustic wave can excite ferromagnetic resonance.[991] Likewise, the transmission of a surface acoustic wave can be damped owing to its resonant coupling to a ferromagnetic resonance in the underlying material.[992] In Chapter 18, the resonant coupling of waves of different natures will be introduced when discussing the coupling of an antiferromagnetic resonance to the electromagnetic waves of a cavity (magnon-polaritons). The coupling of spin waves to acoustic waves is referred to as magnon-polarons. This coupling can generate spin currents in a ferromagnetic thin film.[993] Magnon-polarons were observed in an antiferromagnets.[994] They were also observed in heterostructures combining a ferromagnetic thin film with a ferroelectric or multiferroic thin film.[995–997]

Magnonics

The use of magnons is being explored in the context of *information processing* and drives a field of research called magnonics.[998] One intriguing possibility involves the realization of spin-wave logic gates employing a Mach-Zehnder interferometer configuration.[999] Additionally, magnons can be considered as carriers of angular momentum that connect two magnetic nanostructures. Indeed, magnon propagation has been observed to extend over tens of micrometers in metallic magnets,[1000, 1001] and several millimeters in insulating magnets.[1002]

Similar to the role of diodes in conventional electronics and the significance of spin filtering in spintronics, *non-reciprocity* is a topic of fundamental interest to the field of magnonics. Non-reciprocity refers to the asymmetry in the propagation behavior of waves going in one direction and waves going in the opposite direction.[1003–1005] Control over wave propagation was also obtained by creating spin-wave valves that operate with regard to spin waves like spin valves do with spin-polarized electrons.[986, 987, 1006–1008]

18 Antiferromagnetic Resonance and Polaritons

Edward M. Purcell (1912–1997)

E. M. Purcell received a PhD in physics from Harvard University. During World Ward II, he worked on microwave technology at the MIT Radiation Laboratory. In 1946, he detected magnetic resonance with two of his colleagues, R.V. Pound and H.C. Torrey. In 1952, F. Bloch and E. M. Purcell received the Nobel prize in physics.

Spintronics: A primer

J.-Ph. Ansermet (https://orcid.org/0000-0002-1307-4864)

Chapter 18. Antiferromagnetic Resonance and Polaritons

Antiferromagnets can be characterized by their temperature-dependent susceptibility and spin-flop transition. The uniform antiferromagnetic resonance frequency is predicted by applying the Landau-Lifshitz equation to both sublattices. Magnetic polaritons are coupled modes of the electromagnetic field and magnetization dynamics, akin to the Purcell effect. A classical input-output formalism accounts for this coupling phenomenologically. A quantum mechanical description yields a coupling strength proportional to the square root of the number of correlated moments.

DOI: 10.1201/9781003370017-18

18.1 HISTORICAL INTRODUCTION

Antiferromagnets are of interest in spintronics because their natural dynamics is in the hundreds of GHz.[1009] In a review article, A. N. Slavin mentioned the zero-field resonances of several orthoferrites.[1010] For example, $YFeO_3$ resonate at 299.4 GHz at room temperature in the absence of any applied magnetic field. It has another mode at about 527 GHz.

Antiferromagnets may play a role in cavity spintronics, a field of research that seeks to make use of the interplay of cavity modes and magnetization modes, i.e., polaritons. These polaritons are the magnetic version of cavity quantum dynamics.[1011] Reports on cavity-spintronics devices based on ferromagnets appeared in 2019, after many years of fundamental research.[1012, 1013]

R. W. Sanders, V. Jaccarino and S. M. Rezende showed in 1978 that the coupling of electromagnetic waves with magnetization waves in FeF_2 gave rise to a 50-fold increase in the antiferromagnetic resonance line width.[1014] This demonstrated a concept first identified by J. J. Hopfield for the dielectric response.[1015] Polaritons are coupled modes of, on the one hand, modes of the electromagnetic fields, and on the other hand, normal modes of the material. These can be electric dipole excitations characterized by the dielectric tensor, or magnetic dipole resonances characterized by the permeability tensor. Magnon-photon coupling was first discussed for the case of ferromagnets. [535, 1016] The case of antiferromagnets was treated similarly in 1972.[1017]

In a 1997 article, a few data points suggested anti-crossing dispersion curves caused by a strong coupling between cavity modes and magnetic excitations in a ferromagnet.[1018] In 2010, it was pointed out that the strength of a magnetic polariton is due to the collective nature of the ferromagnetic state. The strength of the coupling increases in proportion to the square root of the number of spins involved in the excitation.[1019] This was experimentally confirmed by the groups of S. Goennenwein and Can-Ming Hu.[1020, 1021]

Elementary excitations in antiferromagnets are magnons. As magnons have integer spins, they are bosons and can undergo a Bose-Einstein condensation (BEC), as reviewed by T. Giamarchi et al.[1022] This bosonic picture provides a unified expression of what occurs in a large variety of systems. It should be noted that the BEC found in 2006 for magnons in ferrimagnets [1023] results from strong pumping. These systems are not in an equilibrium state as was the case in the original BEC phenomenon.

Nowadays, it is possible to detect magnetic resonance in antiferromagnets using a VNA.[854] In the early days, Raman scattering was one of the methods used for detecting magnetic resonance in antiferromagnets. The scattering of light in ferro- and antiferro-magnets was addressed first in the low temperature limit by R. J. Elliot, T. Moriya, and others.[1024–1026] This research was later extended to finite temperatures by Loudon.[1027]

In this chapter, we introduce the notion of antiferromagnetic resonance. We briefly allude to magnons in antiferromagnets. We introduce the concept of magnetic polaritons from a classical point of view at first. Cooperativity effects are analyzed using quantum mechanics.

18.2 QUASI-STATIC RESPONSE OF AN ANTIFERROMAGNET

Above its Néel temperature, an antiferromagnetic material is paramagnetic. The susceptibility follows the Curie-Weiss law with a negative temperature $-T_N$, i.e., the susceptibility has the form $\chi = \chi_0/(T + T_N)$. In other words, a plot of the inverse of the susceptibility as a function of temperature has a zero-intercept at a negative temperature.

In the low temperature phase, the susceptibility depends greatly on whether the field is applied parallel or perpendicular to the magnetization in each sublattice (Fig. 18.1). When the field is perpendicular, the susceptibility is independent of fluctuations, the response

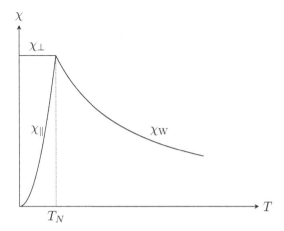

Figure 18.1 Susceptibility of an antiferromagnet as a function of temperature. Above the Néel temperature T_N, a Curie-Weiss behavior is observed. Below the Néel temperature T_N, the susceptibility $\chi_\perp$ is nearly temperature-independent when the field is perpendicular to the magnetization and exerts a torque on the sublattice magnetizations. When the field is parallel, $\chi_\parallel$ is determined by temperature-dependent fluctuations.

being the equilibrium value in the effective field combined with the applied field. When the applied field is parallel to the magnetization of each sublattice, thermal fluctuations give rise to a net average magnetization.

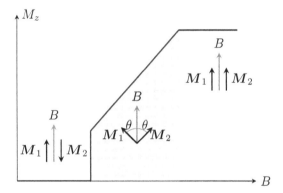

Figure 18.2 Magnetization M_1 and M_2 of both sublattices of an antiferromagnet, as a function of the magnetic field, applied parallel to the anisotropy axis z. The shaded area is the spin-flop region.

If a large magnetic field is applied perpendicular to the magnetization orientation at zero field, a net moment appears, as we would expect. If the field is applied parallel to the magnetization and the magnetic anisotropy is weak, the magnetization undergoes what is known as a ***spin-flop transition*** at some sufficiently high field (Fig. 18.2). See, for example, the data obtained for $Cs_2FeCl_5 \cdot H_2O$.[1028] At still higher fields, the magnetization of both sublattices points in the direction of the applied field. If the magnetic anisotropy is large, there is a direct transition to a state of parallel magnetization.

The spin-flop transition can be understood with a simple zero-temperature model. Consider an antiferromagnet in an applied magnetic field $\boldsymbol{H}$. We note $\boldsymbol{B} = \mu_0 \boldsymbol{H}$. The magnetic enthalpy $V(\boldsymbol{M})$ is a function $V(\theta)$ of the angle θ between the anisotropy axis and sublattice magnetization (Fig. 18.2, right). The value of $V(\boldsymbol{M})$ in this configuration is given by,

$$V(\theta) = -2MB \cos\theta + AM^2 \cos(2\theta) - K \cos^2\theta \qquad (18.1)$$

where A stands for the strength of the exchange energy and K for the magneto-crystalline anisotropy. The energy minimum is given by the condition $\partial V(\theta)/\partial\theta = 0$. Let us write this condition as $\sin\theta\, \partial V(\theta)/\partial\cos\theta = 0$. A non-zero solution is found for $\cos\theta = MB((2AM^2 - K)$. For this value of the angle θ, the energy is,

$$E(B) = -\frac{MB}{2AM^2 - K} \qquad (18.2)$$

For small K, $E(B)$ is a decreasing function of B. When B is large enough, $E(B)$ becomes less than the value $V(\theta = 0) = -AM^2 - K$ of the configuration with antiparallel sublattice magnetizations and a spin flop occurs. While the phenomenon has long been known, it is still being investigated because of the potential applications of antiferromagnets to spintronics. In particular, the dynamics of magnons in the vicinity of the spin-flop transition has been studied.[1029]

When K is large, the slope of $E(B)$ is positive and the minimum is obtained at $\sin\theta = 0$, i.e., either $\theta = 0$ for one sublattice, $\theta = \pi$ for the other, or $\theta = 0$ for both. The stable solution at high fields is characterized by $\theta = 0$ for both sub-lattices.

18.3 RESONANCES IN ANTIFERROMAGNETS

Let us consider an antiferromagnet in which two magnetization sublattices can be distinguished, $\boldsymbol{M}_1$ and $\boldsymbol{M}_2$ with $\boldsymbol{M}_1 = -\boldsymbol{M}_2$, such that the energy V defined in (7.19) is given by,

$$V(\boldsymbol{M}_1, \boldsymbol{M}_2) = \Lambda \boldsymbol{M}_1 \cdot \boldsymbol{M}_2 + K(\sin^2\theta_1 + \sin^2\theta_2). \qquad (18.3)$$

where Λ defines the exchange coupling responsible for the antiferromagnetic configuration at equilibrium ($\Lambda > 0$), K is the anisotropy constant of the uniaxial magnetocrystalline anisotropy (4.2), and θ_i is the angle between the $\hat{z}$ axis and $\boldsymbol{M}_i$, ($i = 1, 2$). Thus, it is assumed that the anisotropy contributions defined for each sublattice add up, i.e., there are no cross-effects. This is valid when the anisotropy is due to one-ion sources of anisotropy.[372] Also, in (18.3) the demagnetizing field is assumed negligible.

Thus, we have two Landau-Lifshitz equations (7.21), one for each magnetization sublattice. Expressing $\sin\theta_i$ in terms of M_x, M_y, and M_z, the derivative of (18.3) yields a term proportional to $\boldsymbol{M}$ that can be discarded since it does not contribute to $\boldsymbol{M} \times \boldsymbol{B}$. Then, only an anisotropy field in the z direction remains. Thus, we find,

$$\begin{aligned}
\frac{d\boldsymbol{M}_1}{dt} &= \gamma \boldsymbol{M}_1 \times \left(\boldsymbol{B} - \Lambda \boldsymbol{M}_2 - \frac{2K}{M_s^2}\hat{z}(\boldsymbol{M}_1 \cdot \hat{z}) \right) \\
\frac{d\boldsymbol{M}_2}{dt} &= \gamma \boldsymbol{M}_2 \times \left(\boldsymbol{B} - \Lambda \boldsymbol{M}_1 - \frac{2K}{M_s^2}\hat{z}(\boldsymbol{M}_2 \cdot \hat{z}) \right)
\end{aligned} \qquad (18.4)$$

These equations were first established by P. Pincus[1030], who omitted the anisotropy term. As can be seen in the analysis below, the anisotropy plays a crucial role in determining the equilibrium and the resonance frequencies (18.21) of antiferromagnets.[372]

Based on the equations (18.4), we now find the susceptibility and deduce from it that an antiferromagnet undergoes a magnetic resonance, even in zero applied magnetic field. The calculation is analogous to what led to the Polder tensor (17.17) describing ferromagnetic resonance.

We notice that, in (18.4), we have opposite anisotropy fields in the sublattices. We note them as $\boldsymbol{B}_A = \pm B_A \hat{\boldsymbol{z}}$. Thus, we write,

$$\frac{d\boldsymbol{M}_1}{dt} = \gamma \boldsymbol{M}_1 \times \left((B_0 + B_A)\hat{\boldsymbol{z}} - B_E \frac{\boldsymbol{M}_2}{M_s} + \boldsymbol{B}_1(t) \right) \tag{18.5}$$

$$\frac{d\boldsymbol{M}_2}{dt} = \gamma \boldsymbol{M}_2 \times \left((B_0 - B_A)\hat{\boldsymbol{z}} - B_E \frac{\boldsymbol{M}_1}{M_s} + \boldsymbol{B}_1(t) \right) \tag{18.6}$$

where $B_E = \Lambda M_s \hat{\boldsymbol{z}}$ is the exchange field experience by the magnetization of both sublattices. We now linearize these equations, assuming that $\boldsymbol{B}_1(t)$ is small and the response is likewise small compared to the saturation magnetization M_s. We assume that a uniform static field $\boldsymbol{B}_0$ is applied along the z axis and a microwave field $\boldsymbol{B}_1(t)$ is also applied. Thus, we write,

$$\boldsymbol{M}_1 = \begin{pmatrix} m_x^{(1)} \\ m_y^{(1)} \\ M_s \end{pmatrix} \exp(i\omega t) \qquad \boldsymbol{M}_2 = \begin{pmatrix} m_x^{(2)} \\ m_y^{(2)} \\ -M_s \end{pmatrix} \exp(i\omega t) \tag{18.7}$$

where $m_x^{(n)}, m_y^{(n)}$ $(n = 1, 2)$ are small. To write the evolution equation for $m_x^{(n)}, m_y^{(n)}$ $(n = 1, 2)$ once for both sublattices, we define,

$$B^{(1)} = B_0 + B_A + B_E \tag{18.8}$$

$$B^{(2)} = B_0 - B_A - B_E \tag{18.9}$$

The signs reflect the opposite directions of the magnetization in the sublattices. Thus, we have,

$$i\omega \begin{pmatrix} m_x^{(n)} \\ m_y^{(n)} \\ 0 \end{pmatrix} = \begin{vmatrix} \hat{\boldsymbol{x}} & m_x^{(n)} & \gamma B_{1x} - \gamma B_E \frac{m_x^{(\bar{n})}}{M_s} \\ \hat{\boldsymbol{y}} & m_y^{(n)} & \gamma B_{1y} - \gamma B_E \frac{m_y^{(\bar{n})}}{M_s} \\ \hat{\boldsymbol{z}} & -(-1)^n M_s & \gamma B^{(n)} \end{vmatrix} \tag{18.10}$$

where $\bar{n} = 1$ if $n = 2$ and $\bar{n} = 2$ if $n = 1$. From (18.10), we deduce,

$$\omega m_+^{(n)} = -m_+^{(n)} \gamma B^{(n)} - (-1)^n M_s \gamma B_+ + (-1)^n \gamma B_E m_+^{(\bar{n})} \tag{18.11}$$

$$\omega m_-^{(n)} = +m_-^{(n)} \gamma B^{(n)} + (-1)^n M_s \gamma B_- - (-1)^n \gamma B_E m_-^{(\bar{n})} \tag{18.12}$$

where we have introduced the rotating fields defined by,

$$m_\pm^{(n)} = m_x^{(n)} \pm i m_y^{(n)} \qquad B_+ = B_x \pm i B_y \tag{18.13}$$

We notice that the equation for $m_-^{(n)}$ is the same as the one for $m_+^{(n)}$, provided we replace ω by $-\omega$. The magnetization response is to be defined for the sum over both sublattices, i.e., we need to calculate $m_+ = m_+^{(1)} + m_+^{(2)}$ on the one hand, and $m_- = m_-^{(1)} + m_-^{(2)}$ on the other. Thus, after some simple algebraic manipulations, we find,

$$m_x + i m_y = c(+\omega)(B_x + i B_y) \tag{18.14}$$

$$m_x - i m_y = c(-\omega)(B_x - i B_y) \tag{18.15}$$

where,

$$c(\omega) = \frac{-2\gamma B_A M_s}{(\omega + \gamma B^{(1)})(\omega + \gamma B^{(2)}) + \gamma^2 B_E^2} \tag{18.16}$$

Therefore, the susceptibility tensor $\chi^{(rf)}$ has the components,

$$\chi_{xx}^{(B)} = \chi_{xx}^{(B)} = \frac{c(\omega) + c(-\omega)}{2} \tag{18.17}$$

$$\chi_{xy}^{(B)} = -\chi_{yx}^{(B)} = i\frac{c(\omega) - c(-\omega)}{2} \tag{18.18}$$

When $B_0 = 0$, combining the above results, with,

$$\chi_{xy} = 0 \tag{18.19}$$

Then χ is diagonal. Using (7.39), we find,

$$\chi_{xx} = \chi_{yy} = \frac{\gamma 2 B_A \mu_0 M_S}{\Omega_0^2 - \omega^2} \tag{18.20}$$

where,

$$\Omega_0^2 = \gamma^2 B_A (B_A + 2B_E) \tag{18.21}$$

Ω_0 is the antiferromagnetic resonance frequency at zero applied magnetic field.

The permeability $\mu \equiv \mu_{xx} = \mu_0(1 + \chi_{xx})$ of the antiferromagnet at zero applied field is given by,

$$\mu = \mu_0 \left(1 + \frac{\gamma^2 B_A \mu_0 M_s}{\Omega_0^2 - \omega^2}\right) \tag{18.22}$$

and the permeability tensor $\boldsymbol{\mu}$ is isotropic, i.e., $\boldsymbol{\mu} = \mu\,\mathbb{1}$.

When a field is applied, the resonance frequency depends on the value of the applied magnetic field. When the field is applied parallel to the equilibrium magnetization vectors, field values in the Tesla range are needed to obtain significant changes in frequency. When the field reaches the spin-flop transition, the sublattice magnetization vectors take on the configuration of a canted antiferromagnet. This is the configuration found in the absence of a magnetic field when the Dzyaloshinskii-Moriya interaction is present. The typical magnetic field dependence of the resonance frequency is shown in Fig 18.3 for a uniaxial antiferromagnet.[372] One mode above the spin-flop transition is called the **quasi-ferromagnetic** resonance because it corresponds to the magnetization precession of a weak ferromagnet, also called canted antiferromagnet. The other mode is predicted by this model to have a zero frequency. In practice, this mode has a non-vanishing frequency because of anisotropies not considered when assuming a strict uniaxial anisotropy.[1031] This mode is represented in Fig. 18.3 as if the magnetization wiggled in the figure plane because the magnetization vectors m_1 and m_2 describe ellipses of very large eccentricities.[1032] This very low frequency mode is the Goldstone mode (§7.6.1) associated with the break in symmetry of the spin-flop transition.

18.4   SPIN WAVES IN ANTIFERROMAGNETS

As D. C. Mattis explains [392], H. Bethe solved the problem of a one-dimensional Heisenberg chain. The solution, however, is so complex that it is difficult to draw much physics from it. D. C. Mattis and E. Lieb found an exact and tractable solution for the so-called XY model, which is a Heisenberg linear chain of half-integer spins, where only in-plane spin components (i.e., components of the plane normal to the chain axis) are considered. Their striking finding is that the magnon dispersion is linear at small wave vectors, i.e., the energy goes linearly to zero in the limit of vanishing small wave vectors. A material for which the linear chain model is relevant is $PrCl_3$. D. Thouless and collaborators previously found a

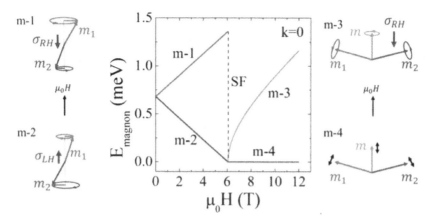

Figure 18.3 Magnetic resonance energy $E_{\text{magnon}} = \hbar\omega(\mu_0 \boldsymbol{H})$ for the antiferromagnet Cr_2O_3, with $\boldsymbol{H}$ parallel to the anisotropy axis. Below the spin-flop transition, the $m-1$ and $m-2$ modes are counter-rotating, as indicated by σ_{RH} and σ_{LH}. Above the spin-flop transition, the canting results in a net magnetization $\boldsymbol{m} = \boldsymbol{m}_1 + \boldsymbol{m}_2$ that either precesses in the basal plane (mode $m-3$), or oscillates parallel to the anisotropy axis (mode $m-4$). Adapted from [1033].

maximum absorption at low temperatures at 3 cm^{-1}.[1034] D. R. Taylor observed electron spin echoes and used relaxation time measurements to infer some of the properties of spin waves in PrCl$_3$ and in praseodymium ethyl sulfate.[1035]

In his treatise on magnetic materials, J. M. D. Coey presents the spin wave spectrum of hematite, an antiferromagnet.[1036] There, too, the dispersion is linear at low momentum values and the energy is vanishingly small (Fig. 18.4). The notion of *magnons* in antiferromagnets was reviewed by Rezende et al.[1032]

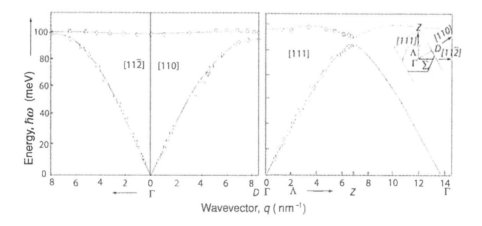

Figure 18.4 Dispersion relation for the antiferromagnet haematite, αFe_2O_3. Adapted from [1037].

18.5 MAGNETIC POLARITONS

In introductory physics, resonances are explained by considering a point mass attached to a spring. First, the natural oscillation of such a harmonic oscillator is observed. Second, a force is applied to the point mass and it can be observed that, when the frequency of the excitation matches the eigenfrequency of the system, a resonance occurs. In this section, we introduce the following question: what happens if we excite a system with a wave that matches the frequency and wave vector of the eigenmodes of an antiferromagnet? To answer this question, we first follow the 1975 publication of Mills and Burnstein.[1038] Then, we use the N-body quantum formalism to describe coupled photons and magnons.

18.5.1 ELECTROMAGNETIC WAVES IN AN ANTIFERROMAGNET

Let us first need to establish a dispersion relation for the electromagnetic waves propagating in the sample. We can then show that, if the permeability depends strongly on frequency, the dispersion relation splits into two branches. In an insulator, the Maxwell equations read, [170]

$$\boldsymbol{\nabla} \times \boldsymbol{H} = \frac{\partial \boldsymbol{D}}{\partial t} \qquad \boldsymbol{\nabla} \times \boldsymbol{E} = \frac{\partial \boldsymbol{B}}{\partial t} \tag{18.23}$$

We consider the linear response of the medium in which electromagnetic waves propagate. We thus write,

$$\boldsymbol{B} = \mu \boldsymbol{H} \qquad \boldsymbol{D} = \epsilon \boldsymbol{E} \tag{18.24}$$

where μ and ϵ are tensors.

Let us consider the propagation of plane waves, characterized by the fields,

$$\boldsymbol{E} \exp(-i\omega t)\exp(i\boldsymbol{k}\boldsymbol{x}) \qquad \boldsymbol{H} \exp(-i\omega t)\exp(i\boldsymbol{k}\boldsymbol{x}) \tag{18.25}$$

Thus, (18.23) yields,

$$\boldsymbol{k} \times \boldsymbol{H} = -\omega\epsilon(\boldsymbol{k},\omega)\boldsymbol{E} \qquad \boldsymbol{k} \times \boldsymbol{E} = \omega\mu(\boldsymbol{k},\omega)\boldsymbol{H} \tag{18.26}$$

From these two relations, we can deduce,

$$\boldsymbol{k} \times \left[\epsilon(\boldsymbol{k},\omega)^{-1}\boldsymbol{k} \times \boldsymbol{H}\right] = -\omega^2\mu(\boldsymbol{k},\omega)\boldsymbol{H} \tag{18.27}$$

At this point, we can simplify the analysis by assuming that the dielectric response is isotropic,

$$\epsilon(\boldsymbol{k},\omega) = \epsilon_0 \mathbf{1} \tag{18.28}$$

with ϵ_0 independent of frequency. Then, we have,

$$(\boldsymbol{k} \cdot \boldsymbol{H})\boldsymbol{k} - k^2\boldsymbol{H} = -\omega^2\epsilon_0\mu(\boldsymbol{k},\omega)\boldsymbol{H} \tag{18.29}$$

From the Maxwell equation $\boldsymbol{\nabla}\boldsymbol{B} = \mathbf{0}$, we have,

$$\boldsymbol{k} \cdot (\mu\boldsymbol{H}) = 0 \tag{18.30}$$

Therefore, in general, $(\boldsymbol{k} \cdot \boldsymbol{H})$ does not vanish. But in the case of zero applied magnetic field, i.e., $B_0 = 0$, the permeability μ is isotropic (18.19), which implies $\boldsymbol{k} \cdot \boldsymbol{H} = \mathbf{0}$. In other words, we have a **transverse magnetic wave**, with the dispersion relation,

$$k^2 = \omega^2\epsilon_0\mu \tag{18.31}$$

18.5.2 MODES SPLITTING

Given the dispersion relation (18.31) and the permeability (18.22), we have,

$$\frac{c^2 k^2}{\omega^2} = 1 + \frac{\Omega_C^2}{\Omega_0^2 - \omega^2} \tag{18.32}$$

where $c = 1/\sqrt{\epsilon_0 \mu_0}$ and,

$$\Omega_C^2 = \gamma^2 B_A \mu_0 M_s \tag{18.33}$$

Equation (18.32) yields a second-degree polynomial equation for ω^2, the solution to which

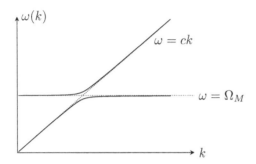

Figure 18.5 Dispersion relation (18.34) for an electromagnetic mode propagating in a material with a strong frequency-dependent susceptibility, such as the one given by (18.22), corresponding to a uniaxial antiferromagnet at zero applied magnetic field.

can be written as,

$$\omega_\pm^2(k) = \frac{1}{2}\left[\omega_{free}^2 + \Omega_M^2 \pm \sqrt{(\omega_{free}^2 - \Omega_M^2)^2 + 4\omega_{free}^2 \Omega_C^2}\right] \tag{18.34}$$

where

$$\omega_{free}^2 = c^2 k^2 \tag{18.35}$$

Our result (18.34) closely resembles the eigenfrequencies of two coupled oscillators. Indeed, consider a mechanical system composed of two harmonic oscillators, for example, two pendula of the same mass m but different lengths. Assume they are coupled by a spring of elastic constant k. Then, the eigenfrequencies are given by,

$$\omega_\pm^2 = \frac{1}{2}\left[\Omega_1^2 + \Omega_2^2 \pm \sqrt{(\Omega_1^2 - \Omega_2^2)^2 + 4\Omega_C^4}\right] \tag{18.36}$$

where $\Omega_C^2 = k/m$ and $\Omega_i^2 = g/L_i + k/m$, i=1,2. The dispersion relations (18.34) are characteristic of the coupling between two wavelike modes. The dispersion $\omega(k)$ is plotted in Fig. 18.5.

Experimentally, we only have access to magnons of very small wavevectors. In free space, the electromagnetic wavevector is comparatively large, but in a cavity, the electromagnetic wave dispersion is folded and there can be crossing points with the magnon dispersion at $k \approx 0$. The resonance condition can be reached by tuning the cavity size. Alternatively, it is possible to go through the resonance condition by changing the temperature, as generally the magnetic resonance frequency is temperature-dependent, whereas the cavity frequency is almost temperature-independent (Fig. 18.6).

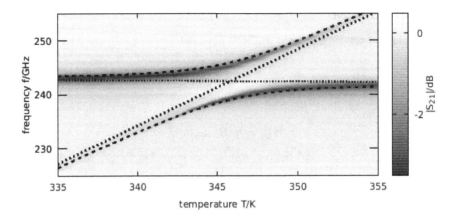

Figure 18.6 Transmission of millimeter waves through a cylindrical cavity containing a small cube of hematite (Fe_2O_3), as a function of the sample temperature and frequency of the incident radiation. The temperature range was chosen so that the AFMR resonance frequency range includes the frequency of a standing wave. Dashed line: fit using (18.48). Dotted lines: temperature dependence of the AFMR and standing wave frequencies if there were no coupling ($G = 0$) between the AFMR and the standing wave mode of the electromagnetic field. Adapted from [1039].

18.5.3 SCATTERING MATRIX MODELING OF MAGNETIC POLARITONS

Let us represent a cavity as a device with $N = 2$ ports, numbered with $i = 1 \ldots N$ (Fig. 18.7). The generalization to $N \neq 2$ is straightforward.[1040] At each port, we distinguish incoming and outgoing waves. The total voltage and total current at each port are given by,

$$V_i = V_i^+ + V_i^- \qquad I_i = I_i^+ + I_i^- \qquad (i = 1 \ldots N) \tag{18.37}$$

One way of characterizing the cavity is to specify the parameters A, B, C, D, defined by,

$$\begin{pmatrix} V_1 \\ I_1 \end{pmatrix} = \begin{pmatrix} A & B \\ C & D \end{pmatrix} \begin{pmatrix} V_2 \\ I_2 \end{pmatrix} \tag{18.38}$$

For the two-port device represented by the impedances as shown in Fig. 18.7, the A, B, C, D parameters are given by,[1040]

$$A = 1 + \frac{Z_1}{Z_3} \quad B = Z_1 + Z_2 + \frac{Z_1 Z_2}{Z_3} \quad C = \frac{1}{Z_3} \quad D = 1 + \frac{Z_2}{Z_3} \tag{18.39}$$

The scattering matrix is defined by,

$$\begin{pmatrix} V_1^- \\ V_2^- \end{pmatrix} = \begin{pmatrix} S_{11} & S_{12} \\ S_{21} & S_{22} \end{pmatrix} \begin{pmatrix} V_1^+ \\ V_2^+ \end{pmatrix} \tag{18.40}$$

In particular, the often-measured S_{21} parameter is given by,

$$S_{21} = \left. \frac{V_2^-}{V_1^+} \right|_{V_2^+ = 0} \tag{18.41}$$

For the circuit of Fig. 18.7, S_{21} can be calculated as,

$$S_{21} = \frac{2}{A + B/Z_0 + C Z_0 + D} \tag{18.42}$$

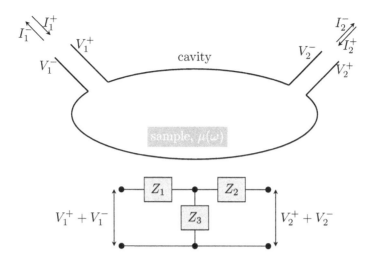

Figure 18.7 Top: cavity as a two-port device. At each port, there are incoming and outgoing waves characterized by voltages and currents. Inside the cavity, there is a piece of matter characterized by its permittivity $\mu(\omega)$. Bottom: two-port circuit with impedances Z_1, Z_2, Z_3 in a "T" configuration.

where Z_0 is the impedance of the transmission lines (or waveguides) at the ports.

Let us apply this formalism to analyze a magnetic material located inside the cavity. We represent the cavity by discrete R, L, C components. We assume that the measurement setup is such that, away from a resonance of the cavity or away from a resonance of the material in the cavity, the signal propagates through the cavity without loss, i.e., $S_{21} = 1$. One possible circuit that would do this would have $Z_1 = 0 = Z_2$ and Z_3 would correspond to the impedance of an inductance L, a capacitance C and a resistance R in series. Then, Z_3 is given by,

$$Z_3 = j\omega L + \frac{1}{j\omega C} + R \tag{18.43}$$

Using (18.42), S_{21} can be calculated,

$$S_{21} = \frac{1}{1 + \frac{Z_0}{2Z_3}} \approx 1 - \frac{Z_0}{2Z_3} \tag{18.44}$$

We assume that the RLC circuit is designed so that the transmission is nearly one at all frequencies, i.e., $Z_3 \gg Z_0$.

We take into account the presence of the material in the cavity by assuming that the inductance L is filled entirely with a medium of effective permeability μ. The permeability μ is written in the form of a Lorentzian line shape, so as to account for the magnetic resonance in the material. In other words, the inductance L is given by $L = \mu L_0$, where L_0 is the inductance of the empty cavity and μ is given by,

$$\mu = 1 + \frac{\Delta\mu\,\omega_m^2}{\omega_m^2 - \omega^2 - j\omega\gamma_m} \tag{18.45}$$

$\Delta\mu$ is a unitless parameter that characterizes the coupling of the material to the cavity. Using Z_3 given by (18.43) and (18.44), we have,

$$S_{21} = 1 - \frac{Z_0 j\omega C/2}{1 - j\omega RC - \omega^2 C L_0\left(1 + \frac{\Delta\mu\,\omega_m^2}{\omega_m^2 - \omega^2 - j\omega\gamma_m}\right)} \tag{18.46}$$

Let us call $\omega_c^2 = 1/(L_0 C)$ the resonance if the cavity were empty. We are interested in what happens near the resonances of this system, namely, the resonance of the cavity and that of the material. Near the resonances, we can make the approximation,

$$\omega^2 - \omega_c^2 = (\omega - \omega_c)(\omega + \omega_c) \approx 2\omega_c(\omega - \omega_c) \tag{18.47}$$

Since the resonances are sharp, ω can be replaced by ω_c when ω is not in a term that leads to a divergence (in the absence of damping). Thus, we can write,[1041]

$$S_{21} = 1 + \cfrac{a}{j(\omega - \omega_c) - \cfrac{\kappa}{2} + \cfrac{G^2}{j(\omega - \omega_m) - \gamma_m/2}} \tag{18.48}$$

where $a = j\omega_c^2 Z_0 C/4$, $G^2 = \omega_c \omega_m \Delta\mu/4$ and $\kappa = \omega_c^2 RC$.

The ***input-output formalism*** is the quantum-mechanical extension of the development presented here. Namely, a quantized incoming electromagnetic field enters one port of a cavity in which a piece of matter has a magnetic or dielectric response, and an outgoing field leaves the cavity at the second port.[1021, 1041–1043]

18.6 CAVITY QUANTUM ELECTRODYNAMICS OF MAGNON POLARITONS

18.6.1 THE PURCELL EFFECT

The ***Purcell effect*** describes the change in the contribution of spontaneous emission to the nuclear spin relaxation rate when the spin is in a cavity instead of in free space.[1044] As we will see, this effect is too small to detect.

Let us start from a classical standpoint and consider the radiation power emitted by an oscillating magnetic dipole $\boldsymbol{m}$. The power emitted is given by:([170] Eq. 7.17)

$$P = \frac{1}{6\pi}\sqrt{\frac{\mu_0}{\epsilon_0}}\frac{\ddot{m}^2}{c^4} \tag{18.49}$$

where c is the speed of light, $\sqrt{\frac{\mu_0}{\epsilon_0}}$ the electromagnetic wave impedance of vacuum, which is $Z_0 = 377$ ohms. If $\omega/2\pi$ is the oscillation frequency, we have $\ddot{m}^2 = \omega^2\|\boldsymbol{m}\|^2$.

To be completely clear about the units, consider a dipole consisting of a current loop of radius a. Then $\|\boldsymbol{m}\| = \pi a^2 I$ where I oscillates with the pulsation ω. Then, $P = Z_0 I^2(\omega^4 a^4/(6\pi c^4))$. As the last parenthesis is unitless, this expression of P has the form of a Joule heating power.

In the following, we will consider that the magnetic dipole has one Bohr magneton, i.e., $\|\boldsymbol{m}\| = 9 \cdot 10^{-24}$ J/T.[1] Thinking of resonance in antiferromagnets, consider a frequency of 300 GHz. One Bohr magneton would emit radiation at this frequency with a power $P = 2.6 \cdot 10^{-30}$ W.

Now, in a manner similar to the energy considerations for the two-level system of §12.2, let us equate P with $W_{\text{spont}}\hbar\omega$ where W_{spont} is the rate of spin flips. This is a spontaneous emission since we are not considering that there are occupied modes of the electromagnetic field surrounding this dipole. Thus, we have,

$$W_{\text{spont}} = \frac{P}{\hbar\omega} = \frac{1}{\hbar\omega}\frac{1}{6\pi}\sqrt{\frac{\mu_0}{\epsilon_0}}\frac{\omega^4 \mu_B^2}{c^4} \tag{18.50}$$

---

[1]From the Lorentz force expression $F = qvB$, we have $T = N/(Cms^{-1}) = Nm^{-1}/A)$. Thus, J/T = (units of magnetic moment) = Am^2, in conformity with $\|\boldsymbol{m}\| = \pi a^2 I$.

With the same numerical values as above, we thus find a rate of 10^{-8}s^{-1}. This corresponds to an extremely long relaxation time. The rate of stimulated emission is given by:[162]

$$W_{\text{sti}} = n(\omega)W_{\text{spont}} \tag{18.51}$$

If the electromagnetic radiation is at equilibrium with a thermal bath at temperature T, $n(\omega)$ is given by the Bose-Einstein distribution,

$$n(\omega) = \frac{1}{\exp\left(\frac{\hbar\omega}{k_B T}\right) - 1} \approx \frac{k_B T}{\hbar\omega} \tag{18.52}$$

In the last step, we used the high-temperature approximation.

Let us now model the number of modes that one may have in a cavity as compared to what we have in free space. We start by expressing the spectral energy density per unit of volume $E(\nu)$ in free space in terms of $n(\omega)$. To do so, we consider the modes of a box of size L^3 where L is very large. The density of points in k-space is given by $L^3/(2\pi)^3$. We have two polarizations per mode. We need to account for the fact that the two-level magnetic system has a Zeeman splitting due to some applied magnetic induction field $\boldsymbol{B}$. We must also account for the fact that the $\boldsymbol{b}$ field of some of the modes do not couple to the spin, so that only 4 out of 6 $\boldsymbol{k}$ directions correspond to active modes.[866] Thus, for the mode at frequency ν, we have,

$$L^3 \int d\nu E(\nu) = (2) \int 4\pi k^2 dk\, n_{\boldsymbol{k}} \hbar\omega_{\boldsymbol{k}} \frac{L^3}{(2\pi)^3} \frac{2}{3} \tag{18.53}$$

The right-hand side can be transformed into an integral over the frequency, using the dispersion $\omega = ck$. This yields,

$$E(\nu) = 8\pi \frac{h}{\lambda^3} \frac{2}{3} n(\nu) \tag{18.54}$$

We adopt the following model for the spectral density function $E_R(\nu)$ inside a cavity. When the photons are at equilibrium with the cavity walls, processes of photon absorption and emission occur. Each event is similar to a photon entering or leaving the cavity (Fig. 18.8). The cavity response is well known and the energy stored in the cavity has a peak at the resonance frequency ν_0. We assume that the resonance spectrum is a Lorentzian function characterized by a quality factor Q. The cavity has a volume V. We impose the normalization:[866]

$$V \int E_R(\nu)d\nu = n(\nu_0)h\nu \tag{18.55}$$

Thus, the energy density function $E_R(\nu)$ is found to be:

$$E_R(\nu) = \frac{1}{\pi} \frac{\nu_0}{2Q} \frac{n(\nu_0)h\nu_0}{(\nu - \nu_0)^2 + (\nu_0/2Q)^2} \frac{1}{V} \tag{18.56}$$

Hence, the spectral density of energy at resonance is given by:

$$E_R(\nu_0) = \frac{1}{\pi} \frac{2Qn(\nu_0)h}{V} \tag{18.57}$$

Therefore, the cavity gives the enhancement,

$$\frac{E_R(\nu_0)}{E(\nu_0)} = \frac{1}{4\pi^2} \frac{Q\lambda^3}{V} \frac{3}{2} \tag{18.58}$$

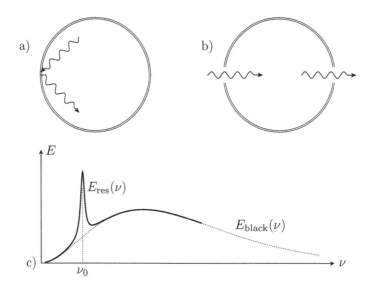

Figure 18.8 a) A cavity with photons at equilibrium with the cavity walls; b) a cavity with incoming and outgoing photons; and c) a spectral density function without (dotted line) and with cavity (black line).

This tells us how much the occupation number $n(\nu_0)$ is enhanced in the cavity. In summary, we find the stimulated emission rate to be,

$$W_{\text{sti}} = \left(\frac{k_B T}{h\nu_0}\right)\left(\frac{1}{4\pi^2}\frac{Q\lambda^3}{V}\frac{3}{2}\right)\left(\frac{1}{\hbar c}\frac{1}{6\pi}\sqrt{\frac{\mu_0}{\epsilon_0}}\frac{\omega^3\mu_B^2}{c^3}\right) \qquad (18.59)$$

where the first factor is the occupation number in free space, the second the cavity enhancement, and the third W_{spont}. Given that the quality factor is by definition $Q = \nu_0/\Delta\nu$ where $\Delta\nu$ is the width of the cavity resonance, (18.59) can be written as,

$$W_{\text{sti}} = \left(\frac{k_B T}{h\Delta\nu_0}\right)\frac{\mu_B^2}{2Vc\hbar}\sqrt{\frac{\mu_0}{\epsilon_0}} \qquad (18.60)$$

Using the same numerical values and assuming a cavity width $\Delta\nu$ of 1 GHz, we thus find $W_{\text{sti}} \approx 4\cdot 10^{-6}$ s^{-1}. This corresponds to a relaxation time far greater than can be expected in an antiferromagnet. We notice that the expression (18.60) does not have an explicit dependence on frequency. However, (18.60) predicts that the smallest possible volume is optimal. Since we are considering the eigenmodes of a cavity, the cavity cannot be smaller than the wavelength. A higher frequency allows us to use a smaller volume for the cavity and hence increase the stimulated emission rate.

Collective effects

As this order-of-magnitude estimate suggests, the Purcell effect, therefore the coupling to cavity modes, of a single two-level system is very small. However, when N two-level systems are coherently coupled in a cavity, the coupling of the **collective excitation** of the two-level systems with the cavity is characterized by a coefficient $g_{\text{eff}} = g\sqrt{N}$, where g is the coupling strength of an isolated two-level system.[1019, 1045] The following two sections explain this enhancement.

The $\sqrt{N}$ enhancement has given the idea to some researchers (see §18.1) that, in a magnetic system, the strength of the coupling might be detectable at temperatures that are not too low and despite the fact that resonances in magnetic materials tend to be rather broad. Experimentally, evidence for a cavity effect was found in an antiferromagnet,[1020] and in ferrimagnets.[1021] Notably, however, this enhancement can be found in classical resonators coupled to one another via their coupling to the cavity. This was exemplified with two yttirum iron garnet (YIG) spheres, of which the resonance frequencies could be tuned individually. As the spheres were located inside a high-quality resonator, the FMR frequencies could be adjusted to match a cavity mode. The strength of the coupling between the cavity mode and the magnetization dynamics was found to be enhanced by a factor $\sqrt{2}$ with respect to the value found for one YIG sphere.[1046]

In Kittel's view,[2] polaritons are one example of a situation where two types of waves are in resonance with one another, the matching condition being in frequency and in wave vector as well.[79] In the next section, we see how to describe with quantum mechanics the coupling between two waves. In the following section, we examine the specific case of the coupling of magnons and photons and show that the coupling strength is enhanced by a factor $\sqrt{N}$ compared to the coupling of an individual spin.[3]

18.6.2 COUPLED MODES IN QUANTUM N-BODY FORMALISM

Let us consider two subsystems, labelled A and B, described by their respective Hilbert spaces $\mathbb{H}_A$ and $\mathbb{H}_B$. The Hilbert space of the whole system is the tensor product $\mathbb{H}_A \otimes \mathbb{H}_B$. Operators corresponding to properties of the A subsystem are of the form $A \otimes \mathbf{1}_B$ and operators of the B subsystem are of the form $\mathbf{1}_A \otimes B$. From now on, we consider two sets of creation and annihilation operators, $A^\dagger, A$ and $B^\dagger, B$, for which the only non-trivial commutation relations are,

$$[A, A^\dagger] = \mathbf{1}_A \quad , \quad [B, B^\dagger] = \mathbf{1}_B \tag{18.61}$$

These commutation relations imply,

$$[A^\dagger A, A] = -A \quad , \quad [A^\dagger A, A^\dagger] = A^\dagger \tag{18.62}$$

and likewise for B and $B^\dagger$. The dynamics of the coupled subsystems is given by the Hamiltonian,

$$\mathcal{H} = \epsilon_A A^\dagger A \otimes \mathbf{1}_B + \epsilon_B \mathbf{1}_A \otimes B^\dagger B + g A^\dagger \otimes B + g^* A \otimes B^\dagger \tag{18.63}$$

where ϵ_A and ϵ_B are real numbers and the coupling strength g is a complex number.

Let us also define the occupation operator,

$$N = N\dagger = A^\dagger A \otimes \mathbf{1}_B + \mathbf{1}_A \otimes B^\dagger B \tag{18.64}$$

We now show that H and N commute. This implies that N is a conserved quantity. Indeed, we have the following commutation relations,

$$[H, A^\dagger A \otimes \mathbf{1}_B] = g[A^\dagger, A^\dagger A] \otimes B + g^*[A, A^\dagger B] \otimes A^\dagger = -g A^\dagger \otimes B + g^* A \otimes B^\dagger$$
$$[H, \mathbf{1}_A \otimes B^\dagger B] = g A^\dagger \otimes [B, B^\dagger B] + g^* A \otimes [B^\dagger, B^\dagger B] = g A^\dagger \otimes B - g^* A \otimes B^\dagger$$

The sum of these two equations yields $[\mathcal{H}, N] = 0$, as announced.

The Hamiltonian $\mathcal{H}$ can be written as,

$$\mathcal{H} = \begin{bmatrix} A^\dagger \otimes \mathbf{1}_B & \mathbf{1}_A \otimes B^\dagger \end{bmatrix} \begin{bmatrix} \epsilon_A & g^* \\ g & \epsilon_B \end{bmatrix} \begin{bmatrix} A \otimes \mathbf{1}_B \\ \mathbf{1}_A \otimes B \end{bmatrix} \tag{18.65}$$

---

[2]See, [79], chapter 10.

[3]The following two sections were developed with the assistance of Dr. F. Reuse, author of [162].

The hermitian 2×2 matrix in (18.65) can be written as,

$$\begin{bmatrix} \epsilon_A & g^* \\ g & \epsilon_B \end{bmatrix} = \epsilon_+ X_+ + \epsilon_- X_- \qquad (18.66)$$

where the eigenvalues ϵ_+ and ϵ_- are real, and the matrices $X_\pm$ are projectors. As such, they have the following properties,

$$X_\pm^\dagger = X_\pm \quad , \quad X_\pm^2 = X_\pm \quad , \quad X_+ X_- = 0 \quad , \quad X_+ + X_- = 1 \qquad (18.67)$$

where 1 is the identity matrix in $\mathbb{C}^2$. The eigenvalues $\epsilon_\pm$ are solutions of,

$$\begin{bmatrix} \epsilon_A & g^* \\ g & \epsilon_B \end{bmatrix} V_\pm = \epsilon_\pm \, V_\pm \qquad (18.68)$$

Hence, the eigenvalues are,

$$\epsilon_\pm = \frac{1}{2}\left(\epsilon_A + \epsilon_B \pm \sqrt{(\epsilon_A - \epsilon_B)^2 + 4gg^*} \right) \qquad (18.69)$$

We note that,

$$\epsilon_A - \epsilon_\pm = \frac{1}{2}\left(\epsilon_A - \epsilon_B \mp \sqrt{(\epsilon_A - \epsilon_B)^2 + 4gg^*} \right) \quad , \quad gg^* + (\epsilon_+ - \epsilon_A)(\epsilon_- - \epsilon_A) = 0 \quad (18.70)$$

The eigenvectors of (18.68) are,

$$V_+ = \alpha_+ \begin{bmatrix} g^* \\ \epsilon_+ - \epsilon_A \end{bmatrix} \quad \text{and} \quad V_- = \alpha_- \begin{bmatrix} g^* \\ \epsilon_- - \epsilon_A \end{bmatrix}$$

The coefficients $\alpha_\pm$ ensure that the eigenvectors are normalized. Hence, we have,

$$\alpha_\pm \alpha_\pm^* = (\epsilon_A - \epsilon_\pm)^2 + gg^* \qquad (18.71)$$

The orthogonal projectors X_+ and X_- are given by,

$$X_\pm = V_\pm V_\pm^\dagger \qquad (18.72)$$

where $V_\pm^\dagger = \alpha_\pm^* [g \quad \epsilon_\pm - \epsilon_A]$. Based on the decomposition (18.66), we can write (18.65) using (18.72) as,

$$\mathcal{H} = \begin{bmatrix} A^\dagger \otimes 1_B & 1_A \otimes B^\dagger \end{bmatrix} \left(\epsilon_+ V_+ V_+^\dagger + \epsilon_- V_- V_-^\dagger \right) \begin{bmatrix} A \otimes 1_B \\ 1_A \otimes B \end{bmatrix} \qquad (18.73)$$

Then, we define the operators,

$$P_\pm^\dagger = \begin{bmatrix} A^\dagger \otimes 1_B & 1_A \otimes B^\dagger \end{bmatrix} V_\pm = \alpha_\pm \left(g^* A^\dagger \otimes 1_B + (\epsilon_\pm - \epsilon_A) 1_A \otimes B^\dagger \right)$$
$$P_\pm = V_\pm^\dagger \begin{bmatrix} A \otimes 1_B \\ 1_A \otimes B \end{bmatrix} = \alpha_\pm^* \left(g A \otimes 1_B + (\epsilon_\pm - \epsilon_A) 1_A \otimes B \right) \qquad (18.74)$$

These operators have the following commutation relations,

$$\begin{bmatrix} P_+, P_+^\dagger \end{bmatrix} = 1 \quad , \quad \begin{bmatrix} P_-, P_-^\dagger \end{bmatrix} = 1 \quad , \quad \begin{bmatrix} P_+, P_-^\dagger \end{bmatrix} = 0 \quad , \quad \begin{bmatrix} P_+, P_- \end{bmatrix} = 0 \qquad (18.75)$$

The Hamiltonian $\mathcal{H}$ can be written in terms of these operators as,

$$\mathcal{H} = \epsilon_+ \, P+^\dagger P_+ + \epsilon_- \, P_-^\dagger P_- \qquad (18.76)$$

This is the diagonal form of the N-body quantum representation of two coupled modes, with energies $\epsilon_\pm$ given by (18.69). In the following section, we examine the case of a collective state of N local moments coupled to electromagnetic modes of a cavity.

18.6.3 COUPLING STRENGTH WHEN ATOMIC SPINS ARE CORRELATED

Let us now consider operators of the Hilbert space $\mathbb{H}_{em}$ of the electromagnetic fields. In the Heisenberg representation, we define electromagnetic fields in the Coulomb gauge that propagate in the z direction, i.e., their wavevector is $\boldsymbol{k} = k\boldsymbol{e}_3$. Here, $(\boldsymbol{e}_1, \boldsymbol{e}_2, \boldsymbol{e}_3)$ is an orthonormal direct basis of unit vectors in the rest frame. Thus, we define the vector potential field,

$$\boldsymbol{A}^H(z,t) = \sqrt{\frac{\hbar}{\varepsilon_0 L^3}} \sum_{k \in \mathcal{L}} \sqrt{\frac{1}{2\omega(k)}} \left[a_k \, \boldsymbol{\epsilon}_{pol} e^{\imath(kz-\omega(k)t)} + a_k^\dagger \, \boldsymbol{\epsilon}_{pol}^* e^{-\imath(kz-\omega(k)t)} \right] \qquad (18.77)$$

where $\mathcal{L}$ is the set of k values defined by,

$$\mathcal{L} = \{ k \neq 0 \,|\, kL = 0 \mod 2\pi \} \qquad (18.78)$$

In (18.77), the polarization vector $\boldsymbol{\epsilon}_{pol}$, defined in $\mathbb{C}^3$, has the properties,

$$\boldsymbol{k} \cdot \boldsymbol{\epsilon}_{pol} = 0 \quad, \quad \boldsymbol{\epsilon}_{pol}^* \cdot \boldsymbol{\epsilon}_{pol} = 1 \qquad (18.79)$$

The electromagnetic wave dispersion is given by,

$$\omega(k) = |k|c \qquad (18.80)$$

The coefficients in (18.77) are such that the creation and annihilation operators satisfy,

$$\begin{aligned} [a_k, \, a_{k'}^\dagger] = \delta_k^{k'} \mathbf{1} \quad &, \quad [a_k, \, a_{k'}] = 0 \\ [a_k a_k^\dagger, \, a_{k'}] = -\delta_k^{k'} \, a_k \quad &, \quad [a_k a_k^\dagger, \, a_{k'}^\dagger] = +\delta_k^{k'} \, a_k^\dagger \end{aligned} \qquad (18.81)$$

We make the assumption that there are no other observables of $\mathbb{H}_{em}$ that are independent of those characterized by the operators a_k and $a_k^\dagger$ ($\forall k$). In other words, if there is a self-adjoint operator $X = X^\dagger$ compatible with a_k and $a_k^\dagger$ ($\forall k$), then,

$$[a_k, X] = 0 \quad \forall k \quad \Rightarrow \quad X = \lambda \mathbf{1} \quad (\lambda \in \mathbb{R}) \qquad (18.82)$$

The electric field can be determined using $\boldsymbol{E}^H(z,t) = -\partial \boldsymbol{A}^H(z,t)/\partial t$. This yields,

$$\boldsymbol{E}^H(z,t) = \sqrt{\frac{\hbar}{\varepsilon_0 L^3}} \sum_{k \in \mathcal{L}} \imath \sqrt{\frac{\omega(k)}{2}} \left[a_k \, \boldsymbol{\epsilon}_{pol} e^{\imath(kz-\omega(k)t)} - a_k^\dagger \, \boldsymbol{\epsilon}_{pol}^* e^{-\imath(kz-\omega(k)t)} \right] \qquad (18.83)$$

The magnetic induction field is derived using $\boldsymbol{B}^H(z,t) = \boldsymbol{\nabla} \times \boldsymbol{A}^H(z,t)$. This yields,

$$\boldsymbol{B}^H(z,t) = \sqrt{\frac{\hbar}{\varepsilon_0 L^3}} \sum_{k \in \mathcal{L}} \sqrt{\frac{1}{2\omega(k)}} \imath \boldsymbol{k} \wedge \left[a_k \, \boldsymbol{\epsilon}_{pol} e^{\imath(kz-\omega(k)t)} - a_k^\dagger \, \boldsymbol{\epsilon}_{pol}^* e^{-\imath(kz-\omega(k)t)} \right] \qquad (18.84)$$

In the Schroedinger representation, these fields are given by,

$$\boldsymbol{A}(z) = \sqrt{\frac{\hbar}{\varepsilon_0 L^3}} \sum_{k \in \mathcal{L}} \sqrt{\frac{1}{2\omega(k)}} \left[a_k \, \boldsymbol{\epsilon}_{pol} e^{\imath kz} + a_k^\dagger \, \boldsymbol{\epsilon}_{pol}^* e^{-\imath kz} \right]$$

$$\boldsymbol{E}(z) = \sqrt{\frac{\hbar}{\varepsilon_0 L^3}} \sum_{k \in \mathcal{L}} \imath \sqrt{\frac{\omega(k)}{2}} \left[a_k \, \boldsymbol{\epsilon}_{pol} e^{\imath kz} - a_k^\dagger \, \boldsymbol{\epsilon}_{pol}^* e^{-\imath kz} \right] \qquad (18.85)$$

$$\boldsymbol{B}(z) = \sqrt{\frac{\hbar}{\varepsilon_0 L^3}} \sum_{k \in \mathcal{L}} \sqrt{\frac{1}{2\omega(k)}} \imath \boldsymbol{k} \wedge \left[a_k \, \boldsymbol{\epsilon}_{pol} e^{\imath kz} - a_k^\dagger \, \boldsymbol{\epsilon}_{pol}^* e^{-\imath kz} \right]$$

Let us now consider the local magnetic moments. We use the Pauli matrices σ_1, σ_2, and σ_3 to define the local magnetic moment as $\mu\boldsymbol{\sigma}$.[4] We recall the commutation relations,

$$[\sigma_i, \sigma_j] = 2\imath\sigma_k \qquad (i, j, k \quad \text{circular permut. of } 1, 2, 3) \qquad (18.86)$$

In the developments below, we will make use of the operators,

$$\sigma_\pm = \frac{1}{2}(\sigma_1 \pm \imath\sigma_2) \qquad\qquad (18.87)$$

These operators have the following commutation relations,

$$[\sigma_3, \sigma_\pm] = \pm\sigma_\pm \quad , \quad [\sigma_+, \sigma_-] = \sigma_3 \qquad\qquad (18.88)$$

Since the excitation is in the xy plane, we write the local moment as,

$$\mu\boldsymbol{\sigma} = \mu(\sigma_1\boldsymbol{e}_1 + \sigma_2\boldsymbol{e}_2) \qquad\qquad (18.89)$$

We can write the local moment as,

$$\mu\boldsymbol{\sigma} = \sqrt{2}\mu(\sigma_+\boldsymbol{e}_- + \sigma_-\boldsymbol{e}_+) \qquad\qquad (18.90)$$

where the vectors $\boldsymbol{e}_\pm$ are defined by,

$$\boldsymbol{e}_\pm = \frac{1}{\sqrt{2}}(\boldsymbol{e}_1 \pm \imath\boldsymbol{e}_2) \quad\Rightarrow\quad \boldsymbol{e}_1 = \frac{\boldsymbol{e}_+ + \boldsymbol{e}_-}{\sqrt{2}} \quad , \quad \boldsymbol{e}_2 = \frac{\boldsymbol{e}_+ - \boldsymbol{e}_-}{\sqrt{2\imath}} \qquad (18.91)$$

The interaction $\mathcal{H}_{\text{int}}^{(1)}$ of one local moment with the magnetic induction field $\boldsymbol{B}(z)$ is given by,

$$\mathcal{H}_{\text{int}}^{(1)} = -\mu\boldsymbol{\sigma} \cdot \boldsymbol{B}(z) = -\mu\Big(\sigma_+(\boldsymbol{e}_- \cdot \boldsymbol{B}(z)) + \sigma_-(\boldsymbol{e}_+ \cdot \boldsymbol{B}(z))\Big) \qquad (18.92)$$

The scalar products in (18.92) can be written as,

$$\boldsymbol{e}_\pm \cdot \boldsymbol{B}(z) = \sum_{k\in\mathcal{L}} B_0(k)\Big[A_\pm(\boldsymbol{\epsilon}_{pol})\, a_k \exp\imath kz + A_\pm(\boldsymbol{\epsilon}_{pol})^*\, a_k^\dagger \exp-\imath kz\Big] \qquad (18.93)$$

where

$$B_0(k) = \sqrt{\frac{\hbar}{\varepsilon_0 L^3 \omega(k)}} \quad , \quad A_\pm(\boldsymbol{\epsilon}_{pol}) = \imath\frac{\boldsymbol{k}}{\|\boldsymbol{k}\|} \cdot (\boldsymbol{\epsilon}_{pol} \wedge \boldsymbol{e}_\pm) \qquad (18.94)$$

Then, the interaction Hamiltonian (18.92) for one local moment reads,

$$\begin{aligned}
\mathcal{H}_{\text{int}}^{(1)} = -\mu\sum_{k\in\mathcal{L}} B_0(k)\Big[&\sigma_+ \otimes \Big(A_-(\boldsymbol{\epsilon}_{pol})\, a_k \exp\imath kz + A_-(\boldsymbol{\epsilon}_{pol})^*\, a_k^\dagger \exp-\imath kz\Big) \\
&+ \sigma_- \otimes \Big(A_+(\boldsymbol{\epsilon}_{pol})\, a_k \exp\imath kz + A_+(\boldsymbol{\epsilon}_{pol})^*\, a_k^\dagger \exp-\imath kz\Big)\Big]
\end{aligned} \qquad (18.95)$$

The coefficients $A_\pm$ defined in (18.94) have the following physical implications concerning which handedness of the electromagnetic wave polarization couples to the spin operators $\sigma_\pm$. If the electromagnetic wave is right-hand circularly polarized,[5] we have,

$$\boldsymbol{\epsilon}_{pol} = \boldsymbol{e}_+ \quad , \quad A_+(\boldsymbol{\epsilon}_{pol}) = 0 \quad , \quad A_-(\boldsymbol{\epsilon}_{pol}) = 1 \qquad (18.96)$$

---

[4]Pauli matrices are given in (19.80). They are introduced in quantum mechanics textbooks.[181]

[5]Using (18.77), it is easy to see that the vector $(\boldsymbol{e}_1 + \imath\boldsymbol{e}_2)\exp-\imath\omega t$ rotates to the right when $k > 0$ and to the left when $k < 0$ (owing to (18.80)). Hence, $\boldsymbol{e}_1 + \imath\boldsymbol{e}_2$ corresponds to a right-handed circular polarization.

Instead, if the wave is left-hand circularly polarized, we have,

$$\epsilon_{pol} = e_- \quad , \quad A_-(\epsilon_{pol}) = 0 \quad , \quad A_+(\epsilon_{pol}) = -1 \tag{18.97}$$

Let us now consider a cavity of size L in the z direction, containing N atoms in a row, at a distance δz from one another. The atoms are located at $z_\nu = \nu\,\delta z$, $\nu = 1, 2, \ldots, N$. The possible values of the wavevector k in (18.85) are,

$$k_n = \frac{2\pi}{L}\, n \quad , \quad n = 1, \ldots, N \tag{18.98}$$

In the following calculations, we are going to encounter functions of the form,

$$\phi_n(\nu) = \frac{1}{\sqrt{N}} \exp \imath k_n z_\nu \tag{18.99}$$

We note here that they have the following properties,

$$\sum_{\nu=1}^{N} \phi_n(\nu)\phi_m(\nu)^* = \delta_{nm} \quad , \quad \sum_{n=1}^{N} \phi_n(\nu)\phi_n(\nu')^* = \delta_{\nu\nu'} \tag{18.100}$$

The Hilbert space $\mathbb{H}_{\mathrm{sw}}$ for the N spins is defined by the tensorial products of N Hilbert spaces for one spin. Hence, we define a one-spin operator as,

$$\boldsymbol{\sigma}^{(\nu)} = \mathbf{1} \otimes \mathbf{1} \otimes \cdots \otimes \mathbf{1} \otimes \boldsymbol{\sigma} \otimes \mathbf{1} \otimes \cdots \otimes \mathbf{1} \otimes \mathbf{1} \quad , \quad \nu = 1, 2, \ldots, N \tag{18.101}$$

These operators have the following commutation relations,

$$\left[\sigma_i^{(\nu)}, \sigma_j^{(\lambda)}\right] = 2\,\imath\,\delta^{(\nu\lambda)}\sigma_k^{(\nu)} \qquad (i, j, k \quad \text{circular permut. of } 1, 2, 3) \tag{18.102}$$

We also define $\sigma_{\mp}^{(\nu)}$ operators,

$$\sigma_{\pm}^{(\nu)} = \frac{1}{\sqrt{2}}\left(\sigma_1^{(\nu)} \pm \imath \sigma_2^{(\nu)}\right) \tag{18.103}$$

The operators $\sigma_{\pm}^{(\nu)}$ have the following commutation relations,

$$\left[\sigma_+^{(\nu)}, \sigma_-^{(\lambda)}\right] = \delta^{(\nu\lambda)}\sigma_3^{(\nu)} \quad , \quad \left[\sigma_3^{(\nu)}, \sigma_{\pm}^{(\lambda)}\right] = \pm\delta^{(\nu\lambda)}\sigma_{\pm}^{(\nu)} \tag{18.104}$$

Let us now examine the coupling between the excitations of the local moments and one mode of the electromagnetic field. Thus, we consider a single wavevector $\boldsymbol{k}_n$ along $\boldsymbol{e}_3$. We also consider a chain of sites located at z_ν, having a transverse spin described by the operator $\boldsymbol{\sigma}_\nu$, $\nu = 1, 2, \ldots, N$. Using the simplifying notation a for a_{k_n}, the interaction Hamiltonian (18.95) allows us to find the interaction Hamiltonian $\mathcal{H}_{\mathrm{int}}$ for this spin chain. It is given by,

$$\begin{aligned}
\mathcal{H}_{\mathrm{int}} = -\mu\, B_0(k)\Big[& A_-(\epsilon_{pol}) \sum_{\nu=1}^{N} \exp\left(\imath k z_\nu\right) \sigma_+^{(\nu)} \otimes a \\
& + A_-(\epsilon_{pol})^* \sum_{\nu=1}^{N} \exp\left(-\imath k z_\nu\right) \sigma_+^{(\nu)} \otimes a^\dagger \\
& + A_+(\epsilon_{pol}) \sum_{\nu=1}^{N} \exp\left(\imath k z_\nu\right) \sigma_-^{(\nu)} \otimes a \\
& + A_+(\epsilon_{pol})^* \sum_{\nu=1}^{N} \exp\left(-\imath k z_\nu\right) \sigma_-^{(\nu)} \otimes a^\dagger \Big]
\end{aligned} \tag{18.105}$$

In view of this result, it is convenient to define the operators,

$$J_\pm(k) = \frac{1}{\sqrt{N}} \sum_{\nu=1}^{N} \exp\left(\pm \imath k z_\nu\right) \sigma_\pm^{(\nu)} \quad , \quad J_3(k) = \sum_{\nu=1}^{N} \sigma_3^{(\nu)} \tag{18.106}$$

The definition of $J_\pm(k)$ implies,

$$J_-(k) = J_+(k)^\dagger \quad , \quad J_+(k) = \left(J_-(k)\right)^\dagger \tag{18.107}$$

The operators $J_\pm(k)$ and J_3 have the following commutation properties,

$$\left[J_+(k), J_-(k)\right] = \frac{1}{N} J_3(k) \quad , \quad \left[J_3(k), J_\pm(k)\right] = \pm J_\pm(k)$$

$$\left[J_+(k)J_-(k), J_+(k)\right] = \frac{-1}{N} J_+(k)J_3(k) \tag{18.108}$$

$$\left[J_+(k)J_-(k), J_-(k)\right] = \frac{1}{N} J_3(k)J_-(k)$$

Using the operators $J_\pm(k)$, we can write $\mathcal{H}_{\text{int}}$ as,

$$\begin{aligned}
\mathcal{H}_{int} = -\mu\, B_0(k)\sqrt{N} \Big[& A_-(\epsilon_{pol})\, J_+(k) \otimes a + A_-(\epsilon_{pol})^*\, J_-(k) \otimes a^\dagger \\
& + A_+(\epsilon_{pol})\, J_-(k) \otimes a + A_+(\epsilon_{pol})^*\, J_+(k) \otimes a^\dagger \Big]
\end{aligned} \tag{18.109}$$

Thus, we find that the coupling strength is proportional to $\sqrt{N}$. If the wave is right-handed circularly polarized, (18.96) implies,

$$\mathcal{H}_{int} = -\mu\, B_0(k)\sqrt{N} \Big[\, J_+(k) \otimes a + J_-(k) \otimes a^\dagger \Big] \tag{18.110}$$

If the wave is left-handed circularly polarized, (18.97) implies,

$$\mathcal{H}_{int} = -\mu\, B_0(k)\sqrt{N} \Big[\, J_-(k) \otimes a + J_+(k) \otimes a^\dagger \Big] \tag{18.111}$$

To complete the parallel with the introduction the introduction to coupled mode of §18.6.2, let us consider the operator $J_+(k)J_-(k)$. Using the definitions (18.106), we have,

$$\begin{aligned}
J_+(k)J_-(k) &= \frac{1}{N} \sum_{\nu=1}^{N} \sum_{\nu'=1}^{N} \exp\left(+\imath k z_\nu\right) \sigma_+^{(\nu)}\, \exp\left(-\imath k z_{\nu'}\right) \sigma_-^{(\nu')} \\
&= \frac{1}{4N} \sum_{\nu=1}^{N} \sum_{\nu'=1}^{N} \exp\left(+\imath k z_\nu\right) \exp\left(-\imath k z_{\nu'}\right) \left(\sigma_1^{(\nu)} + \imath \sigma_2^{(\nu)}\right)\left(\sigma_1^{(\nu')} - \imath \sigma_2^{(\nu')}\right) \\
&= \frac{1}{4N} \sum_{\nu=1}^{N} \sum_{\nu'=1}^{N} \exp\left(+\imath k z_\nu\right) \exp\left(-\imath k z_{\nu'}\right) \\
&\quad \times \left[\left(\sigma_1^{(\nu)}\sigma_1^{(\nu')} + \sigma_2^{(\nu)}\sigma_2^{(\nu')}\right) + \imath\left(\sigma_2^{(\nu)}\sigma_1^{(\nu')} - \sigma_1^{(\nu)}\sigma_2^{(\nu')}\right)\right]
\end{aligned} \tag{18.112}$$

Using the commutation relations (18.104) in (18.112), we find,

$$\begin{aligned}
J_+(k)J_-(k) &= \frac{1}{4N} \sum_{\nu=1}^{N} \sum_{\nu'=1}^{N} \exp\left(+\imath k(z_\nu - z_{\nu'})\right) \left[\sigma_1^{(\nu)}\sigma_1^{(\nu')} + \sigma_2^{(\nu)}\sigma_2^{(\nu')}\right] \\
&\quad + \frac{1}{2N} \sum_{\nu=1}^{N} \sigma_3^{(\nu)}
\end{aligned} \tag{18.113}$$

Likewise, we have,

$$
\begin{aligned}
J_-(k)J_+(k) = &\frac{1}{4N} \sum_{\nu=1}^{N} \sum_{\nu'=1}^{N} \exp\left(+\imath k(z_\nu - z_{\nu'})\right) \left[\sigma_1^{(\nu)}\sigma_1^{(\nu')} + \sigma_2^{(\nu)}\sigma_2^{(\nu')}\right] \\
&- \frac{1}{2N} \sum_{\nu=1}^{N} \sigma_3^{(\nu)}
\end{aligned}
\tag{18.114}
$$

Thus, we have the commutation relation,

$$
\left[J_-(k), J_+(k)\right] = \frac{-1}{2N} \sum_{\nu=1}^{N} \sigma_3^{(\nu)}
\tag{18.115}
$$

We now restrict the dynamics to small excitations near the equilibrium and consider a ferromagnet. Thus, we consider an equilibrium (at zero temperature) where all the spins are aligned in the $-z$ direction. In this case, the equilibrium state is given by,

$$
|\Omega_0> = |-1> \otimes |-1> \otimes \cdots \otimes |-1> \otimes |-1>
\tag{18.116}
$$

where the notation $|-1>$ means,

$$
\sigma_3^{(\nu)}|\Omega_0> = (-1)\,|\Omega_0> \quad , \quad \forall \, \nu = 1, 2, \ldots, N
\tag{18.117}
$$

An excitation corresponds to an increase in the angular moment. We discussed in §7.6.2 that a state where only one spin flipped was not an eigenstate of the Heisenberg exchange Hamiltonian. Instead, the eigenstates (7.59) correspond to the operators $J \pm (k)$ defined in (18.106).

To the extent that the system dynamics is restricted to very small excitations, we have,

$$
\left[J_-(k), J_+(k)\right] \approx 1 \quad , \quad \left[J_+(k)J_-(k), J_\pm(k)\right] \approx \pm J_\pm(k)
\tag{18.118}
$$

In this case, the operators $J_\pm(k)$ in the coupling Hamiltonian $\mathcal{H}_{int}$ (18.111) have the properties assumed for B_+ and B_- in the introductory section §18.6.2. Looking again at (18.110) for example, we see that the annihilation of a photon, expressed by the operator a, is accompanied by the creation of a magnon. Indeed, $J_+(k)$ is associated with σ_+^ν operators which increase the spin, i.e., they correspond to the excitation of the spin wave. Conversely, the second term in (18.110) corresponds to the creation of a photon and the annihilation of a magnon.

We conclude by showing that the Hamiltonian $\mathcal{H}$ for the general case (18.63) can be written has the Hamiltonian of a harmonic oscillator. Here, considering the right-hanced circularly polarized case (18.110), the Hamiltonian $\mathcal{H}$ of the whole system has the form,

$$
\mathcal{H} = \epsilon_{\mathrm{sw}}\, J_+(k)J_-(k) \otimes \mathbb{1}_{\mathrm{em}} + \epsilon_{\mathrm{em}}\, \mathbb{1}_{\mathrm{sw}} \otimes a^\dagger a + g_{\mathrm{eff}} J_+(k) \otimes a + g_{\mathrm{eff}}^* J_-(k) \otimes a^\dagger
\tag{18.119}
$$

with $g_{\mathrm{eff}} = -\mu\, B_0(k)\sqrt{N}$. We define the annihilation operator C (C as in "coupled mode") acting on the Hilbert space $\mathbb{H}_{\mathrm{sw}} \otimes \mathbb{H}_{\mathrm{em}}$,

$$
C = p_{\mathrm{sw}} J_-(k) \otimes \mathbb{1}_{\mathrm{em}} + p_{\mathrm{em}}\, \mathbb{1}_{\mathrm{sw}} \otimes a
\tag{18.120}
$$

where the parameters $p_{\mathrm{sw}}, p_{\mathrm{em}} \in \mathbb{C}$, and are yet to be determined. Owing to (18.107), we have,

$$
C\dagger = p_{\mathrm{sw}}^* J_+(k) \otimes \mathbb{1}_{\mathrm{em}} + p_{\mathrm{em}}^*\, \mathbb{1}_{\mathrm{sw}} \otimes a^\dagger
\tag{18.121}
$$

In view of (18.118), C and $C^\dagger$ have the commutation relation,

$$[C, C^\dagger] = \left(|p_{sw}|^2 + |p_{em}|^2\right) \mathbb{1}_{sw} \otimes \mathbb{1}_{em} \tag{18.122}$$

Assuming

$$\left(|p_{sw}|^2 + |p_{em}|^2\right) = 1 \tag{18.123}$$

and noting $\mathbb{1} = \mathbb{1}_{sw} \otimes \mathbb{1}_{em}$, we obtain the commutation relation,

$$[C, C^\dagger] = \mathbb{1} \tag{18.124}$$

The operator $C^\dagger C$ reads,

$$\begin{aligned}
C^\dagger C = |p_{sw}|^2 J_+(k) J_-(k) \otimes \mathbb{1}_{em} + |p_{em}|^2 \mathbb{1}_{sw} \otimes a^\dagger a \\
+ p_{em} p_{sw}^* \, J_+(k) \otimes a + p_{em}^* p_{sw} \, J_-(k) \otimes a^\dagger
\end{aligned} \tag{18.125}$$

As the analogy between (18.125) and (18.119) suggests, it is indeed possible to write the Hamiltonian $\mathcal{H}$ as,

$$\mathcal{H} = \epsilon \, C^\dagger C \tag{18.126}$$

To do this, we need to impose,

$$|p_{sw}|^2 = \frac{\epsilon_{sw}}{\epsilon} \quad , \quad |p_{em}|^2 = \frac{\epsilon_{em}}{\epsilon} \quad , \quad \epsilon = \epsilon_{sw} + \epsilon_{em} \tag{18.127}$$

The last equality ensures that the condition (18.123) is fulfilled. Then, writing $p_{sw} = |p_{sw}| \exp(\iota\phi_{sw})$ and $p_{em} = |p_{em}| \exp(\iota\phi_{sw})$, we also need to impose,

$$g_{eff} = \sqrt{\epsilon_{sw}\epsilon_{em}} \exp(\iota(\phi_{sw} - \phi_{em})) = \sqrt{\epsilon_{sw}\epsilon_{em}} \tag{18.128}$$

The last step ensures that g_{eff} is real by taking $\phi_{sw} = \phi_{em}$. Since $\mathcal{H}$ in (18.126) describes modes of coupled photons and magnons, it is not surprising that $\mathcal{H}$ can be written as the Hamiltonian for a harmonic oscillator.

18.7   FURTHER READINGS

Dynamics of artificial antiferromagnets

Two ferromagnetic layers separated by a thin non-magnetic metallic layer can align antiferromagnetically at equilibrium, owing to the RKKY coupling (Fig. 1.4). Optical and acoustic spin-wave modes can be excited. The magnetic relaxation of the ferromagnets is modified by the coupling between them. A spin-current mediated damping was identified.[1047]

Spin pumping and antiferromagnets

When the AFMR is excited, the magnetization of both sublattices precesses around the effective field, composed of the anisotropy field B_A and the exchange field B_{ex}. In the case of a uniaxial anisotropy, the AFMR has two modes, often referred to as right-handed and left-handed modes. The precession cone angles θ_1 and θ_2 are not equal; their ratio is given by $\theta_1/\theta_2 \approx (1 + \sqrt{B_A/B_{ex}})^2$.[1048] In view of the simple model of §17.6 for spin pumping, we can expect an ISHE in a heavy metal film capping the antiferromagnet. The ISHE is proportional to $\sin^2\theta_1 - \sin^2\theta_2$. Indeed, the ISHE has been used to detect AFMR.[599, 975, 1048] A more elaborate model would take into account the spin accumulation in the normal metal, which in turn causes a spin backflow into the magnetic material, owing to spin diffusion.[1049, 1050]

Spin cycloids, frustration and magnons

We have seen that the Dzialoshinskii-Moriya interaction (DMI) can give rise to spin canting (§2.7). It can also lead to a long-range structure in the equilibrium magnetization known as a *spin cycloid*. This structure typically extends over many unit cells in rare-earth ferrites. The vibration modes of the cycloid occur at frequencies in the range of hundreds of GHz.[1051–1054] DMI is not a necessary condition for a spin cycloid. This can also happen in frustrated magnets.[1055] *Frustration* typically occurs in a Heisenberg antiferromagnet with a triangular structure. Frustration can also occur when exchange coupling to the nearest neighbor has one sign, and there is also an exchange coupling to the next-nearest neighbors that has the opposite sign.

The use of spin waves in antiferromagnets could greatly advance spintronics, because these spin waves occur at high frequencies. Information processing might require a conversion in electrical signals, which could be achieved in some kind of field-effect-transistor geometry.[1056] The ISHE associated with spin pumping due to AFMR excitation also gives DC signals.[1057]

Spin cavitronics

The realization that magnetization modes can couple to electromagnetic cavity modes gave rise to the research field known as *spin cavitronics*.[1012, 1013, 1058] For example, *strong coupling* of the magnetization of a YIG sphere was obtained in a cavity resonating in the GHz range.[1059, 1060] Putting two YIG spheres into one cavity results in the mutual coupling of the spheres' magnetization. There are also *dark modes* that do not couple to cavity modes. The dark modes have a much longer lifetime than the the magnetization modes that are coupled to electromagnetic cavity modes. When each sphere experiences an adjustable local magnetic induction field, the coupling can be tuned.[1046] The interest in magnetic polaritons stems from their prospective use in information processing, possibly as quantum transducers.[1046, 1061]

When a Hamiltonian has the form (18.126), the eigenstates of the annihilation operator C are *coherent states*. Coherent states minimize the Heisenberg uncertainty relations.[162, 181] In that sense, coherent states most closely resemble a classical state.

Polaritons and parametric pumping

In §18.5, bulk polaritons were described. As one might expect, there can also be surface-magnon polaritons.[1062] In general, dispersions similar to the anticrossing we found for magnetic polaritons are observed in many other coupled wavelike modes. In his famous textbook, Kittel described the coupling of electromagnetic waves with phonons, which he called polaritons.[79] In 1965, C. Henry *at al.* reported this type of polariton in bulk GaP.[1063] The coupling of transverse phonons with magnons was observed in antiferromagnets .[1033] In the 2010s, a coupling was observed between surface Bloch waves and plasmons.[1064, 1065].

In the 1950s, the coupling of waves was thought to provide a means to pump energy from modes of one nature into waves of another nature. The technique is called *parametric pumping*. For example, Comstock explored the possibility of pumping energy into elastic waves by supplying energy to these waves through a microwave excitation of magnetization waves.[1066] In his experiment, instead of a resonance condition that matches both in energy and in momentum, non-linear effects made it possible to consider that the excitation of two waves, characterized by the vectors (k_1, ω_1) and (k_2, ω_2), resulted in a wave characterized by the vector $(k_1 + k_2, \omega_1 + \omega_2)$ (Fig. 18.9). One of these two modes was a polariton due to the coupling of transverse elastic waves of the ferromagnet with electromagnetic modes. The

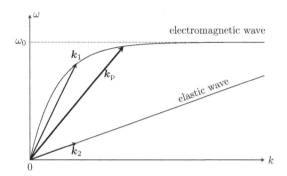

Figure 18.9 Dispersion relation for longitudinal elastic waves and for polaritons of transverse elastic waves coupled to electromagnetic waves. After [1066].

other was a longitudinal acoustic wave. This coupling was possible owing to the magneto-elastic contribution to the energy of the system.[1067]

It is possible to excite magnons with $k \neq 0$ using parametric resonance. When it was first explored, this technique was referred to as ***parallel pumping***.[383, 1068–1070] Evidently, non-linear effects must be involved. Sufficient power can be reached in a cavity under high power. Safonov showed that taking into account the coupling of magnetic resonance to the cavity modes can account for parallel pumping ([391], chapter 4). Parallel pumping was used to excite nuclear spin waves,[1071] and the concomitant excitation of nuclear and electronic spin waves (magnetization waves) was demonstrated.[1072, 1073]

Spin chemistry and self-oscillation

The chemical generation of radio waves is closely related to the physics of spins coupled to a cavity. To understand this connection, imagine first that a magnetic field has polarized a spin population located in a cavity or a radio-frequency circuit ([1074], §4.2), and that a resonant pulse inverts the spins. The slightest fluctuation on any individual spin tilts this spin away from the field. Hence, this spin starts precessing in the applied magnetic field. This induces a current in the cavity or the coil. The current in turn acts on all the other spins, exciting them in their precession. Thus, all spins are coupled to one another via their interaction with the cavity (or resonator) and we can expect ***synchronization*** of the spin precessions. This mechanism is effective only if the decay rate of the induced current is slow compared to the strength of the spin-to-cavity coupling. Also, the effect of the cavity (or coil) fields must be more effective that the intrinsic random fluctuations causing transverse spin relaxation rate $(1/T_2)$. Suppose now that a photochemical reaction generates a negatively-oriented spin polarization. Owing to the process just described, a radio frequency signal can be generated. Chemically-induced radio-frequency emission was first observed in the photochemical reaction of porphyrin with quinone.[1075, 1076]

19 Coherent Spin Dynamics

Charles Pence Slichter (1924–2018)

C. P. Slichter, together with H.S. Gutowsky and D.W. McCall, discovered the indirect coupling of nuclear spins via their coupling to electron spins in molecules. With T.R. Carver, Slichter validated A.W. Overhauser's prediction of dynamic nuclear polarization. His analysis of NMR data obtained by a student, L.C. Hebel, provided the first experimental evidence for J. Bardeen, L. Cooper, and J.R. Schrieffer's theory of superconductivity. Prompted by D. Pines, he succeeded in measuring Pauli spin susceptibility, and later observed antiferromagnetic fluctuations in high-temperature superconductors.

Spintronics: A primer

J.-Ph. Ansermet (https://orcid.org/0000-0002-1307-4864)

Chapter 19. Coherent Spin Dynamics

The Heisenberg representation of spin precession shows the importance of rotations induced by resonant pulses. A spin-operator formalism is applied to describe the spin echo sequence, and to convey the notion that an effective spin Hamiltonian can be generated by applying pulses to the spins. The case of quadrupolar echoes is also analyzed. Double quantum coherence and coherence transfer are described by means of simple introductory examples. An experiment that created an effective entangled state for an ensemble of spins is analyzed. The implementation of a logic gate using coupled spins as qubits is described.

DOI: 10.1201/9781003370017-19

19.1 HISTORICAL INTRODUCTION

In 1938, I.I. Rabi and his collaborators at Columbia University reported on a molecular beam experiment, the purpose of which was to measure nuclear magnetic moments. The beam was 245 cm long. A first magnet, 52 cm in length, spread the beam and a second one, 100 cm long, refocused it. In between both magnets was a solenoid that produced an oscillating field perpendicular to an applied magnetic field of about 6000 gauss. When this field was on resonance with the nuclear spin precession, the refocusing would not work. This effect was detected by scanning the field while monitoring the beam intensity.[793]

In 1950, E.L. Hahn, who was then at the University of Illinois Urbana-Champaign, reported on the successful refocusing of nuclear spins in a liquid. The nuclear induction signal presented a decay caused mainly by the field inhomogeneity. While today, the *spin echo* sequence is commonly thought of as comprising a 90° resonance pulse followed by a 180° resonant pulse, Hahn happened to consider two 90° pulses instead. Using a vector model, he realized the refocusing effect of the second 90° pulse.[1077]

In 1954, H.Y. Carr and E.M. Purcell found that the refocusing could be obtained again by applying a third pulse after the echo.[1078] Later, the technique was refined by S. Meiboom and D. Gill by using different phases for the successive refocusing pulses.[1079] This refocusing strategy, often referred to by the 4 initials of these authors, CPMG, increases the effective *coherence time* of a spin ensemble. The potential application of CPMG to quantum bits is being considered.[1080]

Of course, the free induction decay could arise from T_2 mechanisms. For example, carbon nuclei may relax due to the fluctuating dynamics of nearby protons. In the late 1960s, it was shown that this relaxation mechanism can be quenched by flipping the relaxation-causing spins very fast, a technique called *dynamical decoupling*.[1081–1084] In the case of liquids, quenching the spin-spin coupling among protons via their coupling to carbon nuclei was, demonstrated as early as 1954 by Virginia Royden.[1085]

Among the various forms of double resonance experiments, let us mention *electron-nuclear double resonance* (ENDOR) spectroscopy. For a detailed study of the structure of color centers in potassium chloride, G. Feher carried out an experiment in 1957 that combined electron spin resonance excitation and NMR. This allowed him to improve on the ESR spectral resolution by 4 orders of magnitude.[1086]

The idea of *multidimensional Fourier transform spectroscopy* is attributed to J. Jeneer, who proposed it during an Ampère International Summer School in 1971.[236] In their review article, R. Freeman and G.A. Morris show that R.R. Ernst was the first to appreciate the importance of this method.[1087] R.R. Ernst had reported his solid state NMR experiment in 1969.[1088]

19.2 RESONANT PULSES

We refer back to the single spin precession described in §15.2. Let us consider a field applied on resonance. As can be seen in (15.7), this means $\omega = -\gamma B_0$. In the rotating frame, the magnetic moment evolves according to,

$$\frac{d'\boldsymbol{m}}{dt} = -\gamma B_1 \hat{\boldsymbol{u}} \times \boldsymbol{m} \tag{19.1}$$

Likewise the spin $\hat{\boldsymbol{S}}$ associated with $\boldsymbol{m}$ by $\boldsymbol{m} = \hbar\gamma\hat{\boldsymbol{S}}$ evolves according to,

$$\frac{d'\hat{\boldsymbol{S}}}{dt} = -\gamma B_1 \hat{\boldsymbol{u}} \times \hat{\boldsymbol{S}} \tag{19.2}$$

Thus, we find that in the rotating frame, the spin $\hat{\boldsymbol{S}}$ rotates around the $\hat{\boldsymbol{u}}$ axis at the angular velocity $-\gamma B_1 \hat{\boldsymbol{u}}$.[1]

Suppose that a field $\boldsymbol{B}_1(t)$ is applied for a time T_{90}. In the laboratory frame, it is an oscillating field. In the rotating frame picture, according to (19.2), the spin $\hat{\boldsymbol{S}}$ rotates by an angle $\pi/2$ (90 degrees), provided,

$$\frac{\pi}{2} = T_{90}\,\gamma B_1 \tag{19.3}$$

If the spin is initially along the z axis, after this pulse, it precesses in the Ouv plane (Fig. 15.2). This is the pulse typically used to observe a free induction decay in nuclear magnetic resonance spectroscopy, from which a spectrum can be deduced by a Fourier transform.

19.3 QUANTUM MECHANICAL DESCRIPTION OF SPIN PRECESSION

Let us consider an isolated particle that carries a dipole moment $\boldsymbol{\mu}$ and has the angular moment $\boldsymbol{I}$ such that $\gamma \boldsymbol{I} = \boldsymbol{\mu}$. When this single, isolated dipole is exposed to a magnetic induction field $\boldsymbol{B}$, the Hamiltonian of this system is,

$$\mathcal{H} = -\gamma \boldsymbol{B} \cdot \boldsymbol{I} \tag{19.4}$$

Let us analyze the case when the magnetic induction field is composed of a constant term and a circularly polarized field $\boldsymbol{B}_1$, i.e.,

$$\boldsymbol{B} = B_0 \hat{\boldsymbol{z}} + B_1 \left(\cos(\omega t)\hat{\boldsymbol{x}} + \sin(\omega t)\hat{\boldsymbol{y}}\right) \tag{19.5}$$

where $(\hat{\boldsymbol{x}}, \hat{\boldsymbol{y}}, \hat{\boldsymbol{z}})$ is a direct set of orthogonal unit vectors, fixed in the reference frame and centered at a point O (Fig. 15.1). Thus, the Hamiltonian is,

$$\mathcal{H} = -\hbar\gamma B_0 \hat{I}_z - \hbar\gamma B_1 \left(\hat{I}_x \cos(\omega t) + \hat{I}_y \sin(\omega t)\right) \tag{19.6}$$

Let us define the two unit vectors (Fig. 15.1),

$$\begin{aligned}
\hat{\boldsymbol{u}} &= \cos(\omega t)\hat{\boldsymbol{x}} + \sin(\omega t)\hat{\boldsymbol{y}} \\
\hat{\boldsymbol{v}} &= -\sin(\omega t)\hat{\boldsymbol{x}} + \cos(\omega t)\hat{\boldsymbol{y}}
\end{aligned} \tag{19.7}$$

The operator $\hat{I}_x \cos(\omega t) + \hat{I}_y \sin(\omega t)$ can be thought of as the projection of $\boldsymbol{I}$ on the unit vector $\hat{\boldsymbol{u}}$. For any unit vector $\hat{\boldsymbol{n}}$ rotating at the angular velocity $\boldsymbol{\omega}$, we have,

$$\frac{d\hat{\boldsymbol{n}}}{dt} = \boldsymbol{\omega} \times \hat{\boldsymbol{n}} \tag{19.8}$$

Here, the vectors $\hat{\boldsymbol{u}}, \hat{\boldsymbol{v}}$, and $\boldsymbol{B}_1$ rotate with the angular velocity $\boldsymbol{\omega} = \omega\hat{\boldsymbol{z}}$.

In the Heisenberg representation, the time evolution of an operator A is given by:

$$i\hbar\frac{dA}{dt} = [A, \mathcal{H}] \tag{19.9}$$

The time evolution of the operator $\boldsymbol{I} \cdot \hat{\boldsymbol{n}}$ is given by,

$$\begin{aligned}
i\hbar\frac{d(\boldsymbol{I} \cdot \hat{\boldsymbol{n}})}{dt} &= i\hbar\frac{d\boldsymbol{I}}{dt} \cdot \hat{\boldsymbol{n}} + i\hbar(\boldsymbol{\omega} \times \hat{\boldsymbol{n}}) \cdot \boldsymbol{I} \\
&= [\boldsymbol{I} \cdot \hat{\boldsymbol{n}}, \mathcal{H}] + i\hbar(\boldsymbol{\omega} \times \hat{\boldsymbol{n}}) \cdot \boldsymbol{I}
\end{aligned} \tag{19.10}$$

---

[1]We recall that in classical mechanics, if $\boldsymbol{r}$ designates the position of a point mass, then $d\boldsymbol{r}/dt = \boldsymbol{\Omega} \times \boldsymbol{r}$ is the velocity of the point mass when $\boldsymbol{r}$ undergoes a rotation at the angular velocity $\boldsymbol{\Omega}$. [557]

Since here, $\mathcal{H} = -\hbar\gamma B_0 \hat{I}_z - \hbar\gamma B_1 \hat{I}_u$ where $\hat{I}_u = \hat{\boldsymbol{I}} \cdot \hat{\boldsymbol{u}}$, the time evolution of the spin projection $\hat{\boldsymbol{I}} \cdot \hat{\boldsymbol{n}}$ is given by,

$$i\hbar\frac{d(\hat{\boldsymbol{I}} \cdot \hat{\boldsymbol{n}})}{dt} = -\hbar\gamma B_0 \left[\hat{\boldsymbol{I}} \cdot \hat{\boldsymbol{n}}, \hat{I}_z\right] - \hbar\gamma B_1 \left[\hat{\boldsymbol{I}} \cdot \hat{\boldsymbol{n}}, \hat{I}_u\right] + i\hbar(\boldsymbol{\omega} \times \hat{\boldsymbol{n}}) \cdot \hat{\boldsymbol{I}} \qquad (19.11)$$

When working out the calculation of the commutators in this expression, it is useful to know that,

$$\left[\hat{I}_z, \hat{I}_x\right] = i\hat{I}_y \qquad \text{(and cyclic permutations)} \qquad (19.12)$$

It is easy to verify that, likewise,

$$\left[\hat{I}_z, \hat{I}_u\right] = i\hat{I}_v \qquad \text{(and cyclic permutations)} \qquad (19.13)$$

Thus, applying (19.11) with $\hat{\boldsymbol{n}} = \hat{\boldsymbol{z}}$, we find,

$$i\hbar\frac{d\hat{I}_z}{dt} = -\hbar\gamma B_1 \left[\hat{I}_z, \hat{I}_x \cos(\omega t) + \hat{I}_y \sin(\omega t)\right] = -\hbar\gamma B_1 \left(i\hat{I}_y \cos(\omega t) - i\hat{I}_x \sin(\omega t)\right) \quad (19.14)$$

Applying the definition (19.7) of $\hat{\boldsymbol{v}}$, we have,

$$\frac{d\hat{I}_z}{dt} = -\gamma B_1 \hat{I}_v \qquad (19.15)$$

Applying now (19.11) with $\hat{\boldsymbol{n}} = \hat{\boldsymbol{u}}$, we find,

$$i\hbar\frac{d\hat{I}_u}{dt} = -\hbar\gamma B_0 \left[\hat{I}_u, \hat{I}_z\right] + i\hbar(\boldsymbol{\omega} \times \hat{\boldsymbol{u}}) \cdot \boldsymbol{I} = -\hbar\gamma B_0(-i\hat{I}_v) + i\hbar\omega\hat{I}_v \qquad (19.16)$$

Thus, we find,

$$\frac{d\hat{I}_u}{dt} = (\gamma B_0 + \omega)\hat{I}_v \qquad (19.17)$$

Repeating the calculation with $\hat{\boldsymbol{n}} = \hat{\boldsymbol{v}}$, we find,

$$\begin{aligned} i\hbar\frac{d\hat{I}_v}{dt} &= -\hbar\gamma B_0 \left[\hat{I}_v, \hat{I}_z\right] - \hbar\gamma B_1 \left[\hat{I}_v, \hat{I}_u\right] + i\hbar(\boldsymbol{\omega} \times \hat{\boldsymbol{v}}) \cdot \hat{\boldsymbol{I}} \\ &= -\hbar\gamma B_0(i\hat{I}_u) - \hbar\gamma B_1(-i\hat{I}_z) + i\hbar\omega(-\hat{I}_u) \end{aligned} \qquad (19.18)$$

Thus, we find,

$$\frac{d\hat{I}_v}{dt} = -(\gamma B_0 + \omega)\hat{I}_u + \gamma B_1 \hat{I}_z \qquad (19.19)$$

These results can be summarized by,

$$\frac{d\hat{\boldsymbol{I}}}{dt} = -\gamma \boldsymbol{B}_{\text{eff}} \times \hat{\boldsymbol{I}} \qquad (19.20)$$

where $\boldsymbol{B}_{\text{eff}}$ is the effective field introduced in (15.8) for the classical description of spin precession. We see that the evolution equation for the angular momentum operator has the same form in the Heisenberg representation as in the semi-classical equation for the magnetic moment.

19.4 SPIN-OPERATOR FORMALISM

Many experiments in NMR can be understood by considering the spin evolution projected in the rotating frame. In the rotating frame, the notion that radio-frequency fields rotate spins becomes immediately evident. In this section, we first introduce the formalism expressing the rotations of operators. Second, we apply it to the spin echo obtained with two simple rotations. Third, we examine an echo formed by an arbitrary rotation, relevant for spins with quadrupolar moments. Fourth, we us a spin-operator formalism to analyze a sequence that generates double quantum coherence, and another one that produces a coherence transfer.

19.4.1 ROTATIONS

When addressing spin precession within a quantum mechanical framework, it is useful to know some relations concerning the rotation of angular momenta. For example, we will often use the relation,

$$\hat{I}_x \cos(\omega t) + \hat{I}_y \sin(\omega t) = e^{-i\omega t \hat{I}_z} \hat{I}_x e^{i\omega t \hat{I}_z} \tag{19.21}$$

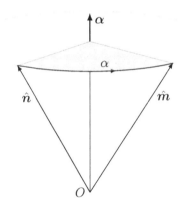

Figure 19.1   A rotation defined by the vector $\boldsymbol{\alpha}$ brings the vector $\hat{n}$ onto the vector $\hat{m}$.

This operator relation expresses the effect of a rotation on a spin operator. It is a particular consequence of a general result that we want to derive here.

Let the unit vector $\hat{n}$ have the image $\hat{m}$ under the action of a rotation defined by the vector $\boldsymbol{\alpha}$. The angle of rotation is $\alpha = |\boldsymbol{\alpha}|$ and the axis of rotation is along the vector $\boldsymbol{\alpha}$ (Fig. 19.1). Let us demonstrate that,

$$\boldsymbol{S} \cdot \hat{m} = e^{-i\boldsymbol{\alpha} \cdot \hat{\boldsymbol{I}}} (\boldsymbol{S} \cdot \hat{n}) e^{i\boldsymbol{\alpha} \cdot \hat{\boldsymbol{I}}} \tag{19.22}$$

To demonstrate this property, we use the fact that the angular momentum is the generator of rotations.[181] Let us designate by $R(\boldsymbol{\omega})$ a rotation characterized by the vector $\boldsymbol{\omega}$. From a geometrical standpoint, we have,

$$R(\boldsymbol{\alpha})R(\boldsymbol{\omega})R(\boldsymbol{\alpha})^{-1} = R(R(\boldsymbol{\alpha})\boldsymbol{\omega}) \tag{19.23}$$

In order to prove (19.23), we note that, first, the composition of rotations is itself a rotation. Second, the vector $R(\boldsymbol{\alpha})\boldsymbol{\omega}$ is invariant under the composition $R(\boldsymbol{\alpha})R(\boldsymbol{\omega})R(\boldsymbol{\alpha})^{-1}$. This can be verified by applying the composition of rotation to this vector:

$$R(\boldsymbol{\alpha})R(\boldsymbol{\omega})R(\boldsymbol{\alpha})^{-1}\left(R(\boldsymbol{\alpha})\boldsymbol{\omega}\right) = R(\boldsymbol{\alpha})R(\boldsymbol{\omega})\boldsymbol{\omega} = R(\boldsymbol{\alpha})\boldsymbol{\omega} \tag{19.24}$$

Therefore, the axis of that rotation is parallel to $R(\boldsymbol{\alpha})\boldsymbol{\omega}$. Finally, the angle of that rotation is $|\boldsymbol{\omega}|$. This can be seen by working out the angle between any vector $\hat{\boldsymbol{n}}$ and its image:

$$\hat{\boldsymbol{n}} \cdot R(\boldsymbol{\alpha})R(\boldsymbol{\omega})R(\boldsymbol{\alpha})^{-1}\hat{\boldsymbol{n}} = R(\boldsymbol{\alpha})^{-1}\hat{\boldsymbol{n}} \cdot R(\boldsymbol{\omega})R(\boldsymbol{\alpha})^{-1}\hat{\boldsymbol{n}} \tag{19.25}$$

which is the cosine of the angle between the vector $R(\boldsymbol{\alpha})^{-1}\hat{\boldsymbol{n}}$ and its image by a rotation $R(\boldsymbol{\omega})$. Therefore, the rotation angle is indeed $|\boldsymbol{\omega}|$. Note that, in (19.25), we used the fact that for an orthogonal matrix R, $\boldsymbol{a} \cdot R\boldsymbol{b} = R^{-1}\boldsymbol{a} \cdot \boldsymbol{b}$. In summary, we have demonstrated that $R(\boldsymbol{\alpha})R(\boldsymbol{\omega})R(\boldsymbol{\alpha})^{-1}$ is a rotation characterized by the vector $R(\boldsymbol{\alpha})\boldsymbol{\omega}$, which is the meaning of expression (19.23).

Since the angular momentum is the generator of rotations,[181] the unitary operator corresponding to the rotation $R(\boldsymbol{\alpha})$ is given by,

$$U(\boldsymbol{\alpha}) = e^{-i\boldsymbol{\alpha}\cdot\hat{\boldsymbol{I}}} \tag{19.26}$$

Thus, we derive from (19.23),

$$e^{-iR(\boldsymbol{\alpha})\boldsymbol{\omega}\cdot\hat{\boldsymbol{I}}} = U(\boldsymbol{\alpha})e^{-i\boldsymbol{\omega}\cdot\hat{\boldsymbol{I}}}U(\boldsymbol{\alpha})^{-1} = e^{-iU(\boldsymbol{\alpha})(\boldsymbol{\omega}\cdot\hat{\boldsymbol{I}})U(\boldsymbol{\alpha})^{-1}} \tag{19.27}$$

The last step can be obtained by developing $e^{-i\boldsymbol{\omega}\cdot\hat{\boldsymbol{I}}}$ as a series, and in each term of the series, replacing $\boldsymbol{\omega} \cdot \hat{\boldsymbol{I}}$ by $U(\boldsymbol{\alpha})(\boldsymbol{\omega} \cdot \hat{\boldsymbol{I}})U(\boldsymbol{\alpha})^{-1}$. From (19.27), we deduce,

$$R(\boldsymbol{\alpha})\boldsymbol{\omega} \cdot \hat{\boldsymbol{I}} = e^{-i\boldsymbol{\alpha}\cdot\hat{\boldsymbol{I}}}\boldsymbol{\omega} \cdot \hat{\boldsymbol{I}}e^{i\boldsymbol{\alpha}\cdot\hat{\boldsymbol{I}}} \tag{19.28}$$

This is equivalent to the statement (19.22). The previous relation (19.21) is a particular example of (19.28).

19.4.2 SIMPLIFIED DENSITY MATRIX

Let us now consider an ensemble of particles that all have the same spin. Thus, each particle has an angular momentum $\boldsymbol{I} = \hbar\hat{\boldsymbol{I}}$. As often happens in practice, these spins are prepared in an initial state, which is the equilibrium of this ensemble in a large magnetic field at a temperature T. In practice, it is very common that the spins experience slightly different magnetic fields distributed around an average value $\boldsymbol{B}_0$. As these field variations $\Delta\boldsymbol{B}$ play a negligible role in establishing the initial state, we can assume that the density matrix for any one spin of this ensemble at equilibrium is given by,

$$\rho_0 = \frac{1}{Z}\exp\left(\frac{\hbar\gamma B_0\hat{I}_z}{k_BT}\right) \approx \frac{1}{Z}\left(1 + \frac{\hbar\gamma B_0\hat{I}_z}{k_BT}\right) \tag{19.29}$$

where $Z = Tr(\rho_0)$.

In analyzing a pulse sequence, we will calculate how the density matrix evolves over time, i.e., we will find $\rho(t)$. We will then calculate the expectation value,

$$\langle \boldsymbol{I} \rangle = Tr\left(\rho(t)\boldsymbol{I}\right) \tag{19.30}$$

We recall that the evolution of the density matrix is given by,

$$-i\hbar\frac{d\rho}{dt} = [\rho, \mathcal{H}] \tag{19.31}$$

There is a sign difference between this equation, which describes the evolution of the density matrix, and equation (19.9),which describes the time evolution of an operator in the Heisenberg picture. If the Hamiltonian $\mathcal{H}$ is time-independent, then the solution to the Heisenberg equation (19.31) is,

$$\rho(t) = e^{-i\mathcal{H}t/\hbar}\rho_0 e^{i\mathcal{H}t/\hbar} \tag{19.32}$$

as can easily be verified.

In (19.29), the first term, i.e., $1/Z$ will be carried out through the steps of the calculations unchanged, provided the Hamiltonian is stepwise time-independent. Consequently, this term will contribute nothing to the trace (19.30). From now on, we will omit this term. We will also omit the coefficient $\hbar\gamma B_0/(k_B T)$. Thus, we will take the liberty of making the substitution,

$$\rho_0 = \frac{1}{Z}\left(1 + \frac{\hbar\gamma B_0}{k_B T}\hat{I}_z\right) \rightarrow \rho_0 = \hat{I}_z \quad \text{(for calculations)} \tag{19.33}$$

Thus, we will focus our attention on how the operator $\hat{I}_z$ evolves during the pulse sequence, keeping in mind that here, $\hat{I}_z$ stands for the density matrix.

19.4.3 SPIN ECHO

Spin echoes result from a simple pulse sequence comprising, for example, one $\pi/2$ pulse on the x-axis followed by a π pulse on the y-axis (Fig. 19.2). Let us now assume that these spins are subjected to two pulses separated by a waiting time τ, during which the spins precess in the field $\boldsymbol{B}_0 + \Delta\boldsymbol{B}_0$ (Fig. 19.2). The term $\Delta\boldsymbol{B}_0$ represents a deviation from a perfectly uniform field $\boldsymbol{B}_0$. It is implicit in the following that the deviation $\Delta\boldsymbol{B}_0$ differs from spin to spin among the spins undergoing the spin echo experiment.

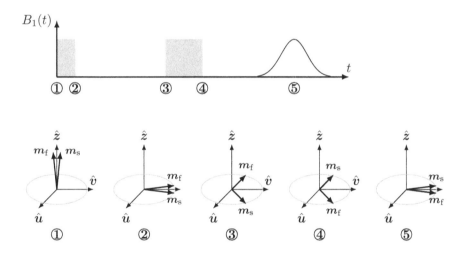

Figure 19.2 Top: schematic representation of a spin echo pulse sequence. Between times 1 and 2, a resonant pulse is applied, with $\boldsymbol{B}_1(t)$ along the x-axis, for a duration such that the spins rotate by 90°. Between times 3 and 4, another pulse is applied along the y-axis for twice the duration, i.e., for a 180° rotation. Bottom: the effect of the pulses is illustrated for two sets of spins, those that are slow in their rotation with respect to the rotating frame (s), and those that are faster (f).

The pulses are assumed to be on resonance, i.e., $\omega = -\gamma B_0$ and the excitation field $\boldsymbol{B}_1(t)$ is so large that $B_1(t) \gg \Delta B_0$. Consequently, ΔB_0 has a negligible effect on the evolution of the spin during the pulses. The first pulse is a 90° pulse applied along the u-axis, and the second pulse is a 180° pulse applied along the v-axis, where u and v are the axes in the frame rotating at the angular frequency $\omega = -\gamma B_0$ (Fig. 15.1). The Hamiltonian for one spin can be written as,

$$\mathcal{H} = -\hbar\gamma B_0 \hat{I}_z - \hbar\gamma\Delta B_0 \hat{I}_z - \hbar\gamma B_1(t)R(\boldsymbol{\omega}t)\hat{\boldsymbol{x}}\cdot\hat{\boldsymbol{I}} \qquad (19.34)$$

when the phase is chosen so that $\boldsymbol{B}_1(t)$ is along x. If $\boldsymbol{B}_1(t)$ is along y, the modification is immediate. In view of (19.28), we can write this as,

$$\mathcal{H} = -\hbar\gamma B_0 \hat{I}_z - \hbar\gamma\Delta B_0 \hat{I}_z - \hbar\gamma B_1(t)e^{-i\omega t\hat{I}_z}(\hat{\boldsymbol{x}}\cdot\hat{\boldsymbol{I}})e^{i\omega t\hat{I}_z} \qquad (19.35)$$

While we used the Heisenberg representation to describe the precession of a spin, here, we want to calculate the evolution of the state, which is given by a density matrix. In order to obtain expressions reminiscent of the rotating frame picture, let us derive an evolution equation for the matrix ρ' defined by,

$$\rho = e^{-i\omega t\hat{I}_z}\rho'e^{i\omega t\hat{I}_z} \Rightarrow \rho' = e^{i\omega t\hat{I}_z}\rho e^{-i\omega t\hat{I}_z} \qquad (19.36)$$

Applying (19.31), we have,

$$\begin{aligned}
-i\hbar\frac{d\rho'}{dt} &= e^{i\omega t\hat{I}_z}\left[\rho, \mathcal{H}\right]e^{-i\omega t\hat{I}_z} + e^{i\omega t\hat{I}_z}\left[\hbar\omega\hat{I}_z, \rho\right]e^{-i\omega t\hat{I}_z} \\
&= e^{i\omega t\hat{I}_z}\left[\rho, \mathcal{H} - \hbar\omega\hat{I}_z\right]e^{-i\omega t\hat{I}_z} \\
&= \left[\rho', \mathcal{H}' - \hbar\omega\hat{I}_z\right]
\end{aligned} \qquad (19.37)$$

where

$$\mathcal{H}' = e^{i\omega t\hat{I}_z}\mathcal{H}e^{-i\omega t\hat{I}_z} \qquad (19.38)$$

These are general results valid for any $\mathcal{H}$. In the particular case of the Hamiltonian (19.35) describing our spin echo experiment, the density matrix ρ' "rotating" at the angular velocity $\omega = -\gamma B_0$, evolves according to (19.37) with the effective Hamiltonian,

$$\mathcal{H}_e = \mathcal{H}' - \hbar\omega\hat{I}_z = -\hbar\gamma\Delta B_0 \hat{I}_z - \hbar\gamma B_1(t)\hat{I}_x \qquad (19.39)$$

We thus find a Hamiltonian that is stepwise time-independent, since $B_1(t)$ in the rotation frame is either "on" or "off."

Let us now proceed to analyze the spin echo sequence. Initially, we have ρ_0 "=" $\hat{I}_z$. After the pulse, we have, according to (19.32),

$$\rho'(t^+) = e^{-i\mathcal{H}_e t/\hbar}\hat{I}_z e^{i\mathcal{H}_e t/\hbar} = e^{i\gamma B_1 t_{90}\hat{I}_x}\hat{I}_z e^{-i\gamma B_1 t_{90}\hat{I}_x} \qquad (19.40)$$

where $\gamma B_1 t_{90} = \pi/2$. In view of (19.28), this means that the effect of the pulse is to rotate $\hat{I}_z$ by $-90°$ around the x-axis. Thus, $\hat{I}_z$ becomes $\hat{I}_y$, i.e., $\rho'(t^+) = \hat{I}_y$. When $B_1(t) = 0$, the spin precesses under the action of ΔB_0, i.e., it rotates around the z-axis by an angle $\gamma\Delta B_0\tau$ from the y-axis. Let us write $\rho'(\tau)$ in the form,

$$\rho'(\tau) = \hat{\boldsymbol{n}}(\pi/2, \pi/2 + \gamma\Delta B_0\tau)\cdot\hat{\boldsymbol{I}} \qquad (19.41)$$

where $\hat{\boldsymbol{n}}(\theta, \phi)$ is a unit vector in the orientation given by the polar angle θ and the azimuthal angle ϕ. The next pulse is a rotation of $180°$ around the y-axis. This changes the sign of $\hat{I}_x$, leaving $\hat{I}_y$ unchanged. Thus,

$$\rho'(\tau^+) = \hat{\boldsymbol{n}}(\pi/2, \pi/2 - \gamma\Delta B_0\tau)\cdot\hat{\boldsymbol{I}} \qquad (19.42)$$

During the next period of evolution, the angular momentum will rotate further with the same angular velocity as before. So, after an additional time τ, the angular momentum will come back to the y-axis, regardless of the value of ΔB_0. If we were to analyze an ensemble average, we would therefore find that the moment is refocused on the y-axis at a time 2τ (Fig. 19.3). This phenomenon is called a ***spin echo***.

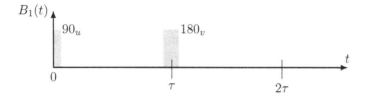

Figure 19.3 Representation of a spin echo sequence. The pulses are assumed to be much shorter than the intervals between pulses. The radio-frequency field $B_1(t)$ is so large that field inhomogeneities can be ignored during the pulses. An echo forms at 2τ, as shown in §19.4.3

We can analyze the spin echo sequence in third way. This will point to an important concept regarding pulse-based spin spectroscopy. Now we write the full evolution as a composition of conjugations like the one shown in (19.32). Thus, the entire evolution is directly expressed as,

$$\rho'(2\tau) = e^{i\gamma\Delta B_0\tau\hat{I}_z}e^{i\pi\hat{I}_y}e^{i\gamma\Delta B_0\tau\hat{I}_z}e^{i\frac{\pi}{2}\hat{I}_x}\hat{I}_z e^{-i\frac{\pi}{2}\hat{I}_x}e^{-i\gamma\Delta B_0\tau\hat{I}_z}e^{-i\pi\hat{I}_y}e^{-i\gamma\Delta B_0\tau\hat{I}_z} \tag{19.43}$$

Once again, we apply our result (19.28) for the rotation of angular momenta operators. First, we simplify the innermost terms.

$$\rho'(2\tau) = e^{i\gamma\Delta B_0\tau\hat{I}_z}e^{i\pi\hat{I}_y}e^{i\gamma\Delta B_0\tau\hat{I}_z}\hat{I}_y e^{-i\gamma\Delta B_0\tau\hat{I}_z}e^{-i\pi\hat{I}_y}e^{-i\gamma\Delta B_0\tau\hat{I}_z} \tag{19.44}$$

Then, we add terms that mutually cancel each other out at both ends of this expression. This will allow us to show how the pulse sequence can be understood as changing the Hamiltonian that determines the evolution of the spin. Thus, we write,

$$\rho'(2\tau) = e^{i\pi\hat{I}_y}e^{-i\pi\hat{I}_y}e^{i\gamma\Delta B_0\tau\hat{I}_z}e^{i\pi\hat{I}_y}e^{i\gamma\Delta B_0\tau\hat{I}_z}\hat{I}_y e^{-i\gamma\Delta B_0\tau\hat{I}_z}e^{-i\pi\hat{I}_y}e^{-i\gamma\Delta B_0\tau\hat{I}_z}e^{i\pi\hat{I}_y}e^{-i\pi\hat{I}_y} \tag{19.45}$$

In this expression, we distinguish the conjugations of operators,

$$e^{-i\pi\hat{I}_y}e^{i\gamma\Delta B_0\tau\hat{I}_z}e^{i\pi\hat{I}_y} \quad \text{and} \quad e^{-i\pi\hat{I}_y}e^{-i\gamma\Delta B_0\tau\hat{I}_z}e^{i\pi\hat{I}_y} \tag{19.46}$$

We apply (19.27) to simplify these expressions. This yields,

$$\rho'(2\tau) = e^{i\pi\hat{I}_y}e^{-i\gamma\Delta B_0\tau\hat{I}_z}e^{i\gamma\Delta B_0\tau\hat{I}_z}\hat{I}_y e^{-i\gamma\Delta B_0\tau\hat{I}_z}e^{i\gamma\Delta B_0\tau\hat{I}_z}e^{-i\pi\hat{I}_y} \tag{19.47}$$

Now we see that the π pulse has the effect of changing the sign of the Hamiltonian that defines the evolution during the time after the π pulse. This is a very simple example where we see that applying pulses can be thought of as changing the effective Hamiltonian of a spin system. Having applied the π pulse, the evolutions during the time before and after the π pulse cancel each other out, meaning that for any field inhomogeneity ΔB_0, its effect cancels out at the time of the echo. Indeed, since $\hat{I}_y$ commutes with the exponential of $\hat{I}_y$, we end up with $\rho'(2\tau) = \hat{I}_y$.

19.4.4 QUADRUPOLAR ECHOES

As an aside, we examine here a type of echo that cannot be analyzed with straightforward geometrical arguments. We consider an ensemble of spins in a uniform magnetic field $\boldsymbol{B}_0$, i.e., $\Delta\boldsymbol{B}_0 = 0$. The spins have a quadrupolar moment that couples to the gradients of the electric field, so that the Hamiltonian for each spin is given by :

$$\mathcal{H} = -\hbar\gamma B_0\hat{I}_z + \mathcal{H}_Q \qquad \mathcal{H}_Q = -\hbar a\hat{I}_z^2 \qquad (19.48)$$

where the coefficient a defines the strength of the quadrupolar effect. The ensemble is characterized by a distribution $f(a)$ of the values of a on each spin.

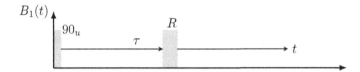

Figure 19.4 Pulse sequence for quadrupolar echo. The second pulse is a rotation R of the spins in the rotating frame.

The pulse sequence in this quadrupolar echo experiment is shown in Fig. 19.4. It is characterized by a 90° pulse that flips the spins from their equilibrium to the Ouv plane. After a time τ of precession, the spins are rotated by a rotation R to be defined. The presence of echoes is monitored during the time t following this second pulse.

In order to examine the evolution in the frame rotating at the angular velocity $\omega = \gamma B_0$, we consider the density matrix ρ' defined in (19.36). The transformed Hamiltonian $\mathcal{H}'$ is readily calculated. Since $[\hat{I}_z^2, \hat{I}_z] = 0$, we have $\mathcal{H}' = \mathcal{H}$. The density matrix immediately after the first pulse is $\rho'(0^+)$. At a time t after the second pulse of the sequence of Fig. 19.4, the density matrix has evolved to,

$$\rho'(\tau,t) = e^{-i\mathcal{H}_Q t/\hbar}Re^{-i\mathcal{H}_Q\tau/\hbar}\rho'(0^+)e^{i\mathcal{H}_Q\tau/\hbar}R^{-1}e^{i\mathcal{H}_Q t/\hbar} \qquad (19.49)$$

It will be convenient to search for the expectation value of $\hat{I}_+ = \hat{I}_x + i\hat{I}_y$, calculated as,

$$<\hat{I}_+>(\tau,t) = Tr\left(\dot{\rho'(\tau,t)}\hat{I}_+\right) \qquad (19.50)$$

In this expression, $<\hat{I}_x>$ and $<\hat{I}_y>$ should be understood as the expectation values of the projections of $\hat{\boldsymbol{I}}$ on the rotating frame, since we are working with ρ', not ρ.

We calculate the trace in (19.50) by using as a basis the eigenstates of $\hat{I}_z$. We make use of,

$$\hat{I}_+|m> = \sqrt{I(I+1) - m(m+1)}|m+1> \qquad (m+1 < I)$$

$$\mathbb{1} = \sum_{m=-I}^{m=+I} |m><m| \qquad (19.51)$$

Let us proceed with the calculation.

$$< \hat{I}_+ > (\tau, t) = \sum_m < m | e^{-i\mathcal{H}_Q t/\hbar} R e^{-i\mathcal{H}_Q \tau/\hbar} \rho'(0^+) e^{i\mathcal{H}_Q \tau/\hbar} R^{-1} e^{i\mathcal{H}_Q t/\hbar} \hat{I}_+ | m >$$

$$= \sum_m \sqrt{I(I+1) - m(m+1)} e^{-im^2 a t} e^{ia(m+1)^2 t}$$

$$< m | R e^{-i\mathcal{H}_Q \tau/\hbar} \rho'(0^+) e^{i\mathcal{H}_Q \tau/\hbar} R^{-1} | m + 1 >$$

$$= \sum_{m,m',m''} \sqrt{I(I+1) - m(m+1)} e^{ia(2m+1)t} \tag{19.52}$$

$$< m | R | m'' >< m'' | e^{-i\mathcal{H}_Q \tau/\hbar} \rho'(0^+) e^{i\mathcal{H}_Q \tau/\hbar} | m' >< m' | R^{-1} | m + 1 >$$

$$= \sum_{m,m',m''} \sqrt{I(I+1) - m(m+1)} e^{ia(2m+1)t} e^{ia(m'^2 - m''^2)\tau}$$

$$< m | R | m'' >< m'' | \rho'(0^+) | m' >< m' | R^{-1} | m + 1 >$$

Since the only operator in $\rho'(0^+)$ is $\hat{I}_x = (\hat{I}_+ + \hat{I}_-)/2$, which couples $|m>$ states in which the m values differ by only 1, the multiple sum can be restricted to,

$$|m' - m''| = 1 \tag{19.53}$$

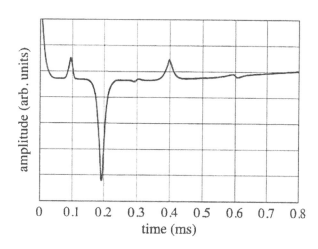

Figure 19.5 Quadrupolar echoes observed in potassium iodine. [1089]

When an ensemble of spins is considered with the distribution $f(a)$ of the quadrupolar coupling, then the signal is vanishingly small except when $< \hat{I}_+ > (\tau, t)$ becomes independent of a, i.e., when,

$$\frac{t}{\tau} = \frac{m'^2 - m''^2}{2m + 1} \tag{19.54}$$

F. Maier used ^{127}I NMR of the compound potassium iodine (KI) and observed 5 echoes (Fig. 19.5).[1089] ^{127}I is a spin 5/2. By trial and error, it is easy to find that the only possible values of the ratio t/τ are $1/2$, 1, $3/2$, 2, 3. The amplitude and the symmetry of the echoes can be accounted for by working out in full the amplitude $< \hat{I}_+ > (\tau, t)$ given by equation (19.52).[1090] Quadrupolar echoes were described in Abragam's textbook.[866] As many as 22 echoes were reported using ^{55}Mn NMR in a manganese ferrite.[1091]

19.4.5 DOUBLE QUANTUM COHERENCE

Let us now consider a spin 1 in a large magnetic field $\boldsymbol{B}_0$, with $\gamma B_0 = \omega_0$, which is subject to the quadrupolar interaction given by,

$$\mathcal{H}_Q = \hbar\omega_Q \hat{I}_z^2 \tag{19.55}$$

This implies that the degenerate Zeeman splittings caused by the static magnetic field $\boldsymbol{B}_0$ change under the influence of the quadrupolar interaction (Fig. 19.6).

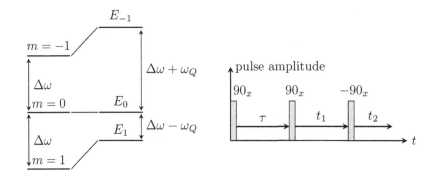

Figure 19.6 Left: Zeeman and quadrupolar splittings. Right: pulse sequence to excite coherence between the states $m = 1$ and $m = -1$.

Adopting once again the rotating picture as in §19.4.3, we have the Hamiltonian

$$\mathcal{H}' = -\hbar\Delta\omega\hat{I}_z - \hbar\gamma B_1(t)\hat{I}_x + \hbar\omega_Q\hat{I}_z^2 \tag{19.56}$$

where $\Delta\omega = \omega_0 + \omega$. We analyze the pulse sequence of Fig. 19.6 and consider that during the pulses, i.e., when $B_1(t)$ is on, the field inhomogeneity $(\omega_0 + \omega)/\gamma$ and the quadrupolar interaction are negligible, i.e., $\gamma B_1 \gg \omega_Q$ and $\gamma B_1 \gg \omega_0 + \omega$.

Following the first 90_x pulse, the density matrix is $\rho'(0^+) = \hat{I}_y$ (in the sense of (19.33)). The matrix elements of $\hat{I}_y$ in the basis set of $\hat{I}_z$ eigenstates can be looked up in a textbook or worked out using the matrix elements of $\hat{I}_\pm$.[181] Hence, $\hat{I}_y$ in the basis $|1>, |0>, |-1>)$ is the matrix,

$$\hat{I}_y = \frac{1}{\sqrt{2}}\begin{pmatrix} 0 & -i & 0 \\ i & 0 & -i \\ 0 & i & 0 \end{pmatrix} \tag{19.57}$$

The eigenstates $|m>$ of $\hat{I}_z$ are eigenstates of the Hamiltonian outside the pulses. Note that here, $\mathcal{H}' = \mathcal{H}$. Therefore, the evolution of the density matrix in this representation can be readily calculated. Thus, we have,

$$\rho'(\tau) = e^{-i\mathcal{H}\tau/\hbar}\rho'(0^+)e^{i\mathcal{H}\tau/\hbar}$$
$$<i|\rho'(t)|j> = e^{-iE_i\tau/\hbar}e^{iE_j\tau/\hbar}<i|\hat{I}_y|j> \tag{19.58}$$

In matrix form, this gives,

$$\rho'(\tau) = \frac{1}{\sqrt{2}}\begin{pmatrix} 0 & A & 0 \\ A^* & 0 & B \\ 0 & B^* & 0 \end{pmatrix} \tag{19.59}$$

where, in view of the splitting illustrated in Fig. 19.6, we have,

$$A = -ie^{i(E_0 - E_1)\tau/\hbar} = -ie^{i(\Delta\omega - \omega_Q)\tau}$$
$$B = -ie^{i(E_{-1} - E_0)/\hbar} = -ie^{i(\Delta\omega + \omega_Q)\tau} \tag{19.60}$$

At the end of this calculation, we will find that the double coherence depends on the difference $A - B$. It is the term $\omega_Q\tau$ that distinguishes A from B. Thus, it is the evolution during the time τ that permits this double coherence to develop.

We are now faced with the problem of calculating the effect of the next pulse on the density matrix $\rho'(\tau)$. We do this with a method known as the ***product-operator formalism***, which consists in using a basis of matrices with which to express $\rho'(\tau)$. We use matrices that are products of angular momentum operators, so the action of the pulse on these matrices can be worked out readily.

We start with the matrix representation of $\hat{I}_x, \hat{I}_y$, and $\hat{I}_z$ on the basis of the eigenstates of $\hat{I}_z$. $\hat{I}_y$ was given in (19.57). The matrices for $\hat{I}_x$ and $\hat{I}_z$ are,

$$\hat{I}_x = \frac{1}{\sqrt{2}} \begin{pmatrix} 0 & 1 & 0 \\ 1 & 0 & 1 \\ 0 & 1 & 0 \end{pmatrix} \qquad \hat{I}_z = \frac{1}{\sqrt{2}} \begin{pmatrix} 1 & 0 & 0 \\ 0 & 0 & 0 \\ 0 & 0 & -1 \end{pmatrix} \tag{19.61}$$

By simple trial and error, it is possible to find the products that have the appropriate matrix forms for our calculations. Two combinations of products connect $m = 1$ and $m = 0$ states. They are,

$$\hat{I}_x + (\hat{I}_x\hat{I}_z + \hat{I}_z\hat{I}_x) = \sqrt{2} \begin{pmatrix} 0 & 1 & 0 \\ 1 & 0 & 0 \\ 0 & 0 & 0 \end{pmatrix}$$
$$\hat{I}_y + (\hat{I}_y\hat{I}_z + \hat{I}_z\hat{I}_y) = \sqrt{2} \begin{pmatrix} 0 & -i & 0 \\ i & 0 & 0 \\ 0 & 0 & 0 \end{pmatrix} \tag{19.62}$$

Two combinations of products connect $m = 0$ and $m = -1$ states. They are,

$$\hat{I}_x - (\hat{I}_x\hat{I}_z + \hat{I}_z\hat{I}_x) = \sqrt{2} \begin{pmatrix} 0 & 0 & 0 \\ 0 & 0 & 1 \\ 0 & 1 & 0 \end{pmatrix}$$
$$\hat{I}_y - (\hat{I}_y\hat{I}_z + \hat{I}_z\hat{I}_y) = \sqrt{2} \begin{pmatrix} 0 & 0 & 0 \\ 0 & 0 & -i \\ 0 & i & 0 \end{pmatrix} \tag{19.63}$$

All the matrices defined above by products of operators connect only two levels and their non-trivial matrix subblocks look like Pauli matrices. When they are expressed in the subspace in which they operate, they are often referred to as the ***fictitious spin 1/2*** matrices.

We can also create Pauli-like matrices that connect the levels $m = 1$ and $m = -1$, which correspond to the double quantum coherences we wish to describe. They are,

$$\hat{I}_x^2 - \hat{I}_y^2 = \begin{pmatrix} 0 & 0 & 1 \\ 0 & 0 & 0 \\ 1 & 0 & 0 \end{pmatrix}$$
$$\hat{I}_x\hat{I}_y + \hat{I}_y\hat{I}_x = \begin{pmatrix} 0 & 0 & -i \\ 0 & 0 & 0 \\ i & 0 & 0 \end{pmatrix} \tag{19.64}$$

To complete the list, we could add the σ_z type of Pauli matrices that connect two pairs of states. Since we do not need these additional matrices in the following, we do not define them here.

We can write $\rho'(\tau)$ in terms of the fictitious spin 1/2 matrices as,

$$\rho'(\tau) = \frac{Re(A)}{\sqrt{2}} \left(\hat{I}_x + (\hat{I}_x \hat{I}_z + \hat{I}_z \hat{I}_x) \right) + \frac{Re(B)}{\sqrt{2}} \left(\hat{I}_x - (\hat{I}_x \hat{I}_z + \hat{I}_z \hat{I}_x) \right)$$
$$- \frac{Im(A)}{\sqrt{2}} \left(\hat{I}_y + (\hat{I}_y \hat{I}_z + \hat{I}_z \hat{I}_y) \right) - \frac{Im(B)}{\sqrt{2}} \left(\hat{I}_y - (\hat{I}_y \hat{I}_z + \hat{I}_z \hat{I}_y) \right) \tag{19.65}$$

The effect of a pulse can be readily computed using the results of §19.4.1 on rotations. When there is a product of angular momentum operators, we can always insert between them a product $e^{i\mathcal{H}t/\hbar}e^{-i\mathcal{H}t/\hbar}$ in order to calculate the rotations of single angular momentum operators. We thus find that, after the second pulse, the density matrix becomes,

$$\rho'(\tau^+) = \frac{Re(A)}{\sqrt{2}} \left(\hat{I}_x + (\hat{I}_x \hat{I}_y + \hat{I}_y \hat{I}_x) \right) + \frac{Re(B)}{\sqrt{2}} \left(\hat{I}_x - (\hat{I}_x \hat{I}_y + \hat{I}_y \hat{I}_x) \right)$$
$$- \frac{Im(A)}{\sqrt{2}} \left(-\hat{I}_z - (\hat{I}_z \hat{I}_y + \hat{I}_y \hat{I}_z) \right) - \frac{Im(B)}{\sqrt{2}} \left(-\hat{I}_z + (\hat{I}_z \hat{I}_y + \hat{I}_y \hat{I}_z) \right) \tag{19.66}$$

Since we are interested in the double quantum coherence, we focus only on the part of the density matrix ρ'_{2Q} that characterizes double quantum coherence, which is given by,

$$\rho'_{2Q}(\tau^+) = \frac{Re(A) - Re(B)}{\sqrt{2}} (\hat{I}_x \hat{I}_y + \hat{I}_y \hat{I}_x) \tag{19.67}$$

Its evolution during the next period is readily calculated, yielding,

$$\rho'_{2Q}(t_1) = \frac{Re(A) - Re(B)}{\sqrt{2}} \begin{pmatrix} 0 & 0 & Q \\ 0 & 0 & 0 \\ Q^* & 0 & 0 \end{pmatrix} \tag{19.68}$$

where $Q = -ie^{i2\Delta\omega t_1}$.

At this stage, we have created a double quantum coherence through the pulses we have applied. This double quantum coherence evolves at twice the offset frequency, as one might expect. One important feature for spectroscopists is that the time evolution during the time (t_1) is independent of ω_Q. Thus, if we were to study a system with a distribution of values of $\Delta\omega$, the double quantum coherence we established by this pulse sequence would be free of this broadening, allowing for high-resolution spectroscopy.

The third pulse will allow the double quantum coherence to become accessible for detection. We write $\rho'_{2Q}(t_1)$ as

$$\rho'_{2Q}(t_1) = Re(Q)(\hat{I}_x^2 - \hat{I}_y^2) - Im(Q)(\hat{I}_x \hat{I}_y + \hat{I}_y \hat{I}_x) \tag{19.69}$$

Therefore, the -90_x pulse produces,

$$\rho'_{2Q}(t_1) = Re(Q)(\hat{I}_x^2 - \hat{I}_z^2) - Im(Q)(\hat{I}_x \hat{I}_z + \hat{I}_z \hat{I}_x) \tag{19.70}$$

The term containing $(\hat{I}_x^2 - \hat{I}_z^2)$ will not contribute to the expectation values of $< \hat{I}_x >$ and $< \hat{I}_y >$. Therefore, from now on, we track only the second term,

$$\rho'_{2Q}(t_1) = -Im(Q)(\hat{I}_x \hat{I}_z + \hat{I}_z \hat{I}_x) = -Im(Q) \begin{pmatrix} 0 & 1 & 0 \\ 1 & 0 & -1 \\ 0 & -1 & 0 \end{pmatrix} \tag{19.71}$$

The t_2-time evolution can be worked out to yield,

$$\rho'_{2Q}(t_2) = \begin{pmatrix} 0 & C & 0 \\ C^* & 0 & D \\ 0 & D^* & 0 \end{pmatrix} \qquad (19.72)$$

where $C = -Im(Q)e^{i(\Delta\omega - \omega_Q)t_2}$ and $D = Im(Q)e^{i(\Delta\omega + \omega_Q)t_2}$. This has the form of $\rho'(\tau)$ in (19.59), whose development in operator products is given in (19.65). Of the four terms, only the one proportional to $\hat{I}_y$ will contribute to the trace that determines the expectation value. Thus, we have,

$$\begin{aligned} <\hat{I}_y>(t_2) &= Tr\left(\hat{I}_y\left(-\frac{Im(C)}{2} - \frac{Im(D)}{2}\right)\hat{I}_y\right) \\ &= Tr(\hat{I}_y^2)\cos(2\Delta\omega t_1)\sin(\omega_Q\tau)\sin(\omega_Q t_2)\cos(\Delta\omega\tau)\cos(\Delta\omega t_2) \end{aligned} \qquad (19.73)$$

Let us now consider what this implies for a system of spins and a distribution of $\Delta\omega$. The signal is small because the phases cancel each other out except at the time $t_2 = \tau$, since then the contribution of $\cos(\Delta\omega\tau)\cos(\Delta\omega t_2)$ adds up to a maximum, i.e., there is an echo. This echo has an amplitude modulated by $\cos(2\Delta\omega t_1)$. We can see how a **two-dimensional spectroscopy** might provide information on both ω_Q and $\Delta\omega$.[1092–1095]

19.4.6 COHERENCE TRANSFER

Using pulses, it is possible to transfer the coherence from an I spin to an S spin, which is coupled to the I spin. The pulse sequence is illustrated in Fig. 19.7. This pulse sequence was first introduced by Mandsley and Ernst in 1977.[1096]

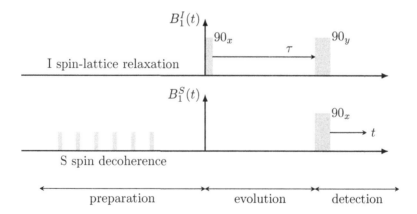

Figure 19.7 Pulse sequence used to transfer coherence from I spins to S spins. The structure of this sequence is typical of quantum mechanical experiments: first, there is a preparation period; then, there is an evolution period, during which the effect of the coupling between the spins builds; and finally, there is a detection period, during which the coherence is manifested in a measurable quantity.

The realization that such a transfer is possible is not only conceptually important, but also useful. For example, the I spins may have a large gyromagnetic moment such that their polarization is high in a given magnetic field. The coherence transfer implies that this strong polarization can be transferred to the S spins, which may carry valuable information about the S spins environment, whereas the I spins may not. Hence, the spectroscopist gains signal intensity compared to doing straightforward S-spin spectroscopy.

As we wish to make the simplest illustration of coherence transfers that use pulses, we will not consider any spread in the Larmor frequencies of the I and S spins. The interaction between the I and S spins will also be made as simple as possible. Thus, the Hamiltonian of the system composed of an I spin coupled to an S spin is written as,

$$\mathcal{H} = -\hbar\gamma_I B_0 \hat{I}_z + \hbar 2a\hat{I}_z\hat{S}_z - \hbar\gamma_S B_0 \hat{S}_z \tag{19.74}$$

The Hilbert space of the system is the tensorial product of the Hilbert spaces of the I and S spins, and the operators should be written as $\mathbb{1} \otimes \hat{S}_z$, $\hat{I}_z \otimes \mathbb{1}$ and $\hat{S}_z \otimes \hat{I}_z$. It is common practice to simplify the notation, as we did in (19.74).

We can move to a double rotating frame picture by applying the transformation (19.36) to both I and S spins. The transformed Hamiltonian becomes

$$\mathcal{H}' = \hbar 2a\hat{I}_z\hat{S}_z \tag{19.75}$$

Let us work out the evolution of the density matrix during the pulses. Only the B_1 field on resonance for one of the spins (I or S spins) needs to be considered. The pulses are said to be selective, i.e., the radio-frequency field which is on-resonance for say, the I spin is ineffective for the S spins. We consider the elements of the density matrix calculated in the basis of the $\hat{S}_z \otimes \hat{I}_z$ eigenstates, namely, the states $|m_S, m_I>$, that we will take in the following order,

$$\{|m_S, m_I>\} = \left\{ |-\frac{1}{2}, -\frac{1}{2}>, |-\frac{1}{2}, +\frac{1}{2}>, |+\frac{1}{2}, -\frac{1}{2}>, |+\frac{1}{2}, +\frac{1}{2}>, \right\} \tag{19.76}$$

The system is initially prepared in a state where the S spins have neither spin coherence nor spin polarization, whereas the I spins have a non-zero polarization but no coherence. Applying the simplification (19.33), we write,

$$\rho'(0) = \begin{pmatrix} A & 0 & 0 & 0 \\ 0 & B & 0 & 0 \\ 0 & 0 & A & 0 \\ 0 & 0 & 0 & B \end{pmatrix} \tag{19.77}$$

In the basis (19.76), $\hat{S}_z$ and $\hat{I}_z$, namely $\mathbb{1} \otimes \hat{S}_z$ and $\hat{I}_z \otimes \mathbb{1}$ in the precise notation, read,

$$\mathbb{1} \otimes \hat{S}_z = \frac{1}{2}\begin{pmatrix} -1 & 0 & 0 & 0 \\ 0 & -1 & 0 & 0 \\ 0 & 0 & 1 & 0 \\ 0 & 0 & 0 & 1 \end{pmatrix} \qquad \hat{I}_z \otimes \mathbb{1} = \frac{1}{2}\begin{pmatrix} -1 & 0 & 0 & 0 \\ 0 & 1 & 0 & 0 \\ 0 & 0 & -1 & 0 \\ 0 & 0 & 0 & 1 \end{pmatrix} \tag{19.78}$$

The form of $\rho'(0)$ expresses the result of the initial preparation, namely no coherence and the polarizations given by,

$$<\hat{S}_z> = Tr(\rho'(0)\hat{S}_z) = 0 \qquad <\hat{I}_z> = Tr(\rho'(0)\hat{I}_z) = B - A \tag{19.79}$$

Now we need to calculate the effect of the 90° pulse on the I spin. We do this using the operator formalism. We recall the Pauli matrices,

$$\mathbb{1} = \begin{pmatrix} 1 & 0 \\ 0 & 1 \end{pmatrix} \qquad \sigma_x = \begin{pmatrix} 0 & 1 \\ 1 & 0 \end{pmatrix} \qquad \sigma_y = \begin{pmatrix} 0 & -i \\ i & 0 \end{pmatrix} \qquad \sigma_z = \begin{pmatrix} 1 & 0 \\ 0 & -1 \end{pmatrix} \tag{19.80}$$

From the definitions (19.80), we deduce,

$$\frac{\mathbb{1}+\sigma_z}{2} = \begin{pmatrix} 1 & 0 \\ 0 & 0 \end{pmatrix} \quad \frac{\mathbb{1}-\sigma_z}{2} = \begin{pmatrix} 0 & 0 \\ 0 & 1 \end{pmatrix}$$
$$\frac{\sigma_x+i\sigma_y}{2} = \begin{pmatrix} 0 & 1 \\ 0 & 0 \end{pmatrix} \quad \frac{\sigma_x-i\sigma_y}{2} = \begin{pmatrix} 0 & 0 \\ 1 & 0 \end{pmatrix} \tag{19.81}$$

Therefore, any matrix ρ given by,

$$\rho = \begin{pmatrix} \rho_{11} & \rho_{12} \\ \rho_{12}^* & \rho_{22} \end{pmatrix} \tag{19.82}$$

can be written as,

$$\rho = \rho_{11}\frac{\mathbb{1}+\sigma_z}{2} + \rho_{22}\frac{\mathbb{1}-\sigma_z}{2} + \rho_{12}\frac{\sigma_x+i\sigma_y}{2} + \rho_{12}^*\frac{\sigma_x-i\sigma_y}{2} \tag{19.83}$$

The effect of a rotation on the angular momentum was worked out in §19.4.1. Thus, after a 90_x pulse, ρ becomes,

$$\rho(90_x) = \rho_{11}\frac{\mathbb{1}+\sigma_y}{2} + \rho_{22}\frac{\mathbb{1}-\sigma_y}{2} + \rho_{12}\frac{\sigma_x-i\sigma_z}{2} + \rho_{12}^*\frac{\sigma_x+i\sigma_z}{2} \tag{19.84}$$

Applying the definitions (19.80) and grouping the terms, we find,

$$\rho(90_x) = \frac{1}{2}\begin{pmatrix} \frac{\rho_{11}+\rho_{22}}{2}+Im(\rho_{12}) & -i\frac{\rho_{11}-\rho_{22}}{2}+Re(\rho_{12}) \\ i\frac{\rho_{11}-\rho_{22}}{2}+Re(\rho_{12}) & \frac{\rho_{11}+\rho_{22}}{2}-Im(\rho_{12}) \end{pmatrix} \tag{19.85}$$

The coherence transfer sequence defined in Fig. 19.7 also contains a 90_y pulse for the I spins, whose effect on the matrix ρ can be calculated the same way. This yields,

$$\rho(90_y) = \frac{1}{2}\begin{pmatrix} \frac{\rho_{11}+\rho_{22}}{2}-Re(\rho_{12}) & -\frac{\rho_{11}-\rho_{22}}{2}+iIm(\rho_{12}) \\ -\frac{\rho_{11}-\rho_{22}}{2}-iIm(\rho_{12}) & \frac{\rho_{11}+\rho_{22}}{2}+Re(\rho_{12}) \end{pmatrix} \tag{19.86}$$

Following the 90_x pulse on the I spins, the density matrix $\rho'(0)$ becomes,

$$\rho'(0+) = \frac{A+B}{2}\begin{pmatrix} 1 & -i\frac{A-B}{A+B} & 0 & 0 \\ i\frac{A-B}{A+B} & 1 & 0 & 0 \\ 0 & 0 & 1 & -i\frac{A-B}{A+B} \\ 0 & 0 & i\frac{A-B}{A+B} & 1 \end{pmatrix} \tag{19.87}$$

In view of the high-temperature approximation (19.33), we keep only the following terms,

$$\rho'(0+) = \begin{pmatrix} 0 & -i(A-B) & 0 & 0 \\ i(A-B) & 0 & 0 & 0 \\ 0 & 0 & 0 & -i(A-B) \\ 0 & 0 & i(A-B) & 0 \end{pmatrix} \tag{19.88}$$

Now the I-S spin pairs evolve under the effect of the coupling (19.75). The density matrix after a time τ is readily calculated using (19.40). Thus, we find,

$$\rho'(\tau) = \begin{pmatrix} 0 & -i(A-B)e^{ia\tau} & 0 & 0 \\ i(A-B)e^{-ia\tau} & 0 & 0 & 0 \\ 0 & 0 & 0 & -i(A-B)e^{-ia\tau} \\ 0 & 0 & i(A-B)e^{ia\tau} & 0 \end{pmatrix} \tag{19.89}$$

After the time τ, a 90_y pulse is applied to the I spins. The effect of the 90_y pulse can be calculated using (19.86). This gives,

$$\rho'(\tau, 90_{(y,I)})$$

$$= (A - B) \begin{pmatrix} -\sin(a\tau) & -i\cos(a\tau) & 0 & 0 \\ i\cos(a\tau) & \sin(a\tau) & 0 & 0 \\ 0 & 0 & \sin(a\tau) & -i\cos(a\tau) \\ 0 & 0 & i\cos(a\tau) & -\sin(a\tau) \end{pmatrix} \qquad (19.90)$$

where we have used,

$$Re(\rho'(\tau)_{12}) = (A - B)\sin(a\tau) \qquad Im(\rho'(\tau)_{12}) = -(A - B)\cos(a\tau) \qquad (19.91)$$

Now we calculate the effect of the 90_x on the S spins. We need to apply (19.85) to the subspace of $m_I = -1/2$ by considering the matrix elements $(1, 1)$, $(1, 3)$, $(3, 1)$, and $(3, 3)$. Then, we apply (19.85) in the subspace $m_I = +1/2$, considering the matrix elements $(2, 2)$, $(2,4)$, $(4, 2)$, and $(4, 4)$. This yields,

$$\rho'(\tau, 90_{(y,I)}, 90_{(x,S)})$$

$$= (A - B) \begin{pmatrix} 0 & -i\cos(a\tau) & i\sin(a\tau) & 0 \\ i\cos(a\tau) & 0 & 0 & -i\sin(a\tau) \\ -i\sin(a\tau) & 0 & 0 & -i\cos(a\tau) \\ 0 & i\sin(a\tau) & i\cos(a\tau) & 0 \end{pmatrix} \qquad (19.92)$$

The system evolves during the time t under the coupling (19.75). This gives,

$$\rho'(\tau, t)$$

$$= (A - B) \begin{pmatrix} 0 & -i\cos(a\tau)e^{iat} & i\sin(a\tau)e^{iat} & 0 \\ i\cos(a\tau)e^{-iat} & 0 & 0 & -i\sin(a\tau)e^{-iat} \\ -i\sin(a\tau)e^{-iat} & 0 & 0 & -i\cos(a\tau)e^{-iat} \\ 0 & i\sin(a\tau)e^{iat} & i\cos(a\tau)e^{iat} & 0 \end{pmatrix} \qquad (19.93)$$

We are now ready to calculate the expectation value $< \hat{S}_+ >= Tr\left(\rho'(\tau,t)\hat{S}_+\right)$. Thus,

$$< \hat{S}_+ > = \sum_{m_I, m_S, m_S'} < m_I, m_S|\rho'(\tau,t)|m_I, m_S' >< m_I, m_S'|\hat{S}_+|m_S >$$

$$=< 1/2|\hat{S}_+| - 1/2 > \sum_{m_I} < m_I, -1/2|\rho'(\tau,t)|m_I, 1/2 > \qquad (19.94)$$

$$=< 1/2|\hat{S}_+| - 1/2 > (\rho'(\tau,t)_{13} + \rho'(\tau,t)_{24})$$

In view of (19.93), which we recall has been renormalized, we obtain a signal,

$$< \hat{S}_+ >\propto (A - B)\sin(a\tau)\sin(at) \qquad (19.95)$$

Thus, we find that the S spin signal is proportional to $< \hat{I}_z >$. This simple experiment uses 3 90° pulses. In practical cases, 180° pulses for I and S spins are inserted at $\tau/2$ and $t/2$ to refocus field inhomogeneities inherent to most practical samples and spectrometers.[1097]

19.5   GENERATING ENTANGLED STATES

Entangled states have properties that are specific to quantum mechanics. In the early days of quantum mechanics, the properties of entangled states signaled a paradox known as the

"EPR paradox."[1098] Their properties are now understood to be inherent to the quantum mechanical nature of microscopic phenomena.[1099] Entanglement comes into play in both quantum cryptography [1100] and quantum computing.[683]

Here, we want to address a practical problem: how can we create an entangled state with spins in a controlled manner, and how can we be sure that the state we created is truly entangled? This was the question raised by M. Mehring and his colleagues in two research papers published in 2003 and 2004.[1101, 1102] In these papers, the authors report on the creation of a quasi-pure entangled state between an electron spin and a nuclear spin. We will see in what sense this state is "quasi"-pure.

By irradiating a crystal of malonic acid with X-rays, some CH_2 groups are converted into ·CH radicals. The unpaired electron spin couples to the nuclear spin, i.e., the proton, via a hyperfine coupling. The choice of material is dictated by the need for well resolved ESR and NMR lines. This means that the lines must be narrow compared to the splittings. The study [1101] on malonic acid was carried out at 30K, while the study reported in [1102] was done with ^{15}N encapsulated in C_{60} cages ($^{15}N@C_{60}$) at 50 K. In both studies, the electron spin resonance was carried out in the X band (near 9.5 GHz). According to ([683], chapter 2), the values for T_1 and T_2 in malonic acid are of about 91 ms and 5 μs, respectively. The splittings are such that the NMR transitions are resolved, a feature that is used extensively to accomplish entanglement by this experiment. The main advantage of using ^{15}N encapsulated in C_{60} cages is that the lines are better resolved than in malonic acid. Another interesting feature of that spin system is that the radical is a spin 3/2, so the energy level diagram more closely resembles what it would look like in practical applications. Let us now analyze the case of a system comprising many repeats of two coupled spin 1/2 (Fig. 19.8).

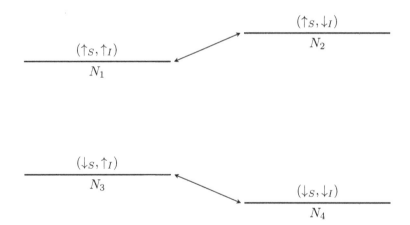

Figure 19.8 Energy level diagram for two coupled spins. The $|i\,j>$ labels represent the eigenstates of spin 1/2 $\hat{S}_z$ and $\hat{I}_z$, projections on the z-axis of the electron spin $\boldsymbol{S}$ and the nuclear spin $\boldsymbol{I}$. The double arrows indicate the nuclear spin transitions. Transitions 1–3 and 2–4 are electron spin transitions.

The entangled states that we wish to create, also known as **Bell states**, are the following:

$$\Psi^{\pm} = \frac{1}{\sqrt{2}} \left(|\uparrow\downarrow> \pm |\downarrow\uparrow> \right) \tag{19.96}$$

$$\Phi^{\pm} = \frac{1}{\sqrt{2}} \left(|\uparrow\uparrow> \pm |\downarrow\downarrow> \right) \tag{19.97}$$

These states present what are called *quantum correlations* or *correlated states*. For example, Cohen-Tanjoudji et al. show in their textbook that a collision between two spins gives rise to correlated states ([169], §10,F).

We have seen that the application of a magnetic excitation pulse rotates those spins that are on resonance. We now make the following assumption. First, we assume that we are capable of creating *selective pulses*, i.e., pulses that rotate the nuclear spins wihtout influencing the electron spins. Second, we assume that we are able to address the nuclear spins coupled to electron spins of a specific direction. Third, we assume that we can address one electron spin transition without changing the nuclear spin. This is possible because the nuclear and electron spin transitions are resolved. In other words, the spectral splittings are larger than the resonance line widths. The pulses we have considered so far covered a frequency band of $1/T_{90}$, where T_{90} is the pulse duration. Such pulses typically cover several lines. How selective pulses are generated is a specialized topic that we need not address here.[1095, 1103–1105]

Assuming selective rotations can be implemented, it is possible to generate an entangled state using selective pulses. For example, starting from the state $|\downarrow\uparrow>$, we can generate Ψ^- with the following pulse sequence,

$$|\downarrow\uparrow> \xrightarrow{P_{34}(\pi/2)} \frac{1}{\sqrt{2}}\left(|\downarrow\uparrow> +|\downarrow\downarrow>\right) \xrightarrow{P_{24}(\pi)} \frac{1}{\sqrt{2}}\left(|\downarrow\uparrow> +|\uparrow\downarrow>\right) = \Psi^+ \tag{19.98}$$

The other Bell states (19.96) can be generated in a similar way.

As the experiment is carried out using a crystal, the system contains an ensemble of spins and should be described with a density matrix. We use for basis the state set,

$$\{|\uparrow\uparrow>, |\uparrow\downarrow>, |\downarrow\uparrow>, |\downarrow\downarrow>\} \equiv \{|1>, |2>, |3>, |4>\} \tag{19.99}$$

with the electron spin magnetic number first, i.e., $|m_S, m_I>$, and the ordering of the states as indicated. At a thermal equilibrium defined by a temperature T, the density matrix ρ_0 is given by,

$$\rho_0 = \frac{1}{Z}\exp\left(\frac{-\mathcal{H}}{k_B T}\right) \approx \frac{1}{4}\left(I_4 - \frac{\mathcal{H}}{k_B T}\right) \tag{19.100}$$

In the last equality, we take into account the high-temperature approximation. I_4 designates the identity matrix in the space spanned by the basis (19.99). In estimating ρ_B, the nuclear spin energy is negligible compared to the electron spin energy (7.5), i.e., $\mathcal{H} \approx g\mu_B B\hat{S}_z \otimes I_2$. Let us look at the matrix components. We have,

$$\rho_0 = \frac{1}{4}\begin{pmatrix} 1-K & 0 & 0 & 0 \\ 0 & 1-K & 0 & 0 \\ 0 & 0 & 1+K & 0 \\ 0 & 0 & 0 & 1+K \end{pmatrix} \tag{19.101}$$

where $K = \mu_B B/(2k_B T)$. Thus, we have equal populations in the states 1 and 2, and in the states 3 and 4 (state numbering consistent with (19.99) and Fig. 19.8). Following [1101], we decompose this matrix into a sum of two terms. The first term is proportional to I_4 and therefore does not contribute to the signal. The second term is K multiplied by a matrix that is written as a density matrix. Hence, we write ρ_0 as,

$$\rho_0 = \left(\frac{1}{4} - \frac{K}{4}\right) I_4 + K\left(\frac{1}{4}I_4 + \frac{1}{4}\begin{pmatrix} -1 & 0 & 0 & 0 \\ 0 & -1 & 0 & 0 \\ 0 & 0 & +1 & 0 \\ 0 & 0 & 0 & +1 \end{pmatrix}\right) \tag{19.102}$$

As in [1101], we now proceed by forgetting, so to speak, the full structure of ρ_0 and focus on the (partial) density matrix ρ_P,

$$\rho_P = \frac{1}{4}I_4 + \frac{1}{4}\begin{pmatrix} -1 & 0 & 0 & 0 \\ 0 & -1 & 0 & 0 \\ 0 & 0 & +1 & 0 \\ 0 & 0 & 0 & +1 \end{pmatrix} \tag{19.103}$$

To the extent that we are not concerned with the factor K in the calculations that follow, we can multiply the K by any constant A and get the same analysis for ρ_P. The definition of ρ_P has changed, however. We now have,

$$\rho_0 = \left(\frac{1}{4} - \frac{K}{4A}\right)I_4 + \frac{K}{A}\left(\frac{1}{4}I_4 + \frac{A}{4}\begin{pmatrix} -1 & 0 & 0 & 0 \\ 0 & -1 & 0 & 0 \\ 0 & 0 & +1 & 0 \\ 0 & 0 & 0 & +1 \end{pmatrix}\right) \tag{19.104}$$

so,

$$\rho_P(0) = \frac{1}{4}I_4 + \frac{A}{4}\begin{pmatrix} -1 & 0 & 0 & 0 \\ 0 & -1 & 0 & 0 \\ 0 & 0 & +1 & 0 \\ 0 & 0 & 0 & +1 \end{pmatrix} = \frac{1}{4}\begin{pmatrix} 1-A & 0 & 0 & 0 \\ 0 & 1-A & 0 & 0 \\ 0 & 0 & 1+A & 0 \\ 0 & 0 & 0 & 1+A \end{pmatrix} \tag{19.105}$$

If we had pure states (not pseudo-pure states), we would enquire about a pulse sequence that produces the pure state $|\downarrow\uparrow>$, of which the corresponding matrix is,

$$\rho_P(final) = \begin{pmatrix} 0 & 0 & 0 & 0 \\ 0 & 0 & 0 & 0 \\ 0 & 0 & +1 & 0 \\ 0 & 0 & 0 & 0 \end{pmatrix} \tag{19.106}$$

Mehring and colleagues arrived at a pseudo-pure state corresponding to the density matrix (19.106) using two selective pulses: one $\pi/2$ pulse acting selectively on the (2–4) electron spin transition, then a second pulse $\pi/2$ acting on the (1–2) transition. We will work out this calculation, but we already know that $(\rho_P)_{(3,3)}$ will remain unchanged. Therefore, we need A such that $1 + A = 4$, i.e., $A = 3$. As a result, we can see that $\rho_{P,init}$ must be,

$$\rho_P(0) = \frac{1}{4}I_4 + \frac{3}{4}\begin{pmatrix} -1 & 0 & 0 & 0 \\ 0 & -1 & 0 & 0 \\ 0 & 0 & +1 & 0 \\ 0 & 0 & 0 & +1 \end{pmatrix} = \frac{1}{4}I_4 + \frac{3}{2}\left(\hat{S}_z \otimes I_2\right) \tag{19.107}$$

The selective pulses do not change I_4. They have an effect on the matrix, truncated from $\rho_P(0)$, which is given by,

$$\rho_P^{(tronc)}(0) = \begin{pmatrix} -1 & 0 & 0 & 0 \\ 0 & -1 & 0 & 0 \\ 0 & 0 & +1 & 0 \\ 0 & 0 & 0 & +1 \end{pmatrix} \tag{19.108}$$

In order to obtain $\rho_P(final)$ as specified in (19.106), the pulses must transform $\rho_P^{(tronc)}(0)$ into a diagonal matrix with the following elements on the diagonal:

$$\rho_P^{(tronc)}(final) = \begin{pmatrix} -1/3 & 0 & 0 & 0 \\ 0 & -1/3 & 0 & 0 \\ 0 & 0 & 1 & 0 \\ 0 & 0 & 0 & -1/3 \end{pmatrix} \tag{19.109}$$

To evaluate the effect of selective pulses, we use the method of fictitious spins $1/2$. [1106–1108] This method consists in expressing the density matrix (or any other matrix of an operator written in a basis like (19.99)) in terms of the Pauli matrices defined for the subspaces corresponding to transitions among the various energy levels of the system. For example, for the (2–4) transition, we can define the Pauli matrix $\sigma_x^{(24)}, \sigma_y^{(24)}, \sigma_z^{(24)}$, and likewise for all other transitions. It is immediately apparent that,

$$\rho_P^{(tronc)}(0) = -\sigma_z^{(24)} - \sigma_z^{(13)} \tag{19.110}$$

A rotation by an angle β around the x-axis is induced by a selective excitation of the (2–4) transition. This rotation is given by the rotation operator,

$$P^{(24)}(\beta) = \cos\frac{\beta}{2}\mathbb{1}_2^{(24)} + i\sin\frac{\beta}{2}\sigma_x^{(24)} \tag{19.111}$$

This rotation does not affect the matrix $\sigma_z^{(13)}$ (as per the definition of selectivity for these pulses). Therefore, after the first pulse that acts on the (2–4) transition, we have,

$$\rho_P^{(tronc)}(1) = -\left(\cos\frac{\beta}{2}\mathbb{1}_2^{(24)} + i\sin\frac{\beta}{2}\sigma_x^{(24)}\right)\sigma_z^{(24)}\left(\cos\frac{\beta}{2}\mathbb{1}_2^{(24)} - i\sin\frac{\beta}{2}\sigma_x^{(24)}\right) - \sigma_z^{(13)} \tag{19.112}$$

Expansion of the product gives several terms,

$$\begin{aligned}\rho_P^{(tronc)}(1) = \ & -\cos^2\frac{\beta}{2}\sigma_z^{(24)} - \sin^2\frac{\beta}{2}\sigma_x^{(24)}\sigma_z^{(24)}\sigma_x^{(24)} \\ & -i\sin\frac{\beta}{2}\cos\frac{\beta}{2}\left[\sigma_x^{(24)}, \sigma_z^{(24)}\right] - \sigma_z^{(13)}\end{aligned} \tag{19.113}$$

The expression (19.113) contains the commutator of $\sigma_x^{(24)}$ and $\sigma_x^{(24)}$, giving a term proportional to $\sigma_y^{(24)}$. Mehring and his colleagues waited until these coherences decayed to zero.[1101] This can be donw with no significant change to the on-diagonal elements because T_2 is so much shorter than T_1. The triple matrix product yields $-\sigma_z^{(24)}$. Using the trigonometric relation $\cos^2 x - \sin^2 x = \cos(2x)$, we find that, after the decay period, the truncated density matrix becomes,

$$\rho_P^{(tronc)}(1) = -\cos\beta\,\sigma_z^{(24)} - \sigma_z^{(13)} \tag{19.114}$$

Thus, after the decay of the coherences, we are left with,

$$\rho_P^{(tronc)}(1) = \begin{pmatrix} -1 & 0 & 0 & 0 \\ 0 & -\cos\beta & 0 & 0 \\ 0 & 0 & 1 & 0 \\ 0 & 0 & 0 & \cos\beta \end{pmatrix} \tag{19.115}$$

Then, the authors of [1101] applied a rotation by an angle θ selectively to the transition (1–2). Let us work out this rotation in matrix form,

$$\begin{pmatrix} \cos\frac{\theta}{2} & -i\sin\frac{\theta}{2} \\ i\sin\frac{\theta}{2} & \cos\frac{\theta}{2} \end{pmatrix}\begin{pmatrix} -1 & 0 \\ 0 & -\cos\beta \end{pmatrix}\begin{pmatrix} \cos\frac{\theta}{2} & i\sin\frac{\theta}{2} \\ -i\sin\frac{\theta}{2} & \cos\frac{\theta}{2} \end{pmatrix} \tag{19.116}$$

$$= \begin{pmatrix} -\cos^2\frac{\theta}{2} - \sin^2\frac{\theta}{2}\cos\beta & i\sin\frac{\theta}{2}\cos\frac{\theta}{2}(1 + \cos\beta) \\ -i\sin\frac{\theta}{2}\cos\frac{\theta}{2}(1 + \cos\beta) & -\sin^2\frac{\theta}{2} - \cos^2\frac{\theta}{2}\cos\beta \end{pmatrix} \tag{19.117}$$

After a delay during which the coherences decay to zero, the truncated density matrix ends up as,

$$\rho_P^{(tronc)}(1) = \begin{pmatrix} -\cos^2\frac{\theta}{2} - \sin^2\frac{\theta}{2}\cos\beta & 0 & 0 & 0 \\ 0 & -\sin^2\frac{\theta}{2} - \cos^2\frac{\theta}{2}\cos\beta & 0 & 0 \\ 0 & 0 & 1 & 0 \\ 0 & 0 & 0 & \cos\beta \end{pmatrix} \quad (19.118)$$

The density matrix $\rho_P^{(tronc)}(1)$ must be equal to the desired form (19.109). This will be the case when,

$$\cos\beta = -1/3 \Rightarrow \beta = 109.47° \quad (19.119)$$

$$\sin^2\frac{\theta}{2} = \cos^2\frac{\theta}{2} = \frac{1}{2} \Rightarrow \theta = 90° \quad (19.120)$$

as can easily be verified. We recall that our accomplishment at this stage is the production of a quasi-pure state defined by the density matrix $\rho_P(1)$ given by (19.106). We call it pseudo-pure because, in fact, the system is characterized by the density matrix (19.104) where $\rho_P(1)$ is weighted by the Boltzmann factor $K = \mu_B B/(2k_B T)$.

Now, we apply the sequence (19.98) of pulses to $\rho_P(1)$. Thus, we start with $\rho_P(1)$, which we write as,

$$\rho_P(1) = \frac{1}{2}\left(\mathbb{1}_2^{(34)} + \sigma_z^{(34)}\right) \quad (19.121)$$

The effect of the NMR pulse is to produce,

$$\rho_P(2) = \frac{1}{2}P_{34}(\pi/2)\left(\mathbb{1}_2^{(34)} + \sigma_z^{(34)}\right)P_{(34)}(\pi/2)^+ = \frac{1}{2}\left(\mathbb{1}_2^{(34)} + \sigma_y^{(34)}\right) \quad (19.122)$$

After the coherences decay, we are left with,

$$\rho_P(3) = \frac{1}{2}\mathbb{1}_2^{(34)} = \frac{1}{2}\begin{pmatrix} 0 & 0 & 0 & 0 \\ 0 & 0 & 0 & 0 \\ 0 & 0 & 1 & 0 \\ 0 & 0 & 0 & 1 \end{pmatrix} = \frac{1}{4}\left(\mathbb{1}_2^{(13)} - \sigma_z^{(13)} + \mathbb{1}_2^{(24)} - \sigma_z^{(24)}\right) \quad (19.123)$$

Now we apply a π pulse to the transition (2–4), which leaves the (1–3) transition untouched. So we get,

$$\rho_P(4) = \frac{1}{4}\left(\mathbb{1}_2^{(13)} - \sigma_z^{(13)} + \mathbb{1}_2^{(24)} + \sigma_z^{(24)}\right) = \frac{1}{2}\begin{pmatrix} 0 & 0 & 0 & 0 \\ 0 & 1 & 0 & 0 \\ 0 & 0 & 1 & 0 \\ 0 & 0 & 0 & 0 \end{pmatrix} = |\Psi^+><\Psi^+| \quad (19.124)$$

In [1101], the reported experiment used the sequence,

- The pulse at 109.41° on the (2–4) electron spin transition,
- The $\pi/2$ pulse on the (1–2) nuclear spin transition,
- The sequence (19.98).

To verify that they had indeed obtained the entangled state Ψ^+, they verified experimentally that they had indeed obtained this entangled state. They called their method **density matrix tomography**. This terminology brings out an important issue of quantum computing; when a computational process leads to coherences, the coherences are most often not readily detectable, as was the case for the double quantum coherence in §19.4.5. Here, we have produced a diagonal matrix and we wish to confirm that its elements represent an entangled state. To do this, the authors of [1101] monitored how the phase of the matrix elements changed under a phase change of the radiation (either rf for NMR or microwave for EPR) in the read-out sequence.

Nowadays, most solid state sources are radio-frequency or microwave sources that remain coherent for the duration of a typical pulsed resonance experiment. The phase coherence of rf and microwave generators can be tested in the following way. If we set a generator at one frequency and combine its output with the output of a second generator, set at the same frequency, the output of the mixer has a low-frequency component that is proportional to the cosine of the phase difference between these two sources. This quasi-DC output will change over a long time scale because the phases of each oscillator change independently of one another over time. The typical time scale for this phase change is very long compared to the time scale of a pulsed resonance experiment. Of course, when the two oscillators are linked, with one signal feeding some sort of trigger signal to the other, then the two sources remain in phase. If a gyrotron (not a gyro-amplifier) were used to make pulses, then the coherence time of this source would have to be good enough. This coherence time can be estimated from the width of the spectrum of the gyrotron output. Some have used a free electron laser to do electron spin resonance.[1109,1110] They controlled the phase by way of a small signal appearing just before the pulse and were able to use this signal to determine the phase of the upcoming pulse.

Mathematically, a phase change of the excitation can be represented by a rotation around the z-axis of the electron and nuclear spin operators defining the excitation. For example, instead of a B_1 field on the u-axis (of the rotating frame), the phase change causes the B_1 field, in the same reference frame, to be at some angle with respect to the u-axis. The operators that generate the rotations of the electron and nuclear spins are, respectively,

$$U_S = \exp\left(-i\phi_S \hat{S}_z\right) \quad \text{and} \quad U_I = \exp\left(-i\phi_I \hat{I}_z\right) \tag{19.125}$$

We apply (19.27) to express the dephasing as rotations. The dephased pulses are given by the operators,

$$U_S P_{24} U_S^+ \quad \text{and} \quad U_I P_{34} U_I^+ \tag{19.126}$$

We recall the rule $(AB)^+ = B^+ A^+$ so, in particular, $(UPU^+)^+ = UP^+U^+$.

In the following, we show that the sequence used in [1101] for the density matrix tomography indeed generates a signal that depends on the phases ϕ_I and ϕ_S. The signal we inspect is the expectation value of σ_x^{24} after acting on the pseudo-pure state Ψ^+ given by $\rho_P(4)$ with a squence of pulses.

First, let us apply a dephased $\pi/2$ pulse on the (3–4) transition to $\rho_P(4)$. To do this, we need to calculate,

$$\rho_P(5) = U_I P_{34} U_I^+ \rho_P(4) U_I P_{34}^+ U_I^+ \tag{19.127}$$

It is best to express $\rho_P(4)$ in terms of the spin operators of the (3–4) and (1–2) transitions, i.e., the nuclear transitions. Thus, we write,

$$\rho_P(5) = U_I P_{34} U_I^+ \frac{1}{4}\left(\mathbb{1}_2^{(12)} - \sigma_z^{(12)} + \mathbb{1}_2^{(34)} + \sigma_z^{(34)}\right) U_I P_{34}^+ U_I^+ \tag{19.128}$$

The innermost conjugation of U_I operators does nothing, since the density matrix is expressed in terms of projections on the z-axis of angular momenta, and U_I is a rotation around the z-axis. We know that the P_{34} conjugation of operators produces a rotation by

$\pi/2$, thus,

$$
\begin{aligned}
\rho_P(5) &= U_I \frac{1}{4} \left(\mathbb{1}_2^{(12)} - \sigma_z^{(12)} + \mathbb{1}_2^{(34)} + \sigma_y^{(34)} \right) U_I^+ \\
&= \frac{1}{4} \left(\mathbb{1}_2^{(12)} - \sigma_z^{(12)} + \mathbb{1}_2^{(34)} + U_I \sigma_y^{(34)} U_I^+ \right)
\end{aligned}
\tag{19.129}
$$

Let us inspect this result in the full 4x4 matrix. We have

$$
\rho_P(5) = \frac{1}{4} \begin{pmatrix} 0 & 0 & 0 & 0 \\ 0 & 2 & 0 & 0 \\ 0 & 0 & 1 & -ie^{-i\phi_I} \\ 0 & 0 & ie^{i\phi_I} & 1 \end{pmatrix}
\tag{19.130}
$$

Now, we apply a pulse on the (2-4) transition,

$$
\rho_P(6) = \frac{1}{4} U_S P_{24} U_S^+ \begin{pmatrix} 0 & 0 & 0 & 0 \\ 0 & 2 & 0 & 0 \\ 0 & 0 & 1 & -ie^{-i\phi_I} \\ 0 & 0 & ie^{i\phi_I} & 1 \end{pmatrix} U_S P_{24}^+ U_S^+
\tag{19.131}
$$

The inner conjugation of the U_S operators leaves the density matrix unchanged (U_S is diagonal in this basis). Rather than solving for this matrix, we move on to the next step and consider the expectation value we are seeking, namely, the measured quantity M given by,

$$
M = Tr \left(\sigma_x^{(24)} \rho_P(6) \right)
\tag{19.132}
$$

We can use cyclic permutations in the trace to write,

$$
\begin{aligned}
M &= Tr \left(\sigma_x^{(24)} \rho_P(6) \right) \\
&= \frac{1}{4} Tr \left(U_S P_{24}^+ U_S^+ \sigma_x^{24} U_S P_{24} U_S^+ \begin{pmatrix} 0 & 0 & 0 & 0 \\ 0 & 2 & 0 & 0 \\ 0 & 0 & 1 & -ie^{-i\phi_I} \\ 0 & 0 & ie^{i\phi_I} & 1 \end{pmatrix} \right)
\end{aligned}
\tag{19.133}
$$

Let us define,

$$
A = \begin{pmatrix} 0 & 0 & 0 & 0 \\ 0 & 2 & 0 & 0 \\ 0 & 0 & 1 & -ie^{-i\phi_I} \\ 0 & 0 & ie^{i\phi_I} & 1 \end{pmatrix}
\tag{19.134}
$$

The conjugation of U_S operators on $\sigma_x^{(24)}$ amounts to a rotation that we can readily write as,

$$
M = \frac{1}{4} Tr \left(U_S P_{(24)}^+ \left(\cos\phi_S \sigma_x^{(24)} + \sin\phi_S \sigma_y^{(24)} \right) P_{24} U_S^+ A \right)
\tag{19.135}
$$

The rotation defined by P_{24} is readily worked out as,

$$
M = \frac{1}{4} Tr \left(U_S \left(\cos\phi_S \sigma_y^{(24)} - \sin\phi_S \sigma_z^{(24)} \right) U_S^+ A \right)
\tag{19.136}
$$

And again, U_S defines a rotation around the z-axis, so we find,

$$
M = \frac{1}{4} Tr \left(\left(\cos\phi_S \left(\cos\phi_S \sigma_y^{(24)} + \sin\phi_S \sigma_x^{(24)} \right) - \sin\phi_S \sigma_z^{(24)} \right) A \right)
\tag{19.137}
$$

Thus, the simple pulse sequence used here to illustrate the tomography concept yields a signal proportional to $\sin\phi$. Refer to [1101] for a sequence that give a signal containing both ϕ_I and ϕ_S.

19.6   QUANTUM LOGIC GATES FOR SPINS

Let us examine a specific example where two spins are used in a logical operation like
a controlled-not (CNOT) gate. Spin 1 is a spin $1/2$. The eigenstates of the $\hat{I}_{z1}$ operator
represent the first input to the CNOT gate. Spin 2 is also a spin $1/2$. The eigenstates of
$\hat{I}_{z2}$ represent the control input of the CNOT gate. Because this example concerns logical
operations, the eigenvalues of $\hat{I}_{z1}$ and $\hat{I}_{z2}$ will be written as 0 or 1.

The CNOT gate is defined as follows. When the control input is 0, the gate does nothing
to the input, i.e., the output is equal to the input. When the control input is 1, then the
gate applies the NOT operation to the input line. We consider this logical gate as operating
on quantum states.[1111] The possible input states are $\{|00\rangle, |01\rangle, |10\rangle, |11\rangle\}$. We consider
these states as the basis to define the input and ouput states. The gate can be thought of
as an operator U_{CNOT} on the states defined in terms of this basis.

In this basis, the operator U_{CNOT} is given by,

$$\begin{pmatrix} 1 & 0 & 0 & 0 \\ 0 & 1 & 0 & 0 \\ 0 & 0 & 0 & 1 \\ 0 & 0 & 1 & 0 \end{pmatrix} \tag{19.138}$$

For example, when the input state is $|10\rangle$, the output state can be calculated as

$$U_{\text{CNOT}}|10\rangle = \begin{pmatrix} 1 & 0 & 0 & 0 \\ 0 & 1 & 0 & 0 \\ 0 & 0 & 0 & 1 \\ 0 & 0 & 1 & 0 \end{pmatrix} \begin{pmatrix} 0 \\ 0 \\ 1 \\ 0 \end{pmatrix} = \begin{pmatrix} 0 \\ 0 \\ 0 \\ 1 \end{pmatrix} = |11\rangle \tag{19.139}$$

Let us assume that the spin pair dynamics is given by the Hamiltonian,

$$\mathcal{H} = \hbar\omega_1\hat{I}_{z1} + \hbar\omega_2\hat{I}_{z2} + \hat{I}_1\mathsf{D}\hat{I}_2 \tag{19.140}$$

where D is a tensor that describes the coupling between the spins. We assume that it is
possible to rotate the spins selectively using resonant pulses. Earlier, when analyzing the
spin echo experiment, we interpreted the evolution given by (19.47) as the spin evolution
that derives from an effective Hamiltonian which resulted from the application of a non-
selective π-pulse. The challenge we face now is to find a sequence of pulses that produces
an effective Hamiltonian $\mathcal{H}_{\text{eff}}$ such that,

$$\exp\left(\frac{-i}{\hbar}\mathcal{H}_{\text{eff}}t\right) = U_{\text{CNOT}} \tag{19.141}$$

We find this pulse sequence by a method that is convenient to apply when considering more
complex gates.[1112] This method uses the concepts of geometric algebra.[1113, 1114] For
any Hilbert space, two idempotent (projection) operators E_+ and E_- are defined, such that,

$$E_+ + E_- = \mathbb{1} \quad E_\pm^2 = E_\pm \quad E_+E_- = \mathbf{0} \tag{19.142}$$

The property of these operators that we will use to solve our problem is the following,

$$\exp(AE_\pm) = e^A E_\pm + E_\mp \quad (\text{when } [A, E_\pm] = \mathbf{0}) \tag{19.143}$$

This can be seen by using the definition of the exponential as a series expansion. For
example,

$$e^{AE_+} = \mathbb{1} + AE_+ + \frac{1}{2!}A^2E_+^2 + \cdots = E_- + E_+\left(1 + A + \frac{1}{2!}A^2 + \cdots\right) = E_- + E_+e^A \tag{19.144}$$

For a system of two coupled spins 1/2, labelled 1 and 2, we have the following projection operators on the subspaces of one or the other eigenvectors of $\hat{I}_{z1}$ and $\hat{I}_{z2}$,

$$E_{+1} = \frac{1}{2}\mathbb{1}_1 + \hat{I}_{z1} \qquad E_{-1} = \frac{1}{2}\mathbb{1}_1 - \hat{I}_{z1}$$
$$E_{+2} = \frac{1}{2}\mathbb{1}_2 + \hat{I}_{z2} \qquad E_{-2} = \frac{1}{2}\mathbb{1}_2 - \hat{I}_{z2} \tag{19.145}$$

Here, the index indicates in which spin Hilbert space the operator acts. For example, in the basis $\{|0\rangle, |1\rangle\}$ of spin 1,

$$E_{+1} = \begin{pmatrix} 1 & 0 \\ 0 & 0 \end{pmatrix} \qquad E_{-1} = \begin{pmatrix} 0 & 0 \\ 0 & 1 \end{pmatrix} \tag{19.146}$$

The matrix U_{CNOT} in (19.138) can be written as,

$$U_{\text{CNOT}} = E_{+1} + 2\hat{I}_{x2}E_{-1} \tag{19.147}$$

The more accurate notation $E_{+1} \otimes \mathbb{1}_2$ is simplified, as is commonly done, to keep the upcoming expressions more legible. Likewise, $E_{-1} \otimes 2\hat{I}_{x2}$ is written $E_{-1}2\hat{I}_{x2} = 2\hat{I}_{x2}E_{-1}$. ($2\hat{I}_{x2}$ is simply the Pauli matrix σ_x for spin 2.) We recall that we are searching for a pulse sequence that transforms the Hamiltonian (19.140) into an effective Hamiltonian equivalent to U_{CNOT}.

In (19.147), we have a sum of operators. When we work out the effect of a pulse sequence, the time evolution is described by a product of operators. Following [1112], we use the properties (19.142) of idempotent operators to write,

$$U_{\text{CNOT}} = E_{+1}E_{+1} + \left(-i2\hat{I}_{x2}\right)(iE_{-1})$$
$$= E_{+1}E_{+1} + \left(-i2\hat{I}_{x2}\right)(iE_{-1}E_{-1}) + iE_{-1}E_{+1} - i2\hat{I}_{x2}E_{-1}E_{+1} \tag{19.148}$$
$$= (E_{+1} + iE_{-1})\left(E_{+1} - i2\hat{I}_{x2}E_{-1}\right)$$

This defines U_{CNOT} as a product of operators that have a structure similar to the linear combinations $m_x \pm im_y$ used in §17.4. Furthermore, we now want to make use of the property (19.143) of the exponential of idempotent operators. To do so, we use a property of the Pauli matrices,

$$e^{i\alpha\sigma_x} = \cos\alpha + i\sigma_x \sin\alpha \tag{19.149}$$

Hence,

$$e^{-i\pi\hat{I}_{x2}} = \cos\frac{\pi}{2} - i2\hat{I}_x \sin\frac{\pi}{2} = -i2\hat{I}_x \tag{19.150}$$

Therefore,

$$\left(E_{+1} - i2\hat{I}_{x2}E_{-1}\right) = \left(E_{+1} + e^{-i\pi\hat{I}_{x2}}E_{-1}\right) = e^{-i\pi\hat{I}_{x2}E_{-1}} \tag{19.151}$$

Likewise,

$$(E_{+1} + iE_{-1}) = \left(E_{+1} + e^{i\pi/2}E_{-1}\right) = e^{i(\pi/2)E_{-1}} \tag{19.152}$$

Applying the definition (19.145) of E_{-1} yields,

$$U_{\text{CNOT}} = e^{i\pi/4}e^{-i(\pi/2)\hat{I}_{z1}}e^{-i(\pi/2)\hat{I}_{x2}}e^{i\pi\hat{I}_{x2}\hat{I}_{z1}} \tag{19.153}$$

We have clearly progressed towards finding the pulse sequence that we are looking for given that this last expression of U_{CNOT} appears as a product of exponentials. There are still a few difficulties. The second exponential looks like the evolution of spin 1 in the applied field, but we cannot make spin 1 evolve in this field without also having spin 2 evolving in the applied field. Furthermore, the last term contains the product $\hat{I}_{x2}\hat{I}_{z1}$ whereas the Hamiltonian (19.140) is composed of the products $\hat{I}_{x1}\hat{I}_{x2}$, $\hat{I}_{y1}\hat{I}_{y2}$, and $\hat{I}_{y1}\hat{I}_{y2}$ when D is isotropic, which is often the case.

Let us continue our search for the pulse sequence under the assumption that the coupled spins are nuclear spins in a large magnetic field. In this case, the Hamiltonian (19.140) can be approximated by keeping only the secular term. Thus, using $D = J\hbar$, we will from now on consider,

$$\mathcal{H} = \hbar\omega_1\hat{I}_{z1} + \hbar\omega_2\hat{I}_{z2} + J\hbar^2\hat{I}_{1z}\hat{I}_{2z} \tag{19.154}$$

If we were instead considering electron spins, this approximation would not be possible. We use it here to keep the algebra as simple as possible and refer to [1112] for that more complex case.

In view of the form of the coupling term in (19.154), we would like to transform the product $\hat{I}_{x2}\hat{I}_{z1}$ appearing in (19.153) into something that contains a product $\hat{I}_{z2}\hat{I}_{z1}$. To achieve this, we need to make the following observations. The term $e^{-i(\pi/2)\hat{I}_{x2}}$ in (19.153) corresponds to a selective $\pi/2$ pulse along the x-axis. It is easy to implement, to the extent that one can apply a selective pulse during which the coupling term can be neglected. To the contrary, the term $e^{-i(\pi/2)\hat{I}_{z1}}$ is problematic. We want to transform it into an exponential that contains $e^{-i(\pi/2)\hat{I}_{x1}}$ or $e^{-i(\pi/2)\hat{I}_{y1}}$.

We can transform the expression (19.153) of U_{CNOT} by making use of the rotation properties expressed in (19.22) and (19.27). From (19.22), we get,

$$\hat{I}_x = e^{-i(\pi/2)\hat{I}_y}\hat{I}_z e^{i(\pi/2)\hat{I}_y} \tag{19.155}$$

$$\hat{I}_z = e^{-i(\pi/2)\hat{I}_x}\hat{I}_y e^{i(\pi/2)\hat{I}_x} \tag{19.156}$$

Applying (19.27), we can then write,

$$e^{i\alpha\hat{I}_x} = e^{-i(\pi/2)\hat{I}_y}e^{i\alpha\hat{I}_z}e^{i(\pi/2)\hat{I}_y} \tag{19.157}$$

$$e^{i\beta\hat{I}_z} = e^{-i(\pi/2)\hat{I}_x}e^{i\beta\hat{I}_y}e^{i(\pi/2)\hat{I}_x} \tag{19.158}$$

As we will see, the evolution under the Hamiltonian (19.154) will provide us an evolution given by $e^{-i\pi(J\hat{I}_{z1}\hat{I}_{z2})}$. We can obtain the correct sign by a rotation of π along the x-axis of the operators in (19.157). Thus we have,

$$e^{i\alpha\hat{I}_x} = e^{i(\pi/2)\hat{I}_y}e^{-i\alpha\hat{I}_z}e^{-i(\pi/2)\hat{I}_y} \tag{19.159}$$

We can apply this to our two-spin system, making use of the fact that the spin operators act on different Hilbert spaces. Hence, we can apply (19.159) to the two-spin system as follows,

$$e^{i\alpha\hat{I}_{x2}\hat{I}_{z1}} = e^{i(\pi/2)\hat{I}_{y2}}e^{-i\alpha\hat{I}_{z2}\hat{I}_{z1}}e^{-i(\pi/2)\hat{I}_{y2}} \tag{19.160}$$

In (19.153), we remove the phase factor $exp(i\pi/4)$; we also replace the $\hat{I}_{z1}$ term using (19.158) and the $\hat{I}_{x2}$ term using (19.160). Thus, we write,

$$U_{\mathrm{CNOT}} = e^{-i(\pi/2)\hat{I}_{x2}}\ e^{-i(\pi/2)\hat{I}_{x1}}e^{-i(\pi/2)\hat{I}_{y1}}e^{i(\pi/2)\hat{I}_{x1}}\ e^{i(\pi/2)\hat{I}_{y2}}e^{-i\pi\hat{I}_{z2}\hat{I}_{z1}}e^{-i(\pi/2)\hat{I}_{y2}}$$
$$\tag{19.161}$$

We have now expressed U_{CNOT} in terms of several selective rotations that can be implemented. There is still one term that is not straightforward. It relies on the spin coupling, but the full Hamiltonian (19.154) contains Zeeman terms. We can use a spin echo strategy to cancel out the Zeeman terms. A non-selective π pulse, i.e., one acting on both nuclei, will reverse the signs of the Zeeman terms, but not the sign of the coupling term. That is to say, we can let the spins evolve for a time $\tau/2$ with $\tau J = \pi$, apply the π pulse, and let the spins evolve again for a time $\tau/2$. Were we to apply another π pulse, we would get an evolution during τ that is equivalent to $e^{i\pi \hat{I}_{z2}\hat{I}_{z1}}$. Indeed, we have,

$$e^{i\pi(\hat{I}_{z1}+\hat{I}_{z2})}e^{-i(\omega_1\hat{I}_{z1}+\omega_2\hat{I}_z2+J\hat{I}_{z1}\hat{I}_{z2})\tau/2}e^{-i\pi(\hat{I}_{z1}+\hat{I}_{z2})}\ e^{-i(\omega_1\hat{I}_{z1}+\omega_2\hat{I}_{z2}+J\hat{I}_{z1}\hat{I}_{z2})\tau/2}$$

$$= e^{-i(\omega_1(-\hat{I}_{z1})+\omega_2(-\hat{I}_{z2})+J\hat{I}_{z1}\hat{I}_{z2})\tau/2}e^{-i(\omega_1\hat{I}_{z1}+\omega_2\hat{I}_{z2}+J\hat{I}_{z1}\hat{I}_{z2})\tau/2} \qquad (19.162)$$

$$= e^{-i\pi(J\hat{I}_{z1}\hat{I}_{z2})}$$

With this last consideration, we now have the pulse sequence that we were looking for. We can read it off (19.161), starting from right to left, taking into account (19.162):

$$90(y2);\text{evolve}(\tau/2);180(x,\text{hard});\text{evolve}(\tau/2);-180(x,\text{hard});$$
$$-90(y2);-90(x1);90(y1);90(x,\text{hard}) \qquad (19.163)$$

The last expression includes a hard $\pi/2$ pulse at the end instead of two juxtaposed selective pulses.

When electron spins are used, considering only the secular form (19.154) is not sufficiently precise. Other terms must be taken into account, and the Hamiltonian reads,

$$\mathcal{H} = \hbar\omega_1\hat{I}_{z1} + \hbar\omega_2\hat{I}_{z2} + J\hbar\hat{I}_{1z}\hat{I}_{2z} + B\hbar\left(\hat{I}_{x1}\hat{I}_{x2} + \hat{I}_{y1}\hat{I}_{y2}\right) \qquad (19.164)$$

If exchange interaction were to dominate, then B would equal J. If dipolar interactions were to play a role, this would no longer be the case. Volkov and Salikhov worked out the pulse sequence in cases where the Hamiltonian has the form (19.164).[1112]

Experimental verification

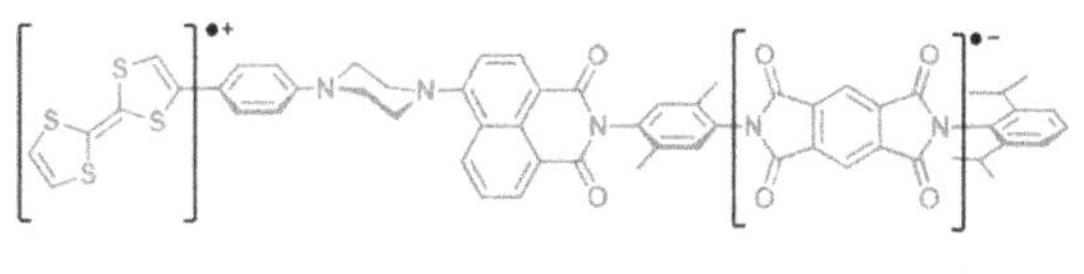

Figure 19.9 Donor tetrathiafulvalene (TTF), chromophore 8-aminonaphthalene-1,8-dicarboximide (ANI), and acceptor pyromellitimide (PI). From [1115]

The prediction (19.163) that is valid for a spin Hamiltonian of the form (19.154) has been demonstrated experimentally.[1113] In that work, ^{13}C-enriched chloroform was used. The spin coupling was due to the dipolar coupling between the protons and the ^{13}C nuclei. The selective pulses were straightforward to implement since the Larmor frequencies of these two nuclear spins are widely different.

Implementation of a CNOT gate using electron spins was also demonstrated.[1116,1117] A radical pair was photogenerated in a molecule that linked together a donor, a chromophore and an acceptor (Fig.19.9).[1115] The phase memory time was 1.8 μs at 85 K. This encouraging result shows that a simple form of supramolecular chemistry can contribute to the implementation of spin-based logic gates.[1118]

19.7 FURTHER READINGS

Rabi oscillations

In NMR, it is possible to follow the spin nutation[2] induced by a resonant radio-frequency pulse. By following the time evolution of this nutation, it is possible to determine a bond length with high precision.[1119] This technique, called ***nutation spectroscopy*** , was proposed as a means to simplify the interpretation of electron spin resonance spectra.[1120] A very simple application of nutation experiments is to measure the inhomogeneity of the applied radio-frequency field. Such inhomogeneity can cause spurious effects that can be avoided by implementing nutation-frequency selective pulses.[1121]

Any two-level system can be mapped onto the up and down states of a fictitious spin 1/2 in a magnetic field. As a consequence, we also speak of Rabi oscillations when dealing with electron charge systems. For example, coherences between the two low-lying states of donor impurities in quantum dots were excited by THz radiation, i.e., rectified laser pulses with a duration of less than 40 ps, and detected by photoconductivity measurements.[1122, 1123]

Very long spin-lattice relaxation times

Nuclei are shielded from the outside world by the electron clouds that surround them. Hence, they tend to have rather long relaxation times, especially in solids at low temperatures, where dipolar coupling fluctuations vanish. This property of nuclei is advantageous for the long-term storage of information in the form of spin orientation, but disadvantageous for writing information quickly. One possible way to overcome this limitation is to use unpaired electron spins, that provide a strong relaxation mechanism for protons, ^{1}H, and hyperpolarize the nuclei by DNP (see §16.3). Proton-proton dipolar couplings are rather large, so spin diffusion is relatively efficient among protons. Far away from the electron spin, this high proton-spin polarization can be transferred to ^{13}C nuclei via cross-polarization. In the absence of resonant irradiation, the ^{1}H and ^{13}C are not on speaking terms, in the sense of (12.160). Since they are far apart, the ^{13}C do not couple to one another either; hence, their relaxation is very slow. Using this strategy, a ^{13}C relaxation time of 16 hours was obtained at 4.2K in a field of only 1 T.[1124]

^{31}P in silicon is a candidate spin-based qubit of interest because of its favorable relaxation properties.[1125] It was also shown to have a long spin lifetime, about 6 s at 300 mK in a field of 1.5T.[1126]

In creating coherences among electron spins, fluctuations of the nuclear spins are the main mechanism that shortens the coherences.[1127] Coordination chemistry strategies may help create specific conformations of organometallic molecular structures that decrease nuclear spin fluctuations.[1128]

Long-lived states, long-lived coherences

In §15.3, we defined the spin-lattice relaxation T_1. There are many circumstances when a long T_1 is desirable. For a given system, how can we store spin polarization for a time longer than T_1? The answer is to create what is now called ***long-lived states*** (LLS). This method uses the spin dynamics tool of magnetic resonance spectroscopy to create spin singlets.[1129, 1130] The point of the technique is that the dipolar coupling among spins is often a dominant relaxation mechanism in liquids and that, for symmetry reasons, the singlet is protected against dipolar field fluctuations. With this technique, relaxation on a time scale of $50T_1$ was obtained.[1131]

---

[2]rotation away from the z-axis seen in the rotating frame

LLS have several potential applications. The relaxation of the LLS can reveal subtleties of the spin dynamics of the system under examination.[1132] Generating LLS can be used to store spin polarization until it is put to use in a magnetic resonance spectroscopy.[1133] Coherences need to be transformed into detectable signals. This requires spin manipulations that have long been known to the NMR spectroscopy field.[1134] NMR can be used to store and retrieve information at specific locations, a technique some authors have coined *NMR photography*.[1135, 1136] This strategy that consists in generating LLS was applied to phosphorous impurities in silicon that could be selectively addressed.[1137]

In quantum sensing, quantum computing and dynamic nuclear polarization (Chapter 16), one is generally called to create superpositions of spin states with long coherence times. It is possible to excite long-lived coherences that can be transformed into measurable signals in dipolar-coupled nuclear systems, even when the spectra are unresolved.[1138] In the more common situation of spin pairs, both the singlet and the $m = 0$ triplet states are notorious in NMR for their insensitivity to field inhomogeneities and fluctuations of intramolecular dipolar couplings.[1139–1141] Thus, these states are referred to as *long lived coherences*. By using multiple pulse sequences, very long relaxation processes have been obtained that were free of modulations caused by chemical shift differences (i.e., atomically local field differences due to shielding by electrons).[1142]

Spin qubit coherence

When using spin-based qubits for quantum computing, coherence time is a limiting factor referred to by some as the *spin bath decoherence*. One of the mechanisms causing decoherence is the fluctuation of the dipolar couplings among spins. Thus, specially engineered materials have been devised, such as isotopically enriched diamond whose *NV centers* are used as qubits.[1143]. Nuclear coherence times in the order of minutes have been obtained.[1144]

Using isotopically enriched silicon, the ^{31}P spin of phosphorous nuclei coherence time reached more than 30 second when the temperature was lowered to 100 mK.[1145] This very long coherence time was produced by means of a refocusing pulse sequence, the so-called Car-Purcell-Meiboom-Gill sequence (CPMG). This pulse sequence consists in a series of carefully phased echo pulses. The phosphorus center holds an unpaired electron that was found to lose coherence in about 0.5 s.[1145] In a quantum dot made with isotopically enriched silicon, a coherence time of 28 ms was observed in a CMOS-based architecture.[1146]

Very long electronic spin lifetimes were observed in *silicon carbide*.[1147] The divacancies in SiC are paramagnetic defects that also present long coherence times.[1148] Likewise, in hexagonal boron nitride, defects were found with relaxation times T_1 of 18 μs and T_2 of 2 μs at room temperature.[1149] These relaxation times increased by orders of magnitude at cryogenic temperatures. Europium dopant in yttrium orthosilicate was found, by optically detected magnetic resonance, to have a coherence time of about 6 hours at 2K![680]

It is not enough to have long coherence times. It is also necessary to be able to control a qubit on a short time scale. Spin-orbit coupling allows for fast spin-qubit control. The spin state of a hole trapped at a boron dopant in a ^{28}Si offers an interesting trade-off in that a coherence time of 10 ms was achieved when a CPMG sequence was applied. The qubit control can be fast, in part because it can be acted on electrically.[1150] Using high-spin nuclei in isotopically enriched silicon allows for the resonant electric excitation of spin qubits.[846] The fast control over a nuclear spin can be ensured via its coupling to an electron spin.[1151]

Coherences can also be manipulated in the case of adatom spins. For example, a pair of titanium atoms were brought in proximity, to the point that they were exchange coupled. The atoms position were specific, owing to the surface lattice constant. The coupling was small compared to the magnetic field difference, so the eigenstates were approximately those of free individual spins. Selective excitation by a scanning tunneling microscope (STM) tip located over one atom generated coherences. As the resonance condition depended on the spin state of the other adatom, this operation was equivalent to a CNOT gate.[1152]

Quantum computing will need multi-qubit devices, which are presently under development.[1153, 1154] Coherences among 10 spins were demonstrated in an implementation comprising an NV-center (one electron spin and one ^{14}N nucleus) and 8 ^{13}C nuclei (Fig. 19.10).[1155] This structure constitutes a qubit register in which 45 coherences can be generated, with a relaxation time of 10 s. Alternatively, the register can be used as memory device, in which the isolated qubit state holds for 75 s at 3.7 K. The issue of fidelity comes to the fore in multi-qubit devices and must be specified along with the coherence time.[1156]

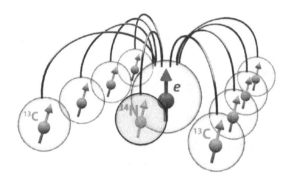

Figure 19.10 The electron spin of a single NV center in diamond acts as a central qubit connected by a two-qubit dynamical control scheme to the intrinsic ^{14}N nuclear spin and 8 ^{13}C nuclear spins surrounding the NV center. Adapted from [1155].

Spin-qubit architecture

Quantum dots are one possible structure for spin-based qubits.[681] Coherent control of coupled electron spins in a semiconductor quantum dot was demonstrated.[1157] A Si/SiGe quantum dot was controlled electrically owing to an artificial spin-orbit field generated with a small magnet.[1158] A Pauli spin-blockade strategy was implemented for the quantum dot readout.[1159]

Single electron transistors (SET) can also be operated as spin qubits. For example, a ^{31}P nucleus located at a SET has a long coherence time because of the spin-orbit coupling is weak in this case. Also, the phosphorous nucleus environment is like a vacuum if the the matrix is isotopically enriched.[686] Alternatively, a SET can drive the Rabi oscillation of ^{31}P located on it.[1160] 97% fidelity in the readout of an SET state was achieved in a measurement time of 1.5 μs.[1161]

Nanowires have been evaluated for their potential use as registers of spin qubits. Pairs of quantum dots in semiconducting nanowires were formed with individual electrical controls.[1162] The spin-orbit coupling in indium antimonide (InSb) is favorable for an electric control of spin qubits.[689] The issue of scalability was addressed by forming a line of quantum dots.[1163]

Supramolecular structure may offer opportunities for making spin qubits.[1164] Indeed, the combination of organic spintronics and spin chemistry may offer valuable opportunities for spin qubits. For example, electron spin polarization transfer from photogenerated radical pairs to a stable radical playing the role of observer could possibly be implemented.[1165] Zero-quantum coherence was studied in a series of covalent spin-correlated pairs.[1166] Operation of a CNOT gate was demonstrated using photogenerated molecular electron-spin qubit pairs.[1116] Entangled spin states were formed in calcium phosphate.[1167] The spin state of purely organic radicals located on a patterned surface can be used as a storage unit controlled by light, magnetic or electric fields.[1168] Room temperature coherences in a molecular qubit were demonstrated.[689] Nuclear coherence time was investigated systematically for a family of complexes in solutions and as dopants in solids.[1169]

CMOS is a highly desirable avenue because it draws on the massive experience and development of a mature technology. The elementary building block is a source-drain channel on which two gates can be formed, thus forming a double quantum dot.[1170, 1171] In view of the CMOS planar technology, the notion of surface codes becomes quite relevant.[1146, 1172] This idea can be traced back to the work of Kitaev on the topological properties of two-dimensional spin arrays.[1173, 1174]

M.Y. Simmons, one of the founding members of the "Centre of Excellence for Quantum Computation and Communication Technology" and founder of Silicon Quantum Computing, the first quantum computing company in Australia, is spearheading the research and development of arrays of addressable spin qubits based on phosphorous atoms embedded in the surface of crystalline silicon.[1175–1178]

List of Abbreviations

AHE: Anomalous Hall effect.
AFMR: Antiferromagnetic resonance.
AMR: Anisotropic magnetoresistance.
ANE: Anomalous Nernst Effect.
AP: Antiparallel (in reference to magnetic configurations).
BEC: Bose-Einstein condensation.
CAP-GMR: Current-at-an-angle magnetoresistance.
CIMS: Current-induced magnetization switching.
CIP-GMR: current-in-plane magnetoresistance.
CISS: Chirality-induced spin selectivity.
CMR: Colossal magnetoresistance.
CMOS: Complementary metal-oxide-semiconductor.
CNOT: Controlled NOT (logical operation) gate.
CPMG: Carr, Purcell, Meiboom, Gill (pulse sequence).
CPP-GMR: current-perpendicular-to-the-plane magnetoresistance.
CW: Continuous wave (method).
DE: Double exchange
DM: Dzyaloshinskii-Moriya (coupling or interaction).
DMI: Dzyaloshinskii-Moriya interaction.
DNP: Dynamic nuclear polarization.
DOS: Density of states.
EDMR: Electrically-detected magnetic resonance.
EPR: Electron paramagnetic resonance.
ESR: Electron spin resonance.
FM: Ferromagnetic.
FMR: Ferromagnetic resonance.
FNR: Ferromagnetic nuclear resonance.
GMR: Giant magnetoresistance.
HH: Heavy hole.
ISHE: Inverse spin Hall effect
HMFM: Half-metallic ferromagnets.
LH: Light hole.
LH: Left handed.
LLG: Landau-Lifshitz-Gilbert (equation).
LSMO: Lanthanum strontium manganese oxide.
MR: magnetoresistance.
MRAM: Magnetoresistive random access memories.
MTEP: Magneto-thermopower.
NMR: Nuclear magnetic resonance.
NOE: Nuclear Overhauser effect.
NOVEL: Nuclear orientation via electron spin locking
NV: Nitrogen vacancy.
ODMR: Optically-detected magnetic resonance.
OHE: Orbital Hall effect.
OPW: Orthogonal plane wave (method).
P: Parallel (in reference to magnetic configurations).

PHIP: Parahydrogen-induced polarization.
PMA: Perpendicular magnetic anisotropy.
QEC: Quantum error correction.
rf: Radio frequency.
RH: Right-handed.
RKKY: Ruderman, Kittel, Kasuya, Yosida (coupling).
SET: Single electron transistor.
SHE: Spin Hall effect.
SOT: Spin-orbit torque.
STM: Scanning tunneling microscope.
STO: Spin-torque oscillator.
TESR: Transmission electron spin resonance.
TMR: Tunnel magnetoresistance.
TREPR: Time-resolved electron paramagnetic resonance.
VNA: Vector network analyzer.
YIG: Yttrium iron garnet.

Glossary

The present glossary takes the alternative approach of defining some terminology in magnetism and spintronics without the rigor of mathematical language, in the hope that these loose definitions give an immediate idea of the concepts being described. A classification in broad categories is proposed, instead of relying on mere alphabetical order. Use the index at the end of this book to find the pages where terms are more formally defined.

Magnetic structures: The magnetic configurations listed here usually define classes of magnetic materials. The possibility of special spin textures is also included.

Antiferromagnet: This large class of materials is characterized by two sublattices of opposing local magnetic moments. Materials are also referred to as antiferromagnetic, where the sublattice magnetic moments do not cancel each other exactly. Most often, the Dzyaloshinskii-Moriya (DM) interaction is responsible for this slight deviation, and we speak of spin canting to refer to this misalignment of opposite moments of nearest neighbors. Thus, the proper term for these materials is canted antiferromagnets.

Ferromagnet: In ferromagnetic materials, in a strict sense, the local magnetic moments are all aligned in the same direction. When speaking of ferromagnets and antiferromagnets, the meaning of the term ferromagnet may include ferrimagnets as well.

Ferrimagnet: In ferrimagnetic materials, the local moments are aligned, but some point in opposite direction to others and the sum of these moments in a unit cell is non-zero.

Helimagnet: Local moments may be subject to a combination of ferromagnetic and antiferromagnetic interactions, resulting in equilibrium configurations that are more complex than a mere alignment or anti-alignment. In a planar helimagnet, the local moments are rotated in a plane compared to their orientation in a nearest parallel plane. In a cycloidal magnet, the local moments of atomic sites located in some crystalline axis form a cone.

Vortices, skyrmions, spin cycloid, merons, hopfions: materials may host stable magnetic configurations extending over a small scale (from ten to hundreds of nanometers), having closed magnetic configurations with specific topological properties. Vortices can be stabilized in a ferromagnet, owing to the shape of the sample. Skyrmions are generally intrinsic to the material, in the sense that a thermodynamic phase can be identified. These topological spin textures have been studied extensively, in part because of their potential applications in information storage. The local spin alignment may be antiferromagnetic locally, but the orientation of the nearest neighbors may rotate over large distances, giving rise to a spin cycloid. More recently, other topological structures have been studied, such as merons, bimerons, and hopfions.

Weak ferromagnet: Spin canting in an antiferromagnet (see "antiferromagnet") is also referred to as weak ferromagnetism. In the case of itinerant, band ferromagnetism, if one spin band is pushed completely below the Fermi energy level, then the term "strong ferromagnet" is used for this material; otherwise, we call the material a "weak ferromagnet".

Magnetization properties: Below are some common features of magnetization in static or quasi-static measurements, *i.e.*, measurements where the applied magnetic field is changed on a much longer time scale than that of magnetization dynamics.

> **Coercive field:** In a material characterized by a hysteresis loop, the coercive field is the applied magnetic field necessary to bring the magnetization to zero. In a nanostructure, the magnetization is likely to switch between two states of opposite magnetization orientations, in that case, we speak of switching field.
>
> **Domain wall:** In bulk magnetic materials, domains with uniform magnetization are observed. Domain walls form between domains. In a Bloch wall, the magnetization rotates around an axis normal to the wall. In a Néel wall, the magnetization rotates in the plane of the wall.
>
> **Exchange biasing:** When an antiferromagnetic material is deposited on top of a ferromagnetic material, the resulting hysteresis is shifted on the applied field axis by an amount called the exchange bias field. This effect is used, for example, in magnetoresistive sensors. This bias mechanism is due to magnetic couplings at the interface between the ferromagnet and the antiferromagnet. The presence of domains in the antiferromagnet plays a role. There are relaxation effects involved, in the sense that running several hysteresis measurements may change the value of the bias field. Temperature also plays a critical role.
>
> **Perpendicular magnetic anisotropy:** In general, the magnetization in a ferromagnetic thin film lies in the plane of the film. But this is not always the case. When the film is covered by oxides or a heavy metal like platinum, the equilibrium magnetization may be perpendicular to the film.
>
> **Proximity effect:** Materials that are non-magnetic but have an electronic structure that suggests they are nearly magnetic may be suspected of turning magnetic when coupled to a ferromagnetic material. It is difficult to evaluate if this is the case because one of the two materials at the interface gives a strong magnetic response, while the proximity effect on the other material is expected to extend only over a few atomic layers.
>
> **Switching field:** When a magnetic material has a hysteresis loop that contains jumps, the field at the jump is called a switching field. At each execution of the hysteresis loop, the value of the switching field might vary. This jump is better characterized by a statistical distribution of the switching field values.

Magnetic resonances and methods: Magnetic resonance spectroscopy may refer to various types of experimental methods depending on what resonates – hence the long list of spectroscopies that follows. In the correlated state of a ferromagnet or an antiferromagnet, it is also possible to excite magnetization waves, commonly called spin waves. Spin waves can be coupled to acoustic waves or to the electromagnetic modes of a cavity.

> **Antiferromagnetic resonance (AFMR):** In an antiferromagnet, the local moments in each sublattice can undergo a magnetic resonance. Unless a magnetic field of several tesla is applied, the main field experienced by the moments in each sublattice is the exchange field corresponding to the nearest-neighbor moments of the other sublattice. Thus, resonances in the range of hundreds of gigahertz are observed without any need for superconducting magnets, as would be the case if dealing with isolated electron spins.

Electron paramagnetic resonance (EPR): The resonance of isolated electron spins in an applied magnetic field is called EPR and sometimes electron spin resonance (ESR). Evidently, EPR is the term used when studying paramagnetic centers, *i.e.*, localized and isolated moments, whereas ESR is the preferred term when exciting the spin resonance of conduction electrons. For several decades, EPR was performed in a continuous wave mode (CW), which is still the method of choice at the higher frequency range (hundreds of gigahertz). Pulsed methods, similar to those routinely used in NMR, are also used for EPR at frequencies below a hundred gigahertz.

Ferromagnetic resonance (FMR): In a ferromagnet, the local magnetic moments are strongly coupled. Exciting magnetic resonance in this case amounts to a collective excitation. One important difference with the resonance of isolated magnetic moments is that the effective magnetic field strength, i.e., the field perceived by each local moment, depends on the orientation of the magnetic moment with respect to the sample and the crystalline axes.

Ferromagnetic nuclear resonance (FNR): This term refers to NMR experiments carried out in ferromagnetic materials. FNR does not require the application of a magnetic field, because the nuclear spins experience the hyperfine field associated with the correlated spins of the electrons.

Nuclear magnetic resonance (NMR): The nuclei of some atoms in the periodic table have a spin. The spin depends on the isotope. Spin one-half nuclei have zero quadrupolar moment, which simplifies the spectrum. In solutions, spectral resolution can reach the part-per-billion (ppb) level using the most advanced spectrometers. For some nuclei, having a magnetic field homogeneity of the order of 10 parts-per-million (ppm) is sufficient. Nuclear spins greater than $1/2$ have a quadrupole moment that couples to the electric field gradients generated by the ions near each spins. The resonance can be driven by applying a magnetic field that couples to the spin at the resonance frequency of the quadrupole moment in the electric field gradients.

Polaritons, polarons: Magnetic polaritons are couples modes of the electromagnetic field and the magnetization dynamics (of a ferromagnet or an antiferromagnet). Thus, they refer to coupled photons and magnons. Magnetic polarons are couples modes involving phonons and magnetization dynamics, in other words, phonons and magnons.

Spin-wave spectroscopy: Spin waves are collective low-lying excitations observed in both ferromagnets and antiferromagnets. In practice, most studies are done in the long-wavelength limit, where exchange energy has practically no influence on the excitation frequency. Only dipolar energies are relevant in this case. The influence of the DM interaction on the dispersion relation has been investigated in ferromagnets and antiferromagnets.

Unusual detection techniques: Optically detected magnetic resonance (ODMR) is highly sensitive. This is particularly the case for the nitrogen-vacancy (NV) centers of diamond. This resonance is so sharp that it can be used to detect spin dynamics in nearby objects, nanostructures, or large molecules. Electrically detected magnetic resonance (EDMR) can detect ESR or FMR of very small samples, provided the samples are electrically connected with leads capable of transmitting high-frequency signals. The spin-diode effect occurs whenever a device has a non-linear electrical current response to an electrostatic potential gradient, owing to any form of magnetoresistance. NMR can be detected via its effect on the EPR of electron spins coupled to nuclear spins (electron-nuclear double resonance or ENDOR). Conversely, electron spin resonance spectra can be obtained in dynamic nuclear polarization (DNP) experiments where the nuclear spin signal amplitude is detected. Scanning-probe microscopy (STM) techniques can be used to detect single, isolated spin resonances.

Spin currents and spin torques: A spin current is a current density of spin polarization, so it contains the information of a current density (a three-dimensional vector) and information about the spin polarization or angular momentum (another three-dimensional property). So, in general, a current density is a tensorial quantity. A spin current in a magnetic layer exerts a torque. Hence, we group here spin currents and spin torques.

Spin transfer torque (charge driven): Specifying that a spin current is charge-driven is usually done only when there might be confusion with, *e.g.*, a heat-driven spin current. A typical spin current can be found in metallic, magnetic nanostructures such as spin valves. In a ferromagnet, there are charge currents associated with majority and minority spins. When there is a privileged spatial direction for the spin polarization, it is possible to describe spin currents with vectors, and the spin current is the difference between the two spin-dependent currents. This spin current acts on the magnetization. Its effect is called a spin transfer torque.

Heat-driven spin transfer torque (conductors): A heat current driven through a conducting, magnetic material gives rise to a heat-driven spin current in bulk materials. As in the case of a charge-driven spin current, there is also a heat-driven spin current in metallic, magnetic nanostructures (such as spin valves) that gives rise to a heat-driven spin transfer torque acting on the magnetization.

Heat-driven spin current (insulators): A temperature gradient imposed on magnetic insulators induces a heat-driven spin current in the bulk of the material.

Spin Seebeck effect: This expression refers to a spin current induced by a temperature gradient in a magnetic material. The material can be an insulator or a conductor. This effect is not to be confused with the spin-dependence of the Seebeck coefficients of majority and minority spins.

Spin-wave spin current: When spin waves are excited in a magnetic material, a spin current is also induced.

Spin pumping: When magnetic resonance is excited in a ferromagnet or an antiferromagnet in the proximity of a heavy-metal, non-magnetic contact, a spin current flows from the magnetic material into the contact. This current is usually detected by the inverse spin Hall effect.

Spin-orbit torque: When a current is driven in-plane through a thin film of a heavy metal, *i.e.*, a metal in which spin-orbit effects play an important role, a spin current is induced perpendicular to the charge current. The spin polarization that this spin current carries is perpendicular to both the charge current and the spin current flow direction.

Transport phenomena: The following list concerns a variety of transport phenomena. When a transport phenomenon is spin dependent, the definitions below extend to the respective spin-dependent cases. Care must be taken to define these spin-dependent transport coefficients precisely, specifying their contributions to the effective experimental quantity that is measured.

Ettingshausen effect: A temperature gradient is observed across a strip of conductor when a charge current is driven along the strip.

Hall effect: A voltage is observed across a conducting strip when a current is driven along the strip.

Ohm's law: A current density proportional to an electrostatic potential gradient is observed under isothermal or adiabatic conditions.

Nernst effect: A voltage is observed across a strip of conductor when a heat current is driven along the strip.

Righi-Leduc effect: A temperature gradient is observed across a strip of conductor when a heat current is driven along the strip.

Seebeck effect: A voltage is observed at the ends of a strip of conductor when a temperature difference is imposed between its ends. A thermodynamic expression of the Seebeck effect shows that it is a volume effect. Confusion may arise from the fact that the path integral of a temperature gradient only depends on the values of the temperatures at both ends of the path.

Soret effect: In fluid containing two types of substances that do not react with one another, typically in suspension of two types of particles, the Soret effect refers to the difference between the chemical potentials of these two substances at the ends of the container holding the fluid.

Thermophoresis: In a fluid containing two types of substances that do not react with one another, typically in a suspension of two types of particles, a temperature gradient induces a diffusive current of one substance with respect to the other. If the fluid is confined and allowed to reach an equilibrium, thermophoresis leads to the Soret effect.

References

1. M. Baibich, J.M. Broto, A. Fert, F. N. Van Dau, F. Petroff, P. Etienne, G. Creuzet, A. Friederich, and J. Chazelas. Giant Magnetoresistance of (001)Fe/(001)Cr Magnetic Superlattices. *Phys. Rev. Lett.*, 61:2472–2475, 1988.

2. G. Binasch, P. Grünberg, F. Saurenbach, and W. Zinn. Enhanced magnetoresistance in layered magnetic structures with antiferromagnetic interlayer exchange. *Phys. Rev. B*, 39:4828–4830(R), 1989.

3. J.S. Moodera, L. R. Kinder, T. M. Wong, and R. Meservey. Large magnetoresistance at room temperature in ferromagnetic thin film tunnel junctions. *Phys. Rev. Lett.*, 74:3273–3276, 1995.

4. S. Tehrani, J.M. Slaughter, E. Chen, M. Durlam, J. Shi, and M. DeHerren. Progress and Outlook for MRAM Technology. *IEEE Trans. Mag.*, 35:2814–2819, 2000.

5. J. A. Katine, F. J. Albert, R. A. Buhrman, E. B. Myers, and D. C. Ralph. Current-Driven Magnetization Reversal and Spin-Wave Excitations in Co/Cu/Co Pillars. *Phys. Rev. Lett.*, 84:3149–3152, 2000.

6. J.-E. Wegrowe, D. Kelly, Y. Jaccard, Ph. Guittienne, and J.-Ph. Ansermet. Current-induced magnetization reversal in magnetic nanowires. *Europhys. Lett.*, 45(5):626–632, 1999.

7. J. Z. Sun. Current-induced magnetic switching device and memory including the same. IBM J. Res. & Dev., 50(1):81–100, 2006.

8. J.-E. Wegrowe, J.-Ph. Ansermet, and S. E. Gilbert. Non-volatile magnetic random access memory. Patent US6172902 (B1), EPFL, 2001.

9. S. I. Kiselev, J. C. Sankey, I. N. Krivorotov, N. C. Emley, R. J. Schoelkopf, R. A. Buhrman, and D.C. Ralph. Microwave oscillations of a nanomagnet driven by a spin-polarized current. *Nature*, 425:380–383, 2003.

10. W. Meiklejohn and C.P. Bean. New Magnetic Anisotropy. *Phys. Rev.*, 105(3):904–913, 1957.

11. J. Noguès and I. K. Schuller. Exchange bias. *J. Magn. Magn. Mater.*, 192:203–232, 1999.

12. M. Kiwi. Exchange bias theory. *J. Magn. Magn. Mater.*, 234(3):584–595, 2001.

13. J. Q. Xiao, J. S. Jiang, and C. L. Chien. Giant magnetoresistance in nonmultilayer magnetic systems. *Phys. Rev. Lett.*, 68:3749–3752, 1992.

14. A. E. Berkowitz, J. R. Mitchell, M. J. Carey, A. P. Young, S. Zhang, F. E. Spada, F. T. Parker, A. Hutten, and G. Thomas. Giant magnetoresistance in heterogeneous Cu-Co alloys. *Phys. Rev. Lett.*, 68:3745–3748, 1992.

15. P. Holody, L. B. Steren, R. Morel, A. Fert, R. Loloee, and P. A. Schroeder. Giant magnetoresistance in hybrid magnetic nanostructures including both layers and clusters. *Phys. Rev. B*, 50:12999, 1994.

16. B. Doudin and J.-Ph. Ansermet. Nanostructuring Materials for Spin Electronics. *Europhys. News*, 28:14–17, 1997.

17. M.A.M. Gijs and G.E.W. Bauer. Perpendicular Giant Magnetoresistance of Magnetic Multilayers. *Adv. Phys.*, 46:285–445, 1997.

18. W. P. Pratt Jr., S.-F. Lee, J. M. Slaughter, R. Loloee, P. A. Schroeder, and J. Bass. Perpendicular giant magnetoresistances of Ag/Co multilayers. *Phys. Rev. Lett.*, 66:3060–3063, 1991.

19. P. Dauguet, P. Gandit, and J. Chaussy. New methods to measure the current perpendicular to the plane magnetoresistance of multilayers. *J. Appl. Phys.*, 79:5823–5825, 1996.

20. M. A. M. Gijs, J. B. Giesbers, and S. K. J. Lenczowski. Perpendicular giant magnetoresistance of microstructured Fe/Cr magnetic multilayers from 4.2 to 300 K. *Phys. Rev. Lett.*, 70:3343–3346, 1993.

21. W. Vavra, S. F. Cheng, A. Fink, J. J. Krebs, and G. A. Prinz. Perpendicular current magnetoresistance in Co/Cu/NiFeCo/Cu multilayered microstructures. *Appl. Phys. Lett.*, 66:2579–2581, 1995.

22. A. Blondel, J. P. Meier, B. Doudin, and J.-Ph. Ansermet. Giant magnetoresistance of nanowires of multilayers. *Appl. Phys. Lett.*, 65:3019–3021, 1994.

23. L. Piraux, J. M. George, J. F. Despres, C. Leroy, E. Ferain, R. Legrasv K. Ounadjela, and A. Fert. Giant Magnetoresistance in Magnetic Multilayered Nanowires. *Appl. Phys. Lett.*, 65(19):2484–2486, 1994.

24. K. Liu, K. Nagodawithana, P. C. Searson, and C. L. Chien. Perpendicular giant magnetoresistance of multilayered Co/Cu nanowires. *Phys. Rev. B*, 51:7381–7384(R), 1995.

25. T. Ono and T. Shinjo. Magnetoresistance of Multilayers Prepared on Microstructured Substrates. *J. Phys. Soc. Jpn.*, 64:363–366, 1995.

26. M. A. M. Gijs, M. T. Johnson, A. Reinders, P. E. Huisman, R. J. M. van de Veerdonk, S. K. J. Lenczowski, and R. M. J. van Gansewinkel. Perpendicular giant magnetoresistance of Co/Cu multilayers deposited under an angle on grooved substrates. *Appl. Phys. Lett.*, 66:1839–1841, 1995.

27. J.C. Slonczewski. Conducance and echange coupling of two ferromagnets separated by a tunneling barrier. *Phys. Rev. B*, 39:6995–7002, 1989.

28. M. Jullière. Tunneling between ferromagnetic films. *Phys. Lett. A*, 54:225–226, 1975.

29. R. Meservey, P.M. Tedrow, and J.S. Moodera. Electron spin polarized tunneling study of ferromagnetic thin films. *J. Magn. Magn. Mater.*, 35(1):1–6, 1983.

30. J. S. Moodera and L. R. Kinder. Ferromagnetic–insulator–ferromagnetic tunneling: Spin-dependent tunneling and large magnetoresistance in trilayer junctions (invited). *Journal of Applied Physics*, 79(8):4724–4729, 04 1996.

31. W. H. Butler, X.-G. Zhang, T. C Schulthess., and J.M. MacLaren. Spin-dependent tunnelling conductance of Fe/MgO/Fe sandwiches. *Phys. Rev. B*, 63:054416, 2001.

32. S. Yuasa, T. Nagahama, A. Fukushima, Y. Suzuki, and K. Ando. Giant room-temperature magnetoresistance in single-crystal Fe/MgO/Fe magnetic tunnel junctions. *Nat. Mater.*, 3:868–871, 2004.

33. W. J. Gallagher and S. S. P. Parkin. Development of the magnetic tunnel junction MRAM at IBM: From first junctions to a 16-Mb MRAM demonstrator chip. *IBM J. Res. Dev.*, 50:5–23, 2006.

34. N. Kobayashi, S. Ohnuma, S. Murakami, T. Masumoto, S. Mitiani, and H. Fujimori. Enhancement of low-field-magnetoresistive response of tunnel-type magnetoresistance in metal–nonmetal granular thin films. *J. Magn. Magn. Mater.*, 188(1–2):30–34, 1998.

35. J. M. D. Coey, M. Viret, and S. von Molnar. Mixed-valence manganites. *Adv. Phys.*, 48(2):167–293, 1999.

36. J. M. D. Coey, A. E. Berkowitz, Ll. Balcells, , F. F. Putris, and A. Barry. Magnetoresistance of Chromium Dioxide Powder Compacts. *Phys. Rev. Lett.*, 80(17):3815–3818, 1998.

37. J. Fontcuberta. Colossal Magnetoresistance. *Phys. World*, 12(2):33–38, 1999.

38. J. Volger. Further experimental investigations on some ferromagnetic oxidic compounds of manganese with perovskite structure. *Physica*, XX:45–66, 1954.

39. S. Jin, T. H. Tiefel, M. McCormack, R. A. Fastnacht, R. Ramesh, and L. H. Chen. Thousand fold Change in Resistivity in magnetoresistive La-Ca-Mn-O Films. *Science*, 264:413–415, 1994.

40. C. Zener. Interaction between the d-Shells in the Transition Metals. II. Ferromagnetic Compounds of Manganese with Perovskite Structure. *Phys. Rev.*, 82:403–405, 1951.

41. P. W. Anderson and H. Hasegawa. Considerations on Double Exchange. *Phys. Rev.*, 100:675–681, 1955.

42. M. Fäth, S. Freisem, A. A. Menovsky, Y. Tomioka, J. Aarts, and J. A. Mydosh. Spatially Inhomogeneous Metal-Insulator Transition in Doped Manganites. *Science*, 285:1540, 1999.

43. A.N. Kocharian and G.R. Reich. Orbital Magnetism and Valence Instabilities in Narrow Bands. *IEEE Trans. Magn.*, 33(1):682–685, 1997.

44. W. Bao, C. Broholm, G. Aeppli, S. A. Carter, P. Dai, T. F. Rosenbaum, J. M. Honig, P. Metcalf, and S. F. Trevino. Magnetic correlations and quantum criticality in the insulating antiferromagnetic, insulating spin liquid, renormalized Fermi liquid, and metallic antiferromagnetic phases of the Mott system V_2O_3. *Phys. Rev. B*, 58:12727–12748, 1998.

45. S. Yu. Ezhov, V. I. Anisimov, D. I. Khomskii, and G. A. Sawatzky. Orbital Occupation, Local Spin, and Exchange Interactions in V_2O_3. *Phys. Rev. Lett.*, 83:4136–4139, 1999.

46. A. Joshi, M. Ma, F. Mila, D. N. Shi, and F. C. Zhang. Elementary Excitation in magnetically ordered systems with orbital dengeneracy. *Phys. Rev. B*, 60(9):6584–6587, 1999.

47. A. Mauger. Indirect exchange in europium chalcogenides. *Phys. Stat. Solidi*, 84(2):761–771, 1977.

48. A. Mauger, C. Godart, M. Escorne, J.C. Achard, and J. P. Desfours. Magnetic properties and instability phenomena in doped EuO. *J. Phys. France*, 39:1125–1133, 1978.

49. A. Mauger and C. Godart. The magnetic, optical, and transport properties of representatives of a class of magnetic semiconductors: The europium chalcogenides. *Phys. Reports*, 141(2-3):51–176, 1986.

50. J.M.D. Coey and C.L. Chien. Half-metallic ferromagnetic oxides. *MRS Bulletin*, 28(10):720–724, 2003.

51. M. Viret, M. Drouet, J. Nassar, J. P. Contour, C. Fermon, and A. Fert. Low-field colossal magnetoresistance in manganite tunnel spin valves. *Europhys. Lett.*, 39(5):545–549, 1997.

52. P.A. Dowben and R. Skomski. Are half-metallic ferromagnets half metals? *J. Appl. Phys.*, 95(11):7453–7458, 2004.

53. M. Bowen, A. Barthélémy, M. Bibes, E. Jacquet, J. P. Contour, A .Fert, D. Wortmann, and S. Blügel. Half-metallicity proven using fully spin-polarized tunnelling. *J. Phys.: Condens. Matter*, 17(41):L407–L409, 2005.

54. M. Bowen, A. Barthélémy, M. Bibes, E. Jacquet, J. P. Contour, A. Fert, F. Ciccacci, L. Duo, and R. Bertacco. Spin-polarized tunneling spectroscopy in tunnel junctions with half-metallic electrodes. *Phys. Rev. Lett.*, 95:137203, 2005.

55. D. Kelly, J.-E. Wegrowe, T.-K. Truong, X. Hoffer, and J.-Ph. Ansermet. Spin-polarized current-induced magnetization reversal in single nanowires. *Phys. Rev. B*, 68:134425, 2003.

56. L. Berger. Emission of spin waves by a magnetic multilayer traversed by a current. *Phys. Rev. B*, 54(13):9353–9358, 1996.

57. J. Slonczewski. Current-driven excitation of magnetic multilayers. *J. Magn. Magn. Mater.*, 159:L1–L7, 1996.

58. M. Tsoi, A. G. M. Jansen, J. Bass, W.-C. Chiang, M. Seck, V. Tsoi, and P. Wyder. Excitation of a magnetic multilayer by an electric current. *Phys. Rev. Lett.*, 80(19):4281–4284, 1998.

59. E. B. Myers, D. C. Ralph, J. A. Katine, R. N. Louie, and R. A. Buhrman. Current-induced switching of domains in magnetic multilayer devices. *Science*, 285:867–870, 1999.

60. F. J. Albert, N. C. Emley, E. B. Myers, D. C. Ralph, and R. A. Buhrman. Quantitative study of magnetization reversal by spin-polarized current in magnetic multilayer nanopillars. *Phys. Rev. Lett.*, 89:226802, 2002.

61. J. C. Slonczewski. Electronic device using magnetic components. Patent US5695864A, IBM, 1995.

62. S. Urazhdin, N. O. Birge, W. P. Pratt Jr., and J. Bass. Current-Driven Magnetic Excitations in Permalloy-Based Multilayer Nanopillars. *Phys. Rev. Lett.*, 91:146803, 2003.

63. A. Fabian, C. Terrier, S. Serrano Guisan, X. Hoffer, M. Dubey, L. Gravier, J.-Ph. Ansermet, and J-E. Wegrowe. Current-induced two-level fluctuations in pseudo-spin-valve (Co/Cu/Co) nanostructures. *Phys. Rev. Lett.*, 91(25):7290, 2003.

64. M. R. Pufall, W. H. Rippard, S. Kaka, S. E. Russek, T. J. Silva, J. Katine, and M. Carey. Current-driven microwave dynamics in magnetic point contacts as a function of applied field angle. *Phys. Rev. B*, 69:214409, 2004.

65. I. N. Krivorotov, N. C. Emley, JC. Sankey, SI. Kiselev, D. C. Ralph, and R. A. Buhrman. Time-domain measurements of nanomagnet dynamics driven by spin-transfer torques. *Science*, 307:228–231, 2005.

66. W. H. Rippard, M. R. Pufall, S. Kaka, T. J. Silva, S. E. Russek, and J. A. Katine. Injection locking and phase control of spin transfer nano-oscillators. *Phys. Rev. Lett.*, 95:067203, 2005.

67. J. C. Sankey, P. M. Braganca, A. G. F. Garcia, I. N. Krivorotov, R. A. Buhrman, and D. C. Ralph. Spin-transfer-driven ferromagnetic resonance of individual nanomagnets. *Phys. Rev. Lett.*, 96:227601, 2006.

68. A. A. Tulapurkar, Y. Suzuki, A. Fukushima, H. Kubota, H. Maehara, K. Tsunekawa, D. D. Djayaprawirav, N. Watanabe, and S. Yuasa. Spin-torque diode effect in magnetic tunnel junctions. *Nature*, 438:339–342, 2005.

69. N. F. Mott. The electrical conductivity of transition metals. *Proc. Roy. Soc.*, 153:699, 1936.

70. N. F. Mott. Electrons in transition metals. *Adv. Phys.*, 13:325–422, 1964.

71. W. Harrison. *Solid State Theory*. Dover Publications Inc., New York, 1979.

72. N. W. Ashcroft and N. D. Mermin. *Solid State Physics*. New York, Chicago, San Francisco, Atlanta, Dallas, Montreal, Toronto, London, Sydney, 1976.

73. A. Matthiessen and M. von Bose. I. On the influence of temperature on the electric conducting power of metals. *Philos. Trans. R. Soc.*, 152:1–27, 1862.

74. S. Reif-Acherman. Augustus Matthiessen: His studies on electrical conductivities and the origins of his 'rule'. *Proc. IEEEE*, 103(4):713–721, 2015.

75. I. A. Campbell, A. Fert, and A. R. Pomeroy. Evidence for Two Current Conduction in Iron. *Phil. Mag.*, 15:977–981, 1967.

76. A. Fert and I. A. Campbell. Two-current Conduction in Nickel. *Phys. Rev. Lett.*, 21(16):1190–1192, 1968.

77. I. A. Campbell and A. Fert. Transport properties of ferromagnets. In *Ferromagnetic Materials*. North-Holland, Amsterdam, 1982.

78. A. Fert and I. A. Campbell. Electrical resistivity of ferromagnetic nickel and iron based alloys. *J. Phys. Metal Phys.*, 6(5):849–871, 1976.

79. C. Kittel. *Introduction to Solid State Physics*. New York, London, Sydney, Toronto, 1976.

80. D. K. C. MacDonald and K. Mendelsohn. Resistivity of pure metals at low temperatures I. The alkali metals. *Proc. Roy. Soc. (London)*, A202:103–126, 1950.

81. R. Hölzle. *Magnetismus von Festkörpern und Grenzflächen*. Forschungszentrum Jülich GmbH, 1993.

82. M. A. M. Gijs, S. K. J. Lenczowski, R. J. M. van de Veerdonk, J. B. Giesbers, M. T. Johnson, and J. B. F. aan de Stegge. Temperature dependence of the spin-dependent scattering in Co/Cu multilayers determined from perpendicular-giant-magnetoresistance experiments. *Phys. Rev. B*, 50:16733–16736, 1994.

83. J. G. Simmons. Electric tunnel effect between dissimilar electrodes separated by a thin insulating film. *J. Appl. Phys.*, 34:2581–2590, 1963.

84. J. G. Simmons. Generalized Thermal J-V characteristic for the electric tunnel effect. *J. Appl. Phys.*, 35:2655–2658, 1964.

85. R. Meservey, P. M. Tedrow, and P. Fulde. Magnetic field splitting of the quasiparticle states in superconducting aluminium films. *Phys. Rev. Lett.*, 25(18):1270–1272, 1978.

86. J. Bardeen. Tunnelling from a many-particle point of view. *Phys. Rev. Lett.*, 6(2):57–59, 1961.

87. J. A. X. Alexander, T. P. Orlando, D. Rainer, and P. M. Tedrow. Theory of Fermi-liquid effects in high-field tunneling. *Phys. Rev. B*, 31(9):5811–5825, 1985.

88. S. T. Chui. Bias Dependence in spin-polarized tunneling. *Phys. Rev. B*, 55(9):5600–5603, 1997.

89. A. Vedyayev, N. Ryzhanova, C. Lacroix, L. Giacomoni, and B. Dieny. Resonance in tunneling through magnetic valve tunnel junctions. *Europhys. Lett.*, 39(2):219–224, 1997.

90. W. H. Butler, X. G. Zhang, X. Wang, J. van Ek, and J. M. MacLaren. Electronic structure of FM/semiconductor/FM spin tunneling structures. *J. Appl. Phys.*, 81(8):5518–5520, 1997.

91. S. F. Alvarado and P. Renaud. Observation of spin-polarized-electron tunneling from a ferromagnet into GaAs. *Phys. Rev. Lett.*, 68:1387, 1992.

92. E. L. Wolf. *Principles of Electron Tunneling Spectroscopy*. Oxford Science Publications, 2nd edition, Oxford, New York, 2012.

93. E.Y. Tsymbal, A. Sokolov, I. F. Sabirianov, and B. Doudin. Resonant Inversion of Tunneling Magnetoresistance. *Phys. Rev. Lett.*, 90:186602, 2003.

94. X.-G. Zhang and W. Butler. Large magnetoresistance in bcc Co/MgO/Co and FeCo/MgO/FeCo tunnel junctions. *Phys. Rev. B*, 70:172407, 2004.

95. R. Wiesendanger, H.-J. Güntherodt, G. Güntherodt, R. Gambino, and R. Ruf. Observation of vacuum tunneling of spin-polarized electrons with the scanning tunneling microscope. *Phys. Rev. Lett.*, 65:247–250, 1990.

96. R. Wiesendanger. Spin mapping at the nanoscale and atomic scale. *Rev. Mod. Phys.*, 81:1495–1550, 2009.

97. P. G. de Gennes and J. Friedel. Anomalies de résistivité dans certains métaux magnétiques. *J. Phys. Chem. Solides*, 4:71–77, 1958.

98. C. Haas. Magnetic semiconductors. *Crit. Rev. Solid State Sci.*, 1(1):47–98, 1970.

99. G.N. Grannemann and L. Berger. Magnon-drag Peltier effect in a Ni-Cu alloy. *Phys. Rev. B*, 13(5):2072–2079, 1976.

100. M. Bailyn. Maximum variational principle for conduction problems in a magnetic field, and the theory of magnon drag. *Phys. Rev.*, 126:2040–2054, 1962.

101. W. M. MacInnes and K. Schroeder. Thermoelectric power of a two-band ferromagnet. *Phys. Rev. B*, 4:4091–4094, 1971.

102. A. Theumann. Effect of magnon drag on electron mass and mobility. *Phys. Rev. B*, 1(11):4400–4405, 1970.

103. F.J. Blatt, D.J. Flood, V. Rowe, P.A. Schroeder, and J.E. Cox. Magnon-drag thermopower in iron. *Phys. Rev. Lett.*, 18(11):395–396, 1967.

104. B. Vuarl and E. E. Thomas. Helicon-spin wave interaction in the magnetic semiconductor $Ag_x\,Cd_{1-x}\,Cr_2\,Se_4$. *Appl. Phys. Lett.*, 12(1):14–17, 1968.

105. J.H. Van Vleck. On The theory of the forward scattering of neutrons by paramagnetic media. *Phys. Rev.*, 55:924–930, 1939.

106. P. G. de Gennes. Inelastic scattering of neutrons in a paramagnetic medium. *C. R. Acad. Sci. Paris*, 244:752–755, 1957.

107. S. von Molnar and S. Methfessel. Giant negative magnetoresistance in ferromagnetic $Eu_{1-x}Gd_xSe$. *J. Appl. Phys.*, 38(3):959–964, 1967.

108. S. Pathak and S. Satpathy. Self-trapped magnetic polaron: Exact solution of a continuum model in one dimension. *Phys. Rev. B*, 63:214413, 2001.

109. D. J. Garcia, K. Hallberg, C. D. Batista, S. Capponi, D. Poilblanc, M. Avignon, and B. Alascio. Charge and spin inhomogeneous phases in the ferromagnetic Kondo lattice model. *Phys. Rev. B*, 65:134444, 2002.

110. M. Umehara. Theory for the bound magnetic polaron in diluted magnetic semiconductors by a modified molecular-field approximation. *Phys. Rev. B*, 61:12209, 2000.

111. Y. U. Dedkov, U. Rüdiger, and G. Güntherodt. Evidence for the half-metallic ferromagnetic state of Fe_3O_4 by spin-resolved photoelectron spectroscopy. *Phys. Rev. B*, 65:064417, 2002.

112. N. Tian, J. Verbeeck, S. Brück, M. Paul, D. Kufer, M. Sing, R. Claessen, and G. van Tendeloo. Interface-induced modulation of charge and polarization in thin film Fe_3O_4. *Adv. Mater.*, 26(3):461–465, 2014.

113. K. Yang, D. H. Kim, and J. Dho. Schottky barrier effect on the electrical properties of Fe_3O_4/ZnO and $Fe_3O_4/Nb:SrTiO_3$ heterostructures. *J. Phys. D: Appl. Phys.*, 44(35):355301, 2011.

114. A. Fert and H. Jaffrès. Conditions for efficient spin injection from a ferromagnetic metal into a semiconductor. *Phys. Rev. B*, 64:184420, 2001.

115. G. Schmidt. Concepts for spin injection into semiconductors – a review. *J. Phys. D: Appl. Phys.*, 38(7):R107–R122, 2005.

116. S. H. Liang, T. T. Zhang, P. Barate, J. Frougier, M. Vidal, P. Renucci, B. Xu, H. Jaffrès, J.-M. George, X. Devaux, M. Hehn, X. Marie, S. Mangin, H. X. Yang, A. Hallal, M. Chshiev, T. Amand, H. F. Liu, D. P. Liu, X. F. Han, Z. G. Wang, and Y. Lu. Large and robust electrical spin injection into GaAs at zero magnetic field using an ultrathin CoFeB/MgO injector. *Phys. Rev. B*, 90:085310, 2014.

117. E. Wada, K. Watanabe, Y. Shirahata, M. Itoh, M. Yamaguchi, and T. Taniyama. Efficient spin injection into GaAs quantum well across Fe_3O_4 spin filter. *App. Phys. Lett.*, 96(10):102510, 2010.

118. R. Fiederling, M. Keim, G. Reuscher, W. Ossau, G. Schmidt, A. Waag, and L. W. Molenkamp. Injection and detection of a spin-polarized current in a light-emitting diode. *Nature*, 402(6763):787–790, 1999.

119. B. Huang and I. Appelbaum. Perpendicular hot electron transport in the spin-valve photodiode. *J. Appl. Phys.*, 100(3):034501, 2006.

120. S. Y. Huang, X. Fan, D. Qu, Y. P. Chen, W. G. Wang, J. Wu, T. Y. Chen, J. Q. Xiao, and C. L. Chien. Transport magnetic proximity effects in platinum. *Phys. Rev. Lett.*, 109:107204, 2012.

121. Y. M. Lu, J. W. Cai, S. Y. Huang, D. Qu, B. F. Miao, and C. L. Chien. Hybrid magnetoresistance in the proximity of a ferromagnet. *Phys. Rev. B*, 87:220409, 2013.

122. T. Lin, C. Tang, and J. Shi. Induced magneto-transport properties at palladium/yttrium iron garnet interface. *Appl. Phys. Lett.*, 103(13):132407, 2013.

123. B. F. Miao, S. Y. Huang, D. Qu, and C. L. Chien. Physical origins of the new magnetoresistance in Pt/YIG. *Phys. Rev. Lett.*, 112:236601, 2014.

124. J. Kirtley. The interaction of tunneling electrons with molecular vibrations. In *Tunneling Spectroscopy: capabilities, applications and new techniques*. Plenum Press New York and London, 1982.

125. G. Herranz, R. Ranchal, M. Bibes, H. Jaffes, E. Jacquet, J. Maurice, K. Bouzehouane, F. Wyczisk, E. Tafra, M. Basletic, A. Hamzic, C. Colliex, JP. Contour, A. Barthélémy, and A. Fert. Co-doped (La,Sr) TiO3-δ: a high Curie Temperature diluted magnetic system with large spin polarization. *Phys. Rev. Lett.*, 96:027207, 2006.

126. R. Meservey, P. M. Tedrow, and R.C. Bruno. Tunneling measurements on spin-paired superconductors with spin-orbit scattering. *Phys. Rev. B*, 11:4224–4235, 1975.

127. B. S. Tao, H. X. Yang, Y. L. Zuo, X. Devaux, G. Lengaigne, M. Hehn, D. Lacour, S. Andrieu, M. Chshiev, T. Hauet, F. Montaigne, S. Mangin, X. F. Han, and Y. Lu. Long-range phase coherence in double-barrier magnetic tunnel junctions with a large thick metallic quantum well. *Phys. Rev. Lett.*, 115:157204, 2015.

128. P. M. Tedrow and R. Meservey. Spin polarization of electrons tunneling from films of Fe, Co, Ni, and Gd. *Phys. Rev. B*, 7:318–326, 1973.

129. I. I. Mazin. How to Define and calculate the degree of spin polarization in ferromagnets. *Phys. Rev. Lett.*, 83:1427–1430, 1999.

130. D. C. Worledge and T. H. Geballe. Maki analysis of spin-polarized tunneling in an oxide ferromagnet. *Phys. Rev. B*, 62:447–451, 2000.

131. I. V. Shvets, R. Wiesendanger, D. Bürgler, G. Tarrach, H.-J. Güntherodt, and J. M. D. Coey. Progress towards spin-polarized scanning tunneling microscopy. *J. Appl. Phys.*, 71(11):5489–5499, 1992.

132. M. Bode, M. Getzlaff, and R. Wiesendanger. Spin-polarized vacuum tunneling into the exchange-split surface state of Gd(0001). *Phys. Rev. Lett.*, 81:4256–4259, 1998.

133. M. Bode, M. Getzlaff, and R. Wiesendanger. Quantitative aspects of spin-polarized scanning tunneling spectroscopy of Gd(0001). *J. Vac. Sci. Technol. A*, 17(4):2228–2232, 1999.

134. O. Pietzsch, A. Kubetzka, M. Bode, and R. Wiesendanger. Real-space observation of dipolar antiferromagnetism in magnetic nanowires by spin-polarized scanning tunneling spectroscopy. *Phys. Rev. Lett.*, 84:5212–5215, 2000.

135. Z. H. Xiong, Di Wu, Z. V. Vardeny, and J. Shi. Giant magnetoresistance in organic spin-valves. *Nature*, 427(6977):821–824, 2004.

136. V. A. Dediu, L.E. Hueso, I. Bergenti, and C. Taliani. Spin routes in organic semiconductors. *Nat. Mater.*, 8(9):707–716, 2009.

137. F. Djeghloul, M. Gruber, E. Urbain, D. Xenioti, L. Joly, S. Boukari, J. Arabski, H. Bulou, F. Scheurer, F. Bertran, P. Le Fèvre, A. Taleb-Ibrahimi, W. Wulfhekel, G. Garreau, S. Hajjar-Garreau, P. Wetzel, M. Alouani, E. Beaurepaire, M. Bowen, and W. Weber. High spin polarization at ferromagnetic metal-organic interfaces: a generic property. *J. Phys. Chem. Lett.*, 7(13):2310–2315, 2016.

138. M. Galbiati, S. Delprat, M. Mattera, S. Manas-Valero, A. Forment-Aliaga, S. Tatay, C. Deranlot, P. Seneor, R. Mattana, and F. Petroff. Recovering ferromagnetic metal surfaces to fully exploit chemistry in molecular spintronics. *AIP Adv.*, 5(05):057131, 2015.

139. M. Galbiati, C. Barraud, S. Tatay, K. Bouzehouane, C. Deranlot, E. Jacquet, A. Fert, P. Seneor, R. Mattana, and F. Petroff. Unveiling self-assembled monolayers' potential for molecular spintronics: spin transport at high voltage. *Adv. Mater.*, 24(48):6429–6432, 2012.

140. K. Watanabe, B. Jinnai, S. Fukami, H. Sato, and H. Ohno. Shape anisotropy revisited in single-digit nanometer magnetic tunnel junctions. *Nat. Commun.*, 9(1):663, 2018.

141. G. Prenat, K. Jabeur, P. Vanhauwaert, G. D. Pendina, F. Oboril, R. Bishnoi, M. Ebrahimi, N. Lamard, O. Boulle, K. Garello, J. Langer, B. Ocker, M. Cyrille, P. Gambardella, M. Tahoori, and G. Gaudin. Ultra-fast and high-reliability SOT-MRAM: from cache replacement to normally-off computing. *IEEE Trans. Multi-Scale Comput. Syst.*, 2(1):49–60, 2016.

142. Z. Luo, A. Hrabec, T. P. Dao, G. Sala, S. Finizio, J. Feng, S. Mayr, J. Raabe, P. Gambardella, and L. J. Heyderman. Current-driven magnetic domain-wall logic. *Nature*, 579(7798):214–218, 2020.

143. N.A. Usov, A.S. Antonov, and A.N. Lagar'kov. Theory of giant magneto-impedance effect in amorphous wires with different types of magnetic anisotropy. *J. Magn. Magn. Mater.*, 185:159–173, 1998.

144. A. Sukstanskii, V. Korenivski, and A. Gromov. Impedance of a ferromagnetic sandwich strip. *J. Appl. Phys.*, 89(1):775–782, 2001.

145. T. Kitho and K. Mohri. Asymmetrical magneto-impedance effect in twisted amorphous wires for sensitive magnetic sensors. *IEEE Trans. Magn.*, 31(6):3137–3139, 1995.

146. M. Noda, L.V. Panina, and K. Mohri. Pulse response bistable magneto-impedance effect in amorphous wires. *IEEE Trans. Magn.*, 31(6):3167–3169, 1995.

147. L. V. Panina, K. Mohri, K. Bushida, and M. Noda. Giant magneto-impedance and magneto-inductive effects in amorphous alloys. *J. Appl. Phys.*, 76:6198–6203, 1994.

148. W. Pauli. Über den Einfluss der Geschwindigkeitabhängigkeit der Elektronenmasse auf den Zeemaneffekt. *Z.Phys.*, 31:373–385, 1925.

149. G. E. Uhlenbeck and S. Goudsmit. Ersetzung der Hypothese vom unmechanischen Zwang durch eine Forderung bezüglich des inneren Verhaltens jedes einzelnen Elektrons. *Naturwissenschaften*, 13:953–954, 1925.

150. G. E. Uhlenbeck and S. Goudsmit. Spinning Electrons and the Structure of Spectra. *Nature*, 117:264–265, 1926.

151. I. G. Kaplan. *The Pauli Exclusion Principle, origin, verifications and applications.* Chichester, 2017.

152. B. Friedrich and D. Herschbach. Stern and Gerlach: How a Bad Cigar Helped Reorient Atomic Physics. *Physics Today*, 56(12):53–59, 2003.

153. R. G. J. Fraser. The effective cross section of the oriented hydrogen atom. *Proc. R. Soc. Lond.*, A114:212–221, 1927.

154. W. Heitler and F. London. Wechselwirkung neutraler Atome und homöopolare Bindung nach der Quantenmechanik. *Z. Phys.*, 44:455–472, 1927.

155. W. Heitler. *Historical Development of Quantum Theory*, volume 6. Archive for the History of Quantum Physics, 1963.

156. R.M. White. *Quantum Theory of Magnetism*, volume 32 of *Springer Series in Solid State Sciences*. Springer Berlin Heidelberg, 2007.

157. H.A. Kramers. L'interaction Entre les Atomes Magnétogènes dans un Cristal Paramagnétique. *Physica*, 1(1-6):182–192, 1934.

158. P. W. Anderson. Antiferromagnetism. Theory of superexchange interaction. *Phys. Rev*, 79(2):350–356, 1950.

159. P. W. Anderson. New approach to the theory of superexchange interactions. *Phys. Rev.*, 115(1):2–13, 1959.

160. P. W. Anderson. Exchange in insulators: superexchange, direct exchange, double exchange. In G.T. Rado, H. Suhl, editor, *Solid State Physics*, volume 1. Academic Press, New York, 1963.

161. P. J. Hay, J. C. Thibeault, and R. Hoffmann. Orbital interactions in metal dimer complexes. *J. Am. Chem. Soc.*, 97(17):4884–4899, 1975.

162. F. Reuse. *Electrodynamique et optique quantique.* Presse Polytechnique Universitaire Romande, Lausanne, 2007.

163. Y. Yafet. g Factors and spin-lattice relaxation of conduction electrons. In Frederick Seitz and David Turnbull, editors, *Solid State Physics*, volume 14, pages 1–98. Academic Press, New York, 1963.

164. J. Friedel, P. Lenghart, and G. Leman. Etude du couplage spin-orbite dans les métaux de transition. Application au platine. *J. Phys. Chem. Solids*, 25:781–800, 1964.

165. R. J. Elliot. Paramagnetic relaxation in Metals. *Phys. Rev.*, 89:689–700, 1953.

166. J. R. Asik, M. A. Ball, and C.P. Slichter. Spin-flip scattering of conduction electrons from impurities. *Phys. Rev. Lett.*, 16(17):740–743, 1966.

167. H. Schwarz. On the spin-forbiddeness of gas-phase ion-molecule reactions: a fruitful intersection of experimental and computational studies. *Int. J. Mass Spectrom.*, 237(1):75–105, 2004.

168. D. L. Coopera and P. B. Karadakov. Spin-coupled descriptions of organic reactivity. *Int. Rev. Phys. Chem.*, 28(2):169–206, 2009.

169. C. Cohen Tannoudji, B. Diu, and F. Laloë. *Mécanique Quantique II*. Hermann, Paris, 1977.

170. F. A. Reuse. *Electrodynamique*. Presses Polytechniques et Universitaires Romandes, Lausanne, 2012.

171. L.H. Thomas. The motion of the Spinning Electron. *Nature*, 117:514–514, 1926.

172. L.H. Thomas. I. The kinematics of an electron with an axis. *Lond. Edinb. Dublin Philos. Mag. J. Sci.*, 3:1–22, 1927.

173. H. C. Corben. Factors of 2 in magnetic moments, spin–orbit coupling, and Thomas precession. *Am. J. Phys.*, 61(6):551–553, 1993.

174. V. Bargmann, L. Michel, and V. L. Telegdi. Precession of the polarization of particles moving in a homogeneous electromagnetic field. *Phys. Rev. Lett.*, 2:435–436, 1959.

175. J. Fröhlich and U. M. Studer. Gauge invariance and current algebra in nonrelativistic many-body theory. *Rev. Mod. Phys.*, 65(3):733–802, 1993.

176. D. C. Mattis. *The Theory of Magnetism I*, volume 17 of *Springer Series in Solid State Sciences*. Springer Berlin Heidelberg New York, 1981.

177. I. Dzialoshinskii. A thermodynamic theory of "weak" ferromagnetism of antiferromagnetics. *J. Chem. Phys. Solids*, 4:241–255, 1958.

178. T. Moriya. Anisotropic superexchange interaction and weak ferromagnetism. *Phys. Rev.*, 120:91–98, 1960.

179. T. Thio. Antisymmetric exchange and its influence on the magnetic structure and conductivity of La_2CuO_4. *Phys. Rev. B*, 38:905–908, 1988.

180. M. Bode, M. Heide, K. von Bergmann, P. Ferriani, S. Heinze, G. Bihlmayer, A. Kubetzka, O. Pietzsch, S. Blügel, and R. Wiesendanger. Chiral magnetic order at surfaces driven by inversion asymmetry. *Nature*, 447:190–193, 2007.

181. L. E. Ballentine. *Quantum Mechanics*. Prentice Hall, Englewood Cliffs, New Jersey, 1990.

182. C.-F. Cheng, J. Hussels, M. Niu, H. L. Bethlem, K. S. E. Eikema, E. J. Salumbides, W. Ubachs, M. Beyer, N. Hölsch, J. A. Agner, F. Merkt, L.-G. Tao, S.-M. Hu, and Ch. Jungen. Dissociation energy of the hydrogen molecule at 10^{-9} accuracy. *Phys. Rev. Lett.*, 121:013001, 2018.

183. A. Aharoni. *Introduction to the Theory of Ferromagnetism*. Oxford Science Publications, Oxford, New York, 2001.

184. R. Skomski. Nanomagnetics. *J. Phys.: Condens. Matter*, 15:R841–R896, 2003.

185. N. Sethulakshmi, A. Mishra, P.M. Ajayan, Y. Kawazoe, A. K. Roy, A. K. Singh, and C. S. Tiwary. Magnetism in two-dimensional materials beyond graphene. *Mater. Today*, 27:107 – 122, 2019.

186. D. Orgassa, H. Fujiwara, T.C. Schulthess, and W. H. Butler. First-principles calculation of the effect of atomic disorder on the electronic structure of the half-metallic ferromagnet NiMnSb. *Phys. Rev. B*, 60:13237–13240, 1999.

187. D. Orgassa, H. Fujiwara, T.C. Schulthess, and W. H. Butler. Disorder dependence of the magnetic moment of the half-metallic ferromagnet NiMnSb from first principles. *J. Appl. Phys.*, 87:5870–5871, 2000.

188. L. M. Sandrastskii. Noncollinear magnetism in itinerant-electron systems: Theory and applications. *Adv. Phys.*, 47:91–160, 1998.

189. L. M. Sandrastskii. Ab initio calculations of exchange interactions, spin-wave stiffness constants, and Curie temperatures of Fe, Co, and Ni. *Phys. Rev. B*, 64:134402, 2001.

190. N. Ishikawa, M. Sugita, T. Ishikawa, S. Koshihara, and Y. Kaizu. Lanthanide double-decker complexes functioning as magnets at the single-molecular level. *J. Am. Chem. Soc.*, 125(29):8694–8695, 2003.

191. L. Bogani and W. Wernsdorfer. Molecular spintronics using single-molecule magnets. *Nat. Mater.*, 7(3):179–186, 2008.

192. A. Chiesa, T. Guidi, S. Carretta, S. Ansbro, G. A. Timco, I. Vitorica-Yrezabal, E. Garlatti, G. Amoretti, R. E. P. Winpenny, and P. Santini. Magnetic exchange interactions in the molecular nanomagnet Mn_{12}. *Phys. Rev. Lett.*, 119:217202, 2017.

193. J. Kishine, K. Inoue, and Y. Yoshida. Synthesis, structure and magnetic properties of chiral molecule-based magnets. *Prog. Theor. Phys. Supp.*, 159(05):82–95, 2005.

194. G. Breit. The effect of retardation on the interaction of two electrons. *Phys. Rev.*, 34:553–573, 1929.

195. G. Breit. The fine structure of He as a test of the spin interactions of two electrons. *Phys. Rev.*, 36:383–397, 1930.

196. G. Breit. Dirac's Equation and the spin-spin interactions of two electrons. *Phys. Rev.*, 39:616–624, 1932.

197. H. A. Bethe and E. E. Salpeter. *Quantum Mechanics of One and Two Atoms*. Plenum Press New York and London, 1977.

198. V. B. Berestetskii, E. M. Lifshitz, and L. P. Pitaevskii. *Quantum Electrodynamics*. Pergamon Press, Oxford, 1980.

199. M. D. Stiles, S. V. Halilov, R. A. Hyman, and A. Zangwill. Spin-other-orbit interaction and magnetocrystalline anisotropy. *Phys. Rev. B*, 64:104430, 2001.

200. I. G. Bostrem, J.Kishine, and A. S. Ovchinnikov. Transport spin current driven by the moving kink crystal in a chiral helimagnet. *Phys. Rev. B*, 77:132405, 2008.

201. Y. Togawa, T. Koyama, K. Takayanagi, S. Mori, Y. Kousaka, J. Akimitsu, S. Nishihara, K. Inoue, A. S. Ovchinnikov, and J. Kishine. Chiral magnetic soliton lattice on a chiral helimagnet. *Phys. Rev. Lett.*, 108:107202, 2012.

202. Y. Togawa, Y. Kousaka, S. Nishihara, K. Inoue, J. Akimitsu, A. S. Ovchinnikov, and J. Kishine. Interlayer magnetoresistance due to chiral soliton lattice formation in hexagonal chiral magnet $CrNb_3S_6$. *Phys. Rev. Lett.*, 111:197204, 2013.

203. J. Kishine and A. S. Ovchinnikov. Theory of Monoaxial Chiral Helimagnet. In R. E. Camley and R. L. Stamps, editors, *Solid State Physics*, volume 66, pages 1 – 130. Academic Press, New York, 2015.

204. Y. Togawa, T. Koyama, Y. Nishimori, Y. Matsumoto, S. McVitie, D. McGrouther, R. L. Stamps, Y. Kousaka, J. Akimitsu, S. Nishihara, K. Inoue, I. G. Bostrem, Vl. E. Sinitsyn, A. S. Ovchinnikov, and J. Kishine. Magnetic soliton confinement and discretization effects arising from macroscopic coherence in a chiral spin soliton lattice. *Phys. Rev. B*, 92:220412, 2015.

205. K. Tsuruta, M. Mito, Y. Kousaka, J. Akimitsu, J. Kishine, Y. Togawa, H. Ohsumi, and K. Inoue. Discrete change in magnetization by chiral soliton lattice formation in the chiral magnet $Cr_{1/3}NbS_2$. *J. Phys. Soc. Jpn.*, 85(1):013707, 2016.

206. Y. Togawa, Y. Kousaka, K. Inoue, and J. Kishine. Symmetry, structure, and dynamics of monoaxial chiral magnets. *J. Phys. Soc. Japan*, 85(11):112001, 2016.

207. S. Muehlbauer, B. Binz, F. Jonietz, C. Pfleiderer, A. Rosch, A. Neubauer, R. Georgii, and P. Boeni. Skyrmion lattice in a chiral magnet. *Science*, 323(5916):915–919, 2009.

208. X. Z. Yu, N. Kanazawa, Y. Onose, K. Kimoto, W. Z. Zhang, S. Ishiwata, Y. Matsui, and Y. Tokura. Near room-temperature formation of a skyrmion crystal in thin-films of the helimagnet FeGe. *Nat. Mater.*, 10(2):106–109, 2011.

209. S Seki, X Z Yu, S Ishiwata, and Y Tokura. Observation of skyrmions in a multiferroic material. *Science*, 336(6078):198–201, 2012.

210. K. Karube, J. S. White, N. Reynolds, J. L. Gavilano, H. Oike, A. Kikkawa, F. Kagawa, Y. Tokunaga, H. M. Rønnow, Y. Tokura, and Y. Taguchi. Robust metastable skyrmions and their triangular–square lattice structural transition in a high-temperature chiral magnet. *Nat. Mater.*, 15(12):1237–1242, 2016.

211. Y. Tokunaga, X. Z. Yu, J. S. White, H. M. Rønnow, D. Morikawa, Y. Taguchi, and Y. Tokura. A new class of chiral materials hosting magnetic skyrmions beyond room temperature. *Nat. Commun.*, 6(1):7638, 2015.

212. J. Sampaio, V. Cros, S. Rohart, A. Thiaville, and A. Fert. Nucleation, stability and current-induced motion of isolated magnetic skyrmions in nanostructures. *Nat. Nanotechnol.*, 8(11):839–844, 2013.

213. N. Romming, C. Hanneken, M. Menzel, J. E. Bickel, B. Wolter, K. von Bergmann, A. Kubetzka, and R. Wiesendanger. Writing and deleting single magnetic skyrmions. *Science*, 341(6146):636–639, 2013.

214. C. Moreau-Luchaire, C. Moutafis, N. Reyren, J. Sampaio, C. A. F. Vaz, N. Van Horne, K. Bouzehouane, K. Garcia, C. Deranlot, P. Warnicke, P. Wohlhüter, J. M. George, M. Weigand, J. Raabe, V. Cros, and A. Fert. Additive interfacial chiral interaction in multilayers for stabilization of small individual skyrmions at room temperature. *Nat. Nanotechnol.*, 11(5):444–448, 2016.

215. S. D. Pollard, J. A. Garlow, J. Yu, Z. Wang, Y. Zhu, and H. Yang. Observation of stable Néel skyrmions in cobalt/palladium multilayers with Lorentz transmission electron microscopy. *Nat. Commun.*, 8(1):14761, 2017.

216. J. C. T Lee, J. J. Chess, S. A. Montoya, X. Shi, N. Tamura, S. K. Mishra, P. Fischer, B. J. McMorran, S. K. Sinha, E. E. Fullerton, S. D. Kevan, and S. Roy. Synthesizing skyrmion bound pairs in Fe-Gd thin films. *Appl. Phys. Lett.*, 109(2), 022402, 2016.

217. S. A. Montoya, S. Couture, J. J. Chess, J. C. T. Lee, N. Kent, D. Henze, S. K. Sinha, M.-Y. Im, S. D. Kevan, P. Fischer, B. J. McMorran, V. Lomakin, S. Roy, and E. E. Fullerton. Tailoring magnetic energies to form dipole skyrmions and skyrmion lattices. *Phys. Rev. B*, 95:024415, 2017.

218. N. Nagaosa and Y. Tokura. Topological properties and dynamics of magnetic skyrmions. *Nat. Nanotechnol.*, 8(12):899–911, 2013.

219. D. Khadka, S. Karayev, and S. X. Huang. Dzyaloshinskii–Moriya interaction in Pt/Co/Ir and Pt/Co/Ru multilayer films. *J. Appl. Phys.*, 123(12):123905, 2018.

220. S. Ding, A. Ross, R. Lebrun, S. Becker, K. Lee, I. Boventer, S. Das, Y. Kurokawa, S. Gupta, J. Yang, G. Jakob, and M. Kläui. Interfacial Dzyaloshinskii-Moriya interaction and chiral magnetic textures in a ferrimagnetic insulator. *Phys. Rev. B*, 100:100406, 2019.

221. K. Miyagawa, A. Kawamoto, Y. Nakazawa, and K. Kanoda. Antiferromagnetic Ordering and Spin Structure in the Organic Conductor, κ-(BEDT-TTF)$_2$Cu[N(CN)$_2$]Cl. *Phys. Rev. Lett.*, 75:1174–1177, 1995.

222. H. H. Wang, K. D. Carlson, U. Geiser, A. M. Kini, A. J. Schultz, J. M. Williams, L. K. Montgomery, W. K. Kwok, U. Welp, K. G. Vandervoort, S. J. Boryschuk, A. V. Strieby Crouch, J. M. Kommers, D. M. Watkins, J. E. Schirber, D. L. Overmeyer, D. Jung, J. J. Novoa, and M.-H. Whangbo. New κ-phase materials, κ-(ET)$_2$Cu[N(CN)$_2$]X. X = Cl, Br and I. The synthesis, structure and superconductivity above 11 K in the Cl (Tc = 12.8 K. 0.3 kbar) and Br (Tc = 11.6 K) salts. *Synth. Met.*, 41–43:1983–1990, 1991.

223. A. M. Kini, U. Geiser, H. H. Wang, K. D. Carlson, J. M. Williams, W. K. Kwok, K. G. Vandervoort, J. E. Thompson, D. L. Stupka, D. Jung, and M.-H. Whangbo. A new ambient-pressure organic superconductor, κ-(ET)2Cu[N(CN)2]Br, with the highest transition temperature yet observed (inductive onset Tc = 11.6 K, resistive onset = 12.5 K). *Inorg. Chem.*, 29:2555–2557, 1990.

224. H. Urayama, H. Yamochi, G. Saito, K. Nozawa, T. Sugano, M. Kinoshitav S. Sato, K. Oshima, A.I Kawamoto, and J. Tanaka. A New Ambient Pressure Organic Superconductor Based on BEDT-TTF with T$_C$ Higher than 10 K (T$_C$=10.4 K). *Chem. Lett.*, 17(1):55–58, 1988.

225. H. Kondo and T. Moriya. Spin Fluctuation-Induced Superconductivity in κ-BEDT-TTF Compounds. *J. Phys. Condens. Matter*, 11:L363–L369, 1999.

226. W. Gilbert. *De Magnete*. Dover Publications Inc., New York, 1958.

227. J. K. Heilbron. *Galileo*. Oxford University Press, Oxford, 2010.

228. C. W. Heaps. The effect of crystal structure on magnetostriction. *Phys. Rev.*, 22:496–501, 1923.

229. I. V. Aleksandrov, B. V. Lidskii, L. G. Mamsurova, M. G. Neigauz, K. S. Pigal'skii, K. K. Pukhov, N. G. Trusevich, and L. G. Shcherbakova. Crystal field effects and the nature of the giant magnetostriction in terbium dititanate. *Sov. Phys. JEPT*, 62(6):1287–1297, 1985.

230. E. du Tremolet de Lacheisserie. *Magnetostriction: theory and applications of magnetoelasticity*. CRC Press Inc., Boca Raton, New York, London, Tokyo, 1993.

231. C. Winter and A. S. Arrott. Irreversible magnetization processes induced by thermal cycling and by ultrasound in Bi-MnBi. *J. Appl. Phys.*, 53:8142–8144, 1982.

232. C. Winter and A. S. Arrott. Ultrasound induced irreversible changes in the magnetization of highly magnetostrictive materials $Tb_{0.27}Dy_{0.73}Fe_2$ and MnBi. *J. Appl. Phys.*, 53:2733–2735, 1982.

233. S. Blundell. *Magnetism in Condensed Matter*. Oxford Master Series in Condensed Matter Physics, Oxford, New York, 2001.

234. J. H. Van Vleck. *The theory of Electric and Magnetic Susceptibilities*. The International Series of Monographs on Physics. Oxford University Press London, 1932.

235. C. Henry and C. P. Slichter. Moments and degeneracy in Optical Spectra. In *Physics of Color Centers*. Academic Press, New York, London, 1968.

236. C. P. Slichter. *Introduction to Magnetic Resonance*, volume 1 of *Springer Series in Solid State Sciences*. Springer-Verlag Berlin Heidelberg New York, 3rd edition, 1989.

237. R. Kubo and Y. Obata. Note on the Paramagnetic susceptibility and the gyromagnetic ratio in metals. *J. Phys. Soc. Jpn.*, 11:547–550, 1956.

238. R. Skomski and J. M. D. Coey. *Permanent Magnetism*. Institute of Physics Publishing, Bristol, Philadelphia, 1999.

239. M. Tachiki. Exchange Contribution to the D-Parameter of Mn2+ in $CoCl_2 \cdot 2H_2O$. *J. Phys. Soc. Jpn.*, 25:686–690, 1968.

240. B. W. Roberts. Kramers degeneracy without eigenvectors. *Phys. Rev. A*, 86(3):8–10, 2012.

241. D.J. Newman and Betty N., editors. *Crystal Field Handbook*. Cambridge University Press, Cambridge, New York, Oakleigh, Madrid, Cape Town, 2009.

242. J. Chen, H. Isshiki, C. Baretzky, T. Balashov, and W. Wulfhekel. Abrupt Switching of Crystal Fields during Formation of Molecular Contacts. *ACS Nano*, 12(4):3280–3286, 2018.

243. X. Zhang, T. Palamarciuc, J.-F. Létard, P.k Rosa, E. V. Lozada, F. Torres, L. G. Rosa, B. Doudin, and P. A. Dowben. The spin state of a molecular adsorbate driven by the ferroelectric substrate polarization. *Chem. Commun.*, 50:2255–2257, 2014.

244. M. A. Halcrow. *Spin-Crossover Materials: Properties and Applications*. Wiley, Chichester, 2013.

245. S. Zhang. Spin-Dependent Surface Screening in Ferromagnets and Magnetic Tunnel Junctions. *Phys. Rev. Lett.*, 83:640–643, 1999.

246. K. T. McCarthy, A. F. Hebard, and S. B. Arnason. Magnetocapacitance: Probe of Spin-Dependent Potentials. *Phys. Rev. Lett.*, 90:117201, 2003.

247. J. L. Rodríguez-López, J. Dorantes-Dávila, and G. M. Pastor. Orbital magnetism at the surfaces of 3d transition metals. *Phys. Rev. B*, 57:1040–1045, 1998.

248. K. W. H. Stevens. *Magnetic Ions in Crystals*. Princeton University Press, Princeton, 1997.

249. H.A. Jahn and E. Teller. Stability of polyatomic molecules in degenerate electronic states I : Orbital degeneracy. *Proc. Roy. Soc.*, A161:220–235, 1937.

250. J. H. Van Vleck. The Jahn-Teller Effect and Crystalline Stark Splitting for Clusters of the Form XY_6. *J. Chem. Phys.*, 7:72–84, 1939.

251. H. Ham. *Jahn-Teller effect in Electron Paramagnetic Resonance Spectra*. Plenum, New York, 1978.

252. C.A. Bates. Jahn-Teller effects in paramagnetic crystals. *Phys. Rep.*, 35:187–304, 1978.

253. Gehring G. A. and Gehring K. A. Cooperative Jahn-Teller Effects. *Rep. Prog. Phys.*, 38:1–89, 1975.

254. P. Bruno. Tight-binding approach to the orbital magnetic moment and magnetocrystalline anisotropy of transition-metal monolayers. *Phys. Rev. B*, 39:865–868, 1989.

255. S. V. Halilov, A. Y. Perlov, P. M. Oppeneer, A. N. Yaresko, and V. N. Antonov. Magnetocrystalline anisotropy energy in cubic Fe, Co, and Ni: Applicability of local-spin-density theory reexamined. *Phys. Rev. B*, 57(16):9557–9560, 1998.

256. Stoehr J. and Siegmann H. C. *Magnetism, from Fundamentals to Nanoscale Dynamics*, volume 152 of *Springer Solid State Series*. Springer-Verlag Berlin Heidelberg, 2006.

257. Néel L. Anisotropie magnétique superficielle et surstructures d'orientation. *J. Phys. Radium*, 15(4):225–239, 1954.

258. P. Gambardella, S. Rusponi, M. Veronese, S. S. Dhesi, C. Grazioli, A. Dallmeyer, I. Cabria, R. Zeller, P. H. Dederichs, K. Kern, C. Carbone, and H. Brune. Giant Magnetic Anisotropy of Single Cobalt Atoms and Nanoparticles. *Science*, 300(5622):1130–3, 2003.

259. E. C. Stoner and E. P. Wohlfarth. A mechanism of magnetic hysteresis in heterogeneous alloys. *Philos. Trans. Royal Soc. A*, 240(826):599–642, 1948.

260. W. T. Coffey, D. S. F. Crothers, J. L. Dormann, Y. P. Kalmykov, E. C. Kennedy, and W. Wernsdorfer. Thermally activated relaxation time of a single domain ferromagnetic particle subjected to a uniform field at an oblique angle to the easy axis: comparison with experimental observations. *Phys. Rev. Lett.*, 80(25):5655–5658, 1998.

261. P. Bruno. Physical origin and theoretical models of magnetic anisotropy. In *Magnetismus von Festkörpern unde Grenzflächen*. IFF-Ferienkurs Jülich, 1993.

262. H. F. J. Jansen. Magnetic anisotropy in density-functional theory. *Phys. Rev. B*, 38:8022–8029, 1988.

263. G. J. Bowden, G. Van Der Laan, T. Hesjedal, and R. J. Hicken. Expanding the Lorentz concept in magnetism. *New J. Phys.*, 21:073063, 2019.

264. H. J. Draaisma and W. J. M. De Jonge. Surface and volume anisotropy from dipole-dipole interactions in ultrathin ferromagnetic films. *J. Appl. Phys.*, 64:3610–3613, 1988.

265. A. Abragam, B. Bleaney. *Electron Paramagnetic Resonance of Transition Ions*. Clarendon Press, New York, 1970.

266. D. F. Smith and C. P. Slichter. Precise Determination of the Orientation of the Dzialoshinskii-Moriya Vector in kappa-$(BEDT\text{-}TTF)_2Cu[N(CN)_2]Cl$. *Phys. Rev. Lett.*, 93:167002, 2004.

267. O. Cepas. *Effets des anisotropies de spins dans les oxydes de basse dimension*. PhD thesis, University Joseph Fournier, Grenoble, France, 2000.

268. D. Amoroso, P. Barone, and S. Picozzi. Spontaneous skyrmionic lattice from anisotropic symmetric exchange in a Ni-halide monolayer. *Nat. Commun.*, 11(1):1–9, 2020.

269. G. van der Laan. Microscopic origin of magnetocrystalline anisotropy in transition metal thin films. *J. Phys. Condens. Matter*, 10(14):3239–3253, 1998.

270. R. C. Handley. *Modern Magnetic Materials: Principles and Applications*. Wiley, 1999.

271. C.-Y. Liang, S. M. Keller, A. E. Sepulveda, A. Bur, W.-Y. Sun, K. Wetzlar, and G. P. Carman. Modeling of magnetoelastic nanostructures with a fully coupled mechanical-micromagnetic model. *Nanotechnology*, 25:435701, 2014.

272. C.-Y. Liang, S. M. Keller, A. E. Sepulveda, W.-Y. Sun, J. Cui, C. S. Lynch, and G. P. Carman. Electrical control of a single magnetoelastic domain structure on a clamped piezoelectric thin film—analysis. *J. Appl. Phys.*, 116(12):123909, 2014.

273. G. Giannopoulos, R. Salikhov, G. Varvaro, V. Psycharis, A. M. Testa, M. Farle, and D. Niarchos. Coherently strained [Fe–Co(C)/Au–Cu]n multilayers: a path to induce magnetic anisotropy in Fe–Co films over large thicknesses. *J. Phys. D: Appl. Phys.*, 51(5):055009, 2018.

274. M. Gruber, F. Ibrahim, S. Boukari, H. Isshiki, L. Joly, M. Peter, M. Studniarek, V. Da Costa, H. Jabbar, V. Davesne, U. Halisdemir, J. Chen, J. Arabski, E. Otero, F. Choueikani, K. Chen, P. Ohresser, W. Wulfhekel, F. Scheurer, M. Alouani, E. Beaurepaire, and M. Bowen. Exchange bias and room-temperature magnetic order in molecular layers. *Nat. Mater.*, 14:981–984, 2015.

275. S. Boukari, H. Jabbar, F. Schleicher, M. Gruber, G. Avedissian, J. Arabski, V. Da Costa, G. Schmerber, P. Rengasamy, B. Vileno, and W. Weber. Disentangling magnetic hardening and molecular spin chain contributions to exchange bias in ferromagnet/molecule Bilayers. *Nano Lett.*, 18(8):4659–4663, 2018.

276. D. Chylarecka, T. K. Kim, K. Tarafder, K. Müller, K. Gödel, I. Czekaj, C. Wäckerlin, M. Cinchetti, Md. E. Ali, C. Piamonteze, F. Schmitt, J. P. Wüstenberg, C. Ziegler, F. Nolting, M. Aeschlimann, P. M. Oppeneer, N. Ballav, and T. A. Jung. Indirect magnetic

coupling of manganese porphyrin to a ferromagnetic cobalt substrate. *J. Phys. Chem. C*, 115(4):1295–1301, 2011.

277. H. Wende, M. Bernien, J. Luo, C. Sorg, N. Ponpandian, J. Kurde, J. Miguel, M. Piantek, X. Xu, Ph. Eckhold, W. Kuch, K. Baberschke, P. M. Panchmatia, B. Sanyal, P. M. Oppeneer, and O. Eriksson. Substrate-induced magnetic ordering and switching of iron porphyrin molecules. *Nat. Mater.*, 6(7):516–520, 2007.

278. J. Girovsky, K. Tarafder, C. Wäckerlin, J. Nowakowski, D. Siewert, T. Hählen, A. Wäckerlin, Armin Kleibert, N. Ballav, T. A. Jung, and P. M. M. Oppeneer. Antiferromagnetic coupling of Cr-porphyrin to a bare Co substrate. *Phys. Rev. B*, 90(22):220404, 2014.

279. O. B. Dor, S. Yochelis, S. P. Mathew, R. Naaman, and Y. Paltiel. A chiral-based magnetic memory device without a permanent magnet. *Nat. Commun.*, 4:2256, 2013.

280. O. Ben Dor, S. Yochelis, A. Radko, K. Vankayala, E. Capua, A. Capua, S.-H. Yang, L. T. Baczewski, S. S. P. Parkin, R. Naaman, and Y. Paltiel. Magnetization switching in ferromagnets by adsorbed chiral molecules without current or external magnetic field. *Nat. Commun.*, 8:14567, 2017.

281. R. Naaman, Y. Paltiel, and D. H. Waldeck. Chiral molecules and the electron spin. *Nat. Rev. Chem.*, 3(4):250–260, 2019.

282. C. Fontanesi, E. Capua, Y. Paltiel, D. H. Waldeck, and R. Naaman. Spin-Dependent Processes Measured without a Permanent Magnet. *Adv. Mater.*, 30:1707390, 2018.

283. P. C. Mondal, P. Roy, D. Kim, E. E. Fullerton, H. Cohen, and R. Naaman. Photospintronics: Magnetic Field-Controlled Photoemission and Light-Controlled Spin Transport in Hybrid Chiral Oligopeptide-Nanoparticle Structures. *Nano Lett.*, 16(4):2806–2811, 2016.

284. S. G. Ray, S. S. Daube, G. Leitus, Z. Vager, and R. Naaman. Chirality-induced spin-selective properties of self-assembled monolayers of DNA on gold. *Phys. Rev. Lett.*, 96(3):1–4, 2006.

285. D. Mishra, T. Z. Markus, R. Naaman, M. Kettner, B. Göhler, H. Zacharias, N. Friedman, M. Sheves, and C. Fontanesi. Spin-dependent electron transmission through bacteriorhodopsin embedded in purple membrane. *Proc. Natl. Acad. Sci.*, 110(37):14872–14876, 2013.

286. C. Blouzon, F. Ott, L. Tortech, D. Fichou, and J.-B. Moussy. Anti-ferromagnetic coupling in hybrid magnetic tunnel junctions mediated by monomolecular layers of alpha-sexithiophene. *Appl. Phys. Lett.*, 103:042417, 2013.

287. F. Al Ma'Mari, T. Moorsom, G. Teobaldi, W. Deacon, T. Prokscha, H. Luetkens, S. W. Lee, G. E. Sterbinsky, D. A. Arena, Donald A. MacLaren, M. Flokstra, M. Ali, M. C. Wheeler, G. Burnell, B. J. Hickey, and O. Cespedes. Beating the Stoner criterion using molecular interfaces. *Nature*, 524:69–73, 2015.

288. G. Avedissian, J. Arabski, J. A. Wytko, J. Weiss, and C. Meny. Probing the growth of organic molecular films embedded between Cobalt and iron electrodes: ferromagnetic nuclear resonance approach. *Adv. Funct. Mater.*, 30:2005605, 2020.

289. R. L. Stamps. Mechanisms for exchange bias. *J. Phys. D: Appl. Phys.*, 33:R247–R268, 2000.

290. F. Radu and H. Zabel. Exchange Bias effect of ferro-/antiferromagnetic heterostructures. In Hartmut Zabel and Samuel D. Bader, editors, *Magnetic Heterostructures: Advances and Perspectives in Spinstructures and Spintransport*, pp. 97–184. Springer Berlin Heidelberg, 2008.

291. N. H. Pham, S. Ohya, M. Tanaka, S. E Barnes, and S. Maekawa. Electromotive force and huge magnetoresistance in magnetic tunnel junctions. *Nature*, 458:489–492, 2009.

292. S. Rusponi, T. Cren, N. Weiss, M. Epple, P. Buluschek, L. Claude, and H. Brune. The remarkable difference between surface and step atoms in the magnetic anisotropy of two-dimensional nanostructures. *Nat. Mater.*, 2:546–551, 2003.

293. S. Baumann, F. Donati, S. Stepanow, S. Rusponi, W. Paul, S. Gangopadhyay, I. G. Rau, G. E. Pacchioni, L. Gragnaniello, M. Pivetta, J. Dreiser, C. Piamonteze, C. P. Lutz, R. M. Macfarlane, B. A. Jones, P. Gambardella, A. J. Heinrich, and H. Brune. Origin of perpendicular magnetic anisotropy and large orbital moment in Fe atoms on MgO. *Phys. Rev. Lett.*, 115:237202, 2015.

294. I. G. Rau, S. Baumann, S. Rusponi, F. Donati, S. Stepanow, L. Gragnaniello, J. Dreiser, C. Piamonteze, F. Nolting, S. Gangopadhyay, O. R. Albertini, R. M. Macfarlane, C. P. Lutz, B. A. Jones, P. Gambardella, A. J. Heinrich, and H. Brune. Reaching the magnetic anisotropy limit of a 3d metal atom. *Science*, 344(6187):988–992, 2014.

295. C. Hübner, B. Baxevanis, A. A. Khajetoorians, and D. Pfannkuche. Symmetry effects on the spin switching of adatoms. *Phys. Rev. B*, 90:155134, 2014.

296. F. Natterer, F. Donati, F. Patthey, and H. Brune. Thermal and magnetic-field stability of holmium single-atom magnets. *Phys. Rev. Lett.*, 121:027201, 2018.

297. H. S. Gutowsky, D. W. McCall, and C. P. Slichter. Nuclear magnetic resonance multiplets in liquids. *J. Chem. Phys.*, 21:279–292, 1953.

298. Ruderman M. A. and C. Kittel. Indirect exchange coupling of nuclear magnetic moments by conduction electrons. *Phys. Rev.*, 96:99–102, 1954.

299. T. Kasuya. A Theory of Metallic ferro- and antiferromagnetism on Zener's model. *Prog. Theo. Phy.*, 16(1):45–57, 1956.

300. R. A. Duine, K.-J. Lee, S. S. P. Parkin, and M. D. Stiles. Synthetic antiferromagnetic spintronics. *Nat. Phys.*, 14:217–219, 2018.

301. F. A. Reuse, K. Maschke, V. de Coulon, J. van der Klink, and J.-Ph. Ansermet. Dissipative evolution of quantum statistical ensembles and nonlinear response to a time-periodic perturbation. *Eur. Phys. J. B.*, 36(4):573–592, 2003.

302. H.J . Ziegler and G. W. Pratt. *Magnetic Interactions in Solids*. Oxford University Press, Oxford, 1973.

303. N. Bloembergen and T. J. Rowland. Nuclear spin exchange in solids: Tl^{203} and Tl^{205} magnetic resonance in thallium and thallic oxide. *Phys. Rev.*, 97:1679–1698, Mar 1955.

304. J.B. Robert and L. Wiesenfeld. Magnetic anisotropic interactions of nuclei in condensed matter. *Phys. Rep.*, 86(7):363–401, 1982.

305. M. P. Ledbetter, G. Saielli, A. Bagno, N. Tran, and M. V. Romalis. Observation of scalar nuclear spin–spin coupling in van der Waals complexes. *Proc. Natl. Acad. Sci.*, 109(31):12393–12397, 2012.

306. M. V. Hosseini, Z. Karimi, and J. Davoodi. Indirect exchange interaction between magnetic impurities in one-dimensional gapped helical states. *J. Phys.: Condens. Matter*, 33:085801, Dec 2020.

307. K. Wang, V. Bheemarasetty, and G. Xiao. Spin textures in synthetic antiferromagnets: challenges, opportunities, and future directions. *APL Mater.*, 11(7):070902, 2023.

308. S. Parkin, Xin Jiang, C. Kaiser, A. Panchula, K. Roche, and M. Samant. Magnetically engineered spintronic sensors and memory. *Proc. IEEE*, 91(5):661–680, 2003.

309. D. Houssameddine, J. F. Sierra, D. Gusakova, B. Delaet, U. Ebels, L. D. Buda-Prejbeanu, M.-C. Cyrille, B. Dieny, B. Ocker, J. Langer, and W. Maas. Spin torque driven excitations in a synthetic antiferromagnet. *Appl. Phys. Lett.*, 96(7):072511, 2010.

310. T. Newhouse-Illige, Y. Liu, M. Xu, Reifsnyder H. D., A. Kundu, H. Almasi, C. Bi, X. Wang, J. W. Freeland, D. J. Keavney, C. J. Sun, Y. H. Xu, M. Rosales, X. M. Cheng, S. Zhang, K. A. Mkhoyan, and W. G. Wang. Voltage-controlled interlayer coupling in perpendicularly magnetized magnetic tunnel junctions. *Nat. Commun.*, 8:15232, 2017.

311. M. Fechner, P. Zahn, S. Ostanin, M. Bibes, and I. Mertig. Switching magnetization by 180° with an electric field. *Phys. Rev. Lett.*, 108:197206, May 2012.

312. L. Wang, X. Zhou, R. Hao, and T. Min. Low-energy picosecond magnetic switching for synthetic ferrimagnetic free layer utilizing the electric-field-tuned RKKY effect. *IEEE Trans. Mag.*, 57:4200407, 2021.

313. L. Trifunovic, O. Dial, M. Trif, J. R. Wootton, R. Abebe, A. Yacoby, and D. Loss. Long-distance spin-spin coupling via floating gates. *Phys. Rev. X*, 2:011006, 2012.

314. J.H. Van Vleck. Models of exchange coupling in ferromagnetic media. *Rev. Mod. Phys.*, 25:211–219, 1953.

315. D. P. Young, D. Hall, M. E. Torelli, Z. Fisk, J. D. Thompson, H. R. Ott, S. B. Oseroff, R. G. Goodrich, and R. Zysler. High-temperature weak ferromagnetism in a low-density free-electron gas. *Nature*, 397:412–414, 1999.

316. H. R. Ott, J. L. Gavilano, B. Ambrosini, P. Vonlanthen, E. Felder, L. Degiorgi, D. P. Young, Z. Fisk, and R. Zysler. Unusual magnetism of hexaborides. *Physica B*, 281–282:423–427, 2000.

317. P. Vonlanthen, S. Paschen, D. Pushin, A. D. Bianchiv, H. R. Ott, J. L. Sarrao, and Z. Fisk. Thermal conductivity of EuB_6. *Phys. Rev. B*, 62:3246–3249, 2000.

318. P. Phillips. *Advanced Solid State Physics*. Westview Press, Boulder, Oxford, 2003.

319. P. Fulde. *Electron Correlations in Molecules and Solids*. Springer Series in Solid State Sciences, Berlin, Heidelberg, New York, 1995.

320. P. Anderson. Localized Magnetic States in Metals. *Phys. Rev.*, 124:41–53, 1961.

321. V. L. Moruzzi, J. F. Janak, and A. R. Williams. *Calculated Electronic Properties of Metals*. Pergamon Press, Oxford, 1978.

322. K. N. Altmann, D. Y. Petrovykh, G. J. Mankey, N. Shannon, N. Gilman, M. Hochstrasser, R. F. Willis, and F. J. Himpsel. Enhanced spin polarization of conduction electrons in Ni explained by comparison with Cu. *Phys. Rev. B*, 61:15661–15665, 2000.

323. P. Coleman. *Introduction to Many-Body Physics*. Cambridge University Press, Cambridge, New York, Melbourne, Delhi, Singapore, 2016.

324. S. N. Kaul, A. Semwal, and H. E. Schaefer. Exchange-enhanced Pauli spin paramagnetism in nanocrystalline Ni_3Al. *Phys. Rev. B*, 62(21):13892–13895, 2000.

325. A.H. Mitchell. Ferromagnetic Relaxation by the Exchange Interaction between Ferromagnetic Electrons and Conduction Electrons. *Phys. Rev.*, 105:1439–1444, 1957.

326. U. Larsen. A simple derivation of the s-d exchange interaction. *J. Phys. C: Solid State Phys.*, 4(13):1835–1836, 1971.

327. M. Viret, D. Vignoles, D. Cole, J. M. D. Coey, W. Allen, D. S. Daniel, and J. F. Gregg. Spin scattering in ferromagnetic thin films. *Phys. Rev. B*, 53:8464–8468, 1996.

328. J.-E. Wegrowe, X. Hoffer, Ph. Guittienne, A. Fabian, L. Gravier, T. Wade, and J.-Ph. Ansermet. Spin-polarized current induced magnetization switch: Is the modulus of the magnetic layer conserved? *J. Appl. Phys.*, 91:6806–6811, 2002.

329. T. Schaepers. *Semiconductor Spintronics*. De Gruyter, Berlin, Boston, 2016.

330. E. Frantzeskakis. *Spectroscopic Studies on Semiconducting Interfaces with Giant Spin Splitting*. PhD thesis, EPFL, 2010.

331. S. LaShell, B. A. McDougall, and E. Jensen. Spin splitting of an Au(111) surface state band observed with angle resolved photoelectron spectroscopy. *Phys. Rev. Lett.*, 77:3419–3422, 1996.

332. L. Petersen and P. A Hedegård. A simple tight-binding model of spin-orbit splitting of sp-derived surface states. *Surf. Sci.*, 459:49–56, 2000.

333. Nicolay G., Reinert F., Hüfiner S., and Blaha P. Spin-orbit splitting of the L-Gap surface state on Au(111) and Ag(111). *Phys. Rev. B*, 65:033407, 2001.

334. C. R. Ast, J. Henk, A. Ernst, L. Moreschini, M. C. Falub, D. Pacilé, P. Bruno, K. Kern, and M. Grioni. Giant Spin Splitting through Surface Alloying. *Phys. Rev. Lett.*, 98:186807, 2007.

335. T. Hirahara, K. Miyamoto, I. Matsuda, T. Kadono, A. Kimura, T. Nagao, G. Bihlmayer, E. V. Chulkov, S. Qiao, K. Shimada, H. Namatame, M. Taniguchi, and S. Hasegawa. Direct observation of spin splitting in bismuth surface states. *Phys. Rev. B*, 76:153305, 2007.

336. L. Moreschini, A. Bendounan, I. Gierz, C. R. Ast, H. Mirhosseini, H. Höchst, K. Kern, J. Henk, A. Ernst, S. Ostanin, F. Reinert, and M. Grioni. Assessing the atomic contribution to the Rashba spin-orbit splitting in surface alloys: Sb/Ag(111). *Phys. Rev. B*, 79:075424, 2009.

337. I. Gierz, T. Suzuki, E. Frantzeskakis, S. Pons, S. Ostanin, A. Ernst, J. Henk, M. Grioni, K. Kern, and C. R. Ast. Silicon Surface with Giant Spin Splitting. *Phys. Rev. Lett.*, 103:046803, 2009.

338. K. Yaji, Y. Ohtsubo, S. Hatta, H. Okuyama, K. Miyamoto, T. Okuda, A. Kimura, H. Namatame, M. Taniguchi, and T. Aruga. Large Rashba spin splitting of a metallic surface-state band on a semiconductor surface. *Nat. Commun.*, 1(1):17, 2010.

339. E. E. Krasovskii. Spin–orbit coupling at surfaces and 2D materials. *J. Phys.: Condens. Matter*, 27(49):493001, 2015.

340. S. Oh and H. J. Choi. Orbital angular momentum analysis for giant spin splitting in solids and nanostructures. *Sci. Rep.*, 7(1):1–10, 2017.

341. L. Petersen and P. Hedegard. A simple tight-binding model of spin-orbit splitting of sp-derived surface states. *Surf. Sci.*, 459:49–56, 2000.

342. P. M. Haney, H.-W. Lee, K.-J. Lee, A. Manchon, and M. D. Stiles. Current-induced torques and interfacial spin-orbit coupling. *Phys. Rev. B*, 88:214417, 2013.

343. P. M. Haney, H.-W. Lee, K. J. Lee, A. Manchon, and M. D. Stiles. Current-induced torques and interfacial spin-orbit coupling. *Phys. Rev. B*, 88(21):1–7, 2013.

344. F Freimuth, S Blügel, and Y Mokrousov. Berry phase theory of Dzyaloshinskii–Moriya interaction and spin–orbit torques. *J. Phys.: Condens. Matter*, 26(10):104202, 2014.

345. H. Kurebayashi, J. Sinova, D. Fang, A. C. Irvine, T. D. Skinner, J. Wunderlich, V. Novák, R. P. Campion, B. L. Gallagher, E. K. Vehstedt, L. P. Zârbo, K. Výborný, A. J. Ferguson, and T. Jungwirth. An antidamping spin–orbit torque originating from the Berry curvature. *Nat. Nanotechnol.*, 9(3):211–217, 2014.

346. P. Gambardella and I. M. Miron. Current-induced spin-orbit torques. *Philos. Trans. Royal Soc. A*, 369(1948):3175–3197, 2011.

347. P. Gambardella, H. Luo, and L. J. Heyderman. Magnetic logic driven by electric current. *Phys. Today*, April:62–63, 2021.

348. A. Manchon, J. Železný, I. M. Miron, T. Jungwirth, J. Sinova, A. Thiaville, K. Garello, and P. Gambardella. Current-induced spin-orbit torques in ferromagnetic and antiferromagnetic systems. *Rev. Mod. Phys.*, 91(3):035004, 2019.

349. D. Xiao, Y. Yao, Z. Fang, and Q. Niu. Berry-phase effect in anomalous thermoelectric transport. *Phys. Rev. Lett.*, 97(2):2–5, 2006.

350. J.D. Jackson. *Classical Electrodynamics*. John Wiley & Sons, New York, 1975.

351. S. J. Cyvin, J. Brunvoll, and B. N. Cyvin. The hunt for concealed non-Kekuléan polyhexes. *J. Math. Chem.*, 4(1):47–54, 1990.

352. O. Gröning, S. Wang, X. Yao, C. A. Pignedoli, G. Borin Barin, C. Daniels, A. Cupo, V. Meunier, X. Feng, A. Narita, K. Müllen, P. Ruffieux, and R. Fasel. Engineering of robust topological quantum phases in graphene nanoribbons. *Nature*, 560(7717):209–213, 2018.

353. S. Mishra, D. Beyer, K. Eimre, S.. Kezilebieke, R. Berger, O. Gröning, C. A. Pignedoli, K. Müllen, P. Liljeroth, P. Ruffieux, X. Feng, and R. Fasel. Topological frustration induces unconventional magnetism in a nanographene. *Nat. Nanotechnol.*, 15(1):22–28, 2020.

354. S. Mishra, D. Beyer, K. Eimre, R. Ortiz, J. Fernández-Rossier, R. Berger, O. Gröning, C. A. Pignedoli, R. Fasel, X. Feng, and P. Ruffieux. Collective All-Carbon Magnetism in Triangulene Dimers. *Angew. Chem., Int. Ed. Engl.*, 59(29):12041–12047, 2020.

355. S. Mishra, X. Yao, Q. Chen, K. Eimre, O. Gröning, R. Ortiz, M. Di Giovannantonio, J. C. Sancho-García, J. Fernández-Rossier, C. A. Pignedoli, K. Müllen, P. Ruffieux, A. Narita, and R. Fasel. Large magnetic exchange coupling in rhombus-shaped nanographenes with zigzag periphery. *Nat. Chem.*, 13(6):581—586, 2021.

356. A. Keerthi, C. Sánchez-Sánchez, O. Deniz, P. Ruffieux, D. Schollmeyer, X. Feng, A. Narita, R. Fasel, and K. Müllen. On-surface Synthesis of a Chiral Graphene Nanoribbon with Mixed Edge Structure. *Chem. Asian J.*, 15(22):3807–3811, 2020.

357. R. Pawlak, X. Liu, S. Ninova, P. D'Astolfo, C. Drechsel, S. Sangtarash, R. Häner, S. Decurtins, H. Sadeghi, C. J. Lambert, U. Aschauer, S.-X. Liu, and E. Meyer. Bottom-up Synthesis of Nitrogen-Doped Porous Graphene Nanoribbons. *J. Am. Chem. Soc.*, 142(29):12568–12573, 2020.

358. T. Ando. Spin-orbit interaction in carbon nanotubes. *J. Phys. Soc. Jpn.*, 69(6):1757–1763, 2000.

359. N. Xu, P. K. Biswas, J. H. Dil, R. S. Dhaka, G. Landolt, S. Muff, C. E. Matt, X. Shi, N. C. Plumb, M. Radović, E. Pomjakushina, K. Conder, A. Amato, S. V. Borisenko, R. Yu, H. M. Weng, Z. Fang, X. Dai, J. Mesot, H. Ding, and M. Shi. Direct observation of the spin texture in SmB_6 as evidence of the topological Kondo insulator. *Nat. Commun.*, 5(1):4566, 2014.

360. Y. Ando. Topological Insulator Materials. *J. Phys. Soc. Jpn.*, 82(10):102001, 2013.

361. C. Kloeffel, M. Trif, and D. Loss. Strong spin-orbit interaction and helical hole states in Ge/Si nanowires. *Phys. Rev. B*, 84:195314, 2011.

362. S. Furthmeier, F. Dirnberger, M. Gmitra, A. Bayer, M. Forsch, J. Hubmann, C. Schüller, E. Reiger, J. Fabian, T.s Korn, and D. Bougeard. Enhanced spin–orbit coupling in core/shell nanowires. *Nat. Commun.*, 7(1):12413, 2016.

363. W.-T. Wang, C. L. Wu, S. F. Tsay, M. H. Gau, I. Lo, H. F. Kao, D. J. Jang, J.-C. Chiang, M.-E. Lee, Y.-C. Chang, C.-N. Chen, and H. C. Hsueh. Dresselhaus effect in bulk wurtzite materials. *Appl. Phys. Lett.*, 91(8):082110, 2007.

364. L. Liang, J. Shan, Q. H. Chen, J. M. Lu, G. R. Blake, T. T. M. Palstra, G. E. W. Bauer, B. J. van Wees, and J. T. Ye. Gate-controlled magnetoresistance of a paramagnetic-insulator—platinum interface. *Phys. Rev. B*, 98:134402, 2018.

365. L. Liang, Q. Chen, J. Lu, W. Talsma, J. Shan, G. R. Blake, T. T. M. Palstra, and J. Ye. Inducing ferromagnetism and Kondo effect in platinum by paramagnetic ionic gating. *Sci. Adv.*, 4(4):eaar2030, 2018.

366. Atesci H. *Towards an ab-axis giant proximity effect using ionic liquid gating*. PhD thesis, Leiden Institute of Physics, 2018.

367. S. Shimizu, K. S. Takahashi, T. Hatano, M. Kawasaki, Y. Tokura, and Y. Iwasa. Electrically Tunable Anomalous Hall Effect in Pt Thin Films. *Phys. Rev. Lett.*, 111:216803, 2013.

368. T. Berlijn, P. C. Snijders, O. Delaire, H.-D. Zhou, T. A. Maier, H.-B. Cao, S.-X. Chi, M. Matsuda, Y. Wang, M. R. Koehler, P. R. C. Kent, and H. H. Weitering. Itinerant Antiferromagnetism in RuO_2. *Phys. Rev. Lett.*, 118:077201, 2017.

369. Z. Sniadecki, M. Werwinski, A. Szajek, U. K. Roessler, and B. Idzikowski. Induced magnetic ordering in alloyed compounds based on Pauli paramagnet YCo_2. *J. Appl. Phys.*, 115:17E129, 2014.

370. G. Rahman, I. G. Kim, H. K. D. H. Bhadeshia, and A. J. Freeman. First-principles investigation of magnetism and electronic structures of substitutional $3d$ transition-metal impurities in BCC Fe. *Phys. Rev. B*, 81:184423, 2010.

371. G. M. Vanacore, G. Berruto, I. Madan, E. Pomarico, P. Biagioni, R. J. Lamb, D. McGrouther, O. Reinhardt, I. Kaminer, B. Barwick, H. Larocque, V. Grillo, E. Karimi, F. J. García de Abajo, and F. Carbone. Ultrafast generation and control of an electron vortex beam via chiral plasmonic near fields. *Nat. Mater.*, 18(6):573–579, 2019.

372. A. G. Gurevich and G. A. Melkov. *Magnetization Oscillations and Waves*. CRC Press, Boca Raton, New York, London, Tokyo, 1996.

373. M.E. Rose. *Elementary Theory of Angular Momentum*. Dover Publications Inc., New York, 1957.

374. S.D. Brechet and J.-Ph. Ansermet. Variational principle for magnetisation dynamics in a temperature gradient. *Europhys. Lett.*, 112:17006, 2015.

375. H. P. Callen. *Thermodynamics*. John Wiley & Sons, New York, 1960.

376. J.-Ph. Ansermet and S. Bréchet. *Principles of Thermodynamics*. Cambridge University Press, Cambridge, New York, Melbourne, New Delhi, Singapore, 2019.

377. S. D. Brechet and J.-Ph. Ansermet. Thermodynamics of continuous media with electric and magnetic dipoles. *Eur. Phys. J. B*, 86:318, 2013.

378. H. Yu, K.O. d'Allivy, V. Cros, R. Bernard, P. Bortolotti, A. Anane, F. Brandl, R. Huber, I. Stasinopoulos, and D. Grundler. Magnetic thin-film insulator with ultra-low spin wave damping for coherent nanomagnonics. *Sci. Rep.*, 4(1):6848, 2014.

379. T. L. Gilbert. A phenomenological theory of damping in ferromagnetic materials. *IEEE Trans. Magn.*, 40:3443–3449, 2004.

380. R. Courant and D. Hilbert. *Methods of Mathematical Physics*. Wiley Interscience, New York, London, Sydney, 1953.

381. C. Liu, J. Chen, T. Liu, F. Heimbach, H. Yu, Y. Xiao, J. Hu, M. Liu, H. Chang, T. Stueckler, S. Tu, Y. Zhang, Y. Zhang, P. Gao, Z. Liao, D. Yu, K. Xia, N. Lei, W. Zhao, and M. Wu. Long-distance propagation of short-wavelength spin waves. *Nat. Commun.*, 9(1):738, 2018.

382. G. Dieterle, J. Förster, H. Stoll, A. S. Semisalova, S. Finizio, A. Gangwar, M. Weigand, M. Noske, M. Fähnle, I. Bykova, J. Gräfe, D. A. Bozhko, H. Yu. Musiienko-Shmarova, V. Tiberkevich, A. N. Slavin, C. H. Back, J. Raabe, G. Schütz, and S. Wintz. Coherent

excitation of heterosymmetric spin waves with ultrashort wavelengths. *Phys. Rev. Lett.*, 122:117202, 2019.

383. V.S. L'vov. *Wave Turbulence Under Parametric Excitation, Applications to Magnets.* Springer Series in Nonlinear Dynamics, Heidelberg, 1994.

384. P. W. Anderson. More is different. *Science*, 177(4047):393–396, 1972.

385. P. W. Anderson. *Concepts in Solids: Lectures on the Theory of Solids.* World Scientific, Singapore, New Jersey, London, Hong Kong, 1963.

386. H. Callen. *Thermodynamics and Introduction to Thermostatistics.* Wiley, New York, Chichester, Brisbane, Toronto, Singapore, 1985.

387. M. Grassi, M.z Geilen, K.A. Oukaci, Y. Henry, D. Lacour, D. Stoeffler, M. Hehn, P. Pirro, and M. Bailleul. Higgs and Goldstone spin-wave modes in striped magnetic texture. *Phys. Rev. B*, 105:094444, 2022.

388. Camara I.S., Tacchi S., Garnier L.-C., Eddrief M., Fortuna F., Carlotti G., and Marangolo M. Magnetization dynamics of weak stripe domains in Fe–N thin films: a multi-technique complementary approach. *J. Phys.: Condens. Matter*, 29(46):465803, 2017.

389. J. M. Radcliffe. Some properties of coherent spin states. *J. Phys. A: Gen. Phys.*, 4(3):313–323, 1971.

390. J. Van Kranendonk and J. H. Van Vleck. Spin waves. *Rev. Mod. Phys.*, 30:1–23, 1958.

391. V. L. Safonov. *Nonequilibrium Magnons, Theory, Experiment, and Applications.* Wiley-VCH, 2013.

392. D. C. Mattis. *The Theory of Magnetism Made Simple.* World Scientific, Singapore, Hackensack New Jersey, London, 2006.

393. M. H. Cohen and F. Keffer. Dipolar Sums in the Primitive Cubic Lattices. *Phys. Rev.*, 99:1128–1134, 1955.

394. Sandercock J.R. and Wettling W. Light scattering from surface and bulk thermal magnons in iron and nickel. *J. Appl. Phys.*, 50:7784–7789, 1979.

395. Kalinikos B. A and Slavin A. N. Theory of of dipole-exchange spin wave spectrum for ferromagnetic films with mixed exchange boundary conditions. *J. Phys. C. Solid State Phys.*, 19:7013–7033, 1986.

396. H. B. Callen. A ferromagnetic dynamical equation. *J. Phys. Chem. Solids*, 4:256–270, 1958.

397. S. D. Brechet and J.-Ph. Ansermet. Thermodynamics of continuous media with intrinsic rotation and magnetoelectric coupling. *Continuum Mech. Thermodyn.*, 26(2):115–142, 2013.

398. M. Sparks. *Ferromagnetic Relaxation Theory.* McGraw-Hill, Inc., New York, San Francisco, Toronto, London, 1964.

399. S.V. Vonsovskii, editor. *Ferromagnetic Resonance.* Pergamon Press, Oxford, 1966.

400. J. B. Sokoloff. Theory of ferromagnetic resonance relaxation in very small solids. *J. Appl. Phys.*, 75(10):6075–6077, 1994.

401. L. Chen, S. Mankovsky, S. Wimmer, M. A. W. Schoen, H. S. Körner, M. Kronseder, D. Schuh, D. Bougeard, H. Ebert, D. Weiss, and C. H. Back. Emergence of anisotropic Gilbert damping in ultrathin Fe layers on GaAs(001). *Nature Physics*, 14(5):490–494, 2018.

402. Y. Li, F. Zeng, S. S.-L. Zhang, H. Shin, H. Saglam, V. Karakas, O.n Ozatay, J. E. Pearson, O. G. Heinonen, Y. Wu, A. Hoffmann, and W. Zhang. Giant Anisotropy of Gilbert Damping in Epitaxial CoFe Films. *Phys. Rev. Lett.*, 122:117203, 2019.

403. K. Gilmore, M. D. Stiles, J. Seib, D. Steiauf, and M. Fähnle. Anisotropic damping of the magnetization dynamics in Ni, Co, and Fe. *Phys. Rev. B*, 81:174414, 2010.

404. D. Steiauf, J. Seib, and M. Fähnle. Unified theory of near-adiabatic magnetization dynamics for collinear and noncollinear magnetization. *Phys. Rev. B*, 78:020410R, 2008.

405. W. B. Yelon and L. Berger. Magnon Heat Conduction and Magnon Scattering Processes in Fe-Ni Alloys. *Phys. Rev. B*, 6:1974–1985, 1972.

406. N. Zhu, H. Chang, A. Franson, T. Liu, X. Zhang, E. Johnston-Halperin, M. Wu, and H. X. Tang. Patterned growth of crystalline $Y_3Fe_5O_{12}$ nanostructures with engineered magnetic shape anisotropy. *Appl. Phys. Lett.*, 110(25):252401, 2017.

407. P. Saraiva, A. Nogaret, J. C. Portal, H. E. Beere, and D. A. Ritchie. Dipolar spin waves of lateral magnetic superlattices. *Phys. Rev. B*, 82:224417, 2010.

408. V. P. Antropov, M. I. Katsnelson, M. van Schilfgaarde, and B. N. Harmon. Ab initio spin dynamics in magnets. *Phys. Rev. Lett.*, 75:729–732, 1995.

409. V. P. Antropov, M. I. Katsnelson, B. N. Harmon, M. van Schilfgaarde, and D. Kusnezov. Spin dynamics in magnets: equation of motion and finite temperature effects. *Phys. Rev. B*, 54:1019–1035, 1996.

410. Q. Niu and L. Kleinman. Spin-wave dynamics in real crystals. *Phys. Rev. Lett.*, 80:2205–2208, 1998.

411. S. Neusser, G. Duerr, H. G. Bauer, S. Tacchi, M. Madami, G. Woltersdorf, G. Gubbiotti, C. H. Back, and D. Grundler. Anisotropic propagation and damping of spin waves in a nanopatterned antidot lattice. *Phys. Rev. Lett.*, 105:067208, 2010.

412. H. Wang, Y. Yang, M. Madami, Y.n Wang, M.n Du, J. Chen, Y. Zhang, L. Sheng, J. Zhang, C. Wen, Y. Zhang, S. Hao, G. Yu, X. Han, G. Gubbiotti, K. Shen, J. Zhang, and H. Yu. Anomalous anisotropic spin-wave propagation in thin manganite films with uniaxial magnetic anisotropy. *Appl. Phys. Lett.*, 120:192402, 2022.

413. T. Yu, Z. Luo, and G. E. W. Bauer. Chirality as generalized spin–orbit interaction in spintronics. *Phys. Rep.*, 1009:1–115, 2023.

414. Cortes-Ortuno D. and Landeros P. Influence of the Dzyaloshinskii–Moriya interaction on the spin-wave spectra of thin films. *J. Phys. Condens. Matter*, 25:156001, 2013.

415. J. H. Moon, S. M. Seo, K. J. Lee, K. W. Kim, J. Ryu, H. Lee, R. D. McMichael, and M. D. Stiles. Spin-wave propagation in the presence of interfacial Dzyaloshinskii-Moriya interaction. *Phys. Rev. B*, 88(18):1–6, 2013.

416. A. Qaiumzadeh, I. A. Ado, R. A. Duine, M. Titov, and A. Brataas. Theory of the Interfacial Dzyaloshinskii-Moriya Interaction in Rashba Antiferromagnets. *Phys. Rev. Lett.*, 120(19):197202, 2018.

417. X. Ma, G. Yu, C. Tang, X. Li, C. He, J Shi, K. L. Wang, and X. Li. Interfacial Dzyaloshinskii-Moriya Interaction: Effect of $5d$ Band Filling and Correlation with Spin Mixing Conductance. *Phys. Rev. Lett.*, 120(15):157204, 2017.

418. H. T. Nembach, J. M. Shaw, M. Weiler, E. Jué, and T. J. Silva. Linear relation between Heisenberg exchange and interfacial Dzyaloshinskii-Moriya interaction in metal films. *Nat. Phys.*, 11(10):825–829, 2015.

419. J. M. Lee, C. Jang, B. C. Min, S. W. Lee, K. J. Lee, and J. Chang. All-Electrical Measurement of Interfacial Dzyaloshinskii-Moriya Interaction Using Collective Spin-Wave Dynamics. *Nano Lett.*, 16(1):62–67, 2016.

420. H. Wang, J. Chen, T. Liu, J. Zhang, K. Baumgaertl, C. Guo, Y. Li, C. Liu, P. Che, S. Tu, Song Liu, P. Gao, X. Han, D. Yu, M. Wu, D. Grundler, and H. Yu. Chiral Spin-Wave Velocities Induced by All-Garnet Interfacial Dzyaloshinskii-Moriya Interaction in Ultrathin Yttrium Iron Garnet Films. *Phys. Rev. Lett.*, 124:027203, 2020.

421. R. Schlitz, S. Vélez, A. Kamra, C.-H. Lambert, M. Lammel, S. T. B. Goennenwein, and P. Gambardella. Control of nonlocal magnon spin transport via magnon drift currents. *Phys. Rev. Lett.*, 126:257201, 2021.

422. C. O. Avci, E. Rosenberg, L. Caretta, F. Büttner, M. Mann, C. Marcus, D. Bono, C. A. Ross, and G. S.D. Beach. Interface-driven chiral magnetism and current-driven domain walls in insulating magnetic garnets. *Nat. Commun.*, 14(6):561–566, 2019.

423. V. G. Bar'yakhtar. Phenomenological description of relaxation processes in magnetic materials. *Sov. Phys. JETP*, 60:683–687, 1984.

424. J.H. Van Vleck. On the Anisotropy of Cubic Ferromagnetic Crystals. *Phys. Rev.*, 52:1178–1198, 1937.

425. H. Brooks. Ferromagnetic Anisotropy and the Itinerant Electron Model. *Phys. Rev.*, 58:909–918, 1940.

426. S. V. Halilov, H. Eschrig, A. Y. Perlov, and P. M. Oppeneer. Adiabatic spin dynamics from spin-density-functional theory: Application to Fe, Co, and Ni. *Phys. Rev. B*, 58:293–302, 1998.

427. D. C. Venerus and H. C. Oettinger. *A Modern Course Course in Transport Phenomena*. Cambridge University Press, Cambridge, New York, Melbourne, New Delhi, Singapore, 2018.

428. M. I. Dyakonov and V. I. Perel. Current-induced spin orientation of electrons in semiconductors. *Phys. Lett. A*, 35(6):459–460, 1971.

429. M. Johnson and R. H. Silsbee. Interfacial charge-spin coupling: Injection and detection of spin magnetization in metals. *Phys. Rev. Lett.*, 55:1790–1793, 1985.

430. M. Johnson and R. H. Silsbee. Thermodynamic analysis of interfacial transport and of the thermomagnetoelectric system. *Phys. Rev. B*, 35(10):4959–4972, 1987.

431. P. C. Van Son and H. Van Kempen and P. Wyder. Boundary Resistance of the Ferromagnetic-Nonferromagnetic Metal Interface. *Phys. Rev. Lett.*, 58(21): 2271–2273, 1987.

432. L. Gravier, S. Serrano-Guisan, F. Reuse, and J.-Ph. Ansermet. Spin-dependent Peltier effect of perpendicular currents in multilayered nanowires. *Phys. Rev. B*, 73:052410, 2006.

433. L. Gravier, S. Serrano-Guisan, F. Reuse, and J.-Ph.Ansermet. Thermodynamic description of heat and spin transport in magnetic nanostructures. *Phys. Rev. B*, 73:024419, 2006.

434. I. Prigogine. *Thermodynamics of Irreversible Processes*. Interscience Publishers, New York, London, Sydney, 1955.

435. J.-Ph. Ansermet. Thermodynamic description of spin mixing in spin-dependent transport. *IEEE Trans. Mag.*, 44(3):329–335, 2008.

436. J. Dubois and J-Ph Ansermet. Second-harmonic response and temperature derivative of noncollinear spin valves. *Phys. Rev. B*, 78(18):184430, 2010.

437. J.-E. Wegrowe, A. Comment, Y. Jaccard, J.-Ph. Ansermet, N. M. Dempsey, and J.-P. Nozières. Spin-dependent scattering of a domain wall of controlled size. *Phys. Rev. B*, 61(18):12216–12220, 2000.

438. G.E.W. Bauer, A.H. MacDonald, and S. Maekawa. Spin Caloritronics. *Solid State Commun.*, 150(11–12):459–460, 2010.

439. G. E. W. Bauer, E. Saitoh, and B. J. Van Wees. Spin caloritronics. *Nat. Mater.*, 11(5):391–399, 2012.

440. L. Piraux, A. Fert, P.A. Schroeder, R. Loloee, and P. Etienne. Large magnetothermoelectric power in Co/Cu, Fe/Cu and Fe/Cr multilayers. *J. Magn. Magn. Mater.*, 110(3):L247–L253, 1992.

441. J. Shi, R. C. Yu, S. S. P. Parkin, and M. B. Salamon. Magnetothermopower of Co/Cu multilayers. *J. Appl. Phys.*, 73:5524–5426, 1993.

442. J. Shi and K. Pettit. Field-dependent thermoelectric power and thermal conductivity in multilayered and granular giant magnetoresistive systems. *Phys. Rev. B*, 54(21):15273–15283, 1996.

443. S.A. Baily, M.B. Salamon, and W. Oepts. Magnetothermopower of cobalt/copper multilayers with gradient perpendicular to planes. *J. Appl. Phys.*, 87(9 II):4855–4857, 2000. cited By 25.

444. O. Tsyplyatyev, O. Kashuba, and V.I. Fal'ko. Thermally excited spin current and giant magnetothermopower in metals with embedded ferromagnetic nanoclusters. *Phys. Rev. B*, 74:132403, 2006.

445. W. M. Saslow. Spin Hall effect and irreversible thermodynamics: center-to-edge transverse current-induced voltage. *Phys. Rev. B*, 91:014401, 2015.

446. H. Yu, S. Bréchet, and J.-Ph Ansermet. Spin Caloritronics, origin and outlook. *Phys. Lett. A*, 381(9):825–837, 2017.

447. M. Johnson and R. H. Silsbee. Ferromagnet-nonferromagnet interface resistance [1]. *Phys. Rev. Lett.*, 60(4):377, 1988.

448. M. R. Sears and W. M. Saslow. Irreversible thermodynamics of transport across interfaces. *Can. J. Phys.*, 89(10):1041–1050, 2011.

449. D. Meier, T. Kuschel, L. Shen, A. Gupta, T. Kikkawa, K. Uchida, E. Saitoh, J.-M. Schmalhorst, and G. Reiss. Thermally driven spin and charge currents in thin $NiFe_2O_4$/Pt films. *Phys. Rev. B*, 87:054421, 2013.

450. K. Uchida, S. Takahashi, K. Harii, J. Ieda, W. Koshibae, K. Ando, S. Maekawa, and E. Saitoh. Observation of the spin Seebeck effect. *Nature*, 455:778–781, 2008.

451. K. Uchida, T. Ota, K. Harii, S. Takahashi, S. Maekawa, Y. Fujikawa, and E. Saitoh. Spin-Seebeck effects in $Ni_{81}Fe_{19}$/Pt films. *Solid State Commun.*, 150(11–12):524–528, 2010.

452. C. M. Jaworski, J. Yang, S. Mack, D. D. Awschalom, J. P. Heremans, and R. C. Myers. Observation of the spin-Seebeck effect in a ferromagnetic semiconductor. *Nat. Mater.*, 9(11):898–903, 2010.

453. K. Uchida, J. Xiao, H. Adachi, J. Ohe, S. Takahashi, J. Ieda, T. Ota, Y. Kajiwara, H. Umezawa, H. Kawai, G.E.W. Bauer, S. Maekawa, and E. Saitoh. Spin Seebeck insulator. *Nature Materials*, 9(11):894–897, 2010.

454. S. Seki, T. Ideue, M. Kubota, Y. Kozuka, R. Takagi, M. Nakamura, Y. Kaneko, M. Kawasaki, and Y. Tokura. Thermal generation of spin current in an antiferromagnet. *Phys. Rev. Lett.*, 115(26), 2015.

455. S.M. Wu, W. Zhang, A. KC, P. Borisov, J.E. Pearson, J.S. Jiang, D. Lederman, A. Hoffmann, and A. Bhattacharya. Antiferromagnetic spin Seebeck effect. *Phys. Rev. Lett.*, 116:097204, 2016.

456. S. D. Brechet and J.-Ph. Ansermet. Heat-driven spin currents on large scales. *Phys. Status Solidi – Rapid Res. Lett.*, 5(12):423–425, 2011.

457. L. J. Cornelissen, K. J.H. Peters, G. E.W. Bauer, R. A. Duine, and B. J. Van Wees. Magnon spin transport driven by the magnon chemical potential in a magnetic insulator. *Phys. Rev. B*, 94:1–16, 2016.

458. M. D. Stiles and A. Zangwill. Anatomy of spin-transfer torque. *Phys. Rev. B*, 66:014407, 2002.

459. V. M. Edelstein. Spin polarization of conduction electrons induced by electric current in two-dimensional asymmetric electron systems. *Solid State Commun.*, 73(3):233 – 235, 1990.

460. R. H. Silsbee. Spin–orbit induced coupling of charge current and spin polarization. *J. Phys.: Condens. Matter*, 16(7):R179–R207, 2004.

461. F. J. Jedema, A. T. Filip, and B. J. van Wees. Electrical spin injection and accumulation at room temperature in an all-metal mesoscopic spin valve. *Nature*, 410(6826):345–348, 2001.

462. H. Idzuchi, Y. Fukuma, and Y. Otani. Spin transport in non-magnetic nano-structures induced by non-local spin injection. *Phys. E: Low-Dimens.*, 68:239–263, 2015.

463. J. Bass and Pratt Jr. W. P. Spin-diffusion lengths in metals and alloys, and spin-flipping at metal/metal interfaces: an experimentalist's critical review. *J. Phys.: Condens. Matter*, 19(18):183201, 2007.

464. Y. Fukuma, L. Wang, H. Idzuchi, S. Takahashi, S. Maekawa, and Y. Otani. Giant enhancement of spin accumulation and long-distance spin precession in metallic lateral spin valves. *Nat. Mater.*, 10(7):527–531, 2011.

465. A. C. Pinon, J. Schlagnitweit, P. Berruyer, A. J. Rossini, M. Lelli, E. Socie, M. Tang, T. Pham, A. Lesage, S. Schantz, and L. Emsley. Measuring nano- to microstructures from relayed dynamic nuclear polarization NMR. *J. Phys. Chem. C*, 121:15993–16005, 2017.

466. A. Slachter, F. L. Bakker, J-P. Adam, and B. J. van Wees. Thermally driven spin injection from a ferromagnet into a non-magnetic metal. *Nat. Phys.*, 6(11):879–882, 2010.

467. L. Fitoussi, F. A. Vetro, C. Caspers, L. Gravier, H. Yu, and J.-Ph. Ansermet. Linear response to a heat-driven spin torque. *Appl. Phys. Lett.*, 106:162401, 2015.

468. T. Nomura, T. Ariki, S. Hu, and T. Kimura. Efficient thermal spin injection in metallic nanostructures. *J. Phys. D: Appl. Phys.*, 50:465003, 2017.

469. J. C. Slonczewski. Initiation of spin-transfer torque by thermal transport from magnons. *Phys. Rev. B*, 82:054403, 2010.

470. H. Yu, S. D. Brechet, P. Che, F. A. Vetro, M. Collet, S. Tu, Y. G. Zhang, Y. Zhang, T. Stueckler, L. Wang, H. Cui, B. M. D. Wang, C. Zhao, P. Bortolotti, A. Anane, J.-Ph. Ansermet, and Zhao W. Thermal spin torques in magnetic insulators. *Phys. Rev. B*, 95:104432, 2017.

471. M. Bailyn. Maximum Variational Principle for Conduction Problems in a Magnetic Field, and the Theory of Magnon Drag. *Phys. Rev.*, 126:2040–2054, 1962.

472. G. D. Mahan. *Many-Particle Physics*. Kluwer Academics, Plenum Publ., New York, 2000.

473. S. R. de Groot and P. Mazur. *Non-Equilibrium Thermodynamics*. Dover Publications Inc., New York, 1962.

474. D. R. Brown, T. Day, K. A. Borup, S. Christensen, B. B. Iversen, and G. J. Snyder. Phase transition enhanced thermoelectric figure-of-merit in copper chalcogenides. *APL Mater.*, 1:052107, 2013.

475. S. N. Guin, J. Pan, A. Bhowmik, D. Sanyal, U. V. Waghmare, and K. Biswas. Temperature dependent reversible p n p type condution switching with colossal change in thermopower of semiconducting AgCuS. *J. Am. Chem. Soc.*, 136:12712–12720, 2014.

476. S. N. Guin, S. Banerjee, D. Sanyal, S. K. Pati, and K. Biswas. Origin of the Order-Disorder Transition and the Associated Anomalous Change of Thermopower in $AgBiS_2$ Nanocrystals: A Combined Experimental and Theoretical Study. *Inorg. Chem.*, 55:6323–6331, 2016.

477. D. Byeon, R. Sobota, K. Delime-Codrin, S. Choi, K. Hirata, M. Adachi, M. Kiyama, T. Matsuura, Y. Yamamoto, M. Matsunami, and T. Takeuchi. Discovery of colossal Seebeck effect in metallic Cu_2Se. *Nat. Commun.*, 10:72, 2019.

478. B. H. Kim, J. S. Kim, T. H. Park, D. S. Le, and Y. W. Park. Magnon drag effect as the dominant contribution to the thermopower in $BiLaSrMnO_3$. *J. Appl. Phys.*, 103:113717, 2008.

479. F.J. Blatt, P.A. Schroeder, C.L. Foiles, and D. Greig. *Thermoelectric power of metals*. Plenum Press New York and London, 1976.

480. D. K. C. MacDonald. *Thermoelectricity, An introduction to the Principles*. Dover Publications Inc., Mineola, New York, 2010, 1962.

481. A.A. Tulapurkar and Y. Suzuki. Contribution of electron-magnon scattering to the spin-dependent Seebeck effect in a ferromagnet. *Solid State Commun.*, 150 (11–12):466–470, 2010.

482. Xiao J., Bauer G. E. W., Uchida K., Saitoh E., and M. Maekawa. Theory of magnonsw-driven spin Seebeck effect. *Phys. Rev. B*, 81:214418, 2010.

483. Harvey S. L. Teaching the photon gas in introductory physics. *Am. J. Math.*, 70:792–797, 2002.

484. G. Baym. *Lectures on Quantum Mechanics*. Benjamin Cummings, London, Amsterdam, Don Mills Ontario, Sydney, Tokyo, 1969.

485. A. Miele, R. Fletcher, E. Zaremba, Y. Feng, C. T. Foxon, and J. J. Harris. Phonon-drag thermopower and weak localisation. *Phys. Rev. B*, 58(19):13181, 1998.

486. A. Fert and I. A. Campbell. Two-Current Conduction in Nickel. *Phys. Rev. Lett.*, 21:1190, 1968.

487. N.F. Mott and J. Jones. *The Theory of the Properties of Metals and Alloys*. Dover Publications Inc., New York, 1958.

488. M. P. Marder. *Condensed Matter Physics*. Wiley Interscience, New York, Chichester, Weinheim, Brisbane, Singapore, Toronto, 2000.

489. A. Matthiessen. Über die elekteische Leitungsfähigkeit der Metalle. *Ann. Phys. (Berlin)*, 179(3):428–434, 1858.

490. R. Daou, R. Frésard, V. Eyert, S. Hébert, and A. Maignan. Unconventional aspects of electronic transport in delafossite oxides. *Sci. Technol. Adv. Mater.*, 18(1):919–938, 2017.

491. J. Kimling, K. Nielsch, K. Rott, and G. Reiss. Field-dependent thermal conductivity and Lorenz number in Co/Cu multilayers. *Phys. Rev. B*, 87:134406, 2013.

492. D. Vu, W. Zhang, C. Sahin, M. E. Flatté, N. Trivedi, and J. P. Heremans. Thermal chiral anomaly in the magnetic-field-induced ideal Weyl phase of Bi(1-x)Sb(x). *Nat. Mater.*, 20(11):1525–1531, 2021.

493. N. Wakeham, A. F. Bangura, X. Xu, J.-F. Mercure, M. Greenblatt, and N. E. Hussey. Gross violation of the Wiedemann–Franz law in a quasi-one-dimensional conductor. *Nat. Commun.*, 2(1):396, 2011.

494. M. Jonson and G. D. Mahan. Mott's formula for the thermopower and the Wiedemann-Franz law. *Phys. Rev. B*, 21:4223–4229, 1980.

495. S. J. Watzman, R. A. Duine, Y. Tserkovnyak, S. R. Boona, H. Jin, A. Prakash, Y. Zheng, and J. P. Heremans. Magnon-drag thermopower and Nernst coefficient in Fe, Co, and Ni. *Phys. Rev. B*, 94:144407, 2016.

496. M. V. Costache, G. Bridoux, I. Neumann, and S. O. Valenzuela. Magnon-drag thermopile. *Nat. Mater.*, 11:199–202, 2012.

497. M. J. Adams, M. Verosky, M. Zebarjadi, and J. P. Heremans. Active Peltier Coolers Based on Correlated and Magnon-Drag Metals. *Phys. Rev. Appl.*, 11:054008, 2019.

498. F. K. Dejene, J. Flipse, and B. J. van Wees. Spin-dependent Seebeck coefficients of $Ni_{80}Fe_{20}$ and Co in nanopillar spin valves. *Phys. Rev. B*, 86:024436, 2012.

499. H.W. Kunert. Thermoelectric effects in spin valves based on layered magnetic structures. *Acta Polinica*, 132:124–132, 2017.

500. N. Liebing, S. Serrano-Guisan, K. Rott, G. Reiss, J. Langer, B. Ocker, and H. W. Schumacher. Determination of spin-dependent Seebeck coefficients of CoFeB/MgO/CoFeB magnetic tunnel junction nanopillars. *J. Appl. Phys.*, 111(7):07C520, 2012.

501. A. Alekhin, I. Razdolski, N. Ilin, J. P. Meyburg, D. Diesing, V. Roddatis, I. Rungger, M. Stamenova, S. Sanvito, U. Bovensiepen, and A. Melnikov. Femtosecond Spin Current Pulses Generated by the Nonthermal Spin-Dependent Seebeck Effect and Interacting with Ferromagnets in Spin Valves. *Phys. Rev. Lett.*, 119:017202, 2017.

502. T. Valet and A. Fert. Theory of the perpendicular magnetoresistance in magnetic multilayers. *Phys. Rev. B*, 48:7099–7113, 1993.

503. M. Stiles and A. Zangwill. Non-collinear Spin Transfer in Co/Cu/Co Multilayers. *J. Appl. Phys.*, 91(10):6812–6817, 2002.

504. B. Doudin, A. Blondel, and J.-Ph. Ansermet. Arrays of multilayered nanowires. *J. Appl. Phys.*, 79:6090–6094, 1996.

505. L. Piraux, S. Dubois, and A. Fert. Perpendicular giant magnetoresistance in magnetic multilayered nanowires. *J. Magn. Magn. Mater.*, 159(3):L287–L292, 1996.

506. L. Piraux, S. Dubois, C. Marchal, J.M. Beuken, L. Filipozzi, J.F. Despres, K. Ounadjela, and A. Fert. Perpendicular magnetoresistance in Co/Cu multilayered nanowires. *J. Magn. Magn. Mater.*, 156(1):317–320, 1996.

507. Q. Yang, P. Holody, SF. Lee, LL. Henry, R. Loloee, P. A. Schroeder, W. P. Pratt Jr., and J. Bass. Spin flip diffusion length and giant magnetoresistance at low temperatures. *Phys. Rev. Lett.*, 72:3274–3277, 1994.

508. S. Dubois, L. Piraux, J. M. George, K. Ounadjela, J. L. Duvail, and A. Fert. Evidence for a short spin diffusion length in permalloy from the giant magnetoresistance of multilayered nanowires. *Phys. Rev. B*, 60:477–484, 1999.

509. W. P. Pratt, S.D. Steenwyk, S. Y. Hsu, W. C. Chiang, A. C. Schaefer, R. Loloee, and J. Bass. Perpendicular-current transport in exchange-biased spin-valves. *IEEE Trans. Mag.*, 33(5):3505–3510, 1997.

510. H. J. Zhang, S. Yamamoto, Y. Fukaya, M. Maekawa, H. Li, A. Kawasuso, T. Seki, E. Saitoh, and K. Takanashi. Current-induced spin polarization on metal surfaces probed by spin-polarized positron beam. *Sci. Rep.*, 4:4844, 2014.

511. P. Kotissek, M. Bailleul, M. Sperl, A. Spitzer, D. Schuh, W. Wegscheider, C. H. Back, and G. Bayreuther. Cross-sectional imaging of spin injection into a semiconductor. *Nat. Phys.*, 3:872–877, 2007.

512. A. J. Drew, J. Hoppler, L. Schulz, F. L. Pratt, P. Desai, P. Shakya, T. Kreouzis, W. P. Gillin, A. Suter, N. A. Morley, V. K. Malik, A. Dubroka, K. W. Kim, H. Bouyanfif, F. Bourqui, C. Bernhard, R. Scheuermann, G. J. Nieuwenhuys, T. Prokscha, and E. Morenzoni. Direct measurement of the electronic spin diffusion length in a fully functional organic spin valve by low-energy muon spin rotation. *Nat. Mater.*, 8:109–114, 2009.

513. J.-Ph. Ansermet. Perpendicular transport of spin-polarized electrons through magnetic nanostructures. *Journal of Physics: Condensed Matter*, 10(27):6027–6050, 1998.

514. J. C. Slater. The Ferromagnetism of Nickel. *Phys. Rev.*, 49:537–545, 1936.

515. L. Pauling. The Nature of the Interatomic Forces in Metals. *Phys. Rev.*, 54:899–904, 1938.

516. J. M. MacLaren, T. C. Schulthess, W. H. Butler, R. Sutton, and M. McHenry. Electronic structure, exchange interactions, and Curie temperature of FeCo. *J. Appl. Phys.*, 85(8):4833–4835, 1999.

517. J.-I. Inoue. GMR, TMR and BMR. In T. Shinjo, editor, *Nanomagnetism and Spintronics*. Elsevier Science, Oxford, Amsterdam, 2009.

518. S. Skaftouros, K. Özdogan, E. Sasloglu, and I. Galanakis. Generalized Slater-Pauling rule for the inverse Heusler compounds. *Phys. Rev. B*, 87:024420, 2013.

519. P. Grünberg and D. Bürgler. Metallic Multilayers: Discovery of Interlayer Exchange Coupling and GMR Metallic Multilayers: Discovery of Interlayer Exchange Coupling and GMR

Metallic Multilayers: Discovery of Interlayer Exchange Coupling and GMR. In Y. Xu; D. D. Awschalom; J. Nitta, editors, *Handbook of Spintronics*. Springer, Dordrecht, Heidelberg, New York, London, 2015.

520. A. Manchon and S. Zhang. Theory of nonequilibrium intrinsic spin torque in a single nanomagnet. *Phys. Rev. B*, 78:212405, 2008.

521. M. Maekawa, S. O. Valenuela, E. Saitoh, and T. Kimuar, editors. *Spin Current*. Oxford University Press, Oxford, 2017.

522. E. H. Hall. On a New Action of the Magnet on Electric Currents. *Am. J. Math.*, 2:287–292, 1879.

523. L. J. van der Pauw. A method of measuring specific resistivity and Hall effect of discs of arbitrary shape. *Philips Res. Repts*, 13:1–9, 1958.

524. L. J. van der Pauw. A method of measuring the resistivity and Hall coefficient on lamellae of arbitrary shape. *Philips Res. Repts*, 20:220–224, 1958.

525. L. Onsager. Reciprocal Relations in Irreversible Processes. I. *Phys. Rev.*, 37:405–426, 1931.

526. L. Onsager. Reciprocal Relations in Irreversible Processes. II. *Phys. Rev.*, 38:2265–2279, 1931.

527. S. D. Brechet. Onsager-Casimir reciprocal relations. *arXiv*, 2210.04289, 2022.

528. Casimir H. B. G. On Onsager's Principle of Microscopic Reversibility. *Rev. Mod. Phys.*, 17:343–350, 1945.

529. S. Tu, T. Ziman, G. Yu, C. Wan, J. Hu, H. Wu, H. Wang, M. Liu, C. Liu, C. Guo, J. Zhang, Marco A. Cabero Z., Y. Zhang, P. Gao, S. Liu, D. Yu, X. Han, I. Hallsteinsen, D. A. Gilbert, P. Woelfle, K. L. Wang, J.-Ph. Ansermet, S. Maekawa, and H. Yu. Record thermopower found in an IrMn-based spintronic stack. *Nat. Commun.*, 11(1):DOI:10.1038/s41467–020–15797–6, 2020.

530. Mitzuguchi M. and Naktsuji S. Interplay of Nernst and Ettinshausen effect. *Sci. Tech. Adv. Mat.*, 20:262–275, 2019.

531. Norwood M. H. Theory of Nernst Generators and Refrigerators. *J. Appl. Phys.*, 34:594–599, 1963.

532. S.J. Watzman, T.M. McCormick, C. Shekhard, S-C. Wu, Y. Sun, A. Prakash, C. Felser, N. Trivedi, and J. P. Heremans. Dirac dispersion generates unusually large Nernst effect in Weyl semimetals. *Phys. Rev. B*, 97:161404(R), 2018.

533. U. Stockert, R. D. Dos Reis, M. O. Ajeesh, S. J. Watzman, M. Schmidt, J. P. Shekhar, C. Heremans, C. Felser, M. Baenitz, and M. Nicklas. Thermopower and thermal conductivity in the Weyl semimetal NbP. *J.Phys.: Cond.Matter*, 29:325701, 2017.

534. H. Smith and H. H. Jensen. *Transport Phenomena*. Oxford Science Publications, New York, 1989.

535. C. Kittel. *Quantum Theory of Solids*. Wiley, New York, 1963.

536. M. Creff, F. Faisant, J. M. Rubì, and J. E. Wegrowe. Surface currents in Hall devices. *J. Appl. Phys.*, 128:054501, 2020.

537. C. Xiao, D. Li, and Z. Ma. Unconventional thermoelectric behaviors and enhancement of figure of merit in Rashba spintronic systems. *Phys. Rev. B*, 93:075150, 2016.

538. Y. Wang, Z. A. Xu, T. Kakeshita, S. Uchida, S. Ono, Y. Ando, and N. P. Ong. Onset of the vortexlike Nernst signal above Tc in $La_{2-x}Sr_xCuO_4$ and $Bi_2Sr_{2-y}La_yCuO_6$. *Phys. Rev. B*, 64:224519, 2001.

539. T. Taniguchi. Phenomenological Spin Transport Theory Driven by Anomalous Nernst Effect. *J. Phys. Soc. Jpn.*, 85(7):074705, 2016.

540. P. M. Chaikin. An introduction to Thermopower for Those Who Might Want to Use It to Study Organic Conductors and Superconductors. In *Organic Superconductivity*. Springer Science+Business Media New York, 1990.

541. W. Koshibae, K. Tsutsui, and S. Maekawa. Thermopower in cobalt oxides. *Phys. Rev. B*, 62:6869–6872, 2000.

542. V. Zlatic and R. Monnier. *Modern Theory of Thermoelectricity*. Oxford University Press, Oxford, 2014.

543. S. M. Wu, A. Luican-Mayer, and A. Bhattacharya. Nanoscale measurement of Nernst effect in two-dimensional charge density wave material 1T-TaS_2. *Appl. Phys. Lett.*, 111(22):223109, 2017.

544. S. Hu and T. Kimura. Anomalous Nernst-Ettingshausen effect in nonlocal spin valve measurement under high-bias current injection. *Phys. Rev. B*, 87:014424, 2013.

545. P. Sengupta, Y. Wen, and J. Shi. Spin-dependent magneto-thermopower of narrow-gap lead chalcogenide quantum wells. *Sci. Rep.*, 8(1):5972, 2018.

546. P. Sheng, Y. Sakuraba, Y.-C. Lau, S. Takahashi, S. Mitani, and M. Hayashi. The spin Nernst effect in tungsten. *Sci. Adv.*, 3:e1701503, 2017.

547. X. Yang, C. H. van der Wal, and B. J. van Wees. Spin-dependent electron transmission model for chiral molecules in mesoscopic devices. *Phys. Rev. B*, 99:024418, 2019.

548. R. Naaman and D. H. Waldeck. Comment on "Spin-dependent electron transmission model for chiral molecules in mesoscopic devices". *Phys. Rev. B*, 101:026403, 2020.

549. D. Sanchez and M. Büttiker. Magnetic-field Asymmetry of Nonlinear Mesoscopic Transport. *Phys. Rev. Lett.*, 93:106802, 2004.

550. A. Gerber, A. Milner, M. Karpovsky, B. Lemke, H.U. Habermeir, J. Tuaillon-Combes, M. Negrier, O. Boisron, P. Melinon, and A. Perez. Extraordinary Hall effect in magnetic films. *J. Magn. Magn. Mater.*, 242-245:90–97, 2002.

551. N. A. Sinitsyn. Semiclassical theories of the anomalous Hall effect. *J. Phys.: Condens. Matter*, 20:023201, 2008.

552. G. Sharma, P. Goswami, and S. Tewari. Nernst and magnetothermal conductivity in a lattice model of Weyl fermions. *Phys. Rev. B*, 93:035116, 2016.

553. G. Sharma, C. Moore, S. Saha, and S. Tewari. Nernst effect in Dirac and inversion-asymmetric Weyl semimetals. *Phys. Rev. B*, 96:195119, 2017.

554. A. A. Burkov. Chiral Anomaly and Diffusive Magnetotransport in Weyl Metals. *Phys. Rev. Lett.*, 113:247203, 2014.

555. J. Hu, M. Caputo, E. B. Guedes, E. Martino, A. Magrez, H. Berger, H. J. Dil, H. Yu, and J.-Ph. Ansermet. Large magnetothermopower and anomalous Nernst effect in HfTe$_5$. *Phys. Rev. B*, 100:115201, 2019.

556. C. A. Mead and D. G. Truhlar. On the determination of Born–Oppenheimer nuclear motion wave functions including complications due to conical intersections and identical nuclei. *J. Chem. Phys.*, 70:2284–2296, 1979.

557. J.-Ph. Ansermet, K. Maschke, and F. Reuse. Spin Evolution in the Born-Oppenheimer. *Isr. J. Chem.*, 62:e2022000, 2022.

558. A. M. Ochs, G. H. Fecher, B. He, W. Schnelle, C. Felser, J. P. Heremans, and J. E. Goldberger. Synergizing a Large ordinary Nernst effect and axis-dependent conduction polarity in flat band KMgBi crystals. *Adv. Mater.*, 36:2308151, 2023.

559. M. I. D'yakonov and V. I. Perel'. Possibility of orienting electron spins with current. *Sov. Phys. JETP Lett.*, 13(11):467–469, 1971.

560. J. Froehlich and P. Werner. Gauge theory of topological phases of matter. *Europhys. Lett.*, 101:47007, 2013.

561. M. Lee, M. O. Hachiya, E. Bernardes, J. C. Egues, and D. Loss. Spin Hall effect due to intersubband-induced spin-orbit interaction in symmetric quantum wells. *Phys. Rev. B*, 80(15):1–7, 2009.

562. Y. K. Kato, R. C. Myers, A. C. Gossard, and D. D. Awschalom. Observation of the spin Hall effect in semiconductors. *Science*, 306:1910–1913, 2004.

563. J. Wunderlich, B. Kaestner, J. Sinova, and T. Jungwirth. Experimental Observation of the spin-hall effect in a two-dimensional spin-orbit coupled semiconductor system. *Phys. Rev. Lett.*, 94(4):1–4, 2005.

564. M. König, H. Buhmann, L. W. Molenkamp, T. Hughes, X. L. Liu, C. X.and Qi, and S. C. Zhang. The quantum spin Hall effect: theory and experiment. *J. Phys. Soc. Jpn.*, 77(3):1–14, 2008.

565. M. V. Berry. Quantal phase factors accompanying adiabatic changes. *Proc. R. Soc. Lond.*, A 392:45–57, 1984.

566. M. Berry. The geometric phase. *Sci. Am.*, 259(6):26–34, 1988.

567. E. Kaxiras and J. D. Joannopoulos. *Quantum Theory of Materials*. Cambridge University Press, Cambridge, New York, New Delhi, Melbourne, Singapore.

568. Z. Jia, C. Li, X. Li, J. Shi, Z. Liao, D. Yu, and X. Wu. Thermoelectric signature of the chiral anomaly in Cd_3As_2. *Nat. Commun.*, 7:13013, 2016.

569. T. Liang, J. Lin, Q. Gibson, T. Gao, M. Hirschberger, M. Liu, R. J. Cava, and N. P. Ong. Anomalous Nernst Effect in the Dirac Semimetal Cd_3As_2. *Phys. Rev. Lett.*, 118:136601, 2017.

570. M. Ikhlas, T. Tomita, T. Koretsune, M.-T. Suzuki, D. Nishio-Hamane, R. Arita, Y. Otani, and S. Nakatsuji. Large anomalous Nernst effect at room temperature in a chiral antiferromagnet. *Nat. Phys.*, 13(11):1085–1090, 2017.

571. N. Kiyohara, T. Tomita, and S. Nakatsuji. Giant anomalous hall effect in the chiral antiferromagnet Mn_3Ge. *Phys. Rev. Appl.*, 5:064009, 2016.

572. K.-S. Kim, H.-J. Kim, and M. Sasaki. Boltzmann approach to anomalous transport in a Weyl metal. *Phys. Rev. B*, 89:195137, 2014.

573. N. Nagaosa, J. Sinova, S. Onoda, A.H. MacDonald, and N.P. Ong. Anomalous hall effect. *Rev. Mod. Phys.*, 82(2):1539–1592, 2010.

574. T. Liang, J. Lin, Q. Gibson, S. Kushwaha, M. Liu, W. Wang, H. Xiong, J. A. Sobota, M. Hashimoto, P. S. Kirchmann, Z. X. Shen, R. J. Cava, and N. P. Ong. Anomalous Hall effect in $ZrTe_5$. *Nat. Phys.*, 14(5):451–455, 2018.

575. N. P. Armitage, E. J. Mele, and A. Vishwanath. Weyl and Dirac semimetals in three-dimensional solids. *Rev. Mod. Physics*, 90(1):15001, 2018.

576. H. Wang, C.-K. Li, H. Liu, J. Yan, J. Wang, J. Liu, Z. Lin, Y. Li, Y. Wang, L. Li, D. Mandrus, X. C. Xie, J. Feng, and J. Wang. Chiral anomaly and ultrahigh mobility in crystalline $HfTe_5$. *Phys. Rev. B*, 93(16):165127, 2016.

577. N. Kumar, C. Shekhar, M. Wang, Y. Chen, H. Borrmann, and C. Felser. Large out-of-plane and linear in-plane magnetoresistance in layered hafnium pentatelluride. *Phys. Rev. B*, 95:155128, 2017.

578. Q. Li, D. E. Kharzeev, C. Zhang, Y. Huang, I. Pletikosić, A. V. Fedorov, R. D. Zhong, J. A. Schneeloch, G. D. Gu, and T. Valla. Chiral magnetic effect in ZrTe 5. *Nat. Phys.*, 12(6):550–554, 2016.

579. M. Hirschberger, S. Kushwaha, Z. Wang, Q. Gibson, S. Liang, C. A. Belvin, B. A. Bernevig, R. J. Cava, and N. P. Ong. The chiral anomaly and thermopower of Weyl fermions in the half-Heusler GdPtBi. *Nat. Mater.*, 15(11):1161–1165, 2016.

580. J. Xiong, S. K Kushwaha, T. Liang, J. W Krizan, M. Hirschberger, W. Wang, R. J. Cava, and N. P. Ong. Evidence for the chiral anomaly in the Dirac semimetal Na_3Bi. *Science*, 350:1314–1317, 2015.

581. X. Huang, L. Zhao, Y. Long, P. Wang, D. Chen, Z. Yang, H. Liang, M. Xue, H. Weng, Z. Fang, X. Dai, and G. Chen. Observation of the chiral-anomaly-induced negative magnetoresistance: In 3D Weyl semimetal TaAs. *Phys. Rev. X*, 5:031023, 2015.

582. H. Li, H. He, H.-Z. Lu, H. Zhang, H. Liu, R. Ma, Z. Fan, S.-Q. Shen, and J. Wang. Negative magnetoresistance in Dirac semimetal Cd_3As_2. *Nat. Commun.*, 7:10301, 2016.

583. H. Li, H. He, H. Z. Lu, H. Zhang, H. Liu, R. Ma, Z. Fan, S. Q. Shen, and J. Wang. Negative magnetoresistance in Dirac semimetal Cd_3As_2. *Nat. Commun.*, 7:10301, 2016.

584. X. Dai, Z. Z. Du, and H. Z. Lu. Negative Magnetoresistance without Chiral Anomaly in Topological Insulators. *Phys. Rev. Lett.*, 119:166601, 2017.

585. D. T. Son and B. Z. Spivak. Chiral anomaly and classical negative magnetoresistance of Weyl metals. *Phys. Rev. B*, 88:104412, 2013.

586. H. B. Nielsen and Ninomiya M. The Adler-Bell-Jackiw anomaly and Weyl fermions in a crystal. *Phys. Lett. B*, 130(6):389–396, 1983.

587. A. Y. Alekseev, V. V. Cheianov, and J. Fröhlich. Universality of transport properties in equilibrium, the Goldstone theorem, and chiral anomaly. *Phys. Rev. Lett.*, 81(16):3503–3506, 1998.

588. S. Murakami, N. Nagaosa, and S.-C. Zhang. Dissipationless quantum spin. *Science (New York, N.Y.)*, 301:1348–1351, 2003.

589. L. E. Ballentine. *Quantum Mechanics, A Modern Development*. World Scientific, 2015.

590. C. Duval and P. A. Horvathy. The exotic Galilei group and the 'Peierls substitution'. *Phys. Lett. B: Nucl.*, 479:284–290, 2000.

591. J. Sinova, D. Culcer, Q. Niu, N. A. Sinitsyn, T. Jungwirth, and A. H. MacDonald. Universal intrinsic spin Hall effect. *Phys. Rev. Lett.*, 92:126603, 2004.

592. J. Sinova, S.O. Valenzuela, J. Wunderlich, C.-H. Back, and T. Jungwirth. Spin-Hall effects. *Rev. Mod. Phys.*, 87:1213–1259, 2015.

593. J. Kessler. *Poarlized Electrons*, volume 1 of *Springer Series on Atoms and Plasma*. Springer Berlin Heidelberg, 1976.

594. V. Sih, R. C. Myers, Y. K. Kato, W. H. Lau, A. C. Gossard, and D. D. Awschalom. Spatial imaging of the spin Hall effect and current-induced polarization in two-dimensional electron gases. *Nat. Phys.*, 1(1):31–35, 2005.

595. V. Sih. A Hall of Spin. *Phys. World*, 18(11):33–36, 2005.

596. Y. Niimi, M. Morota, D. H. Wei, C. Deranlot, M. Basletic, A. Hamzic, A. Fert, and Y. Otani. Extrinsic spin hall effect induced by iridium impurities in copper. *Phys. Rev. Lett.*, 106:126601, 2011.

597. J. Inoue and H. Ohno. Physics: taking the Hall effect for a spin. *Science*, 309(5743):2004–2005, 2005.

598. E. Saitoh, M. Ueda, H. Miyajima, and G. Tatara. Conversion of spin current into charge current at room temperature: Inverse spin-Hall effect. *Appl. Phys. Lett.*, 88(18):182509, 2006.

599. J. Li, C. B. Wilson, R. Cheng, M. Lohmann, M. Kavand, W. Yuan, M. Aldosary, N. Agladze, P. Wei, M. S. Sherwin, and J. Shi. Spin current from sub-terahertz-generated antiferromagnetic magnons. *Nature*, 578:70–74, 2020.

600. R. Lebrun, A. Ross, S. A. Bender, A. Qaiumzadeh, L. Baldrati, J. Cramer, A. Brataas, R. A. Duine, and M. Kläui. Tunable long-distance spin transport in a crystalline antiferromagnetic iron oxide. *Nature*, 561:222–225, 2018.

601. D. Go, D. Jo, C. Kim, and H.-W. Lee. Intrinsic spin and orbital hall effects from orbital texture. *Phys. Rev. Lett.*, 121:086602, 2018.

602. C. Hahn, G. de Loubens, O. Klein, M. Viret, V. V. Naletov, and J. Ben Youssef. Comparative measurements of inverse spin Hall effects and magnetoresistance in YIG/Pt and YIG/Ta. *Phys. Rev. B*, 87:174417, 2013.

603. N. Vlietstra, J. Shan, V. Castel, B. J. van Wees, and J. Ben Youssef. Spin-Hall magnetoresistance in platinum on yttrium iron garnet: Dependence on platinum thickness and in-plane/out-of-plane magnetization. *Phys. Rev. B*, 87:184421, 2013.

604. M. Isasa, A. Bedoya-Pinto, S. Vélez, F. Golmar, F. Sánchez, L.E. Hueso, J. Fontcuberta, and F. Casanova. Spin Hall magnetoresistance at Pt/$CoFe_2O_4$ interfaces and texture effects. *Appl. Phys. Lett.*, 105(14):142402, 2014.

605. Z. Ding, B. L. Chen, J. H. Liang, J. Zhu, J. X. Li, and Y. Z. Wu. Spin Hall magnetoresistance in Pt/Fe_3O_4 thin films at room temperature. *Phys. Rev. B*, 90:134424, 2014.

606. T. Shang, Q. F. Zhan, L. Ma, H. L. Yang, Z. H. Zuo, Y. L. Xie, H. H. Li, L. P. Liu, B. M. Wang, Y. H. Wu, S. Zhang, and R.-W. Li. Pure spin-Hall magnetoresistance in Rh/$Y_3Fe_5O_{12}$ hybrid. *Sci. Rep.*, 5(1):17734, 2015.

607. T. Shang, Q. F. Zhan, H. L. Yang, Z. H. Zuo, Y. L. Xie, Y. Zhang, L. P. Liu, B. M. Wang, Y. H. Wu, S. Zhang, and R.-W. Li. Extraordinary Hall resistance and unconventional magnetoresistance in Pt/$LaCoO_3$ hybrids. *Phys. Rev. B*, 92:165114, 2015.

608. H. Nakayama, M. Althammer, Y.-T. Chen, K. Uchida, Y. Kajiwara, D. Kikuchi, T. Ohtani, S. Geprägs, M. Opel, S. Takahashi, R. Gross, G. E. W. Bauer, S. T. B. Goennenwein, and E. Saitoh. Spin Hall Magnetoresistance Induced by a Nonequilibrium Proximity Effect. *Phys. Rev. Lett.*, 110:206601, 2013.

609. I. M. Miron, K. Garello, G. Gaudin, P.-J. Zermatten, M. V. Costache, S. Auffret, S. Bandiera, B. Rodmacq, A. Schuhl, and P. Gambardella. Perpendicular switching of a single ferromagnetic layer induced by in-plane current injection. *Nature*, 476:189–193, 2011.

610. N. Reynolds, P. Jadaun, J. T. Heron, C. L. Jermain, J. Gibbons, R. Collette, R. A. Buhrman, D. G. Schlom, and D. C. Ralph. Spin Hall torques generated by rare-earth thin films. *Phys. Rev. B*, 95:064412, 2017.

611. L. Liu, O. J. Lee, T. J. Gudmundsen, D. C. Ralph, and R. A. Buhrman. Current-induced switching of perpendicularly magnetized magnetic layers using spin torque from the spin hall effect. *Phys. Rev. Lett.*, 109:096602, 2012.

612. N. Roschewsky, T. Matsumura, S. Cheema, F. Hellman, T. Kato, S. Iwata, and S. Salahuddin. Spin-orbit torques in ferrimagnetic GdFeCo alloys. *Appl. Phys. Lett.*, 109:112403, 2016.

613. N. Roschewsky, C.-H. Lambert, and S. Salahuddin. Spin-orbit torque switching of ultralarge-thickness ferrimagnetic GdFeCo. *Phys. Rev. B*, 96:064406, 2017.

614. K. Ueda, M. Mann, P. W. P. de Brouwer, D. Bono, and G. S. D. Beach. Temperature dependence of spin-orbit torques across the magnetic compensation point in a ferrimagnetic TbCo alloy film. *Phys. Rev. B*, 96:064410, 2017.

615. J. Han, A. Richardella, S. A. Siddiqui, J. Finley, N. Samarth, and L. Liu. Room-temperature spin-orbit torque switching induced by a topological insulator. *Phys. Rev. Lett.*, 119:077702, 2017.

616. Y. Wang, D. Zhu, Y. Wu, Y. Yang, J. Yu, R. Ramaswamy, R. Mishra, S. Shi, M. Elyasi, K.-L. Teo, Y. Wu, and H. Yang. Room temperature magnetization switching in topological insulator-ferromagnet heterostructures by spin-orbit torques. *Nat. Commun.*, 8(1):1364, 2017.

617. D. Xiao, M.-C. Chang, and Q. Niu. Berry phase effects on electronic properties. *Rev. Mod. Phys.*, 82:1959, 2010.

618. D. Xiao, Y. Yao, Z. Fang, and Q. Niu. Berry-phase effect in anomalous thermoelectric transport. *Phys. Rev. Lett.*, 97:026603, 2006.

619. D. Suter, G. C. Chingas, R. A. Harris, and A. Pines. Berry's phase in magnetic resonance. *Mol. Phys.*, 61(6):1327–1340, 1987.

620. J. Ye, Y. B. Kim, A. J. Millis, B. I. Shraiman, P. Majumdar, and Z. Tešanović. Berry phase theory of the anomalous hall effect: application to colossal magnetoresistance manganites. *Phys. Rev. Lett.*, 83:3737–3740, 1999.

621. T. Jungwirth, Qian Niu, and A. H. MacDonald. Anomalous Hall effect in ferromagnetic semiconductors. *Phys. Rev. Lett.*, 88:207208, 2002.

622. Z. Fang, N. Nagaosa, K. S. Takahashi, A. Asamitsu, R. Mathieu, T. Ogasawara, H. Yamada, M. Kawasaki, Y. Tokura, and K. Terakura. The anomalous Hall effect and magnetic monopoles in momentum space. *Science*, 302(5642):92–95, 2003.

623. Y. Yao, L. Kleinman, A. H. MacDonald, J. Sinova, T. Jungwirth, D.-S. Wang, E. Wang, and Q. Niu. First principles calculation of anomalous hall conductivity in ferromagnetic bcc Fe. *Phys. Rev. Lett.*, 92:037204, 2004.

624. W.-L. Lee, S.i Watauchi, V. L. Miller, R. J. Cava, and N. P. Ong. Dissipationless anomalous hall current in the ferromagnetic spinel CuCr2Se4-xBrx. *Science*, 303(5664):1647–1649, 2004.

625. C. Zeng, Y. Yao, Q. Niu, and H. H. Weitering. Linear magnetization dependence of the intrinsic anomalous hall effect. *Phys. Rev. Lett.*, 96:037204, 2006.

626. V. P. Amin, J. Li, M. D. Stiles, and P. M. Haney. Intrinsic spin currents in ferromagnets. *Phys. Rev. B*, 99(22):1–5, 2019.

627. L. Dong, C. Xiao, B. Xiong, and Q. Niu. Berry phase effects in dipole density and the mott relation. *Phys. Rev. Lett.*, 124:066601, 2020.

628. J.-P. Hanke, F. Freimuth, S. Blügel, and Y. Mokrousov. Higher-dimensional wannier interpolation for the modern theory of the dzyaloshinskii–moriya interaction: application to co-based trilayers. *J. Phys. Soc. Jpn.*, 87(4):041010, 2018.

629. V. K. Dugaev, P. Bruno, B. Canals, and C. Lacroix. Berry phase of magnons in textured ferromagnets. *Phys. Rev. B*, 72:024456, 2005.

630. Z. Z. Du, Hai Zhou Lu, and X. C. Xie. Nonlinear Hall effects. *Nat. Rev. Phys.*, 3:744–752, 2021.

631. J. Moody, A. Shapere, and F. Wilczek. *Geometric phases in physics*, chapter Adiabatic Effective Lagrangians, pp. 160–183. World Scientific, Singapore, New Jersey, London, Hong Kong, 1989.

632. H. Weng, R. Yu, X. Hu, X. Dai, and Z. Fang. Quantum anomalous Hall effect and related topological electronic states. *Adv. Phys.*, 64(3):227–282, 2015.

633. R. Yu, W. Zhang, H.-J. Zhang, S.-C. Zhang, X. Dai, and Z. Fang. Quantized anomalous Hall effect in magnetic topological insulators. *Science*, 329(5987):61–64, 2010.

634. C.-Z. Chang, J. Zhang, X. Feng, J. Shen, Z. Zhang, M. Guo, K. Li, Y. Ou, P. Wei, L.-L. Wang, Z.-Q. Ji, Y. Feng, S. Ji, X. Chen, J. Jia, X. Dai, Z. Fang, S.-C. Zhang, K. He, Y. Wang, L. Lu, X.-C. Ma, and Q.-K. Xue. Experimental observation of the quantum anomalous Hall effect in a magnetic topological insulator. *Science*, 340(6129):167–170, 2013.

635. C. Tang, C.-Z. Chang, G. Zhao, Y. Liu, Z. Jiang, C.-X. Liu, M. R. McCartney, D. J. Smith, T. Chen, J. S. Moodera, and J. Shi. Above 400-K robust perpendicular ferromagnetic phase in a topological insulator. *Sci. Adv.*, 3(e1700307), 2017.

636. A. J. Bestwick, E. J. Fox, X. Kou, and Wang K. L. Precise quantization of the anomalous Hall effect near zero magnetic field. *Phys. Rev. Lett.*, 114:187201, 2015.

637. C.-Z. Chang, W. Zhao, D. Y. Kim, H. Zhang, B. A. Assaf, D. Heiman, S.-C. Zhang, C. Liu, M. H. W. Chan, and J. S. Moodera. High-precision realization of robust quantum anomalous Hall state in a hard ferromagnetic topological insulator. *Nat. Mater.*, 14:473–477, 2015.

638. C. L. Kane and E. J. Mele. Quantum spin hall effect in graphene. *Phys. Rev. Lett.*, 95:226801, 2005.

639. Y. P. Mizuta and F. Ishii. Large anomalous Nernst effect in a skyrmion crystal. *Sci. Rep.*, 6(1):28076, 2016.

640. S. Saha and S. Tewari. Anomalous Nernst effect in type-II Weyl semimetals. *Eur. Phys. J. B*, 91(1):4, 2018.

641. M. N. Chernodub, A. Cortijo, and M. A. H. Vozmediano. Generation of a Nernst current from the conformal anomaly in dirac and Weyl semimetals. *Phys. Rev. Lett.*, 120:206601, 2018.

642. J. Noky, J. Gayles, C. Felser, and Y. Sun. Strong anomalous Nernst effect in collinear magnetic Weyl semimetals without net magnetic moments. *Phys. Rev. B*, 97:220405, 2018.

643. T. Yoda, T. Yokoyama, and S. Murakami. Current-induced orbital and spin magnetizations in crystals with helical structure. *Sci. Rep.*, 5:12024, 2005.

644. A. Inui, R. Aoki, Y. Nishiue, K. Shiota, Y. Kousaka, H. Shishido, D. Hirobe, M. Suda, J. I. Ohe, J.n I. Kishine, H. M. Yamamoto, and Y. Togawa. Chirality-induced spin-polarized state of a chiral crystal $CrNb_3 S_6$. *Phys. Rev. Lett.*, 124(16):166602, 2020.

645. K. M. Alam and S. Pramanik. Spin filtering through single-wall carbon nanotubes functionalized with single-stranded DNA. *Adv. Funct. Mater.*, 25(21):3210–3218, 2015.

646. Y. Perlitz and K. Michaeli. Helical liquid in carbon nanotubes wrapped with DNA molecules. *Phys. Rev. B*, 98(19):195405, 2018.

647. V. V. Maslyuk, R. Gutierrez, and G. Cuniberti. Spin–orbit coupling in nearly metallic chiral carbon nanotubes: a density-functional based study. *Phys. Chem. Chem. Phys.*, 19(13):8848–8853, 2017.

648. E. Medina, L. A. González-Arraga, D. Finkelstein-Shapiro, B. Berche, and V. Mujica. Continuum model for chiral induced spin selectivity in helical molecules. *J. Chem. Phys.*, 142:194308, 2015.

649. S. Varela, V. Mujica, and E. Medina. Effective spin-orbit couplings in an analytical tight-binding model of DNA: spin filtering and chiral spin transport. *Phys. Rev. B*, 93:155436, 2016.

650. V. V. Maslyuk, R. Gutierrez, A. Dianat, V. Mujica, and G. Cuniberti. Enhanced magnetoresistance in chiral molecular junctions. *J. Phys. Chem. Lett.*, 18:5453–5459, 2018.

651. N. Lazić, M. Milivojević, T. Vuković, and M. Damnjanović. Double line groups : structure, irreducible representations and spin splitting of the bands. *J. Phys. A: Math. Theor.*, 51:225203, 2018.

652. M. Milivojević, N. Lazić, S. Dmitrović, M. Damnjanović, and T. Vuković. Spin splitting in quasi-one dimensional systems. *Phys. Stat. Solidi (B)*, 255(12):1800184, 2018.

653. J. Fransson. Chirality-Induced Spin Selectivity: The Role of Electron Correlations. *J. Phys. Chem. Lett.*, 10(22):7126–7132, 2019.

654. J. Fransson. *Non-Equilibrium Nano-Physics*, volume 809 of *Lecture Notes in Physics*. Springer Dordrecht, 2010.

655. M. Naka, S. Hayami, H. Kusunose, Y. Yanagi, Y. Motome, and H. Seo. Spin current generation in organic antiferromagnets. *Nat. Commun.*, 10:4305, 2019.

656. Y. Imamura, S. Ten-No, K. Yonemitsu, and Y. Tanimura. Structures and electronic phases of the bis(ethylenedithio)tetrathiafulvalene (BEDT-TTF) clusters and κ-(BEDT-TTF) salts: A theoretical study based on ab initio molecular orbital methods. *J. Chem. Phys.*, 111(13):5986–5994, 1999.

657. N.D. Kushch, M.A. Tanatar, T. Ishiguro, S. Kagoshima, E.B. Yagubskii, V.S. Yefanov, and V.A. Bondarenko. Superconductivity of the κ-(BEDT-TTF)$_2$Cu[N(CN)$_2$]I salt under pressure. *Synth. Met.*, 133-134:177–179, 2003.

658. T. Isono, T. Terashima, K. Miyagawa, K. Kanoda, and S. Uji. Quantum criticality in an organic spin-liquid insulator κ-(BEDT-TTF)$_2$Cu$_2$(CN)$_3$. *Nat. Commun.*, 7(1):13494, 2016.

659. Pines D. and Slichter C. P. Relaxation times in magnetic resonance. *Phys. Rev.*, 100:1014–1020, 1955.

660. R. Kubo, M. Toda, and N. Hashitsume. *Statistical Physics II*, volume 31. Springer Series in Solid State Sciences, Berlin, Heidelberg, New York, 1998.

661. H. B. Callen and T. A. Welton. Irreversibility and generalized noise. *Phys. Rev.*, 83(1):34–40, 1951.

662. H. B. Callen, M. L. Barasch, and J. L. Jackson. Statistical mechanics of irreversibility. *Phys. Rev.*, 88:1382–1386, 1952.

663. R. F. Grene and H. B. Callen. On a theorem of irreversible thermodynamics. II. *Phys. Rev.*, 88:1387–1391, 1952.

664. J. K. Nielsen and J. C. Dyre. Fluctuation-dissipation theorem for frequency-dependent specific heat. *Phys. Rev. B*, 54:15754–15761, 1996.

665. P. S. Hubbard. Quantum-mechanical and semiclassical forms of the density operator theory of relaxation. *Rev. Mod. Phys.*, 33:249–264, 1961.

666. F. Bloch. Dynamical theory of nuclear induction. II. *Phys. Rev.*, 102(1):104–135, 1956.

667. R. K. Wangness and F. Bloch. The dynamics of nuclear induction. *Phys. Rev.*, 89(4):728–739, 1953.

668. A. G. Redfield. The theory of relaxation processes. *Adv. Mag. Res.*, 1:1–32, 1965.

669. A.G. Redfield. On the theory of relaxation processes. *IBM J. Res. Dev.*, 1:19–31, 1957.

670. C. H. Pennington, D. J. Durand, C. P. Slichter, J. P. Rice, E. D. Bukowski, and D. M. Ginsberg. Static and dynamic Cu NMR tensors of YBa$_2$Cu$_3$O$_{7-\delta}$. *Phys. Rev. B*, 39(4):2902–2905, 1989.

671. G. Lindblad. On the generators of quantum dynamical semigroups. *Commun. Math. Phys.*, 48(2):119–130, 1976.

672. Gorini V., Kossakowski A., and Sudarshan E. C. G. Completely positive semigroups of N level systems. *J. Math. Phys.*, 17(6):821, 1976.

673. T. F. Havel. Robust procedures for converting among Lindblad, Kraus and matrix representations of quantum dynamical semigroups. *J. Math. Phys.*, 44(2):534–557, 2003.

674. A. Karabanov, G. Kwiatkowski, and W. Köckenberger. Spin dynamic simulations of solid effect DNP: the role of the relaxation superoperator. *Mol. Phys.*, 112(14):1838–1854, 2014.

675. R. Annabestani and D. G. Cory. Dipolar relaxation mechanism of long-lived states of methyl groups. *Quantum Information Processing*, 17(1):15, 2017.

676. J. Jeener. Superoperators in magnetic resonance, Adv. Magn. Opt. Reson. 10:1–51, 1982

677. Levitt M. H. and Di Bari L. Steady state in magnetic resonance pulse experiments. *Phys. Rev. Lett.*, 69(21):3124–3127, 1992.

678. Levitt M. H. and Di Bari L. The homogeneous master equation and the manipulation of relaxation networks. *Bull. Mag. Res.*, 16:94–114, 1992.

679. C. Bengs and M. H. Levitt. A master equation for spin systems far from equilibrium. *J. Mag. Res.*, 310:106645, 2020.

680. M. Zhong, M. P. Hedges, R. L. Ahlefeldt, J. G. Bartholomew, S. E. Beavan, S. M. Witting, J. J. Longdell, and M. J. Sellars. Optically addressable nuclear spins in a solid with a six-hour coherence time. *Nature*, 517:177–180, 2015.

681. D. Loss and D. P. DiVincenzo. Quantum computation with quantum dots. *Phys. Rev. A*, 57:120–126, 1998.

682. L. M.K. Vandersypen and M. A. Eriksson. Quantum computing with semiconductor spins. *Phys. Today*, 72(8):38–45, 2019.

683. T. Takui, L. Berliner, and G. Hanson, editors. *Electron Spin Resoannce (ESR) Based Quantum Computing*, volume 31 of *Biological Magnetic Resonance*. Springer New York, 2016.

684. P. D. DiVincenzo. Topics in quantum computers. In L. Sohn, L. Kouwenhoven, and G. Schon, editors, *Mesoscopic Electron Transport*. NATO ASI Series E, Curaçao, Netherlands Antilles, 1997.

685. A. M. Tyryshkin, S. A. Lyon, A. V. Astashkin, and A. M. Raitsimring. Electron spin relaxation times of phosphorus donors in silicon. *Phys. Rev. B*, 68:193207, 2003.

686. A. Morello, J. J. Pla, F. A. Zwanenburgv K.K. W. Chan, K. Y. Tan, H. Huebl, M. Moettoenen, C. D. Nugroho, C. Yang, J. A. van Donkelaar, A. D. C. Alves, D. N. Jamieson, C. C. Escott, L. C. L. Hollenberg, R. G. Clark, and A. S. Dzurak. Single-shot readout of an electron spin in silicon. *Nature*, 467:687–691, 2010.

687. Y. He, S. K. Gorman, D. Keith, L.Kranz, J. G.Keizer, and M. Y. Simmons. A two-qubit gate between phosphorus donor electrons in silicon. *Nature*, 571:371–375, 2019.

688. W. S. Warren, N. Gershenfeld, and I. Chuang. The usefulness of NMR quantum computing. *Science*, 277(5332):1688–1690, 1997.

689. K. Bader, D. Dengler, S. Lenz, B. Endeward, S.-D. Jiang, P. Neugebauer, and J. van Slageren. Room temperature quantum coherence in a potential molecular qubit. *Nat. Commun.*, 5:5304, 2014.

690. C. Negrevergne, T. S. Mahesh, C. A. Ryan, M. Ditty, F. Cyr-Racine, W. Power, N. Boulant, T. Havel, D. G. Cory, and R. Laflamme. Benchmarking quantum control methods on a 12-qubit system. *Phys. Rev. Lett.*, 96:170501, 2006.

691. J. Li, Z. Luo, T. Xin, H. Wang, D. Kribs, D. Lu, B. Zeng, and R. Laflamme. Experimental implementation of efficient quantum pseudorandomness on a 12-spin system. *Phys. Rev. Lett.*, 123:030502, 2019.

692. A. Gliesche, K. Maschke, and F. A. Reuse. Irreversible evolution of many-electron systems coupled to a statistical environment. *Phys. Rev. B*, 77(24):1–19, 2008.

693. Gliesche A. and Maschke K. Irreversible evolution of many-electron systems: From the quantum-Boltzmann equation toward the semi-classical Boltzmann equation. *Phys. Rev. B*, 77:214301, 2008.

694. Gliesche A. *Evolution of open many-electron systems from a quantum statistical description towards the semi-classical Boltzmann equation*. PhD thesis, Ecole Polytechnique Fédérale de Lausanne, DOI: 10.5075/epfl-thesis-3828, 2007.

695. N. Bloembergen. On the ferromagnetic resonance in nickel and supermalloy. *Phys. Rev.*, 78:572–580, 1950.

696. C. Kittel. Ferromagnetic relaxation and gyromagnetic anomalies in metals. *Phys. Rev.*, 101:1611–1612, 1956.

697. S. E. Barnes. Theory of electron paramagnetic resonance of ions in metals. *Adv. Phys.*, 30:801–938, 1981.

698. J. Winter. *Magnetic Resonance in Metals*. Oxford at the Clarendon Press, London, 1971.

699. H. Hasegawa. Dynamical properties of s-d interaction. *Prog. Theo. Physics*, 98(2):483–500, 1959.

700. F. J. Dyson. Spin resonance absorption in metals. II. Theory of electron diffusion and the skin effect. *Phys. Rev.*, 98(2):349–359, 1955.

701. J. R. Asik. *Spin-flip scattering of conduction electrons from impurities*. PhD thesis, University of Illinois, Urbana-Champaign, 1966.

702. J.M. Ziman. *Principles of the Theory of Solids*. Cambridge University Press, Cambridge, New York, Melbourne, 1972.

703. D. Y. Smith. Theory of spin-orbit effects in the Il band in alkali halides. *Phys. Rev.*, 137(2A):A574–A582, 1965.

704. E.U. Condon and G.H. Shortley. *The Theory of Atomic Spectra*. Cambridge University Press, London, New York, 1959.

705. R. H. Silsbee, A. Janossy, and P. Monod. Coupling between ferromagnetic and conduction-spin-resonance modes at a ferromagnetic—normal-metal interface. *Phys. Rev. B*, 19:4382–4399, 1979.

706. H. Yu, S. Granville, D. P. ; Yu, and J.-Ph. Ansermet. Evidence for thermal spin-transfer torque. *Phys. Rev. Lett.*, 104:146601, 2010.

707. Ph. Guittienne, J.-E. Wegrowe, D. Kelly, and J.-Ph. Ansermet. Switching time measurements of current-induced magnetization reversal. *IEEE Trans. Magn.*, 37:2126–2128, 2001.

708. T. Kasuya. Effects of s-d interaction on transport phenomena. *Prog. Theo. Physics (Kyoto)*, 22:227–246, 1959.

709. I. Mannari. Electrical resistance of ferromagnetic metals. *Prog. Theo. Phys. (Kyoto)*, 22(2):335–343, 1959.

710. D. A. Goodings. Electrical resistivity of ferromagnetic metals at low temperatures. *Phys. Rev.*, 132(2):542–558, 1963.

711. A. Fert. Two-current conduction in ferromagnetic metals and spin wave electron collisions. *J. Phys. C: Solid State Phys.*, 2(2):1784–1788, 1969.

712. O. Madelung. *Introduction to Solid State Theory.* Springer Series in Solid State Sciences, Berlin, Heidelberg, 1978.

713. B. Rauet, M. Viret, E. Sondergard, O. Cespecdes, and R. Manmy. Electron-magnon scattering and magnetic resistivity in 3d ferromagnets. *Phys. Rev. B*, 66:024433, 2002.

714. J. Fabian and S. Das Sarma. Spin relaxation of conduction electrons. *J. Vac. Sci. Technol.*, B17(4):1708–1715, 1999.

715. R. J. Elliott. Theory of the effect of spin-orbit coupling on magnetic resonance in some semiconductors. *Phys. Rev.*, 96(2):266–279, 1954.

716. F. Beuneu and P. Monod. Conduction-electron spin resonance in cold-worked Al, Cu, and Ag: the spin-flip cross section of dislocations. *Phys. Rev. B*, 13(8):3424–3430, 1976.

717. P. Monod and S. Schultz. Transmission electron spin resonance in dilute copper-chromium alloys. *Phys. Rev.*, 173:645–653, 1968.

718. P. Monod and S. Schultz. Conduction electron spin-flip scattering by impurities in copper. *J. Phys. France*, 43:393–401, 1982.

719. A. W. Overhauser. Paramagnetic relaxation in metals. *Phys. Rev.*, 89:689–700, 1953.

720. P. Boguslawski. Electron-electron spin-flip scattering and spin relaxation in III-V and II-VI semiconductors. *Solid State Comm.*, 33:389–391, 1980.

721. M.I . Dyakonov and V. I. Perel. Spin relaxation of conduction electrons in noncentrosymmetric semiconductors. *Sov. Phys. Solid State, USSR*, 13(12):3023–3026, 1972.

722. J. F. Gregg, W. Allen, S. M. Thompson, M. L. Watson, and G. Gehring. Jitterbug spin channel mixing in heterogeneous giant magnetoresistive materials. *J. Appl. Phys.*, 79(8):5593–5595, 1996.

723. L. E. F. Foa Torres, S. Roche, and J.-C. Charlier. *Introduction to Graphene-Based Nanomaterials.* Cambridge University Press, Cambridge, New York, 2014.

724. N. Tombros, C. Jozsa, M. Popinciuc, H. T. Jonkman, and B. J. van Wees. Electronic spin transport and spin precession in single graphene layers at room temperature. *Nature*, 448(7153):571–574, 2007.

725. Wei Han and R. K. Kawakami. Spin relaxation in single-layer and bilayer graphene. *Phys. Rev. Lett.*, 107:047207, 2011.

726. M. P. Gokhale, A. Ormeci, and D. L. Mills. Inelastic scattering of low-energy electrons by spin excitations on ferromagnets. *Phys. Rev. B*, 46:8978–8993, 1992.

727. M. Plihal and D. L. Mills. Spin-flip exchange scattering of low-energy electrons in ferromagnetic iron. *Phys. Rev. B*, 58:14407–14415, 1998.

728. N. J. Harmon, W. O. Putikka, and R. Joynt. Prediction of extremely long mobile electron spin lifetimes at room temperature in wurtzite semiconductor quantum wells. *Appl. Phys. Lett.*, 98(7):073108, 2011.

729. X. Liu, N. Tang, S. Zhang, X. Zhang, H. Guan, Y. Zhang, X. Qian, Y. Ji, W. Ge, and B. Shen. Effective manipulation of spin dynamics by polarization electric field in InGaN/GaN quantum wells at room temperature. *Adv. Sci.*, 7:1903400, 2020.

730. Y. Ohno, R. Terauchi, T. Adachi, F. Matsukura, and H. Ohno. Spin relaxation in GaAs(110) quantum wells. *Phys. Rev. Lett.*, 83:4196–4199, 1999.

731. S. Krishnamurthy, M. van Schilfgaarde, and N. Newman. Spin lifetimes of electrons injected into GaAs and GaN. *Appl. Phys. Lett.*, 83(9):1761–1763, 2003.

732. S. Ghosh, V. Sih, W. H. Lau, D. D. Awschalom, S.-Y. Bae, S. Wang, S. Vaidya, and G. Chapline. Room-temperature spin coherence in ZnO. *Appl. Phys. Lett.*, 86(23):232507, 2005.

733. A. Balocchi, Q. H. Duong, P. Renucci, B. L. Liu, C. Fontaine, T. Amand, D. Lagarde, and X. Marie. Full electrical control of the electron spin relaxation in GaAs quantum wells. *Phys. Rev. Lett.*, 107:136604, 2011.

734. G. Wang, A. Balocchi, D. Lagarde, C. R. Zhu, T. Amand, P. Renucci, Z. W. Shi, W. X. Wang, B. L. Liu, and X. Marie. Temperature dependent electric field control of the electron spin relaxation in (111)A GaAs quantum wells. *App. Phys. Lett.*, 102:242408, 2013.

735. J. Ishihara, M. Ono, Y. Ohno, and H. Ohno. A strong anisotropy of spin dephasing time of quasi-one dimensional electron gas in modulation-doped GaAs/AlGaAs wires. *Appl. Phys. Lett.*, 102:212402, 2013.

736. S. Haroche. Quantum beats and time-resolved fluorescence spectroscopy. In Shimoda K., editor, *Laser Spectroscopy. Topics in Applied Physics*, volume 13. Springer, Berlin, Heidelberg, 1976.

737. M. Dyakonov, X. Marie, T. Amand, P. Le Jeune, D. Robart, M. Brousseau, and J. Barrau. Coherent spin dynamics of excitons in quantum wells. *Phys. Rev. B*, 56:10412–10422, 1997.

738. C. Kittel and J. K. Galt. Ferromagnetic domain theory. In F. Seitz, D. Turnbull, editors, *Solid State Physics*, volume 3. Academic Press Inc., New York, 1956.

739. E. Kneller. Theorie der Magneitsieringskurve kleiner Kristalle. In S. Fluegge and J. P. J. Wijn, editors, *Handbuch der Physik*, volume XVIII/2. Sprinter-Verlag, Berlin, Heidelberg, New York, 1966.

740. R. E. Camley and R. L. Stamps. Magnetic multilayers: spin configurations, excitations and giant magnetoresistance. *J. Phys. Condens. Matter*, 5:3727–3786, 1983.

741. W. T. Coffey and Y. P. Kalmykov. *The Langevin Equation*. 3rd edition. World Scientific, Singapore, Hackensack New Jersey, London, 2012.

742. W. T. Coffey, D. S. F. Crothers, Y. P. Kalmykov, E. S. Massawe, and J. T. Waldron. Exact analytic formula for the correlation time of a single-domain ferromagnetic particle. *Phys. Rev. E*, 49:1869–1882, 1994.

743. W. T. Coffey, D. S. F. Crothers, Y. P. Kalmykov, E. S. Massawe, and J. T. Waldron. Rotational Brownian motion and dielectric relaxation of polar molecules subjected to a constant bias field: Exact solution. *Phys. Rev. E*, 49:3976–3989, 1994.

744. W. T. Coffey, D. S. F. Crothers, Y. P. Kalmykov, and J. T. Waldron. Constant-magnetic-field effect in Néel relaxation of single-domain ferromagnetic particles. *Phys. Rev. B*, 51:15947–15956, 1995.

745. D. A. Garanin. Intregral relaxation time of single-domain ferromagnetic particles. *Phys. Rev. E*, 54:3250–3256, 1996.

746. D. A. Garanin. Fokker-Planck and Landau-Lifshitz-Bloch Equations for Classical Ferromagnets. *Phys. Rev. B*, 55:3050–3057, 1997.

747. W. Wernsdorfer, B. Doudin, D. Mailly, K. Hasselbach, A. Benoit, J. Meier, J.-Ph. Ansermet, and B. Barbara. Nucleation of magnetization reversal in individual nanosized nickel wires. *Phys. Rev. Lett.*, 77:1873–1876, 1996.

748. M. Mezard, G. Parisi, and M. A. Virasoro. *Spin Glass Theory and Beyond*. World Scientific, Singapore, New Jersey, London, 1987.

749. A. Einstein. Über die von der molekular-kinetischen Theorie der Wärme geforderte Bewegung von in ruhenden Fluessigkeiten suspendierten Teilchen. *Ann. Phys. (Berlin)*, 17:549–560, 1905.

750. C. W. Gardiner. *Handbook of Stochastic Methods*. Springer, Berlin, Heidelberg, New York, 2nd edition, 1985.

751. N. Wax. *Noise and Stochastic Processes*. Dover Publications Inc., New York, 1954.

752. Jr. W. F. Brown. Thermal fluctuations of a single-domain particle. *Phys. Rev.*, 130(5):1677–1686, 1963.

753. H. W. Wyld. *Mathematical Methods for Physics*. Benjamin Cummings. Reading, 1976.

754. J. E. Wegrowe, J. P. Meier, B. Doudin, J.-Ph. Ansermet, W. Wernsdorfer, B. Barbara, W. T. Coffey, Y. P. Kalmykovv, and J.-L. Déjardin. Magnetic relaxation of nanowires - beyond the Néel-Brown activation process. *Europhys. Lett.*, 38(5):329–334, 1997.

755. J.-E. Wegrowe, D. Kelly, A. Franck, and J.-Ph. Ansermet. Magnetoresistance of ferromagnetic nanowires. *Phys. Rev. Lett.*, 82(18):3681–3684, 1999.

756. F. Cayssol, D. Ravelosona, C. Chappert, J. Ferré, and J. P. Jamet. Domain Wall Creep in Magnetic Wires. *Phys. Rev. Lett.*, 92:107202, 2004.

757. D. D. Awschalom, D. P. DiVincenzo, and J. F. Smyth. Macroscopic quantum effects in nanometer-scale magnets. *Science*, 258(5081):414–421, 1992.

758. R. Sessoli, D. Gatteschi, A. Caneschi, and M. A. Novak. Magnetic bistability in a metal-ion cluster. *Nature*, 365(6442):141–143, 1993.

759. B. Barbara, W. Wernsdorfer, L. C. Sampaio, J. G. Park, C. Paulsen, M. A. Novak, R. Ferré, D. Mailly, R. Sessoli, A. Caneschi, K. Hasselbach, A. Benoit, and L. Thomas. Mesoscopic quantum tunneling of the magnetization. *J. Mag. Mag. Mat.*, 140-144:1825–1828, 1995.

760. C. Paulsen, J.-G. Park, B. Barbara, R. Sessoli, and A. Caneschi. Novel features in the relaxation times of $Mn_{12}Ac$. *J. Magn. Magn. Mater.*, 140-144:379 – 380, 1995. International Conference on Magnetism.

761. L. Thomas, F. Lionti, R. Ballou, D. Gatteschi, R. Sessoli, and B. Barbara. Macroscopic quantum tunnelling of magnetization in a single crystal of nanomagnets. *Nature*, 383(6596):145–147, 1996.

762. C. Sangregorio, T. Ohm, C. Paulsen, R. Sessoli, and D. Gatteschi. Quantum tunneling of the magnetization in an iron cluster nanomagnet. *Phys. Rev. Lett.*, 78:4645–4648, 1997.

763. B. Barbara. Quantum nanomagnet. *C. R. Phys.*, 6(9):934–944, 2005.

764. J. R. Friedman, M. P. Sarachik, J. Tejada, and R. Ziolo. Macroscopic measurement of resonant magnetization tunneling in high-spin molecules. *Phys. Rev. Lett.*, 76:3830–3833, 1996.

765. I. P. Moseley, C.-Y. Lin, D. Z. Zee, and J. M. Zadrozny. Synthesis and magnetic characterization of a dinuclear complex of low-coordinate iron(II). *Polyhedron*, 175:114171, 2020.

766. Y.. L. Raikher, V. I. Stepanov, J.-C. Bacri, and R. Perzynski. Orientational dynamics of ferrofluids with finite magnetic anisotropy of the particles: relaxation of magneto-birefringence in crossed fields. *Phys. Rev. E*, 66:021203, 2002.

767. X. Liu, N. Kent, A. Ceballos, R. Streubel, Y. Jiang, Y. Chai, P. Y. Kim, J. Forth, F. Hellman, S. Shi, D. Wang, B. A. Helms, P. D. Ashby, P. Fischer, and T. P. Russell. Reconfigurable ferromagnetic liquid droplets. *Science*, 365(6450):264–267, 2019.

768. Y. L Raikher and V. I. Stepanov. Magneto-orientational behavior of a suspension of antiferromagnetic particles. *J. Phys. Condens. Matter*, 20(20):204120, 2008.

769. E. M. Chudnovsky, W. M. Saslow, and R. A. Serota. Ordering in ferromagnets with random anisotropy. *Phys. Rev. B*, 33:251–261, 1986.

770. H. Gleiter. Nanocrystalline materials. *Prog. Mater. Sci.*, 33(4):223–315, 1989.

771. R.W. Siegel. Cluster-assembled nanophase materials. *Annu. Rev. Mater. Sci.*, 21:559–578, 1991.

772. J. P. Perez, V. Dupuis, J. Tuaillon, A. Perez, V. Paillard, P. Melinon, M. Treilleux, L. Thomas, B. Barbara, and B. Bouchet-Fabre. Magnetic properties of nanostructured iron films obtained by low energy neutral cluster beam deposition. *J. Magn. Magn. Mater.*, 145(1):74–80, 1995.

773. J. F. Löffler, J. P. Meier, B. Doudin, J.-P. Ansermet, and W. Wagner. Random and exchange anisotropy in consolidated nanostructured Fe and Ni: role of grain size and trace oxides on the magnetic properties. *Phys. Rev. B*, 57:2915–2924, 1998.

774. J. Nogues, J. Sort, V. Langlais, V. Skumryev, S. Surinach, J. S. Munoz, and M.D. Baro. Exchange bias in nanostructures. *Phys. Rep.*, 422(3):65–117, 2005.

775. S. Lemerle, J. Ferré, C. Chappert, V. Mathet, T. Giamarchi, and P. Le Doussal. Domain wall creep in an ising ultrathin magnetic film. *Phys. Rev. Lett.*, 80:849–852, 1998.

776. B. Alessandro, C. Beatrice, G. Bertotti, and A. Montorsi. Domain-wall dynamics and Barkhausen effect in metallic ferromagnetic materials. *J. Appl. Phys.*, 68(6):2901–2907, 1990.

777. B. Barbara. Propriétés des parois étroites dans les substances ferromagnétiques à forte anisotropie. *J. Phys. France*, 34(11-12):1039–1046, 1973.

778. B. Barbara and M. Hehara. Anisotropy and coercivity in $SmCo_5$-based compounds. *IEEE Trans. Mag.*, 12(6):997–999, 1976.

779. T. Egami. Theory of Bloch wall tunnelling. *Phys. Status Solidi B*, 57(1):211–224, 1973.

780. G. Tatara and H. Fukuyama. Macroscopic quantum tunneling of a domain wall in a ferromagnetic metal. *Phys. Rev. Lett.*, 72:772–775, 1994.

781. M. Uehara and B. Barbara. Noncoherent quantum effects in the magnetization reversal of a chemically disordered magnet : $SmCo_{3.5}Cu_{1.5}$. *J. Phys. France*, 47(2):235–238, 1986.

782. D. A. Allwood, G. Xiong, C. C. Faulkner, D. Atkinson, D. Petit, and R. P. Cowburn. Magnetic domain-wall logic. *Science*, 309(5741):1688–1692, 2005.

783. G. Catalan, J. Seidel, R. Ramesh, and J. F. Scott. Domain wall nanoelectronics. *Rev. Mod. Phys.*, 84:119–156, 2012.

784. S. Emori, U. Bauer, S.-M. Ahn, E. Martinez, and G. S. D. Beach. Current-driven dynamics of chiral ferromagnetic domain walls. *Nat. Mater.*, 12(7):611–616, 2013.

785. D.-Y. Kim, M.-H. Park, Y.-K. Park, J.-S. Kim, Y.-S. Nam, D.-H. Kim, S.-G. Je, H.-C. Choi, B.-C. Min, and S.-B. Choe. Chirality-induced antisymmetry in magnetic domain wall speed. *NPG Asia Mater.*, 10(1):e464, 2018.

786. D.-H. Kim, D.-Y. Kim, S.-C. Yoo, B.-C. Min, and S.-B. Choe. Universality of Dzyaloshinskii-Moriya interaction effect over domain-wall creep and flow regimes. *Phys. Rev. B*, 99:134401, 2019.

787. O. Gomonay, T. Jungwirth, and J. Sinova. High antiferromagnetic domain wall velocity induced by Néel spin-orbit torques. *Phys. Rev. Lett.*, 117:017202, 2016.

788. L. Caretta, M. Mann, F. Büttner, K. Ueda, B. Pfau, C. M. Günther, P. Hessing, A. Churikova, C. Klose, M. Schneider, D. Engel, C. Marcus, D. Bono, K. Bagschik, S. Eisebitt, and G. S. D. Beach. Fast current-driven domain walls and small skyrmions in a compensated ferrimagnet. *Nat. Nanotechnol.*, 13(12):1154–1160, 2018.

789. S. Mangin, M. Gottwald, C.-H. Lambert, D. Steil, V. Uhlíř, L. Pang, M. Hehn, S. Alebrand, M. Cinchetti, G. Malinowski, Y. Fainman, M. Aeschlimann, and E. E. Fullerton. Engineered materials for all-optical helicity-dependent magnetic switching. *Nat. Mater.*, 13(3):286–292, 2014.

790. J. T. Heron, J. L. Bosse, Q. He, Y. Gao, M. Trassin, L. Ye, J. D. Clarkson, C. Wang, J. Liu, S. Salahuddin, D. C. Ralph, D. G. Schlom, J. Íñiguez, B. D. Huey, and R. Ramesh. Deterministic switching of ferromagnetism at room temperature using an electric field. *Nature*, 516:370–373, 2014.

791. C. D. Stanciu, A. V. Kimel, F. Hansteen, A. Tsukamoto, A. Itoh, A. Kirilyuk, and Th. Rasing. Ultrafast spin dynamics across compensation points in ferrimagnetic GdFeCo: the role of angular momentum compensation. *Phys. Rev. B*, 73:220402, 2006.

792. A. Kirilyuk, A. V. Kimel, and T. Rasing. Ultrafast optical manipulation of magnetic order. *Rev. Mod. Phys.*, 82:2731–2784, 2010.

793. I. I. Rabi, J. R. Zacharias, S. Millman, and P. Kusch. A new method of measuring nuclear magnetic moment. *Phys. Rev.*, 53:318–318, 1938.

794. B. I. Kochelaev, Y. V. Yablokov. *The Beginning of Paramagnetic Resonance*. World Scientific, Singapore, New Jersey, London, Hong Kong, 1995.

795. C. J. Gorter and L. J. F. Broer. Negative result of an attempt to observe nuclear magnetic resonance in solids. *Physica*, IX:591–596, 1942.

796. E. Purcell, H. C. Torrey, and R. V. Pound. Resonance absorption by nuclear magnetic moments in a solid. *Phys. Rev.*, 69:37–38, 1946.

797. F. Bloch, W. W. Hansen, and M. Packard. The nuclear induction experiment. *Phys. Rev.*, 70:474–485, 1946.

798. Zavoisky E. Paramagnetic relaxation of liquid solutions for perpendicular fields. *J. Phys.*, IX(2):211–216, 1945.

799. M. D.E. Forbes, L. E. Jarocha, S. Sim, and V. F. Tarasov. Time-resolved electron paramagnetic resonance spectroscopy: History, technique, and application to supramolecular and macromolecular chemistry. *Adv. Phys. Org. Chem.*, 47:1–83, 2013.

800. M. Rohrer, B. Kinzer, and T. F. Prisner. High-field/high-frequency EPR spectrometer operating in pulsed and continuous-wave mode at 180 GHz. *Appl. Magn. Reson.*, 21:257–274, 2001.

801. Y. Yanagisawa, M. Hamada, K. Hashi, and H. Maeda. Review of recent developments in ultra-high field (UHF) NMR magnets in the Asia region. *Supercond Sci Technol.*, 35:044006, 2022.

802. E. A. Turov and M. P. Petrov. *Nuclear Magnetic Resonance in Ferro- and Antiferromagnets*. John Wiley & Sons, New York, 1972.

803. Panissod P. NMR of nanosized magnetic systems, ultrathin films, and granular systems. In Miller J.S. and Drillon M., editors, *Magnetism: Molecules to Materials III*, chapter 8. Wiley-VCH Weinheim, 2002.

804. J. F. Gregg, W. Allen, K. Ounadjela, M. Viret, M. Hehn, S. M. Thompson, and J. M. D. Coey. Giant magnetoresistive effects in a single element magnetic thin film. *Phys. Rev. Lett.*, 77(8):1580–1583, 1996.

805. J.-Ph. Ansermet. Classical description of spin wave excitation by currents in bulk ferromagnets. *IEEE Trans. Magn.*, 40(2):358–360, 2004.

806. C. J. Hardy, W. A. Edelstein, and D. Vatis. Efficient adiabatic fast passage for NMR population inversion in the presence of radiofrequency field inhomogeneity and frequency offsets. *J. Mag. Res.*, 66:470–482, 1986.

807. J.-Ph. Ansermet. Spintronics with metallic nanowires. In A.V. Narlikar; Y.Y. Fu, editor, *The Oxford Handbook of Nanoscience and Technology*. Oxford University Press, Oxford, 2010.

808. A. Janossy. Transmission electron spin resonance determination of the Fermi liquid exchange parameter of copper. *Solid State Commun.*, 36(4):321–325, 1980.

809. P. Monot, H. Hurdequint, A. Janossy, J. Obert, and J. Chaumont. Giant electron spin-resonance transmission in Cu ion implanted with Mn. *Phys. Rev. Lett.*, 29(19):1327–1330, 1972.

810. A. Comment, J.-Ph. Ansermet, C. P. Slichter, H. Rho, C. S. Snow, and S. L. Cooper. Magnetic properties of pure and Gd-doped EuO probed by NMR. *Phys. Rev. B*, 72:014428, 2005.

811. A. Comment. *Study of pure and gadolinium doped single crystals of europium monoxide by nuclear magnetic resonance*. PhD thesis, EPFL, 2003.

812. V. Heine. Hyperfine structure of paramagnetic ions. *Phys. Rev.*, 107:1002–1003, 1957.

813. A. J. Freeman R. E. Watson. Origin of effective fields in magnetic materials. *Phys. Rev.*, 123:2027–2047, 1961.

814. D. M. Grant and R. K. Harris, editors. *Encyclopedia of Nuclear Magnetic Resonance*, volume 9, chapter Schulman. Wiley Chichester, 2002.

815. R. G. Shulman and V. Jaccarino. Effects of superexchange on the nuclear magnetic resonance of MnF_2. *Phys. Rev.*, 103:1126–1127, 1956.

816. R. G. Shulman and V. Jacarino. Nuclear magnetic resonance in paramagnetic MnF_2. *Phys. Rev.*, 108(5):1219–1231, 1957.

817. R.G. Shulman and S. Sugano. Covalency effects in $KNiF_3$. I: nuclear magnetic resonance studies. *Phys. Rev.*, 130:506–511, 1963.

818. R. D. Willett and C.P. Landee. Ferromagnetism in one dimensional systems: synthesis and structural characterization. *J. Appl. Phys.*, 52(3):2004–2009, 1981.

819. R. S. Drago. *Physical Methods in Chemistry*. W.B. Saunders Co., Philadelphia, London, Toronto, 1977.

820. J. Boyce and C. P. Slichter. Conduction-electron spin density around Fe impurities in Cu above and below the Kondo temperature. *Phys. Rev. B*, 13:379–396, 1976.

821. J. Boyce and C. P. Slichter. Conduction-Electron Spin Density around Fe Impurities in Cu above and below TK. *Phys. Rev. Lett.*, 32:61–64, 1974.

822. F. Bloch and A. Siegert. Magnetic resonance for nonrotating fields. *Phys. Rev.*, 57:522–527, 1940.

823. L. I. Sacolick, F. Wiesinger, I. Hancu, and M. W. Vogel. B_1 mapping by Bloch-Siegert shift. *Magn. Reson. Med.*, 63(5):1315–1322, 2010.

824. E. F. Thenell III. *Rabi Oscillations in Strongly-driven magnetic resonance systems*. PhD thesis, University of Utah, 2017.

825. I. Pietikäinen, S. Danilin, K. S. Kumar, A. Vepsäläinen, D. S. Golubev, J. Tuorila, and G. S. Paraoanu. Observation of the Bloch-Siegert shift in a driven quantum-to-classical transition. *Phys. Rev. B*, 96:020501, 2017.

826. P. Forn-Díaz, J. Lisenfeld, D. Marcos, J. J. García-Ripoll, E. Solano, C. J. P. M. Harmans, and J. E. Mooij. Observation of the Bloch-Siegert shift in a qubit-oscillator system in the ultrastrong coupling regime. *Phys. Rev. Lett.*, 105:237001, 2010.

827. R. H. Dicke. Coherence in spontaneous radiation processes. *Phys. Rev.*, 93:99–110, 1954.

828. R. C. Roundy and M. E. Raikh. Organic magnetoresistance under resonant AC drive. *Phys. Rev. B*, 88:125206, 2013.

829. D. P. Waters, G. Joshi, M. Kavand, M. E. Limes, H. Malissa, P. L. Burn, J. M. Lupton, and C. Boehme. The spin-Dicke effect in OLED magnetoresistance. *Nat. Phys.*, 11(11):910–914, 2015.

830. A. Bienfait, J. J. Pla, Y. Kubo, M. Stern, X. Zhou, C. C. Lo, C. D. Weis, T. Schenkel, M. L.W. Thewalt, D. Vion, D. Esteve, B. Julsgaard, K. Mølmer, J. J.L. Morton, and P. Bertet. Reaching the quantum limit of sensitivity in electron spin resonance. *Nat. Nanotechnol.*, 11(3):253–257, 2016.

831. S. Probst, A. Bienfait, P. Campagne-Ibarcq, J. J. Pla, B. Albanese, J. F. Da Silva Barbosa, T. Schenkel, D. Vion, D. Esteve, K. Mølmer, J. J. L. Morton, R. Heeres, and P. Bertet. Inductive-detection electron-spin resonance spectroscopy with 65 spins/ Hz sensitivity. *Appl. Phys. Lett.*, 111(20):202604, 2017.

832. T. Eickelkamp, S. Roth, and M. Mehring. Electrically detected magnetic resonance in photoexcited fullerenes. *Mol. Phys.*, 95(5):967–972, 1998.

833. K. R. Wald, L. P. Kouwenhoven, P. L. McEuen, N. C. van der Vaart, and C. T. Foxon. Local dynamic nuclear polarization using quantum point contacts. *Phys. Rev. Lett.*, 73:1011–1014, 1994.

834. M. Xiao, I. Martin, E. Yablonovitch, and H. W. Jiang. Electrical detection of the spin resonance of a single electron in a silicon field-effect transistor. *Nature*, 430(6998):435–439, 2004.

835. M. Orrit and J. Bernard. Single pentacene molecules detected by fluorescence excitation in a p-terphenyl crystal. *Phys. Rev. Lett.*, 65:2716–2719, 1990.

836. J. Wrachtrup, C. von Borczyskowski, J. Bernard, M. Orrit, and R. Brown. Optical detection of magnetic resonance in a single molecule. *Nature*, 363(6426):244–245, 1993.

837. A. Gruber, A. Dräbenstedt, C. Tietz, L. Fleury, J. Wrachtrup, and C. von Borczyskowski. Scanning confocal optical microscopy and magnetic resonance on single defect centers. *Science*, 276(5321):2012–2014, 1997.

838. D. Rugar, R. Budakian, H. J. Mamin, and B. W. Chui. Single spin detection by magnetic resonance force microscopy. *Nature*, 430(6997):329–332, 2004.

839. S. Baumann, W. Paul, T. Choi, C. P. Lutz, A. Ardavan, and A. J. Heinrich. Electron paramagnetic resonance of individual atoms on a surface. *Science*, 350(6259):417–420, 2015.

840. F. D. Natterer, K. Yang, W. Paul, P. Willke, T. Choi, T. Greber, A. J. Heinrich, and C. P. Lutz. Reading and writing single-atom magnets. *Nature*, 543(7644):226–228, 2017.

841. T. S. Seifert, S. Kovarik, D. M. Juraschek, N. A. Spaldin, P. Gambardella, and S. Stepanow. Longitudinal and transverse electron paramagnetic resonance in a scanning tunneling microscope. *Sci. Adv.*, 6:eabc5511, 2020.

842. J. M. Kikkawa and D. D. Awschalom. All-optical magnetic resonance in semiconductors. *Science*, 287(5452):473–476, 2000.

843. G. S. Jenkins, A. B. Sushkov, D. C. Schmadel, N. P. Butch, P. Syers, J. Paglione, and H. D. Drew. Terahertz Kerr and reflectivity measurements on the topological insulator Bi_2Se_3. *Phys. Rev. B*, 82:125120, 2010.

844. X. Wang, J. Lian, Y. Huang, Z. Sun, J. Liu, F. Zhang, S. Gao, X. Yu, P. Li, and M. Zhao. Magneto-optical faraday and kerr effects in topological insulators Bi_2Te_3 and Bi_2Se_3. *Jpn. J. Appl. Phys.*, 52(10R):103001, 2013.

845. N. Bloembergen. Linear Stark Effect in magnetic resonance spectra. *Science*, 133:1363–1370, 1961.

846. S. Asaad, V. Mourik, B. Joecker, M. A. I. Johnson, A. D. Baczewski, Hannes R. Firgau, M. T. Madzik, V. Schmitt, J. J. Pla, F. E. Hudson, K. M. Itoh, J. C. McCallum, A. S. Dzurak, A. Laucht, and A. Morello. Coherent electrical control of a single high-spin nucleus in silicon. *Nature*, 579(7798):205–209, 2020.

847. M. Bennati, C. T. Farrar, J. A. Bryant, S. J. Inati, V. Weis, G. J. Gerfen, P. Riggs-Gelasco, J. Stubbe, and R .G. Griffin. Pulsed electron-nuclear double resonance (ENDOR) at 140 GHz. *J. Mag. Res.*, 138(2):232–243, 1999.

848. V. Lang, C. C. Lo, R. E. George, S. A. Lyon, J. Bokor, T. Schenkel, A. Ardavan, and J. J. L. Morton. Electrically detected magnetic resonance in a W-band microwave cavity. *Rev. Sci. Instrum.*, 82(3):034704, 2011.

849. P. A. S. Cruickshank and G. M. Smith. Force detected electron spin resonance at 94 GHz. *Rev. Sci. Instrum.*, 78(1):015101, 2007.

850. J. P. Campbell, J. T. Ryan, P. R. Shrestha, Z. Liu, C. Vaz, J.-H. Kim, V. Georgiou, and K. P. Cheung. Electron spin resonance scanning probe spectroscopy for ultrasensitive biochemical studies. *Anal. Chem.*, 87(9):4910–4916, 2015.

851. P. R. Shrestha, N. Abhyankar, M. A. Anders, K.P. Cheung, R. Gougelet, J. T. Ryan, V. Szalai, and J. P. Campbell. Nonresonant transmission line probe for sensitive interferometric electron spin resonance detection. *Anal. Chem.*, 91(17):11108–11115, 2019.

852. L. E. Switala, B. E. Black, C. A. Mercovich, A. Seshadri, and J. P. Hornak. An electron paramagnetic resonance mobile universal surface explorer. *J. Magn. Res.*, 285:18–25, 2017.

853. E. J. Reijerse. High-frequency EPR instrumentation. *Appl. Magn. Reson*, 37(1):795–818, 2010.

854. C. Caspers, V. Gandhi, A. Magrez, E. De Rijk, and J.-Ph. Ansermet. Sub-terahertz spectroscopy of magnetic resonance in $BiFeO_3$ using a vector network analyzer. *Appl. Phys. Lett.*, 108(24):241104, 2016.

855. A. W. Overhauser. Polarization of nuclei in metals. *Phys. Rev.*, 92(2):411–415, 1953.

856. T. R. Carver and C. P. Slichter. Polarization of nuclar spins in metals. *Phys. Rev.*, 102(4), 975–980, 1956.

857. N. M. Loening, M. Rosay, V. Weis, and R. G. Griffin. Solution-state dynamic nuclear polarization at high magnetic field. *J. Am. Chem. Soc.*, 124:8808–8809, 2002.

858. E. R. McCarney and S. Han. Spin-labeled gel for the production of radical-free dynamic nuclear polarization enhanced molecules for NMR spectroscopy and imaging. *J. Magn. Res.*, 190:307–315, 2007.

859. D. Wolf. *Spin Temperature and Nuclear Spin Relaxation in Matter: Basic Principles and Applications.* The International Series of Monographs on Physics. Oxford University Press London, 1979.

860. E. M. Purcell and R. V. Pound. A nuclear spin system at negative temperature. *Phys. Rev.*, 81:279–280, 1951.

861. W. T. Wenckebach. Thermodynamics of dynamic nuclear polarization. *Nucl. Instrum.*, 356(1):1–4, 1995.

862. T. Wenckebach. *Essentials of Dynamic Nuclear Polarization.* Spindrift Publications, The Netherlands, 2016.

863. E. Beaurepaires, J.C. Merle, A. Daunois, and J.-Y. Bigot. Ultrafast spin dynamics in ferromagnetic nickel. *Phys. Rev. Lett.*, 76(22):4250–4253, 1996.

864. Solomon I. Relaxation processes in a system of two spins. *Phys.Rev.*, 99(2):559–565, 1955.

865. C. D. Jeffries. *Dynamic Nuclear Orientation.* Interscience Publishers, New York, London, Sydney, 1963.

866. A. Abragam. *The Principles of Nuclear Magnetism.* Clarendon Press, Oxford, 1961.

867. E. C. G. Stückelberg de Breidenbach and P. B. Scheurer. *Thermocinétique Phénoménologique Galiléenne.* Birkhäuser Verlag Basel, Basel, Stuttgart, 1974.

868. J. P. Sethna. *Entropy, Order Parameters and Complexity.* Oxford University Press, 2006.

869. R. Koenig. Frontiers in refrigeration and cooling : how to obtain and sustain ultralow temperatures beyond nature's ambience. *Int. J. Refrig.*, 23:577–587, 2000.

870. A. G. Redfield. Nuclear magnetic resonance saturation and rotary saturation in Solids. *Phys. Rev.*, 98:1787–1809, 1955.

871. H. Brunner, R. H. Fritsc, and K. H. Hausser. Cross polarization in electron nuclear double resonance by satisfying the Hartman–Hahn condition. *Z. Naturforsch.*, 42a:1456–1457, 1987.

872. A. Henstra, P. Dirksen, J. Schmidt, and W. Th. Wenckebach. Nuclear spin orientation via electron spin locking (NOVEL). *J. Magn. Res.*, 77:389–393, 1988.

873. D. J. van den Heuvel, A. Henstra, T.-S. Lin, and W. Th.Wenckebach. Transient oscillations in pulsed dynamic nuclear polarization. *Chem. Phys. Lett.*, 188:194–200, 1992.

874. A. Henstra, P. Dirksen, and W. Th. Wenckebach. Enhanced dynamic nuclear polarization by the integrated solid effect. *Phys. Lett. A*, 134(2):134–136, 1988.

875. A. Henstra, T.-S. Lin, J. Schmid, and W.Th. Wenckebach. High dynamic nuclear polarization at room temperature. *Chem. Phys. Lett.*, 165(1):6–10, 1990.

876. W. T. Wenckebach. The solid effect. *Appl. Magn. Reson.*, 34:227–235, 2008.

877. T.R. Eichhorn, M. Haag, B. van den Brandt, P. Hautle, and W.Th. Wenckebach. High proton spin polarization with DNP using the triplet state of pentacene d(14). *Chem. Phys. Letters*, 555:296–299, 2013.

878. A. Abragam and W. G. Proctor. Experiments on spin temperature. *Phys. Rev.*, 106:160–161, 1957.

879. M. Vladimirova, S. Cronenberger, D. Scalbert, I. I. Ryzhov, V. S. Zapasskii, G. G. Kozlov, A. Lemaître, and K. V. Kavokin. Spin temperature concept verified by optical magnetometry of nuclear spins. *Phys. Rev. B*, 97:041301, 2018.

880. S. Braun, J. P. Ronzheimer, M. Schreiber, S. S. Hodgman, T. Rom, I. Bloch, and U. Schneider. Negative absolute temperature for motional degrees of freedom. *Science*, 339(6115):52–55, 2013.

881. L. D. Carr. Negative temperatures? *Science*, 339(6115):42–43, 2013.

882. S. Hilbert, P. Hänggi, and J. Dunkel. Thermodynamic laws in isolated systems. *Phys. Rev. E*, 90:062116, 2014.

883. C. A. Ryan, O. Moussa, J. Baugh, and R. Laflamme. Spin based heat engine: demonstration of multiple rounds of algorithmic cooling. *Phys. Rev. Lett.*, 100:140501, 2008.

884. Raeisi S. and Mosca M. Asymptotic bound for heat-bath algorithmic cooling. *Phys. Rev. Lett.*, 114:100404, 2015.

885. W. G. Clark and G. Feher. Nuclear polarization in InSb by a dc current. *Phys. Rev. Lett.*, 10(4):134–138, 1963.

886. M. Johnson. Dynamic nuclear polarization by spin injection. *Appl. Phys. Lett.*, 77(11):1680–1682, 2000.

887. C. J. Trowbridge, B. M. Norman, Y. K. Kato, D. D. Awschalom, and V. Sih. Dynamic nuclear polarization from current-induced electron spin polarization. *Phys. Rev. B*, 90(8):1–5, 2014.

888. B. M. Norman, C. J. Trowbridge, D. D. Awschalom, and V. Sih. Current-induced spin polarization in anisotropic spin-orbit fields. *Phys. Rev. Lett.*, 112:056601, 2014.

889. Y. K. Kato, R. C. Myers, A. C. Gossard, and D. D. Awschalom. Current-induced spin polarization in strained semiconductors. *Phys. Rev. Lett.*, 93:176601, 2004.

890. F. Meier and B. P. Zakharchenya, editors. *Optical Orientation*. North Holland, Amsterdam, Oxford, New York, Tokyo, 1984.

891. C. Caspers and J.-Ph. Ansermet. Electrical detection of spin hyperpolarization in InP. *Appl. Phys. Lett.*, 105:133110, 2014.

892. C. Caspers, D. Yoon, M. Soundararajan, and J.-Ph. Ansermet. Opto-spintronics in InP using ferromagnetic tunnel spin filters. *New J. Phys.*, 17:022004, 2015.

893. F. Blumenschein, M. Tamski, C. Roussel, E. Z. B. Smolinsky, F. Tassinari, R. Naaman, and J.-Ph. Ansermet. Spin-dependent charge transfer at chiral electrodes probed by magnetic resonance. *Phys. Chem. Chem. Phys.*, 22(3):997–1002, 2020.

894. R. Schirhagl, K. Chang, M. Loretz, and C. L. Degen. Nitrogen-vacancy centers in diamond: nanoscale sensors for physics and biology. *Annu. Rev. Phys. Chem.*, 65(1):83–105, 2014.

895. V. Jacques, P. Neumann, J. Beck, M. Markham, D. Twitchen, J. Meijer, F. Kaiser, G. Balasubramanian, F. Jelezko, and J. Wrachtrup. Dynamic polarization of single nuclear spins by optical pumping of nitrogen-vacancy color centers in diamond at room temperature. *Phys. Rev. Lett.*, 102:057403, 2009.

896. H.-J. Wang, C. S. Shin, C. E. Avalos, S. J. Seltzer, D. Budker, Al. Pines, and V. S. Bajaj. Sensitive magnetic control of ensemble nuclear spin hyperpolarization in diamond. *Nat. Commun.*, 4(1):1940, 2013.

897. A. Ajoy, K. Liu, R. Nazaryan, X. Lv, P. R. Zangara, B. Safvati, G. Wang, D. Arnold, G. Li, A.r Lin, P. Raghavan, E. Druga, S. Dhomkar, D. Pagliero, J. A. Reimer, D. Suter, C. A. Meriles, and A. Pines. Orientation-independent room temperature optical ^{13}C hyperpolarization in powdered diamond. *Sci. Adv.*, 4(5):eaar5492, 2018.

898. D. Abrams, M. E. Trusheim, D. R. Englund, M. D. Shattuck, and C. A. Meriles. Dynamic nuclear spin polarization of liquids and gases in contact with nanostructured diamond. *Nano Lett.*, 14(5):2471–2478, 2014.

899. D. A. Broadway, J.-P. Tetienne, A. Stacey, J. D. A. Wood, D. A. Simpson, L. T. Hall, and L. C. L. Hollenberg. Quantum probe hyperpolarisation of molecular nuclear spins. *Nat. Commun.*, 9(1):1246, 2018.

900. A. L. Falk, P. V. Klimov, V. Ivády, K. Szász, D. J. Christle, W. F. Koehl, Á. Gali, and D. D. Awschalom. Optical polarization of nuclear spins in silicon carbide. *Phys. Rev. Lett.*, 114:247603, 2015.

901. M. W. Dale and C. J. Wedge. Optically generated hyperpolarization for sensitivity enhancement in solution-state NMR spectroscopy. *Chem. Commun.*, 52:13221–13224, 2016.

902. G. Liu, S. H. Liou, N. Enkin, I. Tkach, and M. Bennati. Photo-induced radical polarization and liquid-state dynamic nuclear polarization using fullerene nitroxide derivatives. *Phys. Chem. Chem. Phys.*, 19(47):31823–31829, 2017.

903. S. E. Barrett, R. Tycko, L. N. Pfeiffer, and K. W. West. Directly detected nuclear magnetic resonance of optically pumped GaAs quantum wells. *Phys. Rev. Lett.*, 72:1368–1371, 1994.

904. R. Tycko, S.E. Barrett, G. Dabbagh, L. N. Pfeiffer, and K. W. West. Electronic states in gallium arsenide quantum wells probed by optically pumped NMR. *Science*, 268(5216):1460–1463, 1995.

905. D. Yoon, M. Soundararajan, and J.-Ph. Ansermet. Nuclear polarization by optical pumping in InP:Fe above liquid nitrogen temperature. *Solid State Nucl. Magn. Reson.*, 70:48–52, 2015.

906. P. Maletinsky, A. Badolato, and A. Imamoglu. Dynamics of quantum dot nuclear spin polarization controlled by a single electron. *Phys. Rev. Lett.*, 99:056804, 2007.

907. P. Malentinsky. *Polarization and manipulation of a mesoscopic nuclear spin ensemble using a single confined electron spin*. PhD thesis, ETHZ, 2008.

908. J. Eills, D. Budker, S. Cavagnero, E. Y. Chekmenev, S. J. Elliott, S. Jannin, A. Lesage, J. Matysik, T. Meersmann, T. Prisner, J. A. Reimer, H. Yang, and I. V. Koptyug. Spin hyperpolarization in modern magnetic resonance. *Chem. Rev.*, 123:1417–1551, 2023.

909. V. Denysenkov and T. Prisner. Liquid state Dynamic Nuclear Polarization probe with Fabry-Perot resonator at 9.2 T. *J. Mag. Res.*, 217:1–5, 2012.

910. M. Goldman. *Quantum Description of High-Resolution NMR in Liquids*, volume 15 of *International Series of Monographs on Chemistry*. Clarendon Press, Oxford, 1991.

911. I. P. Gerothanassis, A. Troganis, V. Exarchou, and K. Barbarossou. Nuclear Magnetic Resonance (NMR) spectroscopy: basic principles and phenomena, and their applications to chemistry, biology and medicine. *Chem. Educ. Res. Pract.*, 3:229–252, 2002.

912. Q. Z. Ni, E. Daviso, T. V. Can, E. Markhasin, S. K. Jawla, T. M. Swager, R. J. Temkin, J. Herzfeld, and R. G. Griffin. High frequency dynamic nuclear polarization. *Acc. Chem. Res.*, 46:1933–1941, 2013.

913. K. N. Hu, G. T. Debelouchina, A. A. Smith, and R. G. Griffin. Quantum mechanical theory of dynamic nuclear polarization in solid dielectrics. *J. Chem. Phys.*, 134:125105, 2011.

914. A. S. Lilly Thankamony, J. J. Wittmann, M. Kaushik, and B. Corzilius. Dynamic nuclear polarization for sensitivity enhancement in modern solid-state NMR. *Prog. Nucl. Magn. Reson. Spectrosc.*, 102-103:120–195, 2017.

915. T. V. Can, M. A. Caporini, F. Mentink-Vigier, B. Corzilius, J. J. Walish, M. Rosay, W. E. Maas, M. Baldus, S. Vega, T. M. Swager, and R. G. Griffin. Overhauser effects in insulating solids. *J. Chem. Phys.*, 141(6):064202, 2014.

916. S. Pylaeva, K. L. Ivanov, M. Baldus, D. Sebastiani, and H. Elgabarty. Molecular mechanism of overhauser dynamic nuclear polarization in insulating solids. *J. Chem. Phys. Lett.*, 8(10):2137–2142, 2017.

917. N. Bloembergen, S. Shapiro, P. S. Pershan, and J. O. Artman. Cross-relaxation in spin systems. *Phys. Rev.*, 114:445–459, 1959.

918. P. S. Pershan. Cross relaxation in LiF. *Phys. Rev.*, 117:109–116, 1960.

919. S. R. Hartmann and E. L. Hahn. Nuclear double resonance in the rotating frame. *Phys. Rev.*, 128:2042–2053, 1962.

920. W. Kolodziejski and J. Klinowski. Kinetics of cross-polarization in solid-state NMR: a guide for chemists. *Chemical Reviews*, 102(3):613–628, 2002.

921. R. A. Wind. Distortion-Free ^{13}C NMR Spectroscopy in Coal. In *Magnetic Resonance of Carbonaceous Solids*, Adv. Chem., 229:217–227, 1992.

922. C.T Farrar, D.A Hall, G.J Gerfen, M Rosay, J.-H Ardenkjær-LarsenLarsen, and R.G Griffin. High-frequency dynamic nuclear polarization in the nuclear rotating frame. *J. Mag. Res.*, 144(1):134–141, 2000.

923. L. R. Becerra, G. J. Gerfen, R. J. Temkin, D. J. Singel, and R. G. Griffin. Dynamic nuclear polarization with a cyclotron resonance maser at 5 T. *Phys. Rev. Lett.*, 71:3561–3564, 1993.

924. A. S. Kiryutin, A. N. Pravdivtsev, K. L. Ivanov, Y. A Grishin, H.-M. Vieth, and A. V. Yurkovskaya. A fast field-cycling device for high-resolution NMR: design and application to spin relaxation and hyperpolarization experiments. *J Magn Reson*, 263:79–91, 2016.

925. K. Kouřil, H. Kouřilová, S. Bartram, M. H. Levitt, and B. Meier. Scalable dissolution-dynamic nuclear polarization with rapid transfer of a polarized solid. *Nat. Commun.*, 10(1):1733, 2019.

926. B. Vuichoud, J. Milani, Q. Chappuis, A. Bornet, G. Bodenhausen, and S. Jannin. Measuring absolute spin polarization in dissolution-DNP by Spin Polarimetry Magnetic Resonance (SPY-MR). *J. Magn. Res.*, 260:127–135, 2015.

927. J. H. Ardenkjaer-Larsen, *Introduction to dissolution DNP: overview, instrumentation, and human applications*, eMagRes. 7:63–78, 2018.

928. K. V. Kovtunov, E. V. Pokochueva, O. G Salnikov, S. F. Cousin, D. Kurzbach, B. Vuichoud, S. Jannin, E. Y. Chekmenev, B. M. Goodson, D. A Barskiy, and I. V. Koptyug. Hyperpolarized NMR Spectroscopy: d-DNP, PHIP, and SABRE Techniques. *Chem. Asian J.*, 13(15):1857–1871, 2018.

929. D. Raftery, H. Long, T. Meersmann, P. J. Grandinetti, L. Reven, and A. Pines. High-field NMR of adsorbed xenon polarized by laser pumping. *Phys. Rev. Lett.*, 66:584–587, 1991.

930. B. Schwarzschild. Inhaling hyperpolarized noble gas helps magnetic resonance imaging of lungs. *Phys. Today*, 48(6):17–18, 1995.

931. M. C. Cassidy, H. R. Chan, B. D. Ross, P. K. Bhattacharya, and C. M. Marcus. In vivo magnetic resonance imaging of hyperpolarized silicon particles. *Nat. Nanotechnol.*, 8(5):363–368, 2013.

932. Griffiths J. H. E. Anomalous high-frequency resistance of ferromagnetic metals. *Nature*, 158:670–671, 1946.

933. C. Kittel. Interpretation of anomalous Larmor frequencies in ferromagnetic resonance experiment. *Phys. Rev.*, 71:270–271, 1947.

934. L. R. Walker. Magnetostatic modes in ferromagnetic resonance. *Phys. Rev. B*, 105(2):390–399, 1957.

935. H. J. Juretschke. Electromagnetic theory of dc effects in ferromagnetic resonance. *J. Appl. Phys.*, 31(8):1401–1406, 1960.

936. H. J. Juretschke. DC detection of spin resonance in thin metallic films. *J. Appl. Phys.*, 34:1223, 1963.

937. P. D. Sparks and R. H. Silsbee. Magnetization transport across a ferromagnetic-paramagnetic interface. *Phys. Rev. B*, 35(10):5198–5208, 1987.

938. M. Johnson and R. H. Silsbee. Thermodynamic analysis of interfacial transport and of the thermomagnetoelectric system. *Phys. Rev. B*, 35:4959–4972, 1987.

939. W. M. Saslow. Spin pumping of current in non-uniform conducting magnets. *Phys. Rev. B*, 76:184434, 2007.

940. W. M. Saslow and K. Rivkin. Irreversible thermodynamics of non-uniform insulating ferromagnets. *J. Magn. Magn. Mater.*, 320(21):2622–2628, 2008.

941. W. Heitler and E. Teller. Time effects in the magnetic cooling method. *Proc. Roy. Soc. (London)*, A155(886):629–639, 1936.

942. P. Grünberg. Magnetostatic spin wave modes of a ferromagnetic double layer. *J. Appl. Phys.*, 51(8):4338–4341, 1980.

943. Barnas J. Spin waves in multilayers. In M.G. Gottam, editor, *Linear and Nonlinear Spin Waves in Magnetic Films and Superlattices*. World Scientific, Singapore, River Edge New Jersey, London, 1994.

944. J. A. Stratton. *Electromagnetic Theory*. IEEE Press 2007, Hoboken New Jersey.

945. J. M. Luttinger and C. Kittel. Eine Bemerkunkg zur Quantentheorie der ferromagnetischen Resonanz. *Helv. Phys. Acta*, 21:480–482, 1948.

946. A.H. Morrish. *The Physical Principles of Magnetism*. Academic Press, New York, 1965.

947. R.W. Damon and J. R. Eshbach. Magnetostatic modes of a ferromagnet slab. *J. Phys. Chem. Solids*, 19:308–320, 1961.

948. L. R. Walker. Resonant modes of ferromagnetic spheroids. *J. Appl. Phys.*, 29:318, 1958.

949. Walker L. R. Spin waves and other magnetic modes. In G. T. Rado and H. Suhl, editors, *Magnetism, A Treatise on Modern Theory and Materials*, volume 1, chapter 8, pp. 299–381. Academic Press, New York, 1963.

950. R. Plumier. Magnetostatic modes in a sphere and polarization current corrections. *Physica*, 28(4):423–444, 1962.

951. L. D. Landau, E. M. Lifshitz, and L.-P. Pitaevskii. Electrodynamics of continuous media. In *Landau and Lifshizt Course of Theoretical Physics*, volume 8. Butterworth Heinemann, Oxford, 2000.

952. D. D. Stancil and A. Prabhakar. *Spin Waves, Theory and Applications*. Springer Science and Business Media, LLC New York, 2009.

953. H. Suhl and H. N. Bertram. Localized surface nucleation of magnetization reversal. *J. Appl. Phys.*, 82(12):6128–6137, 1997.

954. C. Kittel. Excitation of spin wave resonance in a ferromagnet by a uniform rf field. *Phys. Rev.*, 110(6):1295–1297, 1958.

955. M.H. Seavey and P.E. Tannewald. Direct observation of spin-wave resonance. *Phys. Rev. Lett.*, 1(5):168–169, 1958.

956. S. D. Brechet, F. A. Reuse, and J.-Ph. Ansermet. Thermodynamics of continuous media with electromagnetic fields. *Eur. Phys. J. B*, 85, 2012.

957. S. Zhang, P. M. Levy, and A. Fert. Mechanisms of spin-polarized current-driven magnetization switching. *Phys. Rev. Lett.*, 88:236601, 2002.

958. A. Shpiro, P. M. Levy, and S. Zhang. Self-consistent treatment of nonequilibrium spin torques in magnetic multilayers. *Phys. Rev. B*, 67:104430, 2003.

959. G. Tatara and H. Kohno. Theory of current-driven domain wall motion: spin transfer versus momentum transfer. *Phys. Rev. Lett.*, 92:086601, 2004.

960. S. Zhang and Z. Li. Roles of nonequilibrium conduction electrons on the magnetization dynamics of ferromagnets. *Phys. Rev. Lett.*, 93:127204, 2004.

961. D. C. Ralph and M. D. Stiles. Spin transfer torques. *J. Magn. Magn. Mater.*, 320(7):1190–1216, 2008.

962. C. Burrowes, A. P. Mihai, D. Ravelosona, J. V. Kim, C. Chappert, L. Vila, A. Marty, Y. Samson, F. Garcia-Sanchez, L. D. Buda-Prejbeanu, I. Tudosa, E. E. Fullerton, and J. P. Attané. Non-adiabatic spin-torques in narrow magnetic domain walls. *Nat. Phys.*, 6(1):17–21, 2010.

963. L. Heyne, J. Rhensius, D. Ilgaz, A. Bisig, U. Rüdiger, M. Kläui, L. Joly, F. Nolting, L. J. Heyderman, J. U. Thiele, and F. Kronast. Direct determination of large spin-torque nonadiabaticity in vortex core dynamics. *Phys. Rev. Lett.*, 105:187203, 2010.

964. M. Eltschka, M. Wötzel, J. Rhensius, S. Krzyk, U. Nowak, M. Kläui, T. Kasama, R. E. Dunin-Borkowski, L. J. Heyderman, H. J. van Driel, and R. A. Duine. Nonadiabatic spin

torque investigated using thermally activated magnetic domain wall dynamics. *Phys. Rev. Lett.*, 105:056601, 2010.

965. S. D. Pollard, L. Huang, K. S. Buchanan, D. A. Arena, and Y. Zhu. Direct dynamic imaging of non-adiabatic spin torque effects. *Nat. Commun.*, 3(1):1028, 2012.

966. Y. Tserkovnyak, A. Brataas, G.E. Bauer, and B.I. Halperin. Nonlocal magnetization dynamics in ferromagnetic heterostructures. *Rev. Mod. Phys.*, 77:1375–1421, 2005.

967. Y. Tserkovnyak, A. Brataas, and G. E.W. Bauer. Enhanced gilbert damping in thin ferromagnetic films. *Phys. Rev. Lett.*, 88:117601, 2002.

968. F. Liang, J. Wang, Y. H. Yang, and K. S. Chan. Detection of spin current by electron spin resonance. *J. Appl. Phys.*, 104(11):113701, 2008.

969. K. i. Uchida, H. Adachi, Y. Kajiwara, S. Maekawa, and E. Saitoh. Spin-wave spin current in magnetic insulators. In M. Wu and A. Hoffmann, editors, *Recent Advances in Magnetic Insulators – From Spintronics to Microwave Applications*, volume 64 of *Solid State Physics*, pp. 1–27. Academic Press, Oxford, Amsterdam, San Diego, Waltham MA, 2013.

970. G. Counil, Joo-Von Kim, T. Devolder, C. Chappert, K. Shigeto, and Y. Otani. Spin wave contributions to the high-frequency magnetic response of thin films obtained with inductive methods. *J. Appl. Phys.*, 95(10):5646–5652, 2004.

971. I. Neudecker, G. Woltersdorf, B. Heinrich, T. Okuno, G. Gubbiotti, and C.H. Back. Comparison of frequency, field, and time domain ferromagnetic resonance methods. *J. Magn. Magn. Mater.*, 307(1):148–156, 2006.

972. C. Bilzer, T. Devolder, P. Crozat, C. Chappert, S. Cardoso, and P. P. Freitas. Vector network analyzer ferromagnetic resonance of thin films on coplanar waveguides: comparison of different evaluation methods. *J. Appl. Phys.*, 101(7):101, 074505, 2007.

973. I. Harward, T. O'Keevan, A. Hutchison, V. Zagorodnii, and Z. Celinski. A broadband ferromagnetic resonance spectrometer to measure thin films up to 70 GHz. *Rev. Sci. Instrum.*, 82(9), 2011. 095115.

974. Y. Li, W. Han, A. G. Swartz, K. Pi, J. J. I. Wong, S. MacK, D. D. Awschalom, and R. K. Kawakami. Oscillatory spin polarization and magneto-optical Kerr effect in Fe_3O_4 thin films on GaAs(001). *Phys. Rev. Lett.*, 105:167203, 2010.

975. S. T. B. Goennenwein, S. W. Schink, A. Brandlmaier, A. Boger, M. Opel, R. Gross, R. S. Keizer, T. M. Klapwijk, A. Gupta, H. Huebl, C. Bihler, and M. S. Brandt. Electrically detected ferromagnetic resonance. *Appl. Phys. Lett.*, 90:162507, 2007.

976. N. Biziere, E. Murè, and J.-Ph. Ansermet. Current-driven, electrically detected ferromagnetic resonance in electrodeposited spin valves. *J. Magn. Magn. Mater.*, 322(9):1357 – 1359, 2010.

977. N. Biziere, E. Murè, and J.-Ph. Ansermet. Microwave spin-torque excitation in a template-synthesized nanomagnet. *Phys. Rev. B*, 79:012404, 2009.

978. Y. S. Gui, S. Holland, N. Mecking, and C.-M. Hu. Resonances in ferromagnetic gratings detected by microwave photoconductivity. *Phys. Rev. Lett.*, 95:056807, 2005.

979. Y. S. Gui, N. Mecking, A. Wirthmann, L. H. Bai, and C.-M. Hu. Electrical detection of the ferromagnetic resonance: Spin-rectification versus bolometric effect. *App. Phys. Lett..*, 91(8):082503, 2007.

980. K. O. Dong, E. L. Cheol, L. Je-Hyoung, and R. Kungwon. Evidence of electron-spin wave coupling in Co_xNb_y magnetic metal thin film. *J. Magn. Magn. Mater.*, 293(3):880–884, 2005.

981. N. Zhu, X. Zhang, I. H. Froning, M. E. Flatté, E. Johnston-Halperin, and H. X. Tang. Low loss spin wave resonances in organic-based ferrimagnet vanadium tetracyanoethylene thin films. *Appl. Phys. Lett.*, 109(8):082402, 2016.

982. H. Liu, C. Zhang, H. Malissa, M. Groesbeck, M. Kavand, R. McLaughlin, S. Jamali, J. Hao, D. Sun, R. A. Davidson, L. Wojcik, J. S. Miller, C. Boehme, and Z. V. Vardeny. Organic-based magnon spintronics. *Nat. Mater.*, 17(4):308–312, 2018.

983. T. Schwarze, J. Waizner, M. Garst, A. Bauer, I. Stasinopoulos, H. Berger, C. Pfleiderer, and D. Grundler. Universal helimagnon and skyrmion excitations in metallic, semiconducting and insulating chiral magnets. *Nat. Mater.*, 14(5):478–483, 2015.

984. S. Pöllath, A. Aqeel, A. Bauer, C. Luo, H. Ryll, F. Radu, C. Pfleiderer, G. Woltersdorf, and C. H. Back. Ferromagnetic Resonance with Magnetic Phase Selectivity by Means of Resonant Elastic X-Ray Scattering on a Chiral Magnet. *Phys. Rev. Lett.*, 123:167201, 2019.

985. M. Weiler, A. Aqeel, M. Mostovoy, A. Leonov, S. Geprägs, R. Gross, H. Huebl, T. T. M. Palstra, and S. T. B. Goennenwein. Helimagnon resonances in an intrinsic chiral magnonic crystal. *Phys. Rev. Lett.*, 119:237204, 2017.

986. J. Lan, W. Yu, R. Wu, and J. Xiao. Spin-wave diode. *Phys. Rev. X*, 5:041049, 2015.

987. K. Wagner, A. Kákay, K. Schultheiss, A. Henschke, T. Sebastian, and H. Schultheiss. Magnetic domain walls as reconfigurable spin-wave nanochannels. *Nat. Nanotechnol.*, 11(5):432–436, May 2016.

988. C. Liu, S. Wu, J. Zhang, J. Chen, J. Ding, J. Ma, Y. Zhang, Y. Sun, S. Tu, H. Wang, P. Liu, C. Li, Y. Jiang, P. Gao, D. Yu, J. Xiao, R. Duine, M. Wu, C.-W. Nan, J. Zhang, and H. Yu. Current-controlled propagation of spin waves in antiparallel, coupled domains. *Nature Nanotechnology*, 14(7):691–697, 2019.

989. H. Yu, J. Xiao, and H. Schultheiss. Magnetic texture based magnonics. *Phys. Rep.*, 905:1–59, 2021.

990. W. M. Lai, D. Rubin, and E. Krempl. *Introduction to Continuum Mechanics*. Butterworth Heinemann Ltd, Woburn MA, 1996.

991. M. Weiler, L. Dreher, C. Heeg, H. Huebl, R. Gross, M. S. Brandt, and S. T.B. Goennenwein. Elastically driven ferromagnetic resonance in nickel thin films. *Phys. Rev. Lett.*, 106:117601, 2011.

992. R. Sasaki, Y. Nii, and Y. Onose. Magnetization control by angular momentum transfer from surface acoustic wave to ferromagnetic spin moments. *Nat. Commun.*, 12:1–7, 2021.

993. M. Weiler, H. Huebl, F. S. Goerg, F. D. Czeschka, R. Gross, and S. T.B. Goennenwein. Spin pumping with coherent elastic waves. *Phys. Rev. Lett.*, 108:176601, 2012.

994. T. P. Lyons, J. Puebla, K. Yamamoto, R. S. Deacon, Y. Hwang, K. Ishibashi, S. Maekawa, and Y. Otani. Acoustically driven magnon-phonon coupling in a layered antiferromagnet. *Phys. Rev. Lett.*, 131(19):196701, 2023.

995. J. Zhang, M. Chen, J. Chen, K. Yamamoto, H. Wang, M. Hamdi, Y. Sun, K. Wagner, W. He, Y. Zhang, J. Ma, P. Gao, X. Han, D. Yu, P. Maletinsky, J.-Ph. Ansermet, S. Maekawa, D. Grundler, C. W. Nan, and H. Yu. Long decay length of magnon-polarons in $BiFeO_3/La_{0.67}Sr_{0.33}MnO_3$ heterostructures. *Nat. Commun.*, 12:7258, 2021.

996. J. Chen, K. Yamamoto, J. Zhang, J. Ma, H. Wang, Y. Sun, M. Chen, J. Ma, S. Liu, P. Gao, D. Yu, J.-Ph. Ansermet, C. W. Nan, S. Maekawa, and H Yu. Hybridized propagation of spin waves and surface acoustic waves in a multiferroic-ferromagnetic heterostructure. *Phys. Rev. Appl.*, 19:024046, 2023.

997. H. Wang, Y. Yang, J. Chen, J. Wang, H. Jia, P. Chen, Y. Zhang, C. Wan, S. Liu, D. Yu, X. Han, J.-Ph. Ansermet, J. Zhang, and H. Yu. Long-distance coherent propagation of magnon polarons in a ferroelectric-ferromagnetic heterostructure. *Phys. Rev. B*, 108(14):144425, 2023.

998. V. V. Kruglyak, S. O. Demokritov, and D. Grundler. Magnonics. *J. Phys. D: Appl. Phys.*, 43(26):264001, 2010.

999. K.-S. Lee and S.-K. Kim. Conceptual design of spin wave logic gates based on a Mach–Zehnder-type spin wave interferometer for universal logic functions. *J. Appl. Phys.*, 104(5):053909, 2008.

1000. M. Bailleul, D. Olligs, C. Fermon, and S. O. Demokritov. Spin waves propagation and confinement in conducting films at the micrometer scale. *Europhys. Lett.*, 56(5):741–747, 2001.

1001. K. Sekiguchi, K. Yamada, S. M. Seo, K. J. Lee, D. Chiba, K. Kobayashi, and T. Ono. Nonreciprocal emission of spin-wave packet in FeNi film. *Appl. Phys. Lett.*, 97(2):022508, 2010.

1002. A. A. Serga, A. V. Chumak, and B. Hillebrands. YIG magnonics. *J. Phys. D: Appl. Phys.*, 43(26):264002, 2010.

1003. K. Di, S. X. Feng, S. N. Piramanayagam, V. L. Zhang, H. S. Lim, S. C. Ng, and M. H. Kuok. Enhancement of spin-wave nonreciprocity in magnonic crystals via synthetic antiferromagnetic coupling. *Sci. Rep.*, 5(1):10153, 2015.

1004. J. H. Kwon, J. Yoon, P. Deorani, J. M. Lee, J. Sinha, K. J. Lee, M. Hayashi, and H. Yang. Giant nonreciprocal emission of spin waves in Ta/Py bilayers. *Sci. Adv.*, 2(7):6–13, 2016.

1005. H. Wang, J. Chen, T. Yu, C. Liu, C. Guo, S. Liu, K. Shen, H. Jia, T. Liu, J. Zhang, M. A. Cabero, Q. Song, S. Tu, M. Wu, X. Han, K. Xia, D. Yu, G. E. W. Bauer, and H. Yu. Nonreciprocal coherent coupling of nanomagnets by exchange spin waves. *Nano. Res.*, 14:2133–2138, 2021.

1006. H. Wu, L. Huang, C. Fang, B. S. Yang, C. H. Wan, G. Q. Yu, J. F. Feng, H. X. Wei, and X. F. Han. Magnon valve effect between two magnetic insulators. *Phys. Rev. Lett.*, 120:097205, 2018.

1007. C. Y. Guo, C. H. Wan, X. Wang, C. Fang, P. Tang, W. J. Kong, M. K. Zhao, L. N. Jiang, B. S. Tao, G. Q. Yu, and X. F. Han. Magnon valves based on YIG/NiO/YIG all-insulating magnon junctions. *Phys. Rev. B*, 98:134426, 2018.

1008. Z.R. Yan, C.H. Wan, and X.F. Han. Magnon blocking effect in an antiferromagnet-spaced magnon junction. *Phys. Rev. Appl.*, 14:044053, 2020.

1009. T. Jungwirth, X. Marti, P. Wadley, and J. Wunderlich. Antiferromagnetic spintronics. *Nat. Nanotechnol.*, 11:231–241, 2016.

1010. S. Gopalan and A.N. Slavin. *High Frequency Processes in Magnetic Materials*. World Scientific, Singapore, New Jersey, London, Hong Kong, 1995.

1011. S. Haroche and D. Kleppner. Cavity qauntum electrodynamics. *Phys. Today*, 42(1):24–30, 1989.

1012. J. T. Hou and L. Liu. Strong coupling between microwave photons and nanomagnet magnons. *Phys. Rev. Lett.*, 123:107702, 2019.

1013. Y. Li, T. Polakovic, Y.-L. Wang, J. Xu, S. Lendinez, Z. Zhang, J. Ding, T. Khaire, H. Saglam, R. Divan, J. Pearson, W.-K. Kwok, Z. Xiao, V. Novosad, A. Hoffmann, and W. Zhang. Strong coupling between magnons and microwave photons in on-chip ferromagnet-superconductor thin-film devices. *Phys. Rev. Lett.*, 123:107701, 2019.

1014. R.W. Sanders, V. Jaccarino, and S.M. Rezende. Magnetic polariton, impurity mode enhancement, and superradiance effects in FeFe$_2$. *Solid State Commun.*, 28:907–910, 1978.

1015. J. J. Hopfield. Aspects of Polaritons, pp. 771-782. In: E. Burstein, C. Weisbuch, (eds) Confined Electrons and Photons. NATO ASI Series, vol 340. Springer, Boston, MA, 1995.

1016. B.A. Auld. Walker modes in large ferrite samples. *J. Appl. Phys.*, 31:1642–1647, 1960.

1017. C. Manohar and G. Venkataraman. Magnon-photon coupling in antiferromagnets. *Phys. Rev. B*, 5:1993–1999, 1972.

1018. M. R. Jensen, S. A. Feiven, T. J. Parker, and R. E. Camley. Experimental observation and interpretation of magnetic polariton modes in FeF$_2$. *J. Phys.: Condens. Matter.*, 9:7233–7247, 1997.

1019. O. O. Soykal and M. E. Flatte. Strong field interactions between a nanomagnet and a photonic cavity. *Phys. Rev. Lett.*, 104:077202, 2010.

1020. H. Huebl, C. W. Zollitsch, J. Lotze, F. Hocke, M. Greifenstein, A. Marx, R. Gross, and S. T. B. Goennenwein. High cooperativity in coupled microwave resonator ferrimagnetic insulator hybrids. *Phys. Rev. Lett.*, 93:127003, 2013.

1021. M. Harder, L. Bai, C. Match, J. Sirker, and C-M. Hu. Study of the cavity-magnon-polariton transmission line shape. *Sci. China: Phys. Mech. Astron.*, 59(11):117511, 2016.

1022. T. Giamarchi, C. Rüegg, and O. Tchernyshyov. Bose-Einstein condensation in magnetic insulators. *Nat. Phys.*, 4:198–204, 2008.

1023. S. O. Demokritov, V. E. Demidov, O. Dzyapko, G. A. Melkov, A. A. Serga, B. Hillebrands, and A. N. Slavin. Bose–Einstein condensation of quasi-equilibrium magnons at room temperature under pumping. *Nature*, 443:430–433, 2006.

1024. R. J. Elliott and R. Loudon. The possible observation of electronic Raman transitions in crystals. *Phys. Lett.*, 3:189–191, 1963.

1025. T. Moriya. Theory of light scattering by magnetic crystals. *J. Phys. Soc. Jpn.*, 23:490–500, 1967.

1026. P. A. Fleury and R. Loudon. Scattering of light by one- and two-magnon excitations. *Phys. Rev.*, 166:514–530, 1968.

1027. R. Loudon. Theory of the temperature dependence of first-order light scattering by ordered spin systems. *J. Phys. C: Solid State Phys.*, 3:872–890, 1970.

1028. R. S. Freitas, A. Paduan-Filho, and C. C. Becerra. Magnetic phase diagram of the low-anisotropy antiferromagnet $Cs_2FeCl_5 \cdot H_2O$. *J. Magn. Magn. Mater.*, 374:307–310, 2015.

1029. F. L. A. Machado, P. R. T. Ribeiro, J. Holand, R. L. Rodriguez-Suarez, A. Azevedo, and S. M. Rezende. Spin-flop transition in the easy-plane antiferromagnet nickel oxide. *Phys. Rev. B*, 95:104418, 2017.

1030. P. Pincus. Theory of magnetic resonance in Fe_2O_3. *Phys. Rev. Lett.*, 5(1):13–15, 1960.

1031. D. Reitz, J. Li, W. Yuan, J. Shi, and Y. Tserkovnyak. Spin Seebeck effect near the antiferromagnetic spin-flop transition. *Phys. Rev. B*, 102(2):20408, 2020.

1032. S. M. Rezende, A. Azevedo, and R. L. Rodríguez-Suárez. Introduction to antiferromagnetic magnons. *J. Appl. Phys.*, 126(15):151101, 2019.

1033. J. Li, H. T. Simensen, D. Reitz, Q. Sun, W. Yuan, C. Li, Y. Tserkovnyak, A. Brataas, and J. Shi. Observation of magnon polarons in a uniaxial antiferromagnetic insulator. *Phys. Rev. Lett.*, 125(21):217201, 2020.

1034. M. Bevis, A. J. Sievers, J. P. Harrison, D. R. Taylor, and D. J. Thouless. Infrared absorption by elementary excitations of the one-dimensional XY system. *Phys. Rev. Lett.*, 41:987–990, 1978.

1035. D. R. Taylor. Magnetic resonance and spin dynamics in a one-dimensional XY system. *Phys. Rev. Lett.*, 42:1302–1305, 1979.

1036. J. M. D. Coey. *Magnetism and Magnetic Materials*. Cambridge University Press, 2014.

1037. E. J. Samuelsen and G. Shirane. Inelastic neutron scattering investigation of spin waves and magnetic interactions in α-Fe_2O_3. *Phys. Stat. Solidi (B)*, 42(1):241–256, 1970.

1038. D. L. Mills and E. Burstein. Polaritons: the electromagnetic modes of media. *Rep. Prog. Phys.*, 37:817–926, 1974.

1039. M. Białek, J. Zhang, H. Yu, and J. Ph Ansermet. Strong coupling of antiferromagnetic resonance with subterahertz cavity fields. *Phys. Rev. Appl.*, 15:42–45, 2021.

1040. D. M. Pozar. *Microwave Engineering*. John Wiley & Sons Inc., Hoboken New Jersey, 2021.

1041. D. I. Schuster, A. P. Sears, E. Ginossar, L. Dicarlo, L. Frunzio, J. J. L. Morton, H. Wu, G. A.D. Briggs, B. B. Buckley, D. D. Awschalom, and R. J. Schoelkopf. High-cooperativity coupling of electron-spin ensembles to superconducting cavities. *Phys. Rev. Lett.*, 105(14):1–4, 2010.

1042. D.F. Walls and G.J. Milburn. *Quantum Optics*. Springer-Verlag Berlin Heidelberg, 2nd edition, 2008.

1043. A. A. Clerk, M. H. Devoret, S. M. Girvin, F. Marquardt, and R. J. Schoelkopf. Introduction to quantum noise, measurement, and amplification. *Rev. Mod. Phys.*, 82(2):1155–1208, 2010.

1044. E. M. Purcell. Spontaneous emission probabilities at radio frequencies. *Phys. Rev.*, 69:681, 1946.

1045. H. J. Carmichael. Quantum fluctuations in absorptive bistability without adiabatic elimination. *Phys. Rev. A*, 33(5):3262–3265, 1986.

1046. X. Zhang, C. L. Zou, N. Zhu, F. Marquardt, L. Jiang, and H. X. Tang. Magnon dark modes and gradient memory. *Nat. Commun.*, 6:8914, 2015.

1047. A. Sud, C. W. Zollitsch, A. Kamimaki, T. Dion, S. Khan, S. Iihama, S. Mizukami, and H. Kurebayashi. Tunable magnon-magnon coupling in synthetic antiferromagnets. *Phys. Rev. B*, 102:100403, 2020.

1048. R. Cheng, J. Xiao, Q. Niu, and A. Brataas. Spin pumping and spin-transfer torques in antiferromagnets. *Phys. Rev. Lett.*, 113:057601, 2014.

1049. Ø. Johansen and A. Brataas. Spin pumping and inverse spin Hall voltages from dynamical antiferromagnets. *Phys. Rev. B*, 95:220408(R), 2017.

1050. I. Boventer, H. T. Simensen, A. Anane, M. Kläui, A. Brataas, and R. Lebrun. Room temperature antiferromagnetic resonance and inverse spin-Hall voltage in canted antiferromagnets. *Phys. Rev. Lett.*, 126:187201, 2021.

1051. R. S. Fishman, N. Furukawa, J. T. Haraldsen, M. Matsuda, and S. Miyahara. Identifying the spectroscopic modes of multiferroic $BiFeO_3$. *Phys. Rev. B*, 86:220402, 2012.

1052. R. S. Fishman, J. T. Haraldsen, N. Furukawa, and S. Miyahara. Spin state and spectroscopic modes of multiferroic $BiFeO_3$. *Phys. Rev. B*, 87:134416, 2013.

1053. M. Białek, A. Magrez, A. Murk, and J.-Ph. Ansermet. Spin-wave resonances in bismuth orthoferrite at high temperatures. *Phys. Rev. B*, 97:054410, 2018.

1054. M. Białek, T. Ito, H. Rønnow, and J.-Ph. Ansermet. Terahertz-optical properties of a bismuth ferrite single crystal. *Phys. Rev. B*, 99:064429, 2019.

1055. O. Prokhnenko, R. Feyerherm, M. Mostovoy, N. Aliouane, E. Dudzik, A. U. B. Wolter, A. Maljuk, and D. N. Argyriou. Coupling of frustrated Ising spins to the magnetic cycloid in multiferroic $TbMnO_3$. *Phys. Rev. Lett.*, 99:177206, 2007.

1056. R. Cheng, M. W. Daniels, J.-G. Zhu, and D. Xiao. Antiferromagnetic spin wave field-effect transistor. *Sci. Rep.*, 6(1):24223, 2016.

1057. P. Ross, M. Schreier, J. Lotze, H. Huebl, R. Gross, and S. T. B. Goennenwein. Antiferromagentic resonance detected by direct current voltages in MnF_2/Pt bilayers. *J. Appl. Phys.*, 118(23):233907, 2015.

1058. C.-M. Hu. Cavity spintronics gets more with less. *Physics*, 12:97, 2019.

1059. X. Zhang, C. L. Zou, L. Jiang, and H. X. Tang. Strongly coupled magnons and cavity microwave photons. *Phys. Rev. Lett.*, 113(15):1–5, 2014.

1060. A. Leo, A. G. Monteduro, S. Rizzato, L. Martina, and G. Maruccio. Identification and time-resolved study of ferrimagnetic spin-wave modes in a microwave cavity in the strong-coupling regime. *Phys. Rev. B*, 101(1):1–8, 2020.

1061. T. S. Parvini, V. A. S. V. Bittencourt, and S. V. Kusminskiy. Antiferromagnetic cavity optomagnonics. *Phys. Rev. Res.*, 2:022027(R), 2020.

1062. E. F. Sarmento and D. R. Tilley. Surface magnon-polaritons. In A.D. Boardman, editor, *Electromagnetic Surface Modes*, chapter 16. John Wiley & Sons, New York, 1982.

1063. C. H. Henry and J. J. Hopfield. Raman Scattering by polaritons. *Phys. Rev. Lett.*, 15(25):964–966, 1965.

1064. M. Liscidini, D. Gerace, D. Sanvitto, and D. Bajoni. Guided Bloch surface wave polaritons. *Appl. Phys. Lett.*, 98(12):121118, 2011.

1065. A. L. Lereu, M. Zerrad, A. Passian, and C. Amra. Surface plasmons and bloch surface waves: Towards optimized ultra-sensitive optical sensors. *Appl. Phys. Lett.*, 111(1):011107, 2017.

1066. R. L. Comstock. Parametric coupling of the magnetization and strain in a ferrimagnet. II. Parametric excitation of magnetic and elastic plane waves. *J. Appl. Phys.*, 34(5):1465–1468, 1963.

1067. R. L. Comstock and B. A. Auld. Parametric coupling of the magnetization and strain in a ferrimagnet. I. parametric excitation of magnetostatic and elastic modes. *J. Appl. Phys.*, 34(5):1461–1464, 1963.

1068. H. Suhl. The theory of ferromagnetic resonance at high signal powers. *J. Phys. Chem. Solids*, 1:209–227, 1957.

1069. E. Schloemann. Fine structure in the decline of the feromagnetic resonance absorption with increasing power level. *Phys. Rev.*, 116(5):828–837, 1959.

1070. V. E. Zakharov, V. S. Lvov, and S. S. Starobinets. Spin wave turbulences beyond the parametric excitation threshold. *Sov. phys., Usp.*, 18:896–919, 1975.

1071. A. V. Andrienko, V. I. Ozhogin, V. L. Safonov, and A. Y. Yakubovskiĭ. Nuclear spin wave research. *Soviet Physics Uspekhi*, 34(10):843–861, 1991.

1072. L. W. Hinderks and P. M. Richards. Excitation of nuclear and electronic spin waves in $RbMnF_3$ by parallel pumping. *J. Appl. Phys.*, 39(2):824–825, 1968.

1073. A.V. Andrienko, V.L. Safonov, and A. Y. Yakubovskii. Investigation of combined resonance of mixed parametric magnon pairs in antiferromagnetic $MnCO_3$. *JETP*, 69(2):363, 1989.

1074. A. L. Buchachenko and E. L. Frankevich. *Chemcial Generation and Reception of Radio-and Microwaves*. VCH Publishers Inc., New York, 1994.

1075. A. G. Zhuravlev, V. L. Berdinskii, and A. L. Buchachenko. Generation of high-frequency current by the products of a photochemical reaction. *JETP Lett.*, 28:140–142, 1978.

1076. Y. B. Zel'dovich, A. L. Buchachenko, and E. L. Frankevich. Magnetic-spin effects in chemistry and molecular physics. *Sov. phys., Usp.*, 31(5):385–408, 1988.

1077. E. L. Hahn. Spin echoes. *Phys. Rev.*, 80(4):580–594, 1950.

1078. H. Y. Carr and E. M. Purcell. Effects of diffusion on free precession in nuclear magnetic resonance experiments. *Phys. Rev.*, 94:630–638, 1954.

1079. S. Meiboom and D. Gill. Modified spin echo method for measuring nuclear relaxation times. *Rev. Sci. Instrum.*, 29(8):688–691, 1958.

1080. G. S. Uhrig. Keeping a quantum bit alive by optimized π-pulse sequences. *Phys. Rev. Lett.*, 98:100504, 2007.

1081. R.R. Ernst. Nuclear magnetic double resonance with an incoherent radio-frequency field. *J. Chem. Phys.*, 45(10):3845–3861, 1966.

1082. J. S. Waugh, L. M. Huber, and U. Haeberlen. Approach to high-resolution NMR in solids. *Phys. Rev. Lett.*, 20:180–182, 1968.

1083. U. Haeberlen and J. S. Waugh. Coherent averaging effects in magnetic resonance. *Phys. Rev.*, 175:453–467, 1968.

1084. H. J. Reich, M. Jautelat, M. T. Messe, F. J. Weigert, and J. D. Roberts. Nuclear magnetic resonance spectroscopy. Carbon-13 spectra of steroids. *J. Am. Chem. Soc.*, 91(26):7445–7454, 1969.

1085. V. Royden. Measurement of the spin and gyromagnetic ratio of C^{13} by the collapse of spin-spin splitting. *Phys. Rev.*, 96:543–544, 1954.

1086. G. Feher. Electronic structure of F centers in KCl by the electron spin double-resonance technique. *Phys. Rev.*, 105:1122–1123, 1957.

1087. Freeman R. and Morris G. A. Two-dimensional fourier transform in NMR. *Bull. Mag. Res.*, 1:5–26, 1979.

1088. W. P. Aue, E. Bartholdi, and R. R. Ernst. Two-dimensional spectroscopy. Application to nuclear magnetic resonance. *J. Chem. Phys.*, 64(5):2229–2246, 2008.

1089. F. Maier. Quadrupolar echoes. Private commun.

1090. J.-Ph. Ansermet. NMR, the quantum mechanics playground. In M. Anselmino, F. Mila, and J. Soffer, editors, *Spin in Physics*, pp. 199–223. Frontier Group, Virginal Ittre, Belgium, 2002.

1091. C. W. Searle, J. Davis, A. Hirai, and K. Fukuda. Multiple spin echoes for Mn^{55} in manganese ferrites. *Phys. Rev. Lett.*, 27(20):1380–1382, 1971.

1092. J. Jeener, B. H. Meir, P. Bachmann, and R. R. Ernst. Investigation of exchange processes by two-dimensional NMR spectroscopy. *J. Chem. Phys.*, 71:4546–4553, 1979.

1093. W. P. Aue, J. Karhan, and R. R. Ernst. Homonuclear broad band decoupling and two-dimensional J-resolved NMR spectroscopy. *J. Chem. Phys.*, 64:4226–4227, 1976.

1094. H. Kessler, M. Gehrke, and C.Griesinger. Two-dimensional NMR Spectroscopy: background and overview of the experiments. *Angew. Chem. Int. Ed.*, 27(4):490–536, 1988.

1095. R.R. Ernst, G. Bodenhausen, and A. Wokaun. *Principles of Nuclear Magnetic Resonance in One and Two Dimensions.* Clarendon Press, Oxford, 1987.

1096. A. A. Maudsley and R. R. Ernst. Indirect detection of magnetic resonance by heteronuclear two-dimensional spectroscopy. *Chem. Phys. Letters*, 50:368, 1977.

1097. M. H. Levitt. *Spin Dynamics*. John Wiley & Sons Ltd, Chichester, 2008.

1098. A. Einstein, B. Podolsky, and N. Rosen. Can quantum-mechanical description of physical reality be considered complete? *Phys. Rev.*, 47:777–780, 1935.

1099. V. Scarani, C. Lynn, and L. S. Yang. *Six Quantum Pieces*. World Scientific, Singapore, 2010.

1100. V. Scarani. *Bell Nonlocality*. Oxford Graduate Texts, Oxford, 2019.

1101. M. Mehring, J. Mende, and W. Scherer. Entanglement between an electron and a nuclear spin 1/2. *Phys. Rev. Lett.*, 90(15):153001, 2003.

1102. M. Mehring, W. Scherer, and A. Weidinger. Pseudoentanglement of spin states in the multilevel ^{15}N@C$_{60}$ system. *Phys. Rev. Lett.*, 93(20):206603, 2004.

1103. L. Emsley and G. Bodenhausen. Optimization of shaped selective pulses for NMR using a quaternion description of their overall propagators. *J. Mag. Res.*, 97(1):135–148, 1992.

1104. A. K. Dorai and K. Anil. Implementing quantum-logic operations, pseudopure states, and the Deutsch-Jozsa algorithm using noncommuting selective pulses in NMR. *Phys. Rev. A*, 61:042306, 2000.

1105. R. Freeman. *Spin Choreography, Basic steps in High Resolution NMR*. Oxford University Press, Oxford, New York, Tokyo, 1998.

1106. S. Vega and A. Pines. Operator formalism for double quantum NMR. *J. Chem. Phys.*, 66:5624–5644, 1977.

1107. S. Vega. Fictitious spin 1/2 operator formalism for multiple quantum NMR. *J. Chem. Phys.*, 68:5518–5527, 1978.

1108. A. Wokaun and R. R. Ernst. Selective excitation and detection in multilevel spin systems: application of single transition operators. *J. Chem. Phys.*, 67:1752–1758, 1977.

1109. S. Takahashi, L. C. Brunel, D. T. Edwards, J. Van Tol, G. Ramian, S. Han, and M. S. Sherwin. Pulsed electron paramagnetic resonance spectroscopy powered by a free-electron laser. *Nature*, 489:409–413, 2012.

1110. D. T. Edwards, Y. Zhang, S. J. Glaser, S. Han, and M. S. Sherwin. Phase cycling with a 240 GHz, free electron laser-powered electron paramagnetic resonance spectrometer. *Physical Chemistry Chemical Physics*, 15:5707–5719, 4 2013.

1111. N. D. Mermin. *Quantum Computer Science*. Cambridge University Press, Cambridge, New York, Melbourne, Madrid, Cape Town, Singapore, São Paulo, 2007.

1112. M. Y. Volkov and K. M. Salikhov. Pulse protocols for quantum computing with electron spins as qubits. *Appl. Magn. Reson.*, 41:145–154, 2011.

1113. M. D. Price, C. S. S. Somoaroo, H. Tseng, J. C. Gore, A.F. Fahmy, T.F. Havel, and D. G. Cory. Construction and implementation of NMR quantum logic gates for two spin systems. *J. Magn. Res.*, 140:371–378, 1999.

1114. S. S. Somaroo, D. G. Cory, and T. F. Havel. Expressing the operations of quantum computing in multiparticle geometric algebra. *Phys. Lett. A*, 240:1–7, 1998.

1115. J. N. Nelson, J. Zhang, J. Zhou, B. K. Rugg, M. D. Krzyaniak, and M. R. Wasielewski. Effect of electron-nuclear hyperfine interactions on multiple-quantum coherences in photo-generated covalent radical (Qubit) pairs. *J. Phys. Chem. A*, 122:9392–9402, 2018.

1116. J. N. Nelson, J. Zhan, J. Zhou, B. K. Rugg, M. D. Krzyaniak, and M. R. Wasielewski. CNOT gate operation on a photogenerated molecular electron spin-qubit pair. *J. Chem. Phys.*, 152:014503, 2020.

1117. G. J. Pazera, M. D. Krzyaniak, and M. R. Wasielewski. Pulse sequences for manipulating the spin states of molecular radical-pair-based electron spin qubit systems for quantum information applications. *J. Chem. Phys.*, 158:204118, 2023.

1118. M. R. Wasielewski. Light-driven spin chemistry for quantum information science. *Phys. Today*, 76:28–34, 2023.

1119. C. S. Yannoni and R. D. Kendrick. NMR nutation spectroscopy: a method for determining molecular geometry in amorphous solids. *J. Chem. Phys.*, 74(1):747–749, 1981.

1120. A.V. Astashkin and A. Schweiger. Electron-spin transient nutation: a new approach to simplify the interpretation of ESR spectra. *Chem. Phys. Letters*, 174(6):595–602, 1990.

1121. K. Aebischer, N. Wili, Z. Tošner, and M. Ernst. Using nutation-frequency-selective pulses to reduce radio-frequency field inhomogeneity in solid-state NMR. *Magnetic Resonance*, 1(2):187–195, 2020.

1122. B. E. Cole, J. B. Williams, B. T. King, M. S. Sherwin, and C. R. Stanley. Coherent manipulation of semiconductor quantum bits with terahertz radiation. *Nature*, 410(6824):60–63, 2001.

1123. H. S. Brandi, A. Latgé, Z. Barticevic, and L. E. Oliveira. Rabi oscillations in two-level semiconductor systems. *Solid State Commun.*, 135(6):386–389, 2005.

1124. X. Ji, A. Bornet, B. Vuichoud, J. Milani, D. Gajan, A. J. Rossini, L. Emsley, G. Bodenhausen, and S. Jannin. Transportable hyperpolarized metabolites. *Nat. Commun.*, 8(1):13975, 2017.

1125. B. E. Kane. A silicon-based nuclear spin quantum computer. *Nature*, 393(6681):133–137, 1998.

1126. J. J. Pla, K. Y. Tan, J. P. Dehollain, W. H. Lim, J. J. L. Morton, F. A. Zwanenburg, D. N. Jamieson, A. S. Dzurak, and A. Morello. High-fidelity readout and control of a nuclear spin qubit in silicon. *Nature*, 496(7445):334–338, 2013.

1127. S. S. Eaton and G. R. Eaton. Relaxation times of organic radicals and transition metal ions. In L. J. Berliner, S. S. Eaton, and G. R. Eaton, editors, *Biological Magnetic Resonance*, volume 19, pp. 29–154. Kluwer Academic/Plenum , New York, 2002.

1128. S. H. Johnson, C. E. Jackson, and J. M. Zadrozny. Programmable nuclear-spin dynamics in Ti(IV) coordination complexes. *Inorg. Chem.*, 59(11):7479–7486, 2020.

1129. M. Carravetta, O. G. Johannessen, and M. H. Levitt. Beyond the T_1 limit: singlet nuclear spin states in low magnetic fields. *Phys. Rev. Lett.*, 92:153003, 2004.

1130. M. Carravetta and M. H. Levitt. Long-Lived Nuclear Spin States in High-Field Solution NMR. *J. Am. Chem. Soc.*, 126(20):6228–6229, 2004.

1131. J.-N. Dumez, J. T. Hill-Cousins, R. C.D. Brown, and G. Pileio. Long-lived localization in magnetic resonance imaging. *J. Mag. Res.*, 246:27–30, 2014.

1132. P. Ahuja, R. Sarkar, P. R. Vasos, and G. Bodenhausen. Molecular properties determined from the relaxation of long-lived spin states. *J. Chem. Phys.*, 127(13):134112, 2007.

1133. P. R. Vasos, A. Comment, R. Sarkar, P. Ahuja, S. Jannin, J.-Ph. Ansermet, J. A. Konter, P. Hautle, B. van den Brandt, and G. Bodenhausen. Long-lived states to sustain hyperpolarized magnetization. *Proc. Natl. Acad. Sci.*, 106(44):18469–18473, 2009.

1134. H. Geen, J. J. Titman, J. Gottwald, and H. W. Spiess. Solid-state proton multiple-quantum NMR spectroscopy with fast magic angle spinning. *Chem. Phys. Lett.*, 227(1):79–86, 1994.

1135. A. K. Khitrin, V. L. Ermakov, and B. M. Fung. Nuclear magnetic resonance molecular photography. *J. Chem. Phys.*, 117(15):6903–6906, 2002.

1136. K. K. Dey, R. Bhattacharyya, and A. Kumar. Use of spatial encoding in NMR photography. *J. Mag. Res.*, 171(2):359–363, 2004.

1137. S. J. Hile, L. Fricke, M. G. House, E. Peretz, C. Y. Chen, Y. Wang, M. Broome, S. K. Gorman, J. G. Keizer, R. Rahman, and M. Y. Y. Simmons. Addressable electron spin resonance using donors and donor molecules in silicon. *Sci. Adv.*, 4(7):eaaq1459, 2018.

1138. A.K. Khitrin, V.L. Ermakov, and B.M. Fung. Information storage using a cluster of dipolar-coupled spins. *Chem. Phys. Lett.*, 360(1):161–166, 2002.

1139. R. Sarkar, P. Ahuja, P. R. Vasos, and G. Bodenhausen. Long-lived coherences for homogeneous line narrowing in spectroscopy. *Phys. Rev. Lett.*, 104:053001, 2010.

1140. R. Sarkar, P. Ahuja, P. R. Vasos, A. Bornet, O. Wagnières, and G. Bodenhausen. Long-lived coherences for line-narrowing in high-field NMR. *Prog. Nucl. Magn. Reson. Spectrosc.*, 59(1):83–90, 2011.

1141. A. Bornet, S. Jannin, J. A. Konter, P. Hautle, B. van den Brandt, and G. Bodenhausen. Ultra high-resolution NMR: sustained induction decays of long-lived coherences. *J. Am. Chem. Soc.*, 133(39):15644–15649, 10 2011.

1142. M. Singh and N. D. Kurur. An improved method for the measurement of lifetimes of long-lived coherences in NMR. *RSC Adv.*, 5:8236–8238, 2015.

1143. G. Balasubramanian, P. Neumann, D. Twitchen, M. Markham, R. Kolesov, N. Mizuochi, J. Isoya, J. Achard, J. Beck, J. Tissler, V. Jacques, P. R Hemmer, F. Jelezko, and J. Wrachtrup. Ultralong spin coherence time in isotopically engineered diamond. *Nat. Mater.*, 8(5):383–387, 2009.

1144. M. Pfender, N. Aslam, H. Sumiya, S. Onoda, P. Neumann, J. Isoya, C. A. Meriles, and J. Wrachtrup. Nonvolatile nuclear spin memory enables sensor-unlimited nanoscale spectroscopy of small spin clusters. *Nat. Commun.*, 8(1):834, 2017.

1145. J. T. Muhonen, J. P. Dehollain, A. Laucht, F. E. Hudson, R. Kalra, T. Sekiguchi, K. M. Itoh, D. N. Jamieson, J. C. McCallum, A. S. Dzurak, and A. Morello. Storing quantum information for 30 seconds in a nanoelectronic device. *Nat. Nanotechnol.*, 9(12):986–991, 2014.

1146. M. Veldhorst, H. G. J. Eenink, C. H. Yang, and A. S. Dzurak. Silicon CMOS architecture for a spin-based quantum computer. *Nat. Commun.*, 8(1):1766, 2017.

1147. K. C. Miao, A. Bourassa, C. P. Anderson, S. J. Whiteley, A. L. Crook, S. L. Bayliss, G. Wolfowicz, G. Thiering, P. Udvarhelyi, V. Ivády, H. Abe, T. Ohshima, Á. Gali, and D. D. Awschalom. Electrically driven optical interferometry with spins in silicon carbide. *Sci. Adv.*, 5,eaay0527, 2019.

1148. D. J. Christle, P. Klimov, C. F. D. L. Casas, K. Sz'asz, V. Ivády, V. Jokubavivcius, J. Hassan, M. Syvajarvi, W. Koehl, T. Ohshima, N. Son, E. Janz'en, A. G. Gali, and D. Awschalom. Isolated spin qubits in SiC with a high-fidelity infrared spin-to-photon interface. *Phys. Rev. X*, 7:021046, 2017.

1149. A. Gottscholl, M. Diez, V. Soltamov, C. Kasper, A. Sperlich, M. Kianinia, C. Bradac, I. Aharonovich, and V. Dyakonov. Room temperature coherent control of spin defects in hexagonal boron nitride. *Sci. Adv.*, 7(14):eabf3630, 2021.

1150. T. Kobayashi, J. Salfi, C. Chua, J. van der Heijden, M. G. House, D. Culcer, W. D. Hutchison, B. C. Johnson, J. C. McCallum, H. Riemann, N. V. Abrosimov, P. Becker, H.-J. Pohl, M. Y. Simmons, and S. Rogge. Engineering long spin coherence times of spin–orbit qubits in silicon. *Nat. Mater.*, 20(1):38–42, 2021.

1151. B. Hensen, W. W. Huang, C.-H. Yang, K. W. Chan, J. Yoneda, T. Tanttu, F. E. Hudson, A. Laucht, K. M. Itoh, T. D. Ladd, A. Morello, and A. S. Dzurak. A silicon quantum-dot-coupled nuclear spin qubit. *Nat. Nanotechnol.*, 15(1):13–17, 2020.

1152. K. Yang, W. Paul, S.-H. Phark, P. Willke, Y. Bae, T. Choi, T. Esat, A. Ardavan, A. J. Heinrich, and C. P. Lutz. Coherent spin manipulation of individual atoms on a surface. *Science*, 366(6464):509–512, 2019.

1153. M. Veldhorst, C. H. Yang, J. C. C. Hwang, W. Huang, J. P. Dehollain, J. T. Muhonen, S. Simmons, A. Laucht, F. E. Hudson, K. M. Itoh, A. Morello, and A. S. Dzurak. A two-qubit logic gate in silicon. *Nature*, 526(7573):410–414, 2015.

1154. C. H. Yang, K. W. Chan, R. Harper, W. Huang, T. Evans, J. C. C. Hwang, B. Hensen, A. Laucht, T. Tanttu, F. E. Hudson, S. T. Flammia, K. M. Itoh, A. Morello, S. D. Bartlett, and A. S. Dzurak. Silicon qubit fidelities approaching incoherent noise limits via pulse engineering. *Nat. Electron.*, 2(4):151–158, 2019.

1155. C. E. Bradley, J. Randall, M. H. Abobeih, R. C. Berrevoets, M. J. Degen, M. A. Bakker, M. Markham, D. J. Twitchen, and T. H. Taminiau. A ten-qubit solid-state spin register with quantum memory up to one minute. *Phys. Rev. X*, 9:031045, 2019.

1156. W. Huang, C. H. Yang, K. W. Chan, T. Tanttu, B. Hensen, R. C. C. Leon, M. A. Fogarty, J. C. C. Hwang, F. E. Hudson, K. M. Itoh, A. Morello, A. Laucht, and A. S. Dzurak. Fidelity benchmarks for two-qubit gates in silicon. *Nature*, 569(7757):532–536, 2019.

1157. J. R. Petta, A. C. Johnson, J. M. Taylor, E. A. Laird, A. Yacoby, M. D. Lukin, C. M. Marcus, M. P. Hanson, and A. C. Gossard. Coherent manipulation of coupled electron spins in semiconductor quantum dots. *Science*, 309(5744):2180–2184, 2005.

1158. E. Kawakami, T. Jullien, P. Scarlino, D. R. Ward, D. E. Savage, M. G. Lagally, V. V. Dobrovitski, M. Friesen, S. N. Coppersmith, M. A. Eriksson, and L. M. K. Vandersypen. Gate fidelity and coherence of an electron spin in an Si/SiGe quantum dot with micromagnet. *Proc. Natl. Acad. Sci.*, 113(42):11738–11743, 2016.

1159. R. Zhao, T. Tanttu, K. Y. Tan, B. Hensen, K. W. Chan, J. C. C. Hwang, R. C. C. Leon, C. H. Yang, W. Gilbert, F. E. Hudson, K. M. Itoh, A. A. Kiselev, T. D. Ladd, A. Morello, A. Laucht, and A. S. Dzurak. Single-spin qubits in isotopically enriched silicon at low magnetic field. *Nat. Commun.*, 10(1):5500, 2019.

1160. J. J. Pla, K. Y. Tan, J. P. Dehollain, Wee H. L., J. J. L. Morton, D. N. Jamieson, A. S. Dzurak, and A. Morello. A single-atom electron spin qubit in silicon. *Nature*, 489(7417):541–545, 2012.

1161. D. Keith, M. G. House, M. B. Donnelly, T. F. Watson, B. Weber, and M. Y. Simmons. Single-shot spin readout in semiconductors near the shot-noise sensitivity limit. *Phys. Rev. X*, 9:041003, 2019.

1162. C. Fasth, A. Fuhrer, M.T. Björk, and L. Samuelson. Tunable double quantum dots in InAs nanowires defined by local gate electrodes. *Nano Lett.*, 5(7):1487–1490, 2005.

1163. C. Jones, M. A. Fogarty, A. Morello, M. F. Gyure, A. S. Dzurak, and T. D. Ladd. Logical qubit in a linear array of semiconductor quantum dots. *Phys. Rev. X*, 8:021058, 2018.

1164. M. R. Wasielewski, M. D. E. Forbes, N. L. Frank, K. Kowalski, G. D. Scholes, J. Yuen-Zhou, M. A. Baldo, D. E. Freedman, R. H. Goldsmith, T. Goodson, M. L. Kirk, J. K. McCusker, J. P. Ogilvie, D. A. Shultz, S. Stoll, and K. B. Whaley. Exploiting chemistry and molecular systems for quantum information science. *Nat. Rev. Chem.*, 4(9):490–504, 2020.

1165. M. T. Colvin, R. Carmieli, T. Miura, S. Richert, D. M. Gardner, A. L. Smeigh, S. M. Dyar, S. M. Conron, M. A. Ratner, and M. R. Wasielewski. Electron spin polarization transfer from photogenerated spin-correlated radical pairs to a stable radical observer spin. *J. Chem. Phys. A*, 117(25):5314–5325, 06 2013.

1166. J. N. Nelson, M. D. Krzyaniak, N. E. Horwitz, B. K. Rugg, B. T. Phelan, and M. R. Wasielewski. Zero quantum coherence in a series of covalent spin-correlated radical pairs. *J. Phys. Chem. A*, 121(11):2241–2252, 2017.

1167. T. C. Player and P. J. Hore. Posner qubits: Spin dynamics of entangled $Ca_9(PO_4)_6$ molecules and their role in neural processing. *J. R. Soc. Interface*, 15: 20180494, 2018.

1168. K. V. Raman, A. M. Kamerbeek, A. Mukherjee, N. Atodiresei, T. K. Sen, P. Lazić, V. Caciuc, R. Michel, D. Stalke, S. K. Mandal, S. Blügel, M. Münzenberg, and J. S. Moodera. Interface-engineered templates for molecular spin memory devices. *Nature*, 493(7433):509–513, 2013.

1169. S. Lenz, K. Bader, H. Bamberger, and J. van Slageren. Quantitative prediction of nuclear-spin-diffusion-limited coherence times of molecular quantum bits based on copper(ii). *Chem. Commun.*, 53:4477–4480, 2017.

1170. J. I. Colless, A. C. Mahoney, J. M. Hornibrook, A. C. Doherty, H. Lu, A. C. Gossard, and D. J. Reilly. Dispersive readout of a few-electron double quantum dot with fast rf gate sensors. *Phys. Rev. Lett.*, 110:046805, 2013.

1171. R. Maurand, X. Jehl, D. Kotekar-Patil, A. Corna, H. Bohuslavskyi, R. Laviéville, L. Hutin, S. Barraud, M. Vinet, M. Sanquer, and S. De Franceschi. A CMOS silicon spin qubit. *Nat. Commun.*, 7(1):13575, 2016.

1172. A. G. Fowler, M. Mariantoni, J. M. Martinis, and A. N. Cleland. Surface codes: Towards practical large-scale quantum computation. *Phys. Rev. A*, 86:032324, 2012.

1173. A. Y. Kitaev. Quantum computations: algorithms and error correction. *Russ. Math. Surv*, 52(6):1191–1249, 1997.

1174. A. Y. Kitaev. Fault-tolerant quantum computation by anyons. *Ann. Phys. (N. Y.)*, 303(1):2–30, 2003.

1175. H. Büch, S. Mahapatra, R. Rahman, A. Morello, and M. Y. Simmons. Spin readout and addressability of phosphorus-donor clusters in silicon. *Nat. Commun.*, 4(1):2017, 2013.

1176. F.N. Krauth, S.K. Gorman, Y. He, M.T. Jones, P. Macha, S. Kocsis, C. Chua, B. Voisin, S. Rogge, R. Rahman, Y. Chung, and M.Y. Simmons. Flopping-mode electric dipole spin resonance in phosphorus donor qubits in silicon. *Phys. Rev. Appl.*, 17:054006, 2022.

1177. B. Voisin, J. Salfi, D. D. St Médar, B. C. Johnson, J. C. McCallum, M. Y. Simmons, and S. Rogge. A solid-state quantum microscope for wavefunction control of an atom-based quantum dot device in silicon. *Nat. Electron.*, 6(6):409–416, 2023.

1178. L. Kranz, S. Roche, S. K. Gorman, J. G. Keizer, and M. Y. Simmons. High-fidelity CNOT gate for donor electron spin qubits in silicon. *Phys. Rev. Appl.*, 19:024068, 2023.

Index

Printed in the United States
by Baker & Taylor Publisher Services